T0214647

Lecture Notes in Computer Science　　11384

Commenced Publication in 1973
Founding and Former Series Editors:
Gerhard Goos, Juris Hartmanis, and Jan van Leeuwen

Editorial Board

David Hutchison
  Lancaster University, Lancaster, UK
Takeo Kanade
  Carnegie Mellon University, Pittsburgh, PA, USA
Josef Kittler
  University of Surrey, Guildford, UK
Jon M. Kleinberg
  Cornell University, Ithaca, NY, USA
Friedemann Mattern
  ETH Zurich, Zurich, Switzerland
John C. Mitchell
  Stanford University, Stanford, CA, USA
Moni Naor
  Weizmann Institute of Science, Rehovot, Israel
C. Pandu Rangan
  Indian Institute of Technology Madras, Chennai, India
Bernhard Steffen
  TU Dortmund University, Dortmund, Germany
Demetri Terzopoulos
  University of California, Los Angeles, CA, USA
Doug Tygar
  University of California, Berkeley, CA, USA

More information about this series at http://www.springer.com/series/7412

Alessandro Crimi · Spyridon Bakas
Hugo Kuijf · Farahani Keyvan
Mauricio Reyes · Theo van Walsum (Eds.)

Brainlesion: Glioma, Multiple Sclerosis, Stroke and Traumatic Brain Injuries

4th International Workshop, BrainLes 2018
Held in Conjunction with MICCAI 2018
Granada, Spain, September 16, 2018
Revised Selected Papers
Part II

 Springer

Editors
Alessandro Crimi
University Hospital of Zurich
Zürich, Switzerland

Spyridon Bakas (iD)
University of Pennsylvania
Philadelphia, PA, USA

Hugo Kuijf
University Medical Center Utrecht
Utrecht, The Netherlands

Farahani Keyvan
National Cancer Institute
Bethesda, MD, USA

Mauricio Reyes (iD)
University of Bern
Bern, Switzerland

Theo van Walsum (iD)
Erasmus University Medical Center
Rotterdam, The Netherlands

ISSN 0302-9743 ISSN 1611-3349 (electronic)
Lecture Notes in Computer Science
ISBN 978-3-030-11725-2 ISBN 978-3-030-11726-9 (eBook)
https://doi.org/10.1007/978-3-030-11726-9

Library of Congress Control Number: 2018967942

LNCS Sublibrary: SL6 – Image Processing, Computer Vision, Pattern Recognition, and Graphics

© Springer Nature Switzerland AG 2019
This work is subject to copyright. All rights are reserved by the Publisher, whether the whole or part of the material is concerned, specifically the rights of translation, reprinting, reuse of illustrations, recitation, broadcasting, reproduction on microfilms or in any other physical way, and transmission or information storage and retrieval, electronic adaptation, computer software, or by similar or dissimilar methodology now known or hereafter developed.
The use of general descriptive names, registered names, trademarks, service marks, etc. in this publication does not imply, even in the absence of a specific statement, that such names are exempt from the relevant protective laws and regulations and therefore free for general use.
The publisher, the authors and the editors are safe to assume that the advice and information in this book are believed to be true and accurate at the date of publication. Neither the publisher nor the authors or the editors give a warranty, express or implied, with respect to the material contained herein or for any errors or omissions that may have been made. The publisher remains neutral with regard to jurisdictional claims in published maps and institutional affiliations.

This Springer imprint is published by the registered company Springer Nature Switzerland AG
The registered company address is: Gewerbestrasse 11, 6330 Cham, Switzerland

Preface

This volume contains articles from the Brain-Lesion Workshop (BrainLes), as well as the (a) International Multimodal Brain Tumor Segmentation (BraTS) challenge, (b) Ischemic Stroke Lesion Segmentation (ISLES) challenge, (c) grand challenge on MR Brain Image Segmentation (MRBrainS18), (d) Computational Precision Medicine (CPM) challenges, and (e) Stroke Workshop on Imaging and Treatment Challenges (SWITCH). All these events were held in conjunction with the Medical Image Computing for Computer Assisted Intervention (MICCAI) conference during September 16–20, 2018, in Granada, Spain.

The papers presented describe research of computational scientists and clinical researchers working on glioma, multiple sclerosis, cerebral stroke, traumatic brain injuries, and white matter hyper-intensities of presumed vascular origin. This compilation does not claim to provide a comprehensive understanding from all points of view; however, the authors present their latest advances in segmentation, disease prognosis, and other applications to the clinical context.

The volume is divided into seven parts: The first part comprises three invited papers summarizing the presentations of the keynote speakers; the second includes the paper submissions to the BrainLes workshop; the third through the seventh parts contain a selection of papers presenting methods that participated at the 2018 challenges of ISLES, MRBrainS, CPM, SWITCH, and BraTS, respectively.

The first chapter in these proceedings describes invited papers from the four keynote speakers of the MICCAI BrainLes 2018 workshop (www.brainlesion-workshop.org). The overarching aim of these papers is to give an updated review of the work done in (a) the domain of machine learning applied in neuro-oncology diagnostics, (b) connectomics of traumatic brain injury and brain tumors, (c) computational/memory considerations for deep learning in medical image analysis, and (d) computed tomography perfusion. The sequence of these papers reflects the order that they were presented during the workshop.

The aim of the second chapter, focusing on the BrainLes workshop submissions, is to provide an overview of new advances in medical image analysis in all of the aforementioned brain pathologies. Bringing together researchers from the medical image analysis domain, neurologists, and radiologists working on at least one of these diseases. The aim is to consider neuroimaging biomarkers used for one disease applied to the other diseases. This session did not have a specific dataset to be used.

The third chapter contains descriptions of a selection of algorithms that participated in the ISLES 2018 challenge. The purpose of this challenge was to directly compare methods for the automatic prediction of stroke lesion outcome from CT-perfusion imaging. A dataset consisting of CT-perfusion image volumes acquired at acute and 3-month follow-up was released for training. A dedicated test set of cases was used for evaluation. Test data were not released, but participants had to submit their segmentation results to: www.isles-challenge.org.

The fourth chapter includes a number of papers from MRBrainS 2018. The purpose of this challenge is to directly compare methods for segmentation of gray matter, white matter, cerebrospinal fluid, and other structures on 3T MRI scans of the brain, and to assess the effect of (large) pathologies on segmentation and volumetry. Over 30 teams participated and the challenge remains open for future submissions. An up-to-date ranking is hosted on: http://mrbrains18.isi.uu.nl/.

The fifth chapter presents a selection of papers from the leading participants in the two CPM 2018 challenges in brain tumors (http://miccai.cloudapp.net/competitions/). The "Combined MRI and Pathology Brain Tumor Classification" challenge used corresponding imaging and pathology data with the task of classifying a cohort of "low-grade" glioma tumors ($n = 52$) into two sub-types of oligodendroglioma and astrocytoma. This challenge presented a new paradigm in algorithmic challenges, where data and analytical tasks related to the management of brain tumors were combined to arrive at a more accurate tumor classification. In the challenge of "Segmentation of Nuclei in Digital Pathology," participants were asked to detect and segment all nuclei in a set of image tiles ($n = 33$) of glioblastoma and lower-grade glioma extracted from whole slide tissue images. Data from both challenges were obtained from The Cancer Genome Atlas/The Cancer Imaging Archive (TCGA/TCGA) repository.

Finally, the sixth chapter of these proceedings contains scientific contributions of the SWITCH workshop, which aims to bring together clinicians and medical imaging experts to discuss challenges and opportunities for medical imaging in stroke care and treatment. In 2018, three clinical keynote speakers addressed various aspects of stroke and ischemic stroke treatment: Prof. Aad van der Lugt discussed imaging biomarkers related to stroke, Prof. Matt Gounis shared his research on the development for stroke devices, and Prof. Roland Wiest presented stroke mimics and chameleons. The scientific contributions of the medical imaging field, addressing topics such as perfusion parameter estimation and the relation between diffusion MRI and microstructural changes in gray matter, were presented at the workshop in oral and poster presentations. All accepted full paper contributions are part of these proceedings.

The seventh chapter focuses on a selection of papers from the BraTS challenge participants. BraTS 2018 made publicly available a large ($n = 542$) manually annotated dataset of pre-operative brain tumor scans from 19 institutions, in order to gauge the current state-of-the-art in automated glioma segmentation using multi-parametric structural MRI modalities and to compare fairly between different methods. To pinpoint and evaluate the clinical relevance of tumor segmentation, BraTS 2018 also included the prediction of patient overall survival, via integrative analyses of radiomic features and machine learning algorithms (www.cbica.upenn.edu/BraTS2018.html).

We heartily hope that this volume will promote further exciting research on brain lesions.

December 2018

Alessandro Crimi
Spyridon Bakas

Organization

Main Organizing Committee
(Lead Organizers from Each Individual Event)

Spyridon Bakas · Center for Biomedical Image Computing and Analytics, University of Pennsylvania, USA
Alessandro Crimi · African Institute for Mathematical Sciences, Ghana
Keyvan Farahani · National Cancer Institute, National Institutes of Health, USA
Hugo Kuijf · University Medical Center Utrecht, The Netherlands
Mauricio Reyes · Biomedical Neuroimage Analysis Group, University of Bern, Switzerland
Theo van Walsum · Biomedical Imaging Group Rotterdam, Erasmus MC, The Netherlands

Challenges Organizing Committees

Brain Tumor Segmentation (BraTS) Challenge

Spyridon Bakas · University of Pennsylvania, USA
Christos Davatzikos · University of Pennsylvania, USA
Keyvan Farahani · National Cancer Institute (NCI), National Institutes of Health, USA
Jayashree Kalpathy-Cramer · Harvard Medical School, USA
Bjoern Menze · Technical University of Munich, Germany

Computational Precision Medicine (CPM) Challenges

Spyridon Bakas · University of Pennsylvania, USA
Hesham Elhalawani · University of Texas MD Anderson Cancer Center, USA
Keyvan Farahani · National Cancer Institute (NCI), National Institutes of Health, USA
John Freymann · Frederick National Laboratory for Cancer Research, USA
David Fuller · University of Texas MD Anderson Cancer Center, USA
Jayashree Kalpathy-Cramer · Harvard Medical School, USA
Justin Kirby · Frederick National Laboratory for Cancer Research, USA
Tahsin Kurc · Stony Brook Cancer Center, USA
Joel Saltz · Stony Brook Cancer Center, USA
Amber Simpson · Memorial Sloan Kettering Cancer Center, USA

Ischemic Stroke Lesion Segmentation (ISLES) Challenge

Søren Christensen	Stanford University, USA
Arsany Hakim	Inselspital Bern, Switzerland
Maarten G. Lansberg	Stanford University, USA
Mauricio Reyes	University of Bern, Switzerland
David Robben	KU Leuven, Belgium
Roland Wiest	Inselspital Bern, Switzerland
Stefan Winzeck	University of Cambridge, UK
Greg Zaharchuk	Stanford University, USA

Grand Challenge on MR Brain Segmentation 2018 (MRBrainS18)

Edwin Bennink	University Medical Center Utrecht, The Netherlands
Hugo Kuijf	University Medical Center Utrecht, The Netherlands

Stroke Workshop on Imaging and Treatment Challenges (SWITCH)

Adrian Dalca	CSAIL, MIT and MGH, Harvard Medical School, USA
Mauricio Reyes	University of Bern, Biomedical Neuroimage Analysis Group, Switzerland
David Robben	KU Leuven, Belgium
Theo van Walsum	Biomedical Imaging Group Rotterdam, Erasmus MC, The Netherlands
Roland Wiest	Support Center of Advanced Neuroimaging, Inselspital Bern, Switzerland

Program Committee

Chincisan Andra	University Hospital of Zurich, Switzerland
Meritxell Bach Cuadra	University of Lausanne, Switzerland
Jacopo Cavazza	Instituto Italiano di Tecnologia, Italy
Adrian Dalca	Massachusetts Institute of Technology, USA
Guray Erus	University of Pennsylvania, USA
Alvaro Gomariz	ETH Zürich, Switzerland
Moises Hernandez	University of Pennsylvania, USA
Ender Konukoglu	ETH-Zurich, Switzerland
Jana Lipkova	Technical University of Munich, Germany
Yusuf Osmanlioglu	University of Pennsylvania, USA
Saima Rathore	University of Pennsylvania, USA
Zahra Riahi Samani	University of Pennsylvania, USA
Aristeidis Sotiras	University of Pennsylvania, USA

Koen Van Leemput	Harvard Medical School, USA
Benedikt Wiestler	Technical University of Munich, Germany
Stefan Winzeck	University of Cambridge, UK

Sponsoring Institutions

Center for Biomedical Image Computing and Analytics, University of Pennsylvania, USA

Contents – Part II

Contents – Part I

Computational Precision Medicine

Stroke Workshop on Imaging and Treatment Challenges

Brain Tumor Image Segmentation

Segmentation of Brain Tumors and Patient Survival Prediction: Methods for the BraTS 2018 Challenge

Leon Weninger[✉], Oliver Rippel, Simon Koppers, and Dorit Merhof

Institute of Imaging & Computer Vision, RWTH Aachen University,
Aachen, Germany
leon.weninger@lfb.rwth-aachen.de

Abstract. Brain tumor localization and segmentation is an important step in the treatment of brain tumor patients. It is the base for later clinical steps, e.g., a possible resection of the tumor. Hence, an automatic segmentation algorithm would be preferable, as it does not suffer from inter-rater variability. On top, results could be available immediately after the brain imaging procedure. Using this automatic tumor segmentation, it could also be possible to predict the survival of patients. The BraTS 2018 challenge consists of these two tasks: tumor segmentation in 3D-MRI images of brain tumor patients and survival prediction based on these images. For the tumor segmentation, we utilize a two-step approach. First, the tumor is located using a 3D U-net. Second, another 3D U-net – more complex, but with a smaller output size – detects subtle differences in the tumor volume, i.e., it segments the located tumor into tumor core, enhanced tumor, and peritumoral edema.

The survival prediction of the patients is done with a rather simple, yet accurate algorithm which outperformed other tested approaches on the train set when thoroughly cross-validated. This finding is consistent with our performance on the test set - we achieved 3rd place in the survival prediction task of the BraTS Challenge 2018.

Keywords: BraTS 2018 · Brain tumor · Automatic segmentation · Survival prediction · Deep learning

1 Introduction

Brain tumors can appear in different forms, shapes and sizes and can grow to a considerable size until they are discovered. They can be distinguished into glioblastoma (GBM/HGG) and low grade glioma (LGG). A common way of screening for brain tumors is with MRI-scans, where even different brain tumor regions can be determined. In effect, MRI scans of the brain are not only the basis for tumor screening, but are even utilized for pre-operative planning. Thus, an accurate, fast and reproducible segmentation of brain tumors in MRI scans is needed for several clinical applications.

© Springer Nature Switzerland AG 2019
A. Crimi et al. (Eds.): BrainLes 2018, LNCS 11384, pp. 3–12, 2019.
https://doi.org/10.1007/978-3-030-11726-9_1

HGG patients have a poor survival prognosis, as metastases often develop even when the initial tumor was completely resected. Whether patient overall survival can be accurately predicted based on pre-operative scans by employing knowing factors such as radiomics features, tumor location and tumor shape, remains an open question.

The BraTS challenge [11] addresses these problems, and is one of the biggest and well known machine learning challenges in the field of medical imaging. Last year around 50 different competitors from around the world took part. This year, the challenge is divided in two parts: First, tumor segmentation based on 3D-MRI images, and second, survival prediction of the brain tumor patients based on only the pre-operative scans and the age of the patients.

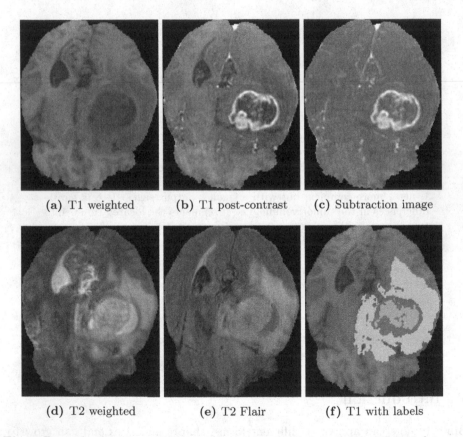

(a) T1 weighted   (b) T1 post-contrast   (c) Subtraction image

(d) T2 weighted   (e) T2 Flair   (f) T1 with labels

Fig. 1. Example of image modalities and groundtruth-labels in the BraTS 2018 dataset. The subtraction image is calculated by subtracting the T1 image (a) from the T1 post-contrast image (b), as described in Sect. 3.1. For the labels, blue indicates the peritumoral edema, green the necrotic and non-enhancing tumor, and red the GD-enhancing core, as described in the BraTS paper [11]. (Color figure online)

Similar to the BraTS 2017 dataset, the BraTS 2018 training dataset consists of MRI-scans of 285 brain tumor patients from 19 different contributors. The dataset includes T1, T1 post-contrast (T1CE), T2, and T2 Fluid Attenuated Inversion Recovery (Flair) volumes, as well as hand-annotated expert labels for each patient [1–3]. An example of a set of images can be seen in Fig. 1.

Motivated by the success of the U-net [14] in biomedical image segmentation, we choose the 3D-adaptation [5] of this architecture to tackle the segmentation part of the BraTS challenge. Two different versions are used, a first one to coarsely locate the tumor, and a second one to accurately segment the located tumor into different areas.

Concerning the survival prediction, we found that complex models using different types of radiomics features such as shape and texture of the tumor and the brain could not outperform a simple linear regressor based on just a few basic features. Using only the patient age and tumor region sizes as features, we achieve competitive results.

The code developed for this challenge is available online: https://github.com/weningerleon/BraTS2018.

2 Related Work

In the last years, deep learning has advanced classification and segmentation in many biomedical imaging applications, and has a preeminent role in current publications.

In the BraTS Challenge last year, all top-ranking approaches of the segmentation task [6,9,16,17] used deep convolutional neural networks. The employed architectures vary substantially among these submission. However, a common ground seems to be the utilization of 3D-architectures instead of 2D-architectures.

One key architecture for biomedical segmentation, which is also heavily used throughout this paper, is the U-Net [14]. Both, 2D as well as 3D-variants [5] have been successfully employed for various biomedical applications, and still achieve competitive results in current biomedical image segmentation challenges [7,8].

3 Methods

3.1 Segmentation

We tackle the segmentation task in a two-step approach: First, the location of the brain tumor is determined. Second, this region is segmented into the three different classes: *peritumoral edema (ed)*, *necrotic tumor (nec)*, and *GD-enhancing core (gde)*.

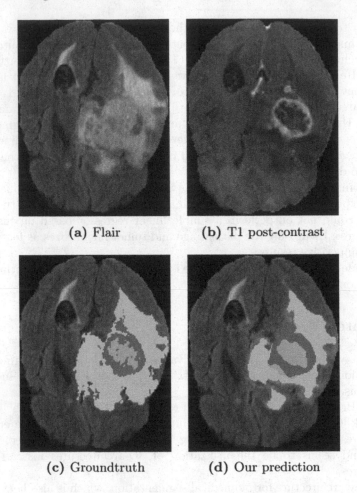

(a) Flair   (b) T1 post-contrast

(c) Groundtruth   (d) Our prediction

Fig. 2. Comparison of our segmentation result with the groundtruth labels.

Preprocessing. We first define a brain mask based on all voxels unequal to zero, on which all preprocessing is carried out. On this brain mask, the mean and standard deviation of the intensity is calculated, and the data normalized accordingly. Since different MRI-scanners and sequences are used, we independently normalize each image and modality based on the obtained values. Non-brain regions remain zero.

The whole tumor is strongly visible in T1, T2 and Flair MRI-images. However, in practice, including all images seems to produce better results even for the whole tumor localization. We also add another image as input, a contrast-agent subtraction image, where the T1 image is subtracted from the T1CE image. This should enhance the contrast-agent sensitive region, as can be seen in Fig. 1c.

We construct a cuboid bounding box around the brain, and crop the non-brain regions to facilitate training. The training target is constructed by merging the three different tumor classes of the groundtruth labels.

For training of the tumor segmentation step, the 3D-images are cropped around a padded tumor area, which is defined as the area of 20 voxels in every direction around the tumor.

Network Architectures and Employed Hardware. For both steps, a 3D U-net [5] with a depth of 4 is employed.

The first U-net uses padding in every convolutional layer, such that the input size corresponds exactly to the output size. Every convolutional layer is followed by a ReLU activation function. 16 feature maps are used in the first layer, and the number of feature maps doubles as the depth increases. For normalization between the different layers, instance-norm layers [15] are used, as they seem to be better suited for normalization in segmentation tasks and for small batch sizes. Testing different training hyperparameters, the Adam optimizer [10] with an initial learning rate of 0.001 together with a binary cross entropy loss was chosen for the tumor localization step. An L2-regularization of 1e−5 is applied to the weights, and the learning rate was reduced by a factor of 0.015 after every epoch. One epoch denotes a training step over every brain.

The U-net utilized in the second step has a similar architecture as the previous one, but with double as many feature maps per layer. To counteract the increased memory usage, no padding is used, which drastically reduces the size of the output as well as the memory consumption of later feature maps.

Here, we apply a multi-class dice loss to the output of our 3D U-net and the labels for training, as described in [12]. A learning rate of 0.005 was chosen, while weight decay and learning rate reduction remain the same as in step 1.

Our contribution to the BraTS challenge was implemented using pyTorch [13]. Training and prediction is carried out on a Nvidia 1080 Ti GPU with a memory size of 11 Gb.

Training. In the first step, we train with complete brain images cropped to the brain mask. The brain mask is determined by all voxels not equal to zero. Using a rather simple U-net, a training pass with a batch-size of one fits on a GPU even for larger brains. Due to the bounding box around the brain, different sizes need to be passed through the network. In practice this is possible using a fully convolutional network architecture and a batch size of one.

For the second step, we choose the input to be fixed to $124 \times 124 \times 124$. Due to the unpadded convolutions, this results in an output shape of $36 \times 36 \times 36$. Hence, the training labels are the $36 \times 36 \times 36$ sized segmented voxels in the middle of the input. Here, a batch-size of two was chosen.

During training, patches are chosen from inside the padded tumor bounding box for each patient. To guarantee a reasonably balanced train set, only training patches which comprise all three tumor classes are kept for training.

Inference. Similar to the training procedure, the first step is carried out directly on a complete 5-channel (T1, T2, Flair, T1CE, and contrast-agent subtraction image) 3D image of the brain.

Before the tumor/non-tumor segmentation of this step is used as basis in the second step, only the largest connected area is kept. Based on the assumption that there is only one tumorous area in the brain, we can suppress false positive voxels in the rest of the brain with this method.

We then predict $36 \times 36 \times 36$ sized patches with the trained unpadded U-net. Patches are chosen so that they cover the tumorous area, and the distance between two neighboring patches was set to 9 in each direction. This results in several predictions per voxel. Finally, a majority vote over these predictions gives the end result.

3.2   Survival Prediction

According to the information given by the segmentation labels, we count the number of voxels of the tumor segmentation. This volume information about the necrotic tumor core, the GD-enhancing tumor and peritumoral edema as well as the distance between the centroids of tumor and brain and the age of the patient were considered as valuable feature for the survival prediction task. We tested single features, as well as combinations of features as input for a linear regressor.

4   Results

4.1   Segmentation

For evaluation on the training dataset, we split the training dataset randomly into 245 training images and 40 test images to evaluate our approach with groundtruth labels. No external data was used for training or pre-training.

Based on our experience with the training dataset, we choose 200 epochs as an appropriate training duration for the first step, and 60 epochs as an appropriate training duration for the second step. We thus train from scratch on all training images for the determined optimal number of epochs, and use the obtained networks for evaluation on the validation set. The results obtained by this method can be seen in Table 1, and an exemplary result is visualized in Fig. 2.

4.2   Survival Prediction

For evaluating our approach on the training dataset, we fit and evaluate our linear regressor with a leave-one-out cross-validation on the training images. We compare the results obtained by solely using the age of the patient versus using the age with a subset of the tumor region sizes as features. On top, we consider the distance between the centroid of the tumor and the centroid of the brain as a feature. Our finding is that all features other than the age of the patient increase

Table 1. Results for the segmentation challenge. Train, validation and test errors according to the online submission system, as available.

Dataset	Dice			Sensitivity			Specificity			Hausdorff 95		
	ET	WT	TC	ET	WT	TC	ET	WT	TC	ET	WT	TC
Train set	0.763	0.860	0.817	0.747	0.784	0.787	0.998	0.998	0.998	5.63	7.01	7.88
Val set	0.712	0.889	0.758	0.757	0.887	0.735	0.998	0.995	0.998	6.28	6.97	10.91
Test set	0.621	0.844	0.728	*	*	*	*	*	*	10.5	8.71	13.3

the error on left-out images. In Tables 2 and 3, we show the exact results for the different input features on the training set (cross-validation) and on the test set (according to the online portal).

In Fig. 3, the survival time in years is plotted against the age for all patients with a resection status of 'gross total resection' in the train dataset. The linear regressor fitted to this data and used for the challenge, is plotted as well. The three classes used during the challenge, dividing the dataset into long, short, and mid-survivors can also be seen.

This age-only linear regressor achieved the 3rd place in the BraTS challenge 2018 [4], with an accuracy of 0.558, a MSE of 277890 and a median SE of 43264 on the test data.

Table 2. Training Data: Mean Squared Error and Median Error for leave-one-out cross-validation of the linear regressor. The different features considered are the age of the patient, the volume in voxels of the enhancing tumor (gde), of the necrotic tumor (nec), of the edema (ed) as well as the distance between the centroid of the tumor and the centroid of the brain ($dist$).

Features	MSE	Median Err.
Age (submitted)	**95082**	216
Age + gde	100941	224
Age + ed	99693	221
Age + nec	98826	216
Age + dist	100928	**215**
Age + gde + ed + nec	109817	222

Table 3. Validation Data: Accuracy metrics according to the online portal.

Features	Accuracy	MSE	Median SE	stdSE	SpearmanR
Age	0.5	**97759.5**	**46120.5**	**139670.7**	**0.267**
Age + gde + ed + nec	**0.536**	101012.0	51006.5	140511.5	0.258

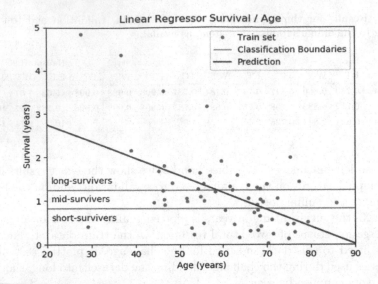

Fig. 3. Our linear regressor (blue) over the age of the tumor patients. The red points are the training data, and the green lines indicate the boundaries between the classes (long, mid, and short survivor), which are used for calculation of the accuracy metric. (Color figure online)

5   Discussion and Conclusion

Our contribution submitted to the BraTS challenge 2018 was summarized in this paper. We used a two-step approach for tumor segmentation and a linear regression for survival prediction.

The segmentation approach already gives promising results. In practice, the two-step framework helps eliminating spurious false-positive classifications in non-tumorous areas, as only the largest connected area is considered as tumor. However, this assumes that there is only one tumorous area in the brain. As there is only one tumorous area in the vast majority of cases, this boosts the accuracy measured. Notwithstanding, it is a simplification that can lead to serious misclassifications in single cases.

This simplification needs to be tackled in future development of the framework. Furthermore, we will evaluate a broader variety of different network architectures, and will also include 3D data-augmentation techniques into our framework.

Our algorithm for the survival analysis task is a straight-forward approach. We considered other, more complex approaches, which were however not able to beat this baseline algorithm.

On the validation set, our survival prediction algorithm ranks among the top submissions, e.g., the age-only approach achieves the lowest MSE and second highest accuracy according to the online portal.

Finally, our top-placement (3rd place) in the challenge underlines the strength of the age as feature for survival prediction. Other teams, using various radiomics and/or deep learning approaches, could not perform much better than our straight-forward approach. Hence, it can be concluded that pre-operative scans are not well suited for survival prediction. However, other datasets could be better suited for survival prediction, e.g., post-operative or follow-up scans of the patient.

References

1. Bakas, S., et al.: Segmentation labels and radiomic features for the pre-operative scans of the TCGA-LGG collection. The Cancer Imaging Archive (2017)
2. Bakas, S., et al.: Segmentation labels and radiomic features for the pre-operative scans of the TCGA-GBM collection. The Cancer Imaging Archive (2017)
3. Bakas, S., et al.: Advancing the cancer genome atlas glioma MRI collections with expert segmentation labels and radiomic features. Scientific Data 4, 170117 EP, September 2017. https://doi.org/10.1038/sdata.2017.117
4. Bakas, S., et al.: Identifying the best machine learning algorithms for brain tumor segmentation, progression assessment, and overall survival prediction in the BRATS challenge. arXiv preprint arXiv:1811.02629
5. Çiçek, Ö., Abdulkadir, A., Lienkamp, S.S., Brox, T., Ronneberger, O.: 3D U-Net: learning dense volumetric segmentation from sparse annotation. In: Ourselin, S., Joskowicz, L., Sabuncu, M.R., Unal, G., Wells, W. (eds.) MICCAI 2016. LNCS, vol. 9901, pp. 424–432. Springer, Cham (2016). https://doi.org/10.1007/978-3-319-46723-8_49
6. Isensee, F., Kickingereder, P., Wick, W., Bendszus, M., Maier-Hein, K.H.: Brain tumor segmentation and radiomics survival prediction: contribution to the BRATS 2017 challenge. CoRR abs/1802.10508 (2018). http://arxiv.org/abs/1802.10508
7. Isensee, F., Kickingereder, P., Wick, W., Bendszus, M., Maier-Hein, K.H.: No new-net. CoRR abs/1809.10483 (2018). http://arxiv.org/abs/1809.10483
8. Isensee, F., et al.: nnU-Net: self-adapting framework for U-Net-based medical image segmentation. CoRR abs/1809.10486 (2018). http://arxiv.org/abs/1809.10486
9. Kamnitsas, K., et al.: Ensembles of multiple models and architectures for robust brain tumour segmentation. In: Crimi, et al. [6], pp. 450–462. https://doi.org/10.1007/978-3-319-75238-9_38
10. Kingma, D.P., Ba, J.: Adam: a method for stochastic optimization. CoRR abs/1412.6980 (2014). http://arxiv.org/abs/1412.6980
11. Menze, B.H., et al.: The multimodal brain tumor image segmentation benchmark (BRATS). IEEE Trans. Med. Imag. 34(10), 1993–2024 (2015). https://doi.org/10.1109/TMI.2014.2377694
12. Milletari, F., Navab, N., Ahmadi, S.: V-Net: Fully convolutional neural networks for volumetric medical image segmentation. CoRR abs/1606.04797 (2016). http://arxiv.org/abs/1606.04797
13. Paszke, A., et al.: Automatic differentiation in PyTorch. In: NIPS-W (2017)
14. Ronneberger, O., Fischer, P., Brox, T.: U-Net: convolutional networks for biomedical image segmentation. In: Navab, N., Hornegger, J., Wells, W.M., Frangi, A.F. (eds.) MICCAI 2015, part III. LNCS, vol. 9351, pp. 234–241. Springer, Cham (2015). https://doi.org/10.1007/978-3-319-24574-4_28

15. Ulyanov, D., Vedaldi, A., Lempitsky, V.S.: Instance normalization: the missing ingredient for fast stylization. CoRR abs/1607.08022 (2016). http://arxiv.org/abs/1607.08022
16. Wang, G., Li, W., Ourselin, S., Vercauteren, T.: Automatic brain tumor segmentation using cascaded anisotropic convolutional neural networks. In: Crimi, et al. [6], pp. 178–190. https://doi.org/10.1007/978-3-319-75238-9_16
17. Yang, T.L., Ou, Y.N., Huang, T.Y.: Automatic segmentation of brain tumor from MR images using SegNet: selection of training data sets. In: Crimi, et al. [6], pp. 450–462. https://doi.org/10.1007/978-3-319-75238-9

Segmenting Brain Tumors from MRI Using Cascaded Multi-modal U-Nets

Michal Marcinkiewicz[1], Jakub Nalepa[1,2(✉)], Pablo Ribalta Lorenzo[2], Wojciech Dudzik[1,2], and Grzegorz Mrukwa[1,2]

[1] Future Processing, Gliwice, Poland
jnalepa@ieee.org
[2] Silesian University of Technology, Gliwice, Poland

Abstract. Gliomas are the most common primary brain tumors, and their accurate manual delineation is a time- consuming and very user-dependent process. Therefore, developing automated techniques for reproducible detection and segmentation of brain tumors from magnetic resonance imaging is a vital research topic. In this paper, we present a deep learning-powered approach for brain tumor segmentation which exploits multiple magnetic-resonance modalities and processes them in two cascaded stages. In both stages, we use multi-modal fully-convolutional neural nets inspired by U-Nets. The first stage detects regions of interests, whereas the second stage performs the multi-class classification. Our experimental study, performed over the newest release of the BraTS dataset (BraTS 2018) showed that our method delivers accurate brain-tumor delineation and offers very fast processing—the total time required to segment one study using our approach amounts to around 18 s.

Keywords: Brain tumor · Segmentation · Deep learning · CNN

1 Introduction

Gliomas are the most common primary brain tumors in humans. They are characterized by different levels of aggressiveness which directly influences prognosis. Due to the gliomas' heterogeneity (in terms of shape and appearance) manifested in multi-modal magnetic resonance imaging (MRI), their accurate delineation is an important yet challenging medical image analysis task. However, manual segmentation of such brain tumors is very time-consuming and prone to human errors. It also lacks reproducibility which adversely affects the effectiveness of patient's monitoring, and can ultimately lead to inefficient treatment.

Therefore, automatic brain tumor *detection* (i.e., which pixels in an input image are tumorous) and *classification* (what is a type of a tumor and/or which

We applied the *sequence-determines-credit* approach for the sequence of authors, and M. Marcinkiewicz, J. Nalepa are joint first authors.

© Springer Nature Switzerland AG 2019
A. Crimi et al. (Eds.): BrainLes 2018, LNCS 11384, pp. 13–24, 2019.
https://doi.org/10.1007/978-3-030-11726-9_2

part of a tumor, e.g., edema, non-enhancing solid core, or enhancing structures a given pixel belongs to; see examples in Fig. 1) from MRI are vital research topics in the pattern recognition and medical image analysis fields. A very wide practical applicability of such techniques encompasses computer-aided diagnosis, prognosis, staging, and monitoring of a patient. In this paper, we propose a deep learning technique to detect and segment gliomas from MRI in a cascaded processing pipeline. These gliomas are further segmented into the enhancing tumor (ET), tumor core (TC), and the whole tumor (WT).

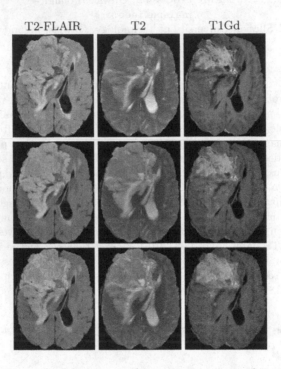

Fig. 1. Different parts of a brain tumor (*detection* is presented in the second row—green parts show the agreement with a human reader) *segmented* using the proposed method (third row) alongside original images (first row): red—peritumoral edema, yellow—necrotic and non-enhancing tumor core, green—GD-enhancing tumor. (Color figure online)

1.1   Contribution

The contribution of this work is multi-fold:

- We propose a deep learning technique for detection and segmentation of brain tumors from MRI. Our deep neural networks (DNNs) are inspired by the U-Nets [28] with considerable changes to the architecture, and they are cascaded—the first DNN performs detection, whereas the second segments a tumor into the enhancing tumor, tumor core, and the whole tumor (Fig. 1).

- To improve generalization capabilities of our segmentation models, we build an ensemble of DNNs trained over different folds of a training set, and average the responses of the base classifiers.
- We show that our approach can be seamlessly applied to the multi-modal MRI analysis, and allows for introducing separate processing pathways for each modality.
- We validate our techniques over the newest release of the Brain Tumor Segmentation dataset (BraTS 2018), and show that they provide high-quality detection and segmentation, and offer instant segmentation.

1.2 Paper Structure

This paper is organized as follows. In Sect. 2, we discuss the current state of the art in brain-tumor delineation. The proposed deep learning-based techniques are presented in Sect. 3. The results of our experiments are analyzed in Sect. 4. Section 5 concludes the paper and highlights the directions of our future work.

2 Related Literature

Approaches for automated brain-tumor delineation can be divided into *atlas-based, unsupervised, supervised,* and *hybrid* techniques (Fig. 2). In the *atlas-based* algorithms, manually segmented images (referred to as *atlases*) are used to segment incoming (previously unseen) scans [25]. These atlases model the anatomical variability of the brain tissue [22]. Atlas images are extrapolated to new frames by warping and applying non-rigid registration techniques. An important drawback of such techniques is the necessity of creating large (and representative) annotated reference sets. It is time-consuming and error prone in practice, and may lead to atlases which cannot be applied to other tumors because they do not encompass certain types of brain tumors [1,6].

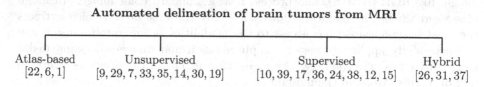

Fig. 2. Automated delineation of brain tumors from MRI—a taxonomy.

Unsupervised algorithms search for hidden structures within unlabeled data [9,19]. In various meta-heuristic approaches, e.g., in evolutionary algorithms [33], brain segmentation is understood as an optimization problem, in which pixels (or voxels) of similar characteristics are searched. It is tackled in a biologically-inspired manner, in which a population of candidate solutions (being the pixel or voxel labels) evolves in time [7]. Other unsupervised algorithms encompass clustering-based techniques [14, 29, 35], and Gaussian modeling [30]. In *supervised* techniques, manually segmented image sets are utilized to train a model. Such algorithms include, among others, decision forests [10, 39], conditional random fields [36], support vector machines [17], and extremely randomized trees [24].

Deep neural networks, which established the state of the art in a plethora of image-processing and image-recognition tasks, have been successful in segmentation of different kinds of brain tissue as well [12, 16, 21] (they very often require computationally intensive data pre-processing). Holistically nested neural nets for MRI were introduced in [38]. White matter was segmented in [11], and convolutional neural networks were applied to segment tumors in [13]. Interestingly, the winning BraTS'17 algorithm used deep neural nets ensembles [15]. However, the authors reported neither training nor inference times of their algorithm which may prevent from using it in clinical practice. *Hybrid* algorithms couple together methods from other categories [26, 31, 37].

We address the aforementioned issues and propose a deep learning algorithm for automated brain tumor segmentation which exploits a new multi-modal fully-convolutional neural network based on U-Nets. The experimental evidence (presented in Sect. 4) obtained over the newest release of the BraTS dataset (BraTS 2018) shows that it can effectively deal with multi-class classification, and it delivers high-quality tumor segmentation in real time.

3   Methods

In this work, we propose an algorithm which utilizes cascaded U-Net-based deep neural networks for detecting and segmenting brain tumors. Our approach for this task is driven by an assumption that the most salient features of a lesion are not contained in a single image modality.

There are multiple ways to exploit all the modalities in deep learning-based engines. One way is to store three (or four) modalities as channels of a single image, like RGB (RGBA), and process it as a standard color image. Although this approach has a significant downside—only the first layer (which extracts the most basic features) has access to the modalities as separate inputs, it can be successfully applied to easier computer-vision and image-processing tasks. Consecutive layers in the network process the outputs of the previous layers—a mix of features from all the modalities.

Hu and Xia processed each modality separately, and merged them at the very end of the processing chain to produce the final segmentation mask, to fully benefit from information manifested in each modality [8]. In this work, we combine both techniques—we use merged modalities for brain-tumor detection, and separate processing pathways for further segmentation of a tumor.

3.1   Detection of Brain Tumors from MRI

The first stage of our image analysis approach involves taking the whole image as an input (i.e., different modalities are stacked together as the channels of an image), and producing a binary mask of the region of interest (therefore, it performs *detection* of a tumor). This binary mask is used to select the voxels of all modalities from the original images (rendering remaining pixels as background). This region is passed to the segmentation unit by the U-Net in the second stage for the final multi-class segmentation.

The architecture of our DNN used for detection is visualized in Fig. 3 (note that we present multiple processing pathways which are exploited in segmentation; for detection, only one pathway is used, and the sigmoid function was applied as the non-linearity). The DNN prediction is binarized using a threshold T_b. The binary mask is post-processed using the 3D connected components analysis—the size of connected components is calculated, and the one with the largest volumes remains. If the next (second) largest connected component is at least T_{cc} (in %) of the volume of the largest, it is kept as well. The binary masks resulting from the first stage are used to produce input to the second stage. More details on the architecture of our deep network itself are presented in the following subsection.

3.2   Segmentation of Detected Brain Tumors

Our DNN for brain tumor *segmentation* separates processing pathways and merges them at the very bottom of the network, where the feature space is compacted the most, and at each bridged connection (Fig. 3). By doing that, we assure that the low- and high-level features are extracted separately for all modalities in the contracting path. Those features can "interact" with each other in the expanding path, producing high-quality segmentations. Our preliminary experiments showed that the pre-contrast T1 modality carries the smallest amount of information, therefore in order to reduce the amount of segmentation time and resources (to make our method easily applicable in a real-life clinical setting), we did not use that modality in our pipeline. However, the proposed U-Net-based architecture is fairly flexible and allows for using any number of input modalities.

Our models are based on a well-known U-Net [28], with considerable changes to the architecture. First, there are separate pathways for each modality, effectively making three contracting paths. In the original architecture the number of filters was doubled at each down-block, whereas in our model it is constant everywhere, except in the very bottom part of the network (where the concatenation and merging of the paths takes place) where it is doubled. The down-block in our model consists of three convolutional layers (48 filters of the size 3×3 each, with stride 1). The second alteration to the original U-Net are the bridged connections, which join (concatenate) activations from each pathway of the contracting paths with their corresponding activations from the expanding path, where they become merged. This procedure allows the DNN to extract high-level features while preserving the context stored earlier. The expanding path is

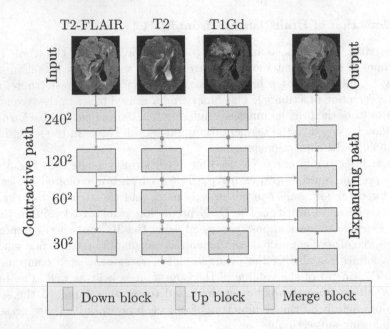

Fig. 3. The proposed deep neural network architecture. Three separate pathways (e.g., for FLAIR, T1c, and T2) are shown as a part of the contractive path. At each level (each set of down blocks) the output is concatenated and sent to a corresponding up block. At the bottom, there is a merging block, where all the features are merged before entering the expanding path. The output layer is a 1 × 1 convolution with one filter for the first stage (detection), and three filters for the second stage (segmentation).

standard—each up-block doubles the size of an activation map by the upsampling procedure, which is followed by two convolutions (48 filters of 3 × 3 size each, with stride 1). In the last layer, there is a 1 × 1 convolution with 1 filter in the detection, and 3 filters in the multi-class classification stages, respectively.

The output of the second stage is an activation map of the size $I_w \times I_h \times 3$, where the last dimension represents the number of classes, and I_w and I_h are the image width and height, respectively. The activation is then passed through a softmax operation, which performs the final multi-class classification.

4   Experimental Validation

4.1   Data

The Brain Tumor Segmentation (BraTS) dataset [2–5, 20] encompasses MRI-DCE data of 285 patients with diagnosed gliomas—210 high-grade glioblastomas (HGG), and 75 low-grade gliomas (LGG). Each study was manually annotated by one to four experienced readers. The data comes in four co-registered modalities: native pre-contrast (T1), post-contrast T1-weighted (T1c), T2-weighted

(T2), and T2 Fluid Attenuated Inversion Recovery (FLAIR). All the pixels have one of four labels attached: healthy tissue, Gd-enhancing tumor (ET), peritumoral edema (ED), the necrotic and non-enhancing tumor core (NCR/NET).

The data was acquired with different clinical protocols, various scanners, and at 19 institutions, therefore the pixel intensity distribution may vary significantly. The studies were interpolated to the same shape ($240 \times 240 \times 155$, hence 155 images of 240×240 size, with voxel size 1 mm^3), and they were pre-processed (skull-stripping was applied). Overall, there are 285 patients in the training set T (210 HGG, 75 LGG), 66 patients in the validation set V (without ground-truth data provided by the BraTS 2018 organizers), and 191 in the test set Ψ (unseen data used for the final verification of the trained models).

4.2   Experimental Setup

The DNN models were implemented using `Python3` with the `Keras` library over CUDA 9.0 and CuDNN 5.1. The experiments were run on a machine equipped with an Intel i7-6850K (15 MB Cache, 3.80 GHz) CPU with 32 GB RAM and NVIDIA GTX Titan X GPU with 12 GB VRAM. The training metric was the DICE score for both stages (detection and segmentation), which is calculated as

$$\mathrm{DICE}(A, B) = \frac{2 \cdot |A \cap B|}{|A| + |B|}, \tag{1}$$

where A and B are two segmentations, i.e., manual and automated. DICE ranges from zero to one (one is the perfect score). The optimizer was Nadam (Adam with Nesterov momentum) with the initial learning rate 10^{-5}, and the optimizer parameters: $\beta_1 = 0.9$, $\beta_2 = 0.999$. The training ran until DICE over the validation set did not increase by at least 0.002 in 10 epochs. The training time for one epoch is around 10 min (similar for both stages). The networks converges in around 20–30 epochs (the complete training for each fold takes 7–8 h). For detection, we used the manually-tuned thresholds: $T_b = 0.5$, and $T_{cc} = 20\%$.

Both networks are relatively small, which directly translates to the low computational requirements during inference—one volume can be processed and classified end-to-end within around 5 s. To exploit the training set completely, and still be able to use validation subset to avoid over-fitting, the final prediction was performed with an ensemble of five models trained on different folds of the training set (we followed the 5-fold cross-validation setting over the training set). Using an ensemble of five models (and averaging their outputs to elaborate the final prediction) was shown to improve the performance, while extending the inference time to around 18 s per full volume.

4.3   Experimental Results

In Table 1, we gather the results (DICE) obtained over the training and validation BraTS 2018 datasets (in the 5-fold cross-validation setting). The whole tumor class represents the performance of the first stage of our classification system (evaluated on all the classes merged into one—exactly as the first stage model is trained). Here, we report the average DICE for 5 non-overlapping folds of the training set T, and the final DICE for validation V, and test Ψ sets obtained using the ensembles of 5 base deep classifiers learned using training examples belonging to different folds. Note that the ground-truth data (pixel labels) for V and Ψ were not known to the participants during the BraTS 2018 challenge, hence they could not be exploited to improve the models.

Table 1. Segmentation performance (DICE) over the BraTS 2018 validation set obtained using our DNNs trained with T1c, T2, and FLAIR images. The scores are presented for whole tumor (WT), tumor core (TC), and enhancing tumor (ET) classes. For the training set, we report the average across 5 non-overlapping folds, whereas for the validation set—the results reported automatically by the BraTS competition server (for validation, we used an ensemble of 5 DNNs trained over different folds).

Dataset	Label	DICE	Sensitivity	Specificity
Training	ET	0.7365	0.8483	0.9981
	WT	0.9268	0.9239	0.9956
	TC	0.8779	0.8891	0.9973
Validation	ET	0.7519	0.8373	0.9972
	WT	0.8980	0.9096	0.9935
	TC	0.8118	0.8142	0.9974

The results show that an ensemble of DNNs manifests fairly good generalization capabilities over the unseen data, and it consistently obtains high-quality classification. Interestingly, we did *not* use any data augmentation techniques in our approach (which can be perceived as an implicit regularization), and even without increasing the size and heterogeneity of the training data, the ensembles were able to accurately delineate brain tumors in unseen scans. It also indicates that data augmentation could potentially further improve the capabilities (both detection and segmentation) of our deep models by providing a large number of artificially created (but visually plausible and anatomically correct) training examples generated using the original T.

In Table 2, we gather the results obtained over the unseen test set Ψ—we report not only DICE, but also the Hausdorff distance (HD) given as

$$\mathrm{HD}(A, B) = \max \left(h(A, B), h(B, A) \right), \qquad (2)$$

where $h(A, B)$ is the directed Hausdorff distance:

$$h(A, B) = \max_{a \in A} \min_{b \in B} ||a - b||, \qquad (3)$$

and $||\cdot||$ is a norm operation (e.g., Euclidean distance) [32]. It can be noted that this metric is quite sensitive to outliers (the lower HD, the higher quality

Table 2. The results reported for the unseen test set Ψ show that our models can generalize fairly well over the unseen data (however, the results are worse when compared to the validation set). We report DICE alongside the Hausdorff distance (HD).

Measure	DICE (ET)	DICE (WT)	DICE (TC)	HD (ET)	HD (WT)	HD (TC)
Mean	0.6493	0.8590	0.7342	7.3117	9.5128	11.5729
Std dev.	0.3021	0.1272	0.2604	12.8380	16.1665	15.1306
Median	0.7770	0.8983	0.8332	3.0000	4.1231	7.3485
25 quantile	0.6023	0.8376	0.6877	2.0000	2.4495	3.4783
75 quantile	0.8423	0.9299	0.9010	5.0500	7.8717	13.9277

segmentation we have in terms of contour similarity). The results show that our deep-network ensemble can generalize quite well over the unseen data, however the DICE values are slightly lowered when compared to the validation set. We can attribute it to the heterogeneity of the testing data (as mentioned earlier, we did not apply any data augmentation to increase the representativeness of the training set). Interestingly, the whole-tumor segmentation remained at the very same level (see DICE in Table 2), and our method delivered high-quality whole-tumor delineation (we can observe the highest decrease of accuracy for the enhancing part of a tumor, and it amounts to more than 0.08 DICE on average). It also leads us to the conclusion that for tumor segmentation (differentiating between different parts of a lesion), the deep models require larger and more diverse sets (perhaps due to subtle tissue differences which cannot be learnt from a limited number of brain-tumor examples) and potentially better regularization.

5   Conclusions

In this paper, we presented an approach for effective detection and segmentation (into different parts of a tumor) of brain lesions from magnetic resonance images which exploits cascaded multi-modal fully-convolutional neural networks inspired by the U-Net architecture. The first deep network in our pipeline performs tumor detection, whereas the second—multi-class tumor segmentation. We cross-validated the proposed technique (in the 5-fold cross-validation setting) over the newest release of the BraTS dataset (BraTS 2018), and the experimental results showed that:

- Our cascaded multi-modal U-Nets deliver accurate segmentation, and ensembling the models (and averaging the response of base classifiers) trained across separate folds allows us to build the final model which generalizes well over the unseen testing data.
- We showed that our networks can be trained fairly fast (7–8 h using 1 GPU), and deliver real-time inference (around 18 s per volume).
- We showed that our models can be seamlessly applied to both two- and multi-class classification (i.e., tumor detection and segmentation, respectively).

Our current research is focused on applying our techniques to different organs and modalities (e.g., lung PET/CT imaging [23]), and developing data augmentation approaches for medical images. Such algorithms (which ideally generate artificial but visually plausible and realistic images) can be perceived as implicit regularizers which help improve the performance of models over the unseen data by introducing new examples into a training set [18,27,34].

Acknowledgments. This research was supported by the National Centre for Research and Development under the Innomed Research and Development Grant No. POIR.01.02.00-00-0030/15.

References

1. Aljabar, P., Heckemann, R., Hammers, A., Hajnal, J., Rueckert, D.: Multi-atlas based segmentation of brain images: atlas selection and its effect on accuracy. NeuroImage **46**(3), 726–738 (2009)
2. Bakas, S., et al.: Advancing the cancer genome atlas glioma MRI collections with expert segmentation labels and radiomic features. Nat. Sci. Data **4**, 1–13 (2017). https://doi.org/10.1038/sdata.2017.117
3. Bakas, S., et al.: Segmentation labels and radiomic features for the pre-operative scans of the TCGA-GBM collection, the Cancer Imaging Archive (2017). https://doi.org/10.7937/K9/TCIA.2017.KLXWJJ1Q
4. Bakas, S., et al.: Segmentation labels and radiomic features for the pre-operative scans of the TCGA-LGG collection, the Cancer Imaging Archive (2017). https://doi.org/10.7937/K9/TCIA.2017.GJQ7R0EF
5. Bakas, S., Reyes, M., et al.: Identifying the best machine learning algorithms for brain tumor segmentation, progression assessment, and overall survival prediction in the BRATS challenge. CoRR abs/1811.02629 (2018). http://arxiv.org/abs/1811.02629
6. Bauer, S., Seiler, C., Bardyn, T., Buechler, P., Reyes, M.: Atlas-based segmentation of brain tumor images using a markov random field-based tumor growth model and non-rigid registration. In: Proceedings of IEEE EMBC, pp. 4080–4083 (2010). https://doi.org/10.1109/IEMBS.2010.5627302
7. Chander, A., Chatterjee, A., Siarry, P.: A new social and momentum component adaptive PSO algorithm for image segmentation. Expert Syst. Appl. **38**(5), 4998–5004 (2011)
8. Dong, H., Yang, G., Liu, F., Mo, Y., Guo, Y.: Automatic brain tumor detection and segmentation using U-Net based fully convolutional networks. CoRR abs/1705.03820 (2017). http://arxiv.org/abs/1705.03820
9. Fan, X., Yang, J., Zheng, Y., Cheng, L., Zhu, Y.: A novel unsupervised segmentation method for MR brain images based on fuzzy methods. In: Liu, Y., Jiang, T., Zhang, C. (eds.) CVBIA 2005. LNCS, vol. 3765, pp. 160–169. Springer, Heidelberg (2005). https://doi.org/10.1007/11569541_17
10. Geremia, E., Clatz, O., Menze, B.H., Konukoglu, E., Criminisi, A., Ayache, N.: Spatial decision forests for MS lesion segmentation in multi-channel magnetic resonance images. NeuroImage **57**(2), 378–390 (2011)
11. Ghafoorian, M., et al.: Location sensitive deep convolutional neural networks for segmentation of white matter hyperintensities. CoRR abs/1610.04834 (2016). http://arxiv.org/abs/1610.04834

12. Ghafoorian, M., et al.: Transfer learning for domain adaptation in MRI: application in brain lesion segmentation. In: Proceedings of MICCAI, pp. 516–524 (2017)
13. Havaei, M., Dutil, F., Pal, C., Larochelle, H., Jodoin, P.-M.: A convolutional neural network approach to brain tumor segmentation. In: Crimi, A., Menze, B., Maier, O., Reyes, M., Handels, H. (eds.) BrainLes 2015. LNCS, vol. 9556, pp. 195–208. Springer, Cham (2016). https://doi.org/10.1007/978-3-319-30858-6_17
14. Ji, S., Wei, B., Yu, Z., Yang, G., Yin, Y.: A new multistage medical segmentation method based on superpixel and fuzzy clustering. Comp. Math. Meth. Med. **2014**, 747549:1–747549:13 (2014)
15. Kamnitsas, K., et al.: Ensembles of multiple models and architectures for robust brain tumour segmentation. In: Crimi, A., Bakas, S., Kuijf, H., Menze, B., Reyes, M. (eds.) BrainLes 2017. LNCS, vol. 10670, pp. 450–462. Springer, Cham (2018). https://doi.org/10.1007/978-3-319-75238-9_38
16. Korfiatis, P., Kline, T.L., Erickson, B.J.: Automated segmentation of hyperintense regions in FLAIR MRI using deep learning. Tomogr. J. Imaging Res. **2**(4), 334–340 (2016). https://doi.org/10.18383/j.tom.2016.00166
17. Ladgham, A., Torkhani, G., Sakly, A., Mtibaa, A.: Modified support vector machines for MR brain images recognition. In: Proceedings of CoDIT, pp. 032–035 (2013). https://doi.org/10.1109/CoDIT.2013.6689515
18. Lorenzo, P.R., Nalepa, J.: Memetic evolution of deep neural networks. In: Proceedings of the Genetic and Evolutionary Computation Conference GECCO 2018, pp. 505–512. ACM, New York (2018)
19. Mei, P.A., de Carvalho Carneiro, C., Fraser, S.J., Min, L.L., Reis, F.: Analysis of neoplastic lesions in magnetic resonance imaging using self-organizing maps. J. Neurol. Sci. **359**(1–2), 78–83 (2015)
20. Menze, B.H., et al.: The multimodal brain tumor image segmentation benchmark (BRATS). IEEE Trans. Med. Imaging **34**(10), 1993–2024 (2015). https://doi.org/10.1109/TMI.2014.2377694
21. Moeskops, P., Viergever, M.A., Mendrik, A.M., de Vries, L.S., Benders, M.J.N.L., Isgum, I.: Automatic segmentation of MR brain images with a convolutional neural network. IEEE Trans. Med. Imaging **35**(5), 1252–1261 (2016). https://doi.org/10.1109/TMI.2016.2548501
22. Park, M.T.M., et al.: Derivation of high-resolution MRI atlases of the human cerebellum at 3T and segmentation using multiple automatically generated templates. NeuroImage **95**, 217–231 (2014)
23. Pawełczyk, K., et al.: Towards detecting high-uptake lesions from lung CT scans using deep learning. In: Battiato, S., Gallo, G., Schettini, R., Stanco, F. (eds.) ICIAP 2017. LNCS, vol. 10485, pp. 310–320. Springer, Cham (2017). https://doi.org/10.1007/978-3-319-68548-9_29
24. Pinto, A., Pereira, S., Correia, H., Oliveira, J., Rasteiro, D.M.L.D., Silva, C.A.: Brain tumour segmentation based on extremely rand. forest with high-level features. In: Proceedings of IEEE EMBC, pp. 3037–3040 (2015). https://doi.org/10.1109/EMBC.2015.7319032
25. Pipitone, J., et al.: Multi-atlas segmentation of the whole hippocampus and subfields using multiple automatically generated templates. NeuroImage **101**, 494–512 (2014)
26. Rajendran, A., Dhanasekaran, R.: Fuzzy clustering and deformable model for tumor segmentation on MRI brain image: a combined approach. Procedia Eng. **30**, 327–333 (2012). https://doi.org/10.1016/j.proeng.2012.01.868
27. Rezaei, M., et al.: Conditional adversarial network for semantic segmentation of brain tumor. CoRR abs/1708.05227, pp. 1–10 (2017)

28. Ronneberger, O., Fischer, P., Brox, T.: U-Net: convolutional networks for biomedical image segmentation. CoRR abs/1505.04597 (2015)
29. Saha, S., Bandyopadhyay, S.: MRI brain image segmentation by fuzzy symmetry based genetic clustering technique. In: Proceedings of IEEE CEC, pp. 4417–4424 (2007)
30. Simi, V., Joseph, J.: Segmentation of glioblastoma multiforme from MR images - a comprehensive review. Egypt. J. Radiol. Nucl. Med. **46**(4), 1105–1110 (2015)
31. Soltaninejad, M., et al.: Automated brain tumour detection and segmentation using superpixel-based extremely randomized trees in FLAIR MRI. Int. J. of Comp. Assist. Radiol. Surg. **12**(2), 183–203 (2017)
32. Taha, A.A., Hanbury, A.: Metrics for evaluating 3D medical image segmentation: analysis, selection, and tool. BMC Med. Imaging **15**(1), 29 (2015)
33. Taherdangkoo, M., Bagheri, M.H., Yazdi, M., Andriole, K.P.: An effective method for segmentation of MR brain images using the ant colony optimization algorithm. J. Digit. Imaging **26**(6), 1116–1123 (2013)
34. Varghese, A., Mohammed, S., Sai, C., Ganapathy, K.: Generative adversarial networks for brain lesion detection. In: Proceedings of SPIE, vol. 10133, p. 10133 (2017)
35. Verma, N., Cowperthwaite, M.C., Markey, M.K.: Superpixels in brain MR image analysis. In: Proceedings of IEEE EMBC, pp. 1077–1080 (2013). https://doi.org/10.1109/EMBC.2013.6609691
36. Wu, W., Chen, A.Y.C., Zhao, L., Corso, J.J.: Brain tumor detection and segmentation in a CRF (conditional random fields) framework with pixel-pairwise affinity and superpixel-level features. Int. J. of Comp. Assist. Radiol. Surg. **9**(2), 241–253 (2014)
37. Zhao, X., Wu, Y., Song, G., Li, Z., Zhang, Y., Fan, Y.: A deep learning model integrating FCNNs and CRFs for brain tumor segmentation. CoRR abs/1702.04528 (2017)
38. Zhuge, Y., et al.: Brain tumor segmentation using holistically nested neural networks in MRI images. Med. Phys. **44**(10), 5234–5243 (2017). https://doi.org/10.1002/mp.12481
39. Zikic, D., et al.: Decision forests for tissue-specific segmentation of high-grade gliomas in multi-channel MR. In: Ayache, N., Delingette, H., Golland, P., Mori, K. (eds.) MICCAI 2012. LNCS, vol. 7512, pp. 369–376. Springer, Heidelberg (2012). https://doi.org/10.1007/978-3-642-33454-2_46

Automatic Brain Tumor Segmentation by Exploring the Multi-modality Complementary Information and Cascaded 3D Lightweight CNNs

Jun Ma and Xiaoping Yang(✉)

Department of Mathematics, Nanjing University of Science and Technology,
Nanjing 210094, China
yangxp@mail.njust.edu.cn

Abstract. Accurate segmentation of brain tumors is critical for clinical quantitative analysis and decision making for glioblastoma patients. Convolutional neural networks (CNNs) have been widely used for this task. Most of the existing methods integrate the multi-modality information by merging them as multiple channels at the input of the network. However, explicitly exploring the complementary information among different modalities has not been well studied. In fact, radiologists rely heavily on the multi-modality complementary information to manually segment each brain tumor substructure. In this paper, such a mechanism is developed by training the CNNs like the annotation process by radiologists. Besides, a 3D lightweight CNN is proposed to extract brain tumor substructures. The dilated convolutions and residual connections are used to dramatically reduce the parameters without loss of the spatial resolution and the number of parameters is only 0.5M. In the BraTS 2018 segmentation task, experiments with the validation dataset show that the proposed method helps to improve the brain tumor segmentation accuracy compared with the common merging strategy. The mean Dice scores on the validation and testing dataset are (0.743, 0.872, 0.773) and (0.645, 0.812, 0.725) for enhancing tumor core, whole tumor, and tumor core, respectively.

Keywords: Brain tumor · 3D lightweight CNN ·
Complementary information · Segmentation · Multi-modality

1 Introduction

Glioblastoma is the most common primary malignant brain tumor [17]. Medical imaging technologies play an important role in the diagnosis, preoperative planning, intraoperative navigation, and postoperative evaluation of the brain cancer. Magnetic Resonance Imaging (MRI) is the most frequently used imaging method in the clinical routine of brain tumors, because it is noninvasive and free of radiation.

© Springer Nature Switzerland AG 2019
A. Crimi et al. (Eds.): BrainLes 2018, LNCS 11384, pp. 25–36, 2019.
https://doi.org/10.1007/978-3-030-11726-9_3

Brain tumor segmentation in multi-modality MRI scans is crucial for the quantitative analysis in clinic. However, it is time-consuming and labor-intensive for radiologists to manually delineate brain tumors. Automatic segmentation of brain tumors in multi-modality MRI scans has a potential to provide a more effective solution, but due to the highly heterogeneous appearance and various shapes of brain tumors, it is one of the most challenging tasks in medical image analysis. Figure 1 presents a brain tumor case and the corresponding label in the BraTS 2018 training dataset.

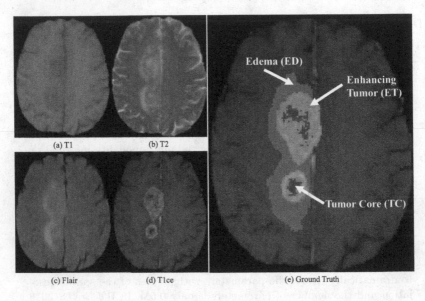

Fig. 1. A brain tumor example (named "Brats18_2013_2_1") in BraTS 2018 dataset. (a–d) show four slices with the same position (107th slice) in different MRI scans. The manual segmentation results of the different substructures are shown in (e).

In recent years, convolutional neural networks (CNNs) have emerged as a powerful tool for medical image segmentation tasks, including organ and lesion segmentation, and achieved unprecedented accuracy. Benefiting from the multimodal brain tumor segmentation challenge [15] which is long-term and competitive, many CNN architectures have been proposed and also achieved state-of-the-art performance. In [13], Kamnitsas et al. constructed an Ensemble of Multiple Models and Architectures (EMMA) for robust brain tumor segmentation including two deepMedic models, three 3D FCNs, and two 3D U-Nets. Wang et al. [19] developed a cascade of fully convolutional neural networks to decompose the multi-class segmentation problem into a sequence of three binary segmentation problems according to the brain tumor substructures hierarchy and proposed anisotropic networks to deal with 3D images as a trade-off among the receptive field, model complexity and memory consumption. The multi-view fusion was used to further reduce noises in the segmentation results.

Isensee et al. [12] modified the U-Net to maximize brain tumor segmentation performance. The architecture consisted of a context aggregation pathway which was used to encode increasingly abstract representations of the input and a localization pathway which was designed to transfer the low level features to a higher spatial resolution.

Most of the existing multi-modality brain tumor segmentation methods use an early-fusion strategy which integrates the multi-modality information from the original MRI scans. For example, four MRI modalities (T1, T2, T1ce, and Flair) are simply merged as four channels at the input of the network [12,13,19]. However, as argued in [18] in the context of multi-modal learning, it is difficult to discover highly non-linear relationships among the low-level features of different modalities. Besides, early-fusion methods implicitly assume that the relationship among different modalities is simple (e.g., linear) and the importance among these modalities is equal for the segmentation of different brain tumor substructures. In fact, when radiologists manually segment tumor substructures, they pay different attention to different modalities. For example, when segmenting the tumor core, radiologists will pay more attention to T1ce modality rather than Flair or T2 modalities. Thus, the importance of different modalities is not the same when segmenting a specific tumor substructure; The complementary information among these modalities plays an important role to the final brain tumor labels. As far as we know, explicitly exploring the complementary information among different modalities has not been well studied for brain tumor substructures segmentation.

In this paper, we train the networks like the manual segmentation process by radiologists to explicitly explore the complementary information among different MRI modalities. Specifically, the pipeline design of the brain tumor segmentation is guided by clinical brain tumor annotation protocol. In addition, we propose a novel 3D lightweight Convolutional Neural Network (CNN) architecture which captures high-level features from a large receptive field without the loss of resolution of the feature maps. The proposed lightweight CNN makes a good balance between the 3D receptive field and model complexity. It has only ten hidden layers and the number of parameters is only $0.5M$. We evaluate the proposed lightweight CNN architecture on the BraTS 2018 validation and testing dataset and achieve the promising segmentation results. Besides, experiments show that an improvement of segmentation accuracy is achieved by exploring the complementary information among different modalities.

2   Methods

2.1   MRI Modality Analysis and Selection of Brain Tumors

The MRI modality selection method is inspired by how radiologists segment the brain tumor substructures. From [15], it can be found that different brain tumor substructures are annotated by different strategies in clinic. Specifically, the edema (belongs to the whole tumor) is segmented primarily from T2 images.

Flair images are used to cross-check the extension of the edema. The enhancing tumor and the tumor core are identified from T1ce images. Motivated by this annotation protocol, different modalities are selected for the segmentation of different brain tumor substructures. Table 1 presents an overview of the used modalities for different substructures segmentation. Briefly, like annotation process by radiologists, we mainly use the Flair and the T2 modalities to segment the whole tumor and use the T1ce modality to segment the enhancing tumor and the tumor core.

Table 1. Overview of the used modalities for the segmentation of different brain tumor substructures.

Substructures	Used modalities
Whole tumor	Flair and T2
Tumor core	T1ce
Enhancing tumor	T1ce

2.2 Proposed 3D Lightweight CNN Architecture

Although for 3D volume data segmentation, traditional 3D architectures such as 3D U-Net and FCN, have high memory consumption in the training phase, the 3D context information would be degenerated if changing the inputs as 2D or 2.5D slices to relieve the computational burden. As a trade-off between memory consumption and 3D context information, a 3D lightweight CNN architecture (Fig. 2) is proposed for 3D brain tumor segmentation which integrates the dilated convolution with different dilated rates and residual connections. Table 2 presents the detailed configurations of the proposed architecture.

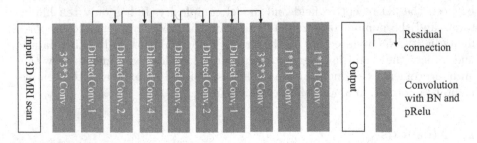

Fig. 2. The proposed 3D lightweight CNN architecture. The number 1, 2, and 4 denote the corresponding dilated rates.

Dilated Convolution with Increasing and Decreasing Dilated Rates. Dilated convolutions have been verified as a very effective structure in deep neural networks [6,21]. The main idea of dilated convolution is to insert "holes" among pixels in traditional convolutional kernels to enlarge the respective field. In order to obtain multi-scale semantic information, we employ different dilation factors in the proposed architecture. The dilation factors are set to 1, 2, and 4 with the increasing and decreasing sequences which can avoid the gridding effect of the standard dilated convolution [8,20].

Residual Connections. To train deep CNNs more effectively, residual connections were first introduced by He et al. [10]. The main idea of residual connections is to learn residual functions through the use of identity-based skip connections which ease the flow of information across units. Our proposed lightweight architecture adds residual connections to each dilated convolutional layer. In addition, each convolutional layer is associated with a batch normalization layer [11] and an element-wise parametric rectified linear Unit (prelu) layer [9] to speed up the convergence of the training process.

Table 2. Configurations of the proposed lightweight CNN architecture. Note that each "Conv" corresponds the sequence Conv-BN-ReLU and a residual connection is added to each "Dilated Conv".

Layers	Configurations (kernel size, channel number)
Conv	$(3 * 3 * 3)$, 8
Dilated Conv	$(3 * 3 * 3)$, 16, dilated factor = 1
Dilated Conv	$(3 * 3 * 3)$, 32, dilated factor = 2
Dilated Conv	$(3 * 3 * 3)$, 64, dilated factor = 4
Dilated Conv	$(3 * 3 * 3)$, 64, dilated factor = 4
Dilated Conv	$(3 * 3 * 3)$, 64, dilated factor = 2
Dilated Conv	$(3 * 3 * 3)$, 64, dilated factor = 1
Conv	$(3 * 3 * 3)$, 64
Conv	$(1 * 1 * 1)$, 64
Conv	$(1 * 1 * 1)$, 2 or 3 for binary/triple segmentation, respectively

2.3 Two-Stage Cascaded Framework

Cascaded strategy has been proved to be an effective way for brain tumor substructures segmentation [19] in the BraTS 2017. Inspired by this work, we deal with the task with a two-stage cascaded framework. Figure 3 presents the whole pipeline. The lightweight CNN architecture is iteratively used to sequentially segment brain tumor substructures. In the first stage, the whole tumor is segmented from Flair and T2 modalities. The segmentation results of Flair and

T2 modalities are merged by simply making an union to generate a bounding box of the whole tumor region of interest (ROI). Besides, an extension with 5 pixels is applied to the whole tumor ROI bounding box so as to avoid possible under-segmentation. Specifically, each side of the bounding box is relaxed by 5 pixels. The whole tumor segmentation from the Flair modality is used as the final whole tumor segmentation result. Besides, we also try to use the union segmentation from Flair and T2 modalities segmentation results as the final whole tumor segmentation result, but there is no improvement of the accuracy. In the second stage, the corresponding T1ce images in the ROI are used to train a new 3D lightweight CNN to make a triple prediction for the enhancing tumor and the tumor core segmentation.

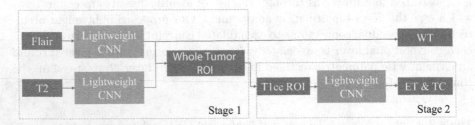

Fig. 3. The two-stage cascaded framework for brain tumor substructures segmentation.

3   Experiments and Results

3.1   Preprocessing

The proposed method was evaluated on the Brain Tumor Segmentation Challenge (BraTS 2018) dataset [2–4]. The training dataset consisted of 210 cases with high grade glioma (HGG), 75 cases with low grade glioma (LGG) and the corresponding manual segmentation. Each case had four 3D MR modalities (T1, T2, Flair, and T1ce).

Table 3. Data preprocessing methods.

Modality	Preprocessing methods
Flair	z-score, histogram equalization, and scale to $[0, 1]$
T2	z-score and scale to $[0, 1]$
T1ce	z-score and scale to $[0, 1]$

To enforce the MR volume data to be more uniform, the following preprocessing strategies (Table 3) were applied to the used modalities. It can be seen that the Flair modality is added an additional histogram equalization compared to the preprocessing methods for the remained modalities. This was because the intensity distributions of the Flair images vary considerably across different cases. Figure 4 presents two examples of Flair modalities in the training dataset. Obviously, the intensity distributions of the two cases still differed remarkably after the z-score normalization. Therefore, the histogram equalization was further applied to make them share similar intensity distribution. For the T2 and T1ce modalities, however, there were no such significant intensity differences among different cases, so a simple z-score preprocessing was enough.

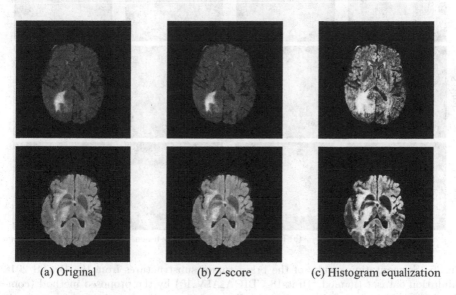

      (a) Original   (b) Z-score   (c) Histogram equalization

Fig. 4. Preprocessing results of two Flair images. The first row is the case named "Brats18_TCIA02_135_1 (78th slice)" and the second row is the case named "Brats18_TCIA02_283_1 (78th slice)". After z-score normalization, there is still a great difference between the two images (the 2nd column). Further, the histogram equalization is applied to make them share similar intensity distributions (the 3rd column).

3.2 Implementation Details

The BraTS 2018 training dataset was randomly divided into training data (80%), validation data (10%), and test data (10%) to find the proper parameters. After that, all the training data were employed to train the final models which were used for the official validation and testing dataset.

The proposed networks were implemented in tensorflow [1] and NiftyNet [7,14]. The input 3D volume data was resized to $64 * 64 * 64$ by the first order spline interpolation. The predicted segmentation was also resized in the same way to retrieve the original 3D volume. The batch size was set to 2 and the maximum number of iterations was 10k. The optimizer was the adam with an initial learning rate 0.001. The loss function was Dice coefficient [16] which can deal with the data imbalance. A $L2$ weight decay of 10^{-5} was used. No external data was used and data augmentation included random rotation, random spatial scaling, and random flipping. The whole training process cost about 30 h on a desktop with an Intel Core i7 CPU and a NVIDIA 1080Ti GPU.

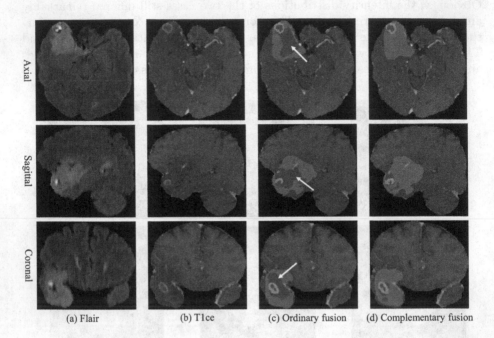

(a) Flair   (b) T1ce   (c) Ordinary fusion   (d) Complementary fusion

Fig. 5. Segmentation results of the brain tumor substructures from the BraTS 2018 validation dataset (named "Brats18_CBICA_ALV_1") by the proposed method (complementary fusion) and its variant (ordinary fusion). Green: edema; red: necrotic and the non-enhancing tumor core; yellow: enhancing tumor core. The obvious mis-segmentations of the non-enhancing tumor core are highlighted by white arrows. (Color figure online)

3.3   Segmentation Results

We test our framework on the BraTS 2018 validation dataset with 66 new cases. To evaluate whether the proposed method (termed as *complementary fusion*) could improve the brain tumor segmentation results, we compare it with the ordinary strategy which merges four MR modalities as four channels at the input of the network. The whole pipeline is also a two-stage cascaded way and we refer to it as *ordinary fusion*. Except the difference at the input of the network, all the hyper-parameters of the *ordinary fusion* are the same with the proposed complementary fusion strategy during the training process.

Table 4 presents quantitative evaluations of the proposed method (*complementary fusion*) and its variant (*complementary fusion*) on the BraTS 2018 validation dataset. For the *ordinary fusion*, the Dice scores are 0.709, 0.851, and 0.751 for enhancing tumor core, whole tumor, and tumor core respectively.

Table 4. Mean values of Dice and 95th percentile Hausdorff measurements of the proposed method on the BraTS 2018 Validation dataset. EN, WT, and TC denote enhancing tumor core, whole tumor and tumor core, respectively. The ordinary fusion denotes that four modalities are simply merged as four channels at the input of the network. The complementary fusion denotes that the proposed method which explicitly explores the complementary information among different modalities.

Dataset	Dice_ET	Dice_WT	Dice_TC	Hdff95_ET	Hdff95_WT	Hdff95_TC
Ordinary fusion	0.709	0.851	0.751	5.65	8.64	13.6
Complementary fusion	0.743	0.872	0.773	4.69	6.12	10.4

For the proposed *complementary fusion*, an improvement is achieved, and the Dice scores are 0.743, 0.872, and 0.773 for these substructures respectively.

Figure 5 shows an example for the brain tumor substructures segmentation from the BraTS 2018 validation dataset. Three views are presented, including the axial view, the sagittal view, and the coronal view. For the simplicity of visualization, only the Flair and T1ce images are shown, because the two modalities can clearly display the whole tumor, enhancing tumor, and tumor core. The first and the second columns present the input images from Flair and T1ce modalities, respectively. We have compared the proposed method with its variant that employed the ordinary fusion method at the input. The third and the fourth columns show the ordinary fusion and the complementary fusion segmentation results, respectively. The green, red, and yellow colors show the edema, tumor core, and enhancing tumor, respectively. It can be observed that the predictions by the *ordinary fusion* seem to have an over segmentation (highlighted by white arrows) of the tumor core. When using the *complementary fusion*, the segmentation results are more accurate.

Table 5 presents quantitative evaluations with the BraTS 2018 testing dataset. It shows the mean values, standard deviations, medians, Dice, and 25 and 75 quantiles of the 95th Hausdorff distance. Compared with the performance on the validation dataset, the performance on the testing dataset is lower, with average Dice scores of 0.645, 0.812, and 0.725 for enhancing tumor core, whole

Table 5. Dice and the 95th percentile Hausdorff measure of the proposed method on the BraTS 2018 Testing dataset. EN, WT, and TC denote enhancing tumor core, whole tumor and tumor core, respectively.

Dataset	Dice_ET	Dice_WT	Dice_TC	Hdff95_ET	Hdff95_WT	Hdff95_TC
Mean	0.645	0.812	0.725	41.1	10.0	28.6
StdDev	0.300	0.175	0.291	105	15.7	78.8
Median	0.768	0.875	0.855	3.00	5.39	5.20
25quantile	0.541	0.829	0.678	1.73	3.74	2.83
75quantile	0.844	0.910	0.921	10.2	8. 22	13.6

tumor, and tumor core, respectively. The higher median values show that good segmentation results are achieved for most cases, and some outliers lead to the lower average scores. The ranking analysis is reported in [5].

4   Discussion and Conclusion

There are several advantages of the proposed framework. Firstly, the complementary information among different modalities is explicitly explored to segment brain tumor substructures which can avoid the interference from other confusing modalities as well as reducing the complexity compared with using all the modalities as inputs simultaneously. Besides, the proposed 3D lightweight CNN effectively uses the dilated convolutions to enlarge the receptive fields and to aggregate the global information. The increasing and decreasing arrangement of the dilate factors can alleviate the gridding effect caused by the standard dilated convolutions. The architecture is very compact and computation efficient. Finally, the cascaded CNNs, which have been proved to be an effective strategy, can separate the complex multiple class segmentation into simper problems and reduce false positives by spatial constrains of brain tumor anatomical structures.

In conclusion, we explicitly explore the complementary information among different modalities according to the clinical annotation protocol. In addition, a compact 3D lightweight CNN architecture is proposed and the number of parameters is only $0.5M$. The proposed approach achieves a promising performance on the BraTS 2018 validation and testing dataset. Experiments with the BraTS 2018 validation dataset show that the complementary fusion strategy helps to improve the brain tumor segmentation accuracy compared with the ordinary fusion method.

Acknowledgements. This work is supported by National Nature Science Foundation of China (No: 11531005). And we also would like to thank the NiftyNet team, they developed the open source convolutional neural networks platform for medical image analysis, which made us more efficiently to build our model. Last but not least, we gratefully thank the BraTS organizers and data contributors for their efforts on hosting the excellent challenge.

References

1. Abadi, M., et al.: TensorFlow: large-scale machine learning on heterogeneous distributed systems. arXiv preprint arXiv:1603.04467 (2016)
2. Bakas, S., et al.: Advancing the cancer genome atlas glioma MRI collections with expert segmentation labels and radiomic features. Sci. Data **4**, 170117 (2017)
3. Bakas, S., et al.: Segmentation labels and radiomic features for the pre-operative scans of the TCGA-GBM collection. Cancer Imaging Arch. (2017)
4. Bakas, S., et al.: Segmentation labels and radiomic features for the pre-operative scans of the TCGA-LGG collection. Cancer Imaging Arch. (2017)

5. Bakas, S., Reyes, M., Menze, B., et al.: Identifying the best machine learning algorithms for brain tumor segmentation, progression assessment, and overall survival prediction in the brats challenge. arXiv preprint arXiv:1811.02629 (2018)
6. Chen, L., Papandreou, G., Kokkinos, I., Murphy, K., Yuille, A.L.: DeepLab: semantic image segmentation with deep convolutional nets, atrous convolution, and fully connected CRFs. IEEE Trans. Pattern Anal. Mach. Intell. **40**(4), 834–848 (2018)
7. Gibson, E., et al.: NiftyNet: a deep-learning platform for medical imaging. Comput. Methods Programs Biomed. **158**, 113–122 (2018)
8. Hamaguchi, R., Fujita, A., Nemoto, K., Imaizumi, T., Hikosaka, S.: Effective use of dilated convolutions for segmenting small object instances in remote sensing imagery. In: 2018 IEEE Winter Conference on Applications of Computer Vision, WACV 2018, Lake Tahoe, NV, USA, 12–15 March 2018, pp. 1442–1450 (2018)
9. He, K., Zhang, X., Ren, S., Sun, J.: Delving deep into rectifiers: surpassing human-level performance on imagenet classification. In: 2015 IEEE International Conference on Computer Vision, ICCV 2015, Santiago, Chile, 7–13 December 2015, pp. 1026–1034 (2015)
10. He, K., Zhang, X., Ren, S., Sun, J.: Deep residual learning for image recognition. In: IEEE Conference on Computer Vision and Pattern Recognition, pp. 770–778 (2016)
11. He, K., Zhang, X., Ren, S., Sun, J.: Identity mappings in deep residual networks. In: Leibe, B., Matas, J., Sebe, N., Welling, M. (eds.) ECCV 2016. LNCS, vol. 9908, pp. 630–645. Springer, Cham (2016). https://doi.org/10.1007/978-3-319-46493-0_38
12. Isensee, F., Kickingereder, P., Wick, W., Bendszus, M., Maier-Hein, K.H.: Brain tumor segmentation and radiomics survival prediction: contribution to the BRATS 2017 challenge. In: Crimi, A., Bakas, S., Kuijf, H., Menze, B., Reyes, M. (eds.) BrainLes 2017. LNCS, vol. 10670, pp. 287–297. Springer, Cham (2018). https://doi.org/10.1007/978-3-319-75238-9_25
13. Kamnitsas, K., et al.: Ensembles of multiple models and architectures for robust brain tumour segmentation. In: Crimi, A., Bakas, S., Kuijf, H., Menze, B., Reyes, M. (eds.) BrainLes 2017. LNCS, vol. 10670, pp. 450–462. Springer, Cham (2018). https://doi.org/10.1007/978-3-319-75238-9_38
14. Li, W., Wang, G., Fidon, L., Ourselin, S., Cardoso, M.J., Vercauteren, T.: On the compactness, efficiency, and representation of 3D convolutional networks: brain parcellation as a pretext task. In: Niethammer, M., et al. (eds.) IPMI 2017. LNCS, vol. 10265, pp. 348–360. Springer, Cham (2017). https://doi.org/10.1007/978-3-319-59050-9_28
15. Menze, B.H., et al.: The multimodal brain tumor image segmentation benchmark (BRATS). IEEE Trans. Med. Imaging **34**(10), 1993–2024 (2015)
16. Milletari, F., Navab, N., Ahmadi, S.: V-Net: fully convolutional neural networks for volumetric medical image segmentation. In: Fourth International Conference on 3D Vision, 3DV 2016, Stanford, CA, USA, 25–28 October 2016, pp. 565–571 (2016)
17. Ostrom, Q.T., et al.: CBTRUS statistical report: primary brain and central nervous system tumors diagnosed in the united states in 2008–2012. Neuro-oncology **17**(suppl-4), iv1–iv62 (2015)
18. Srivastava, N., Salakhutdinov, R.: Multimodal learning with deep boltzmann machines. J. Mach. Learn. Res. **15**(1), 2949–2980 (2014)

19. Wang, G., Li, W., Ourselin, S., Vercauteren, T.: Automatic brain tumor segmentation using cascaded anisotropic convolutional neural networks. In: Crimi, A., Bakas, S., Kuijf, H., Menze, B., Reyes, M. (eds.) BrainLes 2017. LNCS, vol. 10670, pp. 178–190. Springer, Cham (2018). https://doi.org/10.1007/978-3-319-75238-9_16
20. Wang, P., et al.: Understanding convolution for semantic segmentation. In: 2018 IEEE Winter Conference on Applications of Computer Vision, WACV 2018, Lake Tahoe, NV, USA, 12–15 March 2018, pp. 1451–1460 (2018)
21. Yu, F., Koltun, V.: Multi-scale context aggregation by dilated convolutions. arXiv preprint arXiv:1511.07122 (2015)

Deep Convolutional Neural Networks Using U-Net for Automatic Brain Tumor Segmentation in Multimodal MRI Volumes

Adel Kermi[1]([⊠]) [iD], Issam Mahmoudi[1], and Mohamed Tarek Khadir[2]

[1] LMCS Laboratory, National Higher School of Computer Sciences (ESI), BP.68M, 16309 Oued-Smar, El-Harrach, Algiers, Algeria
{a_kermi, di_mahmoudi}@esi.dz
[2] LabGed Laboratory, Department of Computer Sciences, University Badji-Mokhtar of Annaba, BP.12, 23000 Annaba, Algeria
khadir@labged.net

Abstract. Precise 3D computerized segmentation of brain tumors remains, until nowadays, a challenging process due to the variety of the possible shapes, locations and image intensities of various tumors types. This paper presents a fully automated and efficient brain tumor segmentation method based on 2D Deep Convolutional Neural Networks (DNNs) which automatically extracts the whole tumor and intra-tumor regions, including enhancing tumor, edema and necrosis, from pre-operative multimodal 3D-MRI. The network architecture was inspired by U-net and has been modified to increase brain tumor segmentation performance. Among applied modifications, Weighted Cross Entropy (WCE) and Generalized Dice Loss (GDL) were employed as a loss function to address the class imbalance problem in the brain tumor data. The proposed segmentation system has been tested and evaluated on both, BraTS'2018 training and validation datasets, which include a total of 351 multimodal MRI volumes of different patients with HGG and LGG tumors representing different shapes, giving promising and objective results close to manual segmentation performances obtained by experienced neuro-radiologists. On the challenge validation dataset, our system achieved a mean enhancing tumor, whole tumor, and tumor core dice score of 0.783, 0.868 and 0.805 respectively. Other quantitative and qualitative evaluations are presented and discussed along the paper.

Keywords: Brain tumor segmentation · 3D-MRI · Machine learning· Deep learning · Convolutional Neural Networks · U-net· BraTS'2018 challenge

1 Introduction

Brain tumor segmentation in multimodal Magnetic Resonance Imaging (MRI) is widely used as a vital process for surgical planning and simulation, treatment planning prior to radiation therapy, therapy evaluation [1–5], and intra-operative neuro navigation and image neurosurgery [6–8]. However, segmenting brain tumor manually is

© Springer Nature Switzerland AG 2019
A. Crimi et al. (Eds.): BrainLes 2018, LNCS 11384, pp. 37–48, 2019.
https://doi.org/10.1007/978-3-030-11726-9_4

not only a challenging task, but also a time-consuming one, favoring therefore, the emergence of computerized approaches.

Despite considerable research works and encouraging results in the medical imaging domain, fast and precise 3D computerized brain tumors segmentation remains until now a challenging process and a very difficult task to achieve because brain tumors may appear in different size, shape, location and image intensity [2–5]. Many recent research adopted deep learning methods [9], specifically Convolutional Neural Networks (CNNs) [9–12]. CNN shown their effectiveness and were proved successful to automatically classify the normal and pathological brain MRI scans in the past few BraTS challenges as well as other semantic and medical segmentation problems.

This paper proposes an automated and efficient segmentation method of whole tumor and intra-tumor structures, including enhancing tumor, edema and necrosis, in multimodal 3D-MRI. It is based on 2D Deep Convolutional Neural Networks (DNNs) using a modified U-net architecture [10]. The proposed DNN model is trained to segment both High Grade Glioma (HGG) and Lower Grade Glioma (LGG) volumes.

The rest of the paper is organized as follows. First, Sect. 2 presents an overview of the proposed segmentation method. Experimental results with their evaluations are given in Sect. 3. Finally, a conclusion and future work are presented in Sect. 4.

2   The Proposed Method

The proposed segmentation system is entirely automated. The brain tumor segmentation process is based on deep learning more precisely on 2D Convolutional Neural Networks. It includes the main following steps: pre-processing of the 3D-MRI data, training using a U-net architecture, and brain tumoral structures prediction.

2.1   Data and Pre-processing

The BraTS'2018 challenge training dataset consists of 210 pre-operative multimodal MRI scans of subjects with HGG and 75 scans of subjects with LGG, and the BraTS'2018 challenge validation dataset includes 66 different multimodal 3D-MRI [13–16]. Images were acquired at 19 different centers using MR scanners from different vendors and with 3T field strength. They comprise co-registered native (T1) and contrast-enhanced T1-weighted (T1Gd) MRI, as well as T2-weighted (T2) and T2 Fluid Attenuated Inversion Recovery (FLAIR) MRI. All 3D-MRI of BraTS'2018 dataset have a volume dimension of $240 \times 240 \times 155$. They are distributed, co-registered to the same anatomical template and interpolated to the same resolution (1 mm^3). All MRI volumes have been segmented manually, by one to four raters, and their annotations were approved by experienced neuro-radiologists. Each tumor was segmented into edema, necrosis and non-enhancing tumor and active/enhancing tumor.

First, a minimal pre-processing of MRI data is applied, as in [11]. The 1% highest and lowest intensities were removed, then each modality of MR images was normalized by subtracting the mean and dividing by the standard deviation of the intensities within the slice. To address the class imbalance problem in the data, data augmentation technique [17] were employed. This consists in adding new synthetic images by performing operations and transformations on data and the corresponding manual tumors segmentation images obtained by human experts (i.e., ground truth data). The transformations comprise rotation, translation, and horizontal flipping and mirroring.

2.2 Network Architecture and Training

The CNN used in this study has a similar architecture as that of U-net [10]. Our network architecture can be seen in Fig. 1. It consists of a contracting path (left side) and an expanding path (right side). The contracting path consists of 3 pre-activated residual blocks, as in [18, 19], instead of plain blocks in the original U-net. Each block has two convolution units each of which comprises a Batch Normalization (BN) layer, an activation function, called Parametric Rectified Linear Unit (PReLU) [20],instead of ReLU function used in the original architecture [10], and a convolutional layer, like in [12], instead of using Maxpooling [10], with Padding = 2, Stride = 1 and a 3×3 size filter. For down sampling, a convolution layer with a 2×2 filter and a stride of 2 is applied. At each down sampling step, the number of feature channels is doubled. The contracting path is followed by a fourth residual unit that acts as a bridge to connect both paths. In the same way, the expanding path is built using 3 residual blocks. Prior to each block, there is an upsampling operation which increases the feature map size by 2, followed by a 2×2 convolution and a concatenation with the feature maps corresponding to the contracting path. In the last layer of the expanding path, a 1×1 convolution with the Softmax activation function is used to map the multi-channel feature maps to the desired number of classes.

In total, the proposed network model contains 7 residual blocks, 25 convolution layers, 15 layers of BN and 10159748 parameters to optimize.

The designed network was trained with axial slices extracted from training MRI set, including HGG and LGG cases, and the corresponding ground truth segmentations. The goal is to find the network parameters (weights and biases) that minimize a loss function. In this work, this can be achieved by using Stochastic Gradient Descent algorithm (SGD) [17], at each iteration SGD updates the parameters towards the opposite direction of the gradients. In our network model, we used a loss function that adds Weighted Cross Entropy (WCE) [17] and Generalized Dice (GDL) [21] to address the class imbalance problem present in brain tumor data. So, the two components of the loss function are defined as follows:

$$WCE = -\frac{1}{K}\sum_{k}\sum_{i}^{L} W_{i}g_{ik}\log(p_{ik}) \tag{1}$$

$$GDL = 1 - 2\frac{\sum_{i}^{L} w_{i}\sum_{k} g_{ik}p_{ik}}{\sum_{i}^{L} w_{i}\sum_{k}(g_{ik} + p_{ik})} \tag{2}$$

where L is the total number of labels, K denotes the batch size. w_{i} represents the weight assigned to the ith label. As in [21], we set wi to $\frac{1}{\left(\sum_{k} g_{ik}\right)^{2}}$. p_{ik} and g_{ik} representing the value of the (ith, kth) pixel of the segmented binary image and of the binary ground truth image, respectively.

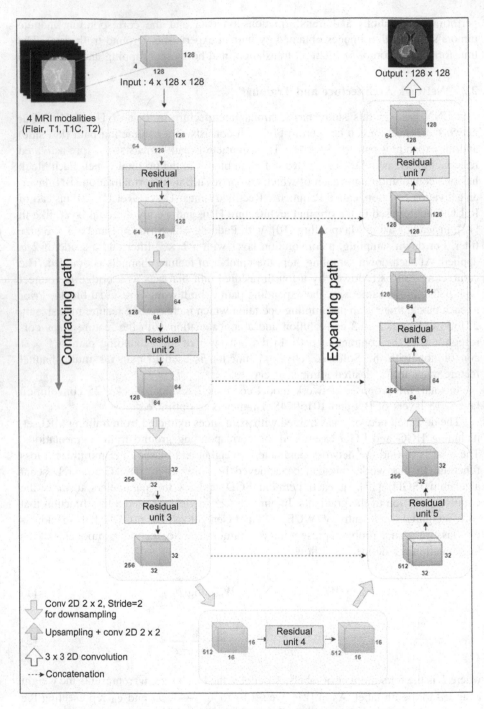

Fig. 1. Architecture of the proposed Deep Convolutional Neural Network.

2.3   Brain Tumoral Structures Prediction

After network training, prediction may be performed. This step consists to provide the network with the four MRI modalities of an unsegmented volume that it has never processed or encountered before, and it must be able to return a segmented image.

3   Experimental Results and Discussion

In this study, we have tested and evaluated our segmentation system on pre-operative multimodal MRI scans of both the training/testing and the validation datasets of the BraTS'2018 challenge [22]. The results of automatically segmented tumors, denoted by A, can be compared with manually segmented tumors by human experts, denoted by B, which are considered as ground truth for evaluation. The results presented in subsequent sections, were obtained during the BraTS'2018 challenge, for training and validation [22]. The top 63 approaches are further compared in terms of results (*Dice*, *Sensitivity* and *Specificity*) and one surface measure based on the Hausdorff distance (*HD*), in [23] on 191 cases. These measures allow to assess the segmentation results accuracy, as well as measuring the similarity between the segmentations A and B [2, 24]. The Dice metric is computed as a performance metric. It measures the similarity between two volumes A and B, corresponding to the output segmentation of the model and clinical ground truth annotations, respectively. The Sensitivity metric measures the proportion of positive voxels of the real brain tumor that are correctly segmented as such, while Specificity metric indicates how well the true negatives are predicted. Employing Sensitivity and Specificity can provide a good assessment of the segmentation result. The *HD* metric indicates the segmentation quality at the tumor's border by evaluating the greatest distance between the two segmentation surfaces A and B, and is independent of the tumor size.

3.1   Performance on 20% of BraTS'2018 Training Dataset (Testing Set)

Preliminary segmentation results for the 285 3D-MRI of the BraTS'2018 training data set have been obtained using 80% of this data set (i.e., 228 subjects) for training and the remaining 20% (i.e., 57 subjects) for validation purposes. Results obtained by our automated system for 10 sample cases are shown in Figs. 2 and 3. Figure 2 shows segmentation results from 5 multimodal MRI where HGG tumors are present and Fig. 3 shows other segmentation results from other 5 MRI with LGG tumors. In these figures, each row represents one clinical case. In the first four columns from left to right, images show one axial slice of MRI acquired in Flair, T1, T1C and T2 modality, respectively, used as input channels to our CNN model. In the fifth and the sixth columns, images show the ground truth (GT) and the prediction labels respectively, where we can distinguish intra-tumoral regions by color-code: enhancing tumor (yellow), peritumoral edema (green) and necrotic and non-enhancing tumor (red). As it can be seen, tumors in the brain MRI of the 10 cases vary in size, shape, position and intensity. By visual inspection, the proposed method usually generates segmentations (Prediction) sensibly similar to the ones obtained by the experts (GT).

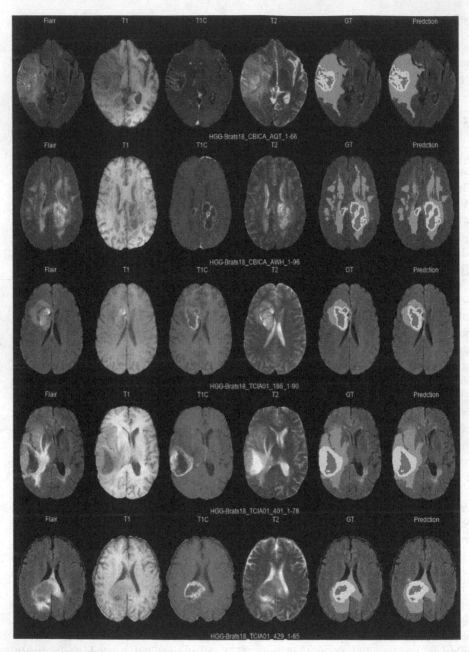

Fig. 2. Intra-tumoral structures segmentation results from 5 multimodal 3D-MRI with HGG of BraTS'2018 training dataset corresponding to 5 different subjects. (Color figure online)

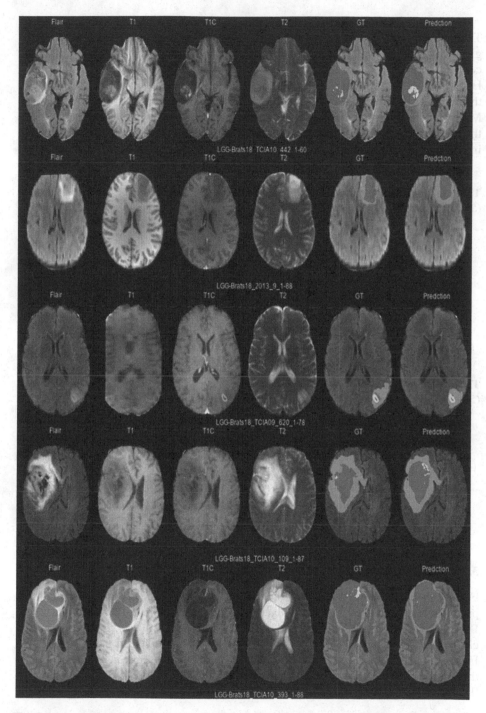

Fig. 3. Intra-tumoral structures segmentation results from 5 other multimodal 3D-MRI with LGG of BraTS'2018 training dataset corresponding to 5 different subjects. (Color figure online)

A quantitative evaluation of segmentation results of Enhancing Tumor (ET), Whole Tumor (WT) and Tumor Core (TC) using the four previously mentioned metrics is given in Tables 1 and 2. Mean, standard deviation, median and 25th and 75th percentile are given for Dice and Sensitivity metrics in Table 1 and for Specificity and Hausdorff distance in Table 2. Values presented in Table 1 show high performance on the Dice metric for WT region, but lower performance for ET and TC regions. Moreover, the proposed method showed good performances for the segmentation of the three intra-tumoral regions. However, the effectiveness, of the approach, is high-lighted in the case of HGG tumors, when compared with the LGG ones.

Table 1. Quantitative evaluation of segmentation results on 20% of BraTS'2018 training dataset (57 MRI scans) using *Dice* and *Sensitivity* metrics.

	Dice			Sensitivity		
	ET	WT	TC	ET	WT	TC
Mean	0.717	0.867	0.798	0.778	0.907	0.84
Std. Dev.	0.275	0.078	0.226	0.3	0.107	0.223
Median	0.831	0.887	0.889	0.912	0.943	0.927
25 quantile	0.726	0.839	0.808	0.777	0.896	0.847
75 quantile	0.859	0.926	0.935	0.951	0.974	0.961

Table 2. Quantitative evaluation of segmentation results on 20% of BraTS'2018 training dataset (57 MRI scans) using *Specificity* and *Hausdorff distance* metrics.

	Specificity			Hausdorff95		
	ET	WT	ET	WT	ET	WT
Mean	0.999	0.998	0.999	4.742	8.706	6.4
Std. Dev.	0.001	0.001	0.002	2.079	2.822	3.685
Median	1	0.999	1	4.123	8.062	5.099
25 quantile	0.999	0.998	0.999	3	6.442	3.871
75 quantile	1	0.999	1	6.633	10.951	8.303

3.2 Performance on BraTS'2018 Validation Dataset

For our participation to BraTS'2018 competition, 100% of the training, including the previous testing dataset (i.e., 285 subjects) is used for training. The performance on BraTS'2018 validation dataset, which is composed of 66 subjects, is diffused in the challenge leaderboard Web site[1] and presented with more statistics in Tables 3, 4, Figs. 4 and 5. In this context, we can compare the obtained segmentation results with those of other participants. The method achieved a mean ET, WT, and TC dice score of 0.783, 0.868 and 0.805 respectively. These scores are close to those obtained by the top

---

[1] https://www.cbica.upenn.edu/BraTS18/lboardValidation.html.

performing methods. Also, an average *HD* scores of 3.728, 8.127 and 9.84 for ET, WT, and TC, respectively were obtained. In addition, it was observed that our DNN model maintains similar WT scores on both, 20% of BraTS'2018 training/testing set used for validation and the final validation dataset proposed for the competition purposes. However, a slight increase in performance on the validation dataset was observed in the ET and TC compartments. It should be noted that this performance was not obtained by overfitting the validation data (i.e., our DNN model has not previously trained on MRI volumes of BraTS'2018 validation dataset).

Table 3. Quantitative evaluation of segmentation results on BraTS'2018 validation dataset (66 MRI scans) by using *Dice* and *Sensitivity* metrics.

	Dice			*Sensitivity*		
	ET	WT	TC	ET	WT	TC
Mean	0.783	0.868	0.805	0.826	0.895	0.807
Std. Dev.	0.216	0.101	0.199	0.241	0.149	0.222
Median	0.846	0.898	0.891	0.901	0.955	0.895
25 quantile	0.769	0.855	0.756	0.82	0.901	0.71
75 quantile	0.893	0.919	0.928	0.969	0.971	0.965

Table 4. Quantitative evaluation of segmentation results on BraTS'2018 validation dataset (66 MRI scans) by using *Specificity* and *Hausdorff distance* metrics.

	Specificity			*Hausdorff95*		
	ET	WT	ET	WT	ET	WT
Mean	0.997	0.991	0.997	3.728	8.127	9.84
Std. Dev.	0.004	0.007	0.003	4.471	10.426	15.385
Median	0.998	0.993	0.998	2.236	4.243	5.431
25 quantile	0.997	0.988	0.997	1.637	3	2.871
75 quantile	0.999	0.996	0.999	3.317	7.778	10.728

This performance can be explained by the fact that the number of learned cases (training dataset) used later for the segmentation of the validation dataset is larger than the one used for the BraTS'2018 segmentation on the training/testing dataset. This represents 285 and 228 cases for both trainings respectively. It is also possible that the slight improvements obtained on the validation dataset, are due to the fact that this latter contains more MRI with HGG tumors than MRI with LGG tumors. Indeed, the segmentation efficiency obtained using the proposed network, is more evident on HGG volumes when compared to LGG ones.

Boxplots showing the dispersion of Dice and Sensitivity scores are represented in Fig. 4 and boxplots of the dispersion of Specificity and HD scores are represented in Fig. 5. In these figures, boxplots show quartile ranges of the scores; whiskers and dots '●' indicate outliers; and 'x' indicates the mean score.

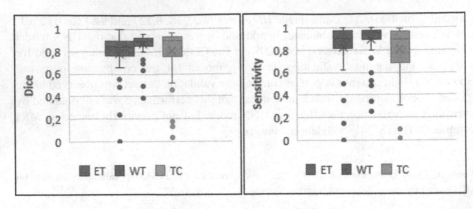

Fig. 4. Dispersion of *Dice* and *Sensitivity* scores for results segmentation of ET, WT, and TC in multimodal MRI scans of the 66 subjects of BraTS'2018 validation dataset.

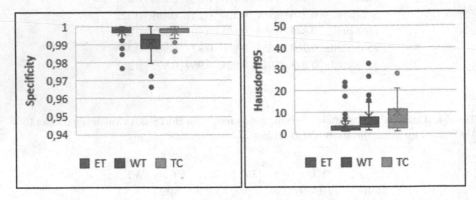

Fig. 5. Dispersion of *Specificity* and *Hausdorff distance* scores for results segmentation of ET, WT, and TC in multimodal MRI scans of the 66 subjects of BraTS'2018 validation dataset.

4   Conclusion and Future Work

In this paper, a fully automatic and accurate method for segmentation of whole brain tumor and intra-tumoral regions using a 2D deep convolutional network based on a well-known architecture in medical imaging called "U-net" is proposed. The constructed DNN model was trained to segment both HGG and LGG volumes.

The proposed method was tested and evaluated quantitatively on both BraTS'2018 training and challenge validation datasets. The total learning computation time of the 285 multimodal MRI volumes of BraTS'2018 training dataset is 185 h on a Cluster machine with Intel Xeon E5-2650 CPU@ 2.00 GHz (64 GB) and NVIDIA Quadro 4000–448 Core CUDA (2 GB) GPU. The average segmentation time of a brain tumor and its components from a given MRI volume is about 62 s on the same GPU. The different tests showed that the segmentation results were very satisfactory, and the evaluation measures confirm that our results are very similar to those manually obtained by the experts, although the proposed method can be further improved.

As future work, a more powerful GPU to further accelerate learning phase of DNN is planned. Thus, a larger number of CNN topologies as well other data augmentation methods may be tested. Also, other interesting perspective consists to use ensemble learning methods, like Stacking and Blending, to improve segmentation performance in tumor core and active tumor regions. Finally, a future work possibility may focus on refining the segmentation results by reducing the false-positive rate using post-processing techniques, such as: applying a conditional random field (CRF) and removing small connected components.

Acknowledgments. This work was in part financially supported by an Algerian research project (CNEPRU) funded by the Ministry of Higher Education and Scientific Research (Project title and number: "PERFORM", B*04120140014). We would like to thank Bakas, S., Ph.D. and Post-doctoral researcher at SBIA of Perelman School of Medicine University of Pennsylvania – USA for providing the entire BraTS'2018 datasets employed in this study. We also gratefully acknowledge the support of CERIST, the Algerian Research Center for Scientific and Technical Information, for allowing us the use of "IBNBADIS" Cluster, without which we could not have performed these tests.

References

1. Dass, R., Priyanka, Devi, S.: Image segmentation techniques. Int. J. Electron. Commun. Technol. (IJECT) **3**, 66–70 (2012)
2. Kermi, A., Andjouh, K., Zidane, F.: Fully automated brain tumour segmentation system in 3D-MRI using symmetry analysis of brain and level sets. IET Image Process. **12**, 1964–1971 (2018)
3. Khotanlou, H., Colliot, O., Atif, J., Bloch, I.: 3D brain tumor segmentation in MRI using fuzzy classification, symmetry analysis and spatially constrained deformable models. Fuzzy Sets Syst. **160**, 1457–1473 (2009)
4. Prastawaa, M., Bullitt, E., Geriga, G.: Simulation of brain tumors in MR images for evaluation of segmentation efficacy. Med. Image Anal. **13**, 297–311 (2009)
5. Sharma, N., Aggarwal, L.M.: Automated medical image segmentation techniques. J. Med. Phys. **35**, 3–14 (2010)
6. Despotovic, I., Goossens, B., Philips, W.: MRI segmentation of the human brain: challenges, methods, and applications. Comput. Math. Methods Med. **2015**, 23 (2015)
7. Wirtz, C.R., et al.: The benefit of neuronavigation for neurosurgery analyzed by its impact on glioblastoma surgery. Neurol. Res. **22**, 354–360 (2000)
8. Yanyun, L., Zhijian, S.: Automated brain tumor segmentation in magnetic resonance imaging based on sliding-window technique and symmetry analysis. Chin. Med. J. **127**, 462–468 (2014)
9. Işın, A., Direkoğlu, C., Şah, M.: Review of MRI-based brain tumor image segmentation using deep learning methods. Procedia Comput. Sci. **102**, 317–324 (2016)
10. Ronneberger, O., Fischer, P., Brox, T.: U-Net: convolutional networks for biomedical image segmentation. In: Navab, N., Hornegger, J., Wells, W.M., Frangi, A.F. (eds.) MICCAI 2015. LNCS, vol. 9351, pp. 234–241. Springer, Cham (2015). https://doi.org/10.1007/978-3-319-24574-4_28
11. Havaei, M., et al.: Brain tumor segmentation with Deep Neural Networks. Med. Image Anal. **35**, 18–31 (2017)

12. Milletari, F., Navab, N., Ahmadi, S.A.: V-Net: fully convolutional neural networks for volumetric medical image segmentation. In: 2016 Fourth International Conference on 3D Vision (3DV), pp. 565–571 (2016)
13. Bakas, S., et al.: Advancing The Cancer Genome Atlas glioma MRI collections with expert segmentation labels and radiomic features. Sci. Data **4**, 170117 (2017)
14. Bakas, S., et al.: Segmentation labels and radiomic features for the pre-operative scans of the TCGA-GBM collection. The Cancer Imaging Archive (2017)
15. Bakas, S., et al.: Segmentation labels and radiomic features for the pre-operative scans of the TCGA-LGG collection. The Cancer Imaging Archive (2017)
16. Menze, B.H., et al.: The multimodal brain tumor image segmentation benchmark (BRATS). IEEE Trans. Med. Imaging **34**, 1993–2024 (2015)
17. Goodfellow, I., Bengio, Y., Courville, A.: Deep Learning. MIT Press, Cambridge (2016)
18. He, K., Zhang, X., Ren, S., Sun, J.: Deep residual learning for image recognition. In: 2016 IEEE Conference on Computer Vision and Pattern Recognition (CVPR), pp. 770–778 (2016)
19. He, K., Zhang, X., Ren, S., Sun, J.: Identity mappings in deep residual networks. In: Leibe, B., Matas, J., Sebe, N., Welling, Max (eds.) ECCV 2016. LNCS, vol. 9908, pp. 630–645. Springer, Cham (2016). https://doi.org/10.1007/978-3-319-46493-0_38
20. He, K., Zhang, X., Ren, S., Sun, J.: Delving deep into rectifiers: surpassing human-level performance on ImageNet classification. In: Proceedings of the 2015 IEEE International Conference on Computer Vision (ICCV), pp. 1026–1034. IEEE Computer Society (2015)
21. Sudre, C.H., Li, W., Vercauteren, T., Ourselin, S., Jorge Cardoso, M.: Generalised dice overlap as a deep learning loss function for highly unbalanced segmentations. In: Cardoso, M.J., et al. (eds.) DLMIA/ML-CDS -2017. LNCS, vol. 10553, pp. 240–248. Springer, Cham (2017). https://doi.org/10.1007/978-3-319-67558-9_28
22. Kermi, A., Mahmoudi, I., Khadir, M.T.: Brain tumor segmentation in multimodal 3D-MRI of BraTS'2018 datasets using Deep Convolutional Neural Networks. In: Pre-conference Proceedings of the 7th International MICCAI BraTS'2018 Challenge, Granada, Spain, pp. 252–263 (2018)
23. Bakas, S., et al.: Identifying the Best Machine Learning Algorithms for Brain Tumor Segmentation, Progression Assessment, and Overall Survival Prediction in the BRATS Challenge. arXiv preprint arXiv:1811.02629 (2018)
24. Castillo, L.S., Daza, L.A., Rivera, L.C., Arbelàez, P.: Volumetric multimodality neural network for brain tumor segmentation. In: Proceedings of the 6th MICCAI BraTS Challenge, Quebec, Canada, pp. 34–41 (2017)

Multimodal Brain Tumor Segmentation Using Cascaded V-Nets

Rui Hua[1,2], Quan Huo[2], Yaozong Gao[2], Yu Sun[1], and Feng Shi[2(✉)]

[1] School of Biomedical Engineering, Southeast University, Nanjing, China
[2] United Imaging Intelligence, Shanghai, China
feng.shi@united-imaging.com

Abstract. In this work, we propose a novel cascaded V-Nets method to segment brain tumor substructures in multimodal brain magnetic resonance imaging (MRI). Although V-Net has been successfully used in many segmentation tasks, we demonstrate that its performance could be further enhanced by using a cascaded structure and ensemble strategy. Briefly, our baseline V-Net consists of four levels with encoding and decoding paths and intra- and inter-path skip connections. Focal loss is chosen to improve performance on hard samples as well as balance the positive and negative samples. We further propose three preprocessing pipelines for multimodal MRI images to train different models. By ensembling the segmentation probability maps obtained from these models, segmentation result is further improved. In other hand, we propose to segment the whole tumor first, and then divide it into tumor necrosis, edema, and enhancing tumor. Experimental results on BraTS 2018 online validation set achieve average Dice scores of 0.9048, 0.8364 and 0.7748 for whole tumor, tumor core and enhancing tumor, respectively. The corresponding values for BraTS 2018 online testing set are 0.8761, 0.7953 and 0.7364, respectively. We further make a prediction of patient overall survival by ensembling multiple classifiers for long, mid and short groups, and achieve accuracy of 0.519, mean square error of 367239 and Spearman correlation coefficient of 0.168.

Keywords: Deep learning · Brain tumor · Segmentation · V-Net

1 Introduction

Gliomas are the most common brain tumors and comprise about 30% of all brain tumors. Gliomas occur in the glial cells of the brain or the spine [1]. They can be further categorized into low-grade gliomas (LGG) and high-grade gliomas (HGG) according to their pathologic evaluation. LGG are well-differentiated and tend to exhibit benign tendencies and portend a better prognosis for the patients. HGG are undifferentiated and tend to exhibit malignant and usually lead to a worse prognosis. With the development of the Magnetic Resonance Imaging (MRI), multimodal MRI plays an important role in disease diagnosis. Different MRI modalities are developed sensitive to different tissues. For example, T2-weighted (T2) and T2 Fluid Attenuation Inversion Recovery (FLAIR) are sensitive to peritumoral edema, and post-contrast T1-weighted (T1Gd) is sensitive to necrotic core and enhancing tumor core. Thus, they can provide complementary information about gliomas.

© Springer Nature Switzerland AG 2019
A. Crimi et al. (Eds.): BrainLes 2018, LNCS 11384, pp. 49–60, 2019.
https://doi.org/10.1007/978-3-030-11726-9_5

Segmentation of brain tumor is a prerequisite while essential task in disease diagnosis, surgical planning and prognosis [2]. Automatic segmentation provides quantitative information that is more accurate and has better reproducibility than conventional qualitative image review. Moreover, the following task of brain tumor classification heavily relies on the results of brain tumor segmentation. Automatic segmentation is considered as a powered engine and empower other intelligent medical application. However, the segmentation of brain tumor in multimodal MRI scans is one of the most challenging tasks in medical imaging analysis due to their highly hetero-geneous appearance, and variable localization, shape and size.

As the rapid development of deep leaning techniques, state-of-the-art performance on brain tumor segmentation have been achieved. For example, in [3], an end-to-end training using fully convolutional network (FCN) showed a satisfactory performance in the localization of the tumor, and patch-wise convolutional neural network (CNN) was used to segment the intra-tumor structure. In [4], a cascaded anisotropic CNN was designed to segment three sub-regions with three Nets, and the segmentation result from previous net was used as receptive field in the next net.

Inspired by the good performance of V-Net in segmentation tasks and the cascaded strategy, we propose a cascaded V-Nets method to segment brain tumor into three substructures and background. In particular, the cascaded V-Nets not only take advantage of residual connection but also use the extra coarse localization and ensemble of multiple models to boost the performance.

2   Method

2.1   Dataset and Preprocessing

The data used in experiments come from BraTS 2018 training set and validation set [5–8]. The training set includes totally 210 HGG patients and 75 LGG patients. The validation set includes 66 patients. Each patient has five MRI modalities including T1-weighted (T1), T2, T1Gd, FLAIR, and a ground truth label of tumor substructures. We use 80% of the training data as our training set, other 20% of the training data as our local testing set. All data used in the experiments are preprocessed with special designed procedures. A flow chart of the proposed preprocessing procedures is shown in Fig. 1, as follows:

(1) Apply bias field correction N4 [9] to T1 and T1Gd images, normalize each modality using histogram matching with respect to a MNI template image, and rescale the images intensity value into range of −1 to 1.
(2) Apply bias field correction N4 to all modalities, compute the standardized z-scores for each image and rescale 0–99.9 percentile intensity values into range of −1 to 1.
(3) Follow the first method, and further apply affine alignment to co-register each image to the MNI template image.

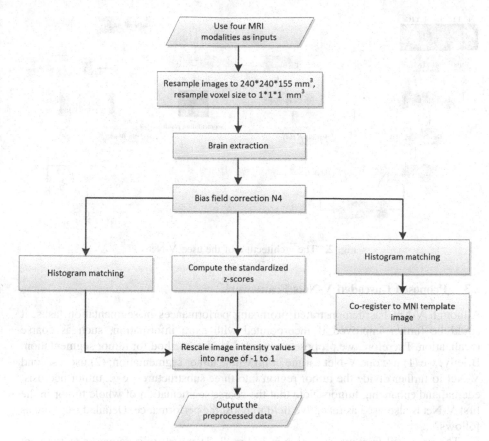

Fig. 1. The flow chart of the preprocessing procedures.

2.2 V-Net Architecture

V-Net was initially proposed to segment prostate by training an end-to-end CNN on MRI [10]. The architecture of our V-Net is shown in Fig. 2. The left side of V-Net reduces the size of the input by down-sampling, and the right side of V-Net recovers the semantic segmentation image that has the same size with input images by applying de-convolutions. The detailed parameters about V-Net is shown in Table 1. By means of introducing residual function and skip connection, V-Net has better segmentation performance compared with classical CNN. By means of introducing the 3D kernel with a size of 1 * 1 * 1, the numbers of parameters in V-Net is decreased and the memory consumption is greatly reduced.

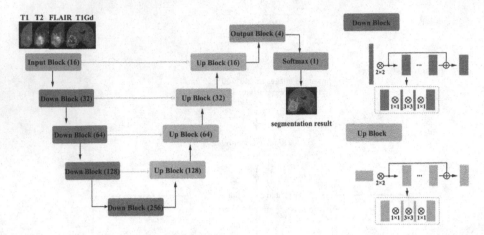

Fig. 2. The architecture of the used V-Net.

2.3 Proposed Cascaded V-Nets Framework

Although V-Net has demonstrated promising performances in segmentation tasks, it could be further improved if incorporated with extra information, such as coarse localization. Therefore, we propose a cascaded V-Nets method for tumor segmentation. Briefly, we (1) use one V-Net for the brain whole tumor segmentation; (2) use a second V-Net to further divide the tumor region into three substructures, e.g., tumor necrosis, edema, and enhancing tumor. Note that the coarse segmentation of whole tumor in the first V-Net is also used as receptive field to boost the performance. Detailed steps are as follows.

The proposed framework is shown in Fig. 3. There are two networks to segment substructures of brain tumors sequentially. The first network (V-Net 1) includes models 1–3, designed to segment the whole tumor. These models are trained by three kinds of preprocessed data mentioned in part of 2.1, respectively. V-Net 1 uses four modalities MR images as inputs, and outputs the mask of whole tumor (WT). The second network (V-Net 2) includes models 4-5, designed to segment the brain tumor into three sub-structures: tumor necrosis, edema, and enhancing tumor. These models are trained by the first two kinds of preprocessed data mentioned in part of 2.1, respectively. V-Net 2 also uses four modalities MR images as inputs, and outputs the segmented mask with three labels. Note that the inputs of V-Net 2 have been processed by using the mask of WT as region of interest (ROI). In other words, the areas out of the ROI are set as background. Finally, we combine the segmentation results of whole tumor obtained by V-Net 1 and the segmentation results of tumor core (TC, includes tumor necrosis and enhancing tumor) obtained by V-Net 2 to achieve more accurate results about the three substructures of brain tumor. In short, the cascaded V-Nets take advantage of segmenting the brain tumor and three substructures sequentially, and ensemble of multiple models to boost the performance and achieve more accurate segmentation results.

Table 1. The detailed parameters of the used V-Net, as shown in Fig. 2. The symbol '-' means the output dimensions are the same with input dimensions.

Blocks	Sub-blocks or layers		Input dimensions	Output dimensions
Input Block	Conv(k = 3, p = 1, s = 1) + BN + ReLU		96 * 96 * 96 * 4	96 * 96 * 96 * 16
Down Block 1	Conv(k = 2, p = 0, s = 2) + BN + ReLU		96 * 96 * 96 * 16	48 * 48 * 48 * 32
	Residual Block	Conv(k = 3, p = 1, s = 1) + BN	48 * 48 * 48 * 32	-
		(input + output) + ReLU	48 * 48 * 48 * 32	-
Down Block 2	Conv(k = 2, p = 0, s = 2) + BN + ReLU		48 * 48 * 48 * 32	24 * 24 * 24 * 64
	Residual Block	Conv Block * 2	24 * 24 * 24 * 64	-
		(input + output) + ReLU	24 * 24 * 24 * 64	-
Down Block 3	Conv(k = 2, p = 0, s = 2) + BN + ReLU		24 * 24 * 24 * 64	12 * 12 * 12 * 128
	Residual Block	Conv Block * 3	12 * 12 * 12 * 128	-
		(input + output) + ReLU	12 * 12 * 12 * 128	-
Down Block 4	Conv(k = 2, p = 0, s = 2) + BN + ReLU		12 * 12 * 12 * 128	6 * 6 * 6 * 256
	Residual Block	Conv Block * 3	6 * 6 * 6 * 256	-
		(input + output) + ReLU	6 * 6 * 6 * 256	-
Up Block 1	Conv(k = 2, p = 0, s = 2) + BN + ReLU		6 * 6 * 6 * 256	12 * 12 * 12 * 128
	Residual Block	Cat(output, skip)	12 * 12 * 12 * 128	12 * 12 * 12 * 256
		Conv Block * 3	12 * 12 * 12 * 256	-
		(input + output) + ReLU	12 * 12 * 12 * 256	-
Up Block 2	Conv(k = 2, p = 0, s = 2) + BN + ReLU		12 * 12 * 12 * 256	24 * 24 * 24 * 64
	Residual Block	Cat(output + skip)	24 * 24 * 24 * 64	24 * 24 * 24 * 128
		Conv Block * 3	24 * 24 * 24 * 128	-
		(input + output) + ReLU	24 * 24 * 24 * 128	-
Up Block 3	Conv(k = 2, p = 0, s = 2) + BN + ReLU		24 * 24 * 24 * 128	48 * 48 * 48 * 32
	Residual Block	Cat(output + skip)	48 * 48 * 48 * 32	48 * 48 * 48 * 64
		Conv(k = 3, p = 1, s = 1) + BN + ReLU	48 * 48 * 48 * 64	-
		Conv(k = 3, p = 1, s = 1) + BN	48 * 48 * 48 * 64	-
		(input + output) + ReLU	48 * 48 * 48 * 64	-
Up Block 4	Conv(k = 2, p = 0, s = 2) + BN + ReLU		48 * 48 * 48 * 64	96 * 96 * 96 * 16
	Residual Block	Cat(output + skip)	96 * 96 * 96 * 16	96 * 96 * 96 * 32
		Conv(k = 3, p = 1, s = 1) + BN	96 * 96 * 96 * 32	-
		(input + output) + ReLU	96 * 96 * 96 * 32	-
Out Block	Conv(k = 1, p = 0, s = 1) + BN + ReLU		96 * 96 * 96 * 32	96 * 96 * 96 * 4
	Softmax		96 * 96 * 96 * 4	96 * 96 * 96 * 1

Note: Each Conv sub-block contains three convolution layers: Conv1(k = 1, p = 0, s = 1), Conv2(k = 3, p = 1, s = 1), and Conv3(k = 1, p = 0, s = 1). k, kernel size; p, padding; s, stride.

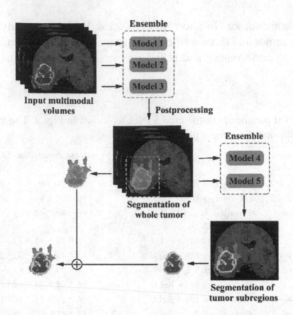

Fig. 3. The proposed framework of cascaded V-Nets for brain tumor segmentation.

2.4   Ensemble Strategy

Our ensemble strategy is simple but efficient. It works by averaging the probability maps obtained from different models. We use ensemble strategy twice in the two-step segmentation of the brain tumor substructures. For example, in V-Net 1, the probability maps of WT obtained from Model 1, Model 2, and Model 3 are averaged to get the final probability map of WT. In V-Net 2, the probability maps of tumor necrosis, edema, and enhancing tumor obtained from Model 4 and Model 5 are averaged to get final probability maps of brain tumor substructures, respectively.

2.5   Network Implementation

Our cascaded V-Nets are implemented in the deep learning framework PyTorch. In our network, we initialize weights with kaiming initialization [11], and use focal loss [12] illustrated in formula (1) as loss function. Adaptive Moment Estimation (Adam) [13] is used as optimizer with learning rate of 0.001, and batch size of 8. Experiments are performed with a NVIDIA Titan Xp 12 GB GPU.

$$\text{Focal Loss } (p_t) = -\alpha(1 - p_t)^r \log(p_t) \tag{1}$$

where, α denotes the weight to balance the importance of positive/negative samples, and r denotes the factor to increase the importance of correcting misclassified samples. p_t is the probability of the ground truth.

In order to reduce the memory consumption in the training process, 3D patches with a size of 96 * 96 * 96 are used. And the center of the patch is confined to the bounding box of the brain tumor. Therefore, every patch used in training process contains both tumor and background. The training efficiency of the network has been greatly improved.

2.6 Post-processing

The predicted segmentation results are post-processed using connected component analysis. We consider that the isolated segmentation labels with small size are prone to artifacts and thus remove them. After the V-Net 1, the components with total voxel number below a threshold (T = 1000) are discarded and these over a threshold (T = 15000) are retained in the binary whole tumor map. For others, their average segmentation probabilities are calculated, and will be retained if over 0.85. After the V-Net 2, masks of different labels are used in the connected component analysis. Moreover, if all the connected components are less than 1000 voxels, we will retain the largest connected component.

2.7 Prediction of Patient Overall Survival

Overall survival (OS) is a direct measure of clinical benefit to a patient. Generally, brain tumor patients could be classified into long-survivors (e.g., >15 months), mid-survivors (e.g., between 10 and 15 months), and short-survivors (e.g., <10 months). From the multimodal MRI data, we propose to use our tumor segmentations and generate imaging markers through Radiomics method to predict the patient OS groups.

From the training data, we extract 40 hand-crafted features and 945 radiomics features in total. The detailed extracted features are shown in Table 2. All features are normalized into range of 0 to 1. Pearson correlation coefficient is used for feature selection. We use support vector machine (SVM), multilayer perceptrons (MLP), XGBoost, decision tree classifier, linear discriminant analysis (LDA) and random forest (RF) as our classifiers in an ensemble strategy. F1-score is used as the evaluation standard. The final result is determined by the vote on all classification results. In order to reduce the bias, a ten-fold cross-validation is used. For the validation and testing data, these selected features are extracted and prediction is made using the above model.

Table 2. Selected features in the training data for the prediction of patient overall survival.

Features	Number of features
Age	1
Volume of whole brain	1
Volume of whole tumor	1
Volumes of three tumor substructures	3
Ratio of the whole tumor in whole brain	1
Ratios of three tumor substructures in whole tumor	3
Extent of lesion in x, y, z directions	3
Center coordinates of the whole tumor	3
Means and variances of three tumor substructures in four MR modalities	24
First order statistics features of three tumor substructures	411
Shape-based features of three tumor substructures	78
Gray level cooccurence matrix features of three tumor substructures	180
Gray level run length matrix features of three tumor substructures	96
Neigbouring gray tone difference matrix features of three tumor substructures	96
Gray level dependence matrix features of three tumor substructures	84

3   Experimental Results

3.1   Segmentation Results on Local Testing Set

We use 20% of all data as our local testing set, which includes 42 HGG patients and 15 LGG patients. Representative segmentation results are shown in Fig. 4. The green shows the edema, the red shows the tumor necrosis, and the yellow shows the enhancing tumor. In order to evaluate the preliminary experimental results, we calculate the average Dice scores, sensitivity and specificity for whole tumor, tumor core and enhancing tumor, respectively. The results are shown in Table 3. The segmentation of whole tumor achieves best results with average Dice score of 0.8505.

3.2   Segmentation Results on MICCAI BraTS 2018 Validation Set of 66 Subjects

The segmentation results on BraTS 2018 online validation set achieve average Dice scores of 0.9048, 0.8364, 0.7768 for whole tumor, tumor core and enhancing tumor, respectively. That performance is slightly better than that in local testing set, while the whole tumor still has best results and enhancing tumor is the most challenging one. The details are shown in Table 4.

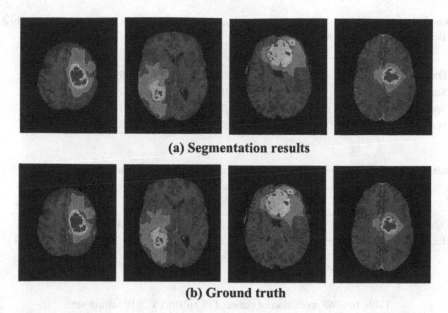

(a) Segmentation results

(b) Ground truth

Fig. 4. The comparison of segmentation results and ground truth on four representative cases from local testing set. (a) The segmentation results of brain tumor. (b) The ground truth of the brain tumor. (Color figure online)

Table 3. Dice, Sensitivity and Specificity measurements of the proposed method on local testing set.

	Whole tumor	Tumor core	Enhancing tumor
Dice mean ± SD	0.8505 ± 0.0972	0.7842 ± 0.1919	0.7426 ± 0.2080
Sensitivity mean ± SD	0.9180 ± 0.1091	0.7596 ± 0.2199	0.7174 ± 0.2337
Specificity mean ± SD	0.9981 ± 0.0012	0.9996 ± 0.0008	0.9997 ± 0.0003

3.3 Segmentation and Prediction Results on MICCAI BraTS 2018 Testing Set of 191 Subjects

The segmentation results on BraTS 2018 online testing set achieve average Dice scores of 0.8761, 0.7953, 0.7364 for whole tumor, tumor core and enhancing tumor, respectively. Compared with the Dice scores on MICCAI BraTS 2018 validation set, the numbers are slightly dropped. The details are shown in Table 5. The prediction of patient OS on BraTS 2018 testing set achieve accuracy of 0.519 and mean square error (MSE) of 367239. The details are shown in Table 6. The BraTS 2018 ranking of all participating teams in the testing data for both tasks has been summarized in [14], where our team listed as "LADYHR" and ranked 18 out of 61 in the segmentation task and 7 out of 26 in the prediction task.

Table 4. Dice, Sensitivity, Specificity and Hausdorff95 measurements of the proposed method on BraTS 2018 validation set.

	Whole tumor	Tumor core	Enhancing tumor
Dice mean ± SD	0.9048 ± 0.0648	0.8364 ± 0.1609	0.7768 ± 0.2355
Sensitivity mean ± SD	0.9146 ± 0.0949	0.8453 ± 0.1781	0.8166 ± 0.2382
Specificity mean ± SD	0.9945 ± 0.0041	0.9971 ± 0.0041	0.9977 ± 0.0032
Hausdorff95 mean ± SD (mm)	5.1759 ± 7.3622	6.2780 ± 7.7681	3.5123 ± 4.5407

Table 5. Dice and Hausdorff95 measurements of the proposed method on BraTS 2018 testing set.

	Whole tumor	Tumor core	Enhancing tumor
Dice mean ± SD	0.8761 ± 0.1247	0.7953 ± 0.2543	0.7364 ± 0.2592
Hausdorff95 mean ± SD (mm)	7.0514 ± 11.5935	6.7262 ± 11.8852	3.9217 ± 6.1934

Table 6. The prediction of patient OS on BraTS 2018 testing set.

	Scores
Accuracy	0.519
Mean squared error (MSE)	367239.974
Median square error (Median SE)	38416
Standard deviation square error	945593.877
Spearman R	0.168

4 Discussion

In this paper, we propose a cascaded V-Nets framework to segment brain tumor. The V-Nets are trained only using provided data, data augmentation and a focal loss formulation. We achieve state-of-the-art results on BraTS 2018 validation set. The experimental results on BraTS 2018 online validation set achieve average Dice scores of 0.9048, 0.8364, 0.7768 for whole tumor, tumor core and enhancing tumor respectively. The corresponding values for BraTS 2018 online testing set are 0.8761, 0.7953 and 0.7364, respectively. Generally, all the three average Dice scores degenerate in testing set compared with validation set. Three are two possible reasons: (1) the testing set includes more cases than validation set, and (2) the thresholds in post-processing maybe more suitable for validation set. Therefore, our future work is to make the models to be more robust.

There are several benefits of using a cascaded framework. First, the cascaded framework breaks down a difficult segmentation task into two easier subtasks. Therefore, a simple network V-Net can have excellent performance. In fact, in our experiment, V-Net does have better performance when segment the tumor substructures step by step than segment background and all the three tumor substructures together. Second, the segmentation results of V-Net 1 helps to reduce the receptive field from whole brain to only whole tumor. Thus, some false positive results can be avoid.

In addition to cascaded framework, ensemble strategy contributes to the segmentation performance. In our cascaded framework, V-Net 1 includes models 1–3 and V-Net 2 includes models 4–5. Every model uses the same network structure V-Net. However, the training data is preprocessed with different pipelines mentioned in part of 2.1. According to our experimental experience, the Dice scores will greatly decrease due to the false positive results. While we did try several ways to change the preprocessing procedures for the training data, or change the model used in the segmentation task, the false positive results always appear. Interestingly, the false positive results appear in different areas in terms of different models. Therefore, ensemble strategy works by averaging probability maps obtained from different models.

Moreover, we find three interesting points in the experiment. Firstly, for multimodal MR images, the combination of data preprocessing procedures is important. In other words, different MRI modalities should be preprocessed independently. For example, in our first preprocessing pipeline, bias field correction only applied to T1 and T1Gd images. The reason is that the histogram matching approach may remove the high intensity information of tumor structure that has negative impact to the segmentation task. Secondly, we use three kinds of preprocessing methods to process the training and validation data, and compared their segmentation results. As a result, there is almost no difference between preprocessing methods in the three average Dice scores for whole tumor, tumor core and enhancing tumor, respectively. However, after the ensemble of the multiple models, the three average Dice scores all rose at least 2%. This suggests that data preprocessing methods is not the most important factor for the segmentation performance, while different data preprocessing methods are complementary and their combination can boost segmentation performance. Thirdly, the postprocessing method is also important that it could affect the average Dices scores largely. If the threshold is too big, some of small clusters will be discarded improperly. If the threshold is too small, some false positive results will be retained. In order to have a better performance, we test a range of thresholds and choose the most suitable two thresholds as the upper and the lower bounds. For the components between upper and lower bounds, their average segmentation probabilities are calculated as a second criterion. Of course, these thresholds may not be suitable for all cases.

5 Conclusions

In conclusion, we propose a cascaded V-Nets framework to segment brain tumor into three substructures of brain tumor and background. The experimental results on BraTS 2018 online validation set achieve average Dice scores of 0.9048, 0.8364, 0.7768 for whole tumor, tumor core and enhancing tumor, respectively. The corresponding values for BraTS 2018 online testing set are 0.8761, 0.7953 and 0.7364, respectively. The state-of-the-art results demonstrate that V-Net is a promising network for 3D medical imaging segmentation tasks, and the cascaded framework and ensemble strategy are efficient for boosting the segmentation performance.

References

1. Mamelak, A.N., Jacoby, D.B.: Targeted delivery of antitumoral therapy to glioma and other malignancies with synthetic chlorotoxin (TM-601). Expert Opin. Drug Deliv. **4**, 175–186 (2007)
2. Bakas, S., et al.: Advancing the cancer genome atlas glioma MRI collections with expert segmentation labels and radiomic features. Sci. Data **4**, 170117 (2017)
3. Cui, S., Mao, L., Jiang, J., Liu, C., Xiong, S.: Automatic semantic segmentation of brain gliomas from MRI images using a deep cascaded neural network. J. Healthc. Eng. **2018**, 4940593 (2018)
4. Crimi, A., Bakas, S., Kuijf, H., Menze, B., Reyes, M. (eds.): BrainLes 2017. LNCS, vol. 10670. Springer, Cham (2018). https://doi.org/10.1007/978-3-319-75238-9
5. Menze, B.H., et al.: The multimodal brain tumor image segmentation benchmark (BRATS). IEEE Trans. Med. Imaging **34**, 1993–2024 (2015)
6. Bakas, S., et al.: Advancing the cancer genome atlas glioma MRI collections with expert segmentation labels and radiomic features. Nat. Sci. Data **4**, 170117 (2017)
7. Bakas, S., et al.: Segmentation labels and radiomic features for the pre-operative scans of the TCGA-GBM collection. The Cancer Imaging Archive (2017)
8. Bakas, S., et al.: Segmentation labels and radiomic features for the pre-operative scans of the TCGA-LGG collection. The Cancer Imaging Archive (2017)
9. Tustison, N.J., et al.: N4ITK: improved N3 bias correction. IEEE Trans. Med. Imaging **29**, 1310–1320 (2010)
10. Milletari, F., Navab, N., Ahmadi, S.: V-Net: fully convolutional neural networks for volumetric medical image segmentation. In: Fourth International Conference on 3D Vision (3DV), Stanford, CA, USA, pp. 565–571 (2016)
11. He, K., Zhang, X., Ren, S., Sun, J.: Delving deep into rectifiers: surpassing human-level performance on ImageNet classification. In: 2015 IEEE International Conference on Computer Vision (ICCV), Santiago, Chile, pp. 1026–1034 (2015)
12. Lin, T., Goyal, P., Girshick, R., He, K., Dollar, P.: Focal loss for dense object detection. In: 2017 IEEE International Conference on Computer Vision (ICCV), Venice, Italy, pp. 2999–3007 (2018)
13. Kingma, D., Ba, J.: Adam, a method for stochastic optimization. In: International Conference on Learning Representations (ICLR), vol. 5 (2014)
14. Bakas, S., et al.: Identifying the Best Machine Learning Algorithms for Brain Tumor Segmentation, Progression Assessment, and Overall Survival Prediction in the BRATS Challenge. arXiv preprint arXiv:1811.02629 (2018)

Automatic Brain Tumor Segmentation Using Convolutional Neural Networks with Test-Time Augmentation

Guotai Wang[1,2]($\boxtimes$), Wenqi Li[1,2], Sébastien Ourselin[1], and Tom Vercauteren[1,2]

[1] School of Biomedical Engineering and Imaging Sciences,
King's College London, London, UK
guotai.1.wang@kcl.ac.uk
[2] Wellcome/EPSRC Centre for Interventional and Surgical Sciences,
University College London, London, UK

Abstract. Automatic brain tumor segmentation plays an important role for diagnosis, surgical planning and treatment assessment of brain tumors. Deep convolutional neural networks (CNNs) have been widely used for this task. Due to the relatively small data set for training, data augmentation at training time has been commonly used for better performance of CNNs. Recent works also demonstrated the usefulness of data augmentation at test time, in addition to training time, for achieving more robust predictions. We investigate how test-time augmentation can improve CNNs' performance for brain tumor segmentation. We used different underpinning network structures and augmented the image by 3D rotation, flipping, scaling and adding random noise at both training and test time. Experiments with BraTS 2018 training and validation set show that test-time augmentation can achieve higher segmentation accuracy and obtain uncertainty estimation of the segmentation results.

Keywords: Brain tumor · Convolutional neural network · Segmentation · Data augmentation

1 Introduction

Gliomas are the most common primary brain tumors that start in the glial cells of the brain in adults. They can be categorized according to their grade: Low-Grade Gliomas (LGG) exhibit benign tendencies and portend a better prognosis for the patient, while High-Grade Gliomas (HGG) are malignant and lead to a worse prognosis [22]. Medical imaging of brain tumors plays an important role for evaluating the progression of the disease before and after treatment. Currently the most widely used imaging modality for brain tumors is Magnetic Resonance Imaging (MRI) with different sequences, such as T1-weighted, contrast enhanced T1-weighted (T1ce), T2-weighted and Fluid Attenuation Inversion Recovery (FLAIR) images. These sequences provide complementary information for different subregions of brain tumors [24]. For example, the tumor

© Springer Nature Switzerland AG 2019
A. Crimi et al. (Eds.): BrainLes 2018, LNCS 11384, pp. 61–72, 2019.
https://doi.org/10.1007/978-3-030-11726-9_6

region and peritumoral edema can be highlighted in FLAIR and T2 images, and the tumor core region without peritumoral edema is more visible in T1 and T1ce images.

Automatic segmentation of brain tumors and substructures from medical images has a potential for accurate and reproducible delineation of the tumors, which can help more efficient and better diagnosis, surgical planning and treatment assessment of brain tumors [5,24]. However, accurate automatic segmentation of the brain tumors is a challenging task for several reasons. First, the boundary between brain tumor and normal tissues is often ambiguous due to the smooth intensity gradients, partial volume effects, and bias field artifacts. Second, the brain tumors vary largely across patients in terms of size, shape, and localization. This prohibits the use of strong priors on shape and localization that are commonly used for robust segmentation of many other anatomical structures, such as the heart [12] and the liver [30].

In recent years, deep Convolutional Neural Networks (CNNs) have achieved the state-of-the-art performance for multi-modal brain tumor segmentation [16, 28]. As a type of machine learning approach, they require a set of annotated training images for learning. Compared with traditional machine learning approaches they do not rely on hand-crafted features and can learn features automatically. In [13], a CNN was proposed to exploit both local and global features for robust brain tumor segmentation. It replaces the final fully connected layer used in traditional CNNs with a convolutional implementation that obtains 40 fold speed up. This approach employs a two-phase training procedure and a cascade architecture to tackle difficulties related to the imbalance of tumor labels. Despite the better performance than traditional methods, this approach works on individual 2D slices without considering 3D contextual information. DeepMedic [17] uses a dual pathway 3D CNN with 11 layers to make use of multi-scale features for brain tumor segmentation. For post-processing, it uses a 3D fully connected Conditional Random Field (CRF) [20] that helps to remove false positives. DeepMedic achieved better performance than using 2D CNNs. However, it works on local image patches and therefore has a relatively low inference efficiency. In [28], a triple cascaded framework was proposed for brain tumor segmentation. The framework uses three networks to hierarchically segment whole tumor, tumor core and enhancing tumor core sequentially. It uses a network structure with anisotropic convolution to deal with 3D images, taking advantage of dilated convolution [31], residual connection [7] and multi-scale fusion [29]. It demonstrated an advantageous trade-off between receptive field, model complexity and memory consumption. This method also fuses the output of CNNs in three orthogonal views for more robust segmentation of brain tumors. In [16], an ensemble of multiple models and architectures including DeepMedic [17], 3D Fully Convolutional Networks (FCN) [21] and U-Net [2,26] was used for robust brain tumor segmentation. The ensemble method reduces the influence of the meta-parameters of individual CNN models and the risk of overfitting the configuration to a specific training dataset. However, it requires much more computational resources to train and run a set of models.

Training with a large dataset plays an important role for the good performance of deep CNNs. For medical images, collecting a very large training set is usually time-consuming and challenging. Therefore, many works have used data augmentation to partially compensate this problem. Data augmentation applies transformations to the samples in a training set to create new ones, so that a relatively small training set can be enlarged to a larger one. Previous works have used different types of transformations such as flipping, cropping, rotation and scaling training images [2]. In [32], a simple and data-agnostic data augmentation routine termed *mixup* was proposed for training neural networks. Recently, several studies have empirically found that the performance of deep learning-based image recognition methods can be improved by combining predictions of multiple transformed versions of a test image, such as in pulmonary nodule detection [15] and skin lesion classification [23]. In [14], test images were augmented by mirroring for brain tumor segmentation. In [27], a mathematical formulation was proposed for test-time augmentation, where a distribution of the prediction was estimated by Monte Carlo simulation with prior distributions of parameters in an image acquisition model. That work also proposed a test-time augmentation-based *aleatoric* uncertainty estimation method that can help to reduce overconfident predictions. The framework in [27] has been validated with binary segmentation tasks, while its application to multi-class segmentation has yet to be demonstrated.

In this paper, we extend the work of [27,28], and apply test-time augmentation to automatic multi-class brain tumor segmentation. For a given input image, instead of obtaining a single inference, we augment the input image with different transformation parameters to obtain multiple predictions from the input, with the same network and associated trained weights. The multiple predictions help to obtain more robust inference of a given image. We explore the use of different CNNs as the underpinning network structures. Experiments with BraTS 2018 training and validation set showed that an improvement of segmentation accuracy was achieved by test-time augmentation, and our method can provide uncertainty estimation for the segmentation output.

2   Methods

2.1   Network Structures

We explore three network configurations as underpinning CNNs for the brain tumor segmentation task: (1) 3D UNet [2], (2) the cascaded networks in [28] where a WNet, TNet and ENet was used to segment whole tumor, tumor core and enhancing tumor core respectively, and (3) adapting WNet [28] for one-pass multi-class prediction without using cascaded prediction, which is referred to as multi-class WNet.

The 3D U-Net has a downsampling and an upsampling path each with four resolution steps. In the downsampling path, each layer has two $3 \times 3 \times 3$ convolutions each followed by a Rectified Linear Unit (ReLU) activation function, and then a $2 \times 2 \times 2$ max pooling layer was used for downsampling. In the upsampling path, each layer uses a deconvolution with kernel size $2 \times 2 \times 2$, followed by two $3 \times 3 \times 3$ convolutions with ReLU. The network has shortcut connections between corresponding layers with the same resolution in the downsampling path and the upsampling path. In the last layer, a $1 \times 1 \times 1$ convolution is used to reduce the number of output channels to the number of segmentation labels, i.e., 4 for the brain tumor segmentation task in the BraTS challenge.

The WNet proposed in [28] is an anisotropic network that considers a trade-off between receptive field, model complexity and memory consumption. It employs dilated convolution [31], residual connection [7] and multi-scale prediction [29] to improve segmentation performance. The network uses 20 intra-slice convolution layers and four inter-slice convolution layers with two 2D down-sampling layers. Since the anisotropic convolution has a small receptive field in the through-plane direction, multi-view fusion was used to take advantage of the 3D contextual information, where the network was applied in axial, sagittal and coronal views respectively. For the multi-view fusion, the softmax outputs in these three views were averaged. In [28], WNet is used to segment the whole tumor. TNet for tumor core segmentation uses the same structure as WNet, and ENet for enhancing core segmentation is a variant of WNet that uses only one down-sampling layer. Compared with multi-label prediction, the cascaded networks require longer time for training and testing. To improve the training efficiency, we compare the cascaded networks [28] with the use of multi-class WNet, where a single WNet for multi-label prediction is employed without using TNet and ENet. Therefore, for this variant we change the output channel number from 2 to 4. Multi-view fusion is also used for this multi-class WNet.

2.2   Data Augmentation for Training and Testing

From the point view of image acquisition, an observed image is only one of many possible observations of the underlying anatomy that can be observed with different spatial transformations and noise. Direct inference with the observed image may lead to a biased result affected by the specific transformation and noise associated with that image. To obtain a more robust prediction, we consider different transformations and noise during the test time. Let β and e represent the parameters for spatial transformation and intensity noise respectively. We assume that β is a combination of f_l, r and s, where f_l is a random variable for flipping along each 3D axis, r is the rotation angle along each 3D axis, s is a scaling factor. We consider these parameters following some prior distributions: $f_l \sim Bern(0.5)$, $r \sim U(0, 2\pi)$, $s \sim U(0.8, 1.2)$. For the intensity noise, we assume $e \sim N(0, 0.05)$ according to the reduced standard deviation of a median-filtered version of a normalized image [27].

For data augmentation, we randomly sample β and e from the above prior distributions and use them to transform the image. We use the same distributions of augmentation parameters at both training and test time for a given CNN. For test-time augmentation, we obtain N samples from the distributions of β and e by Monte Carlo simulation, and the resulting transformed version of the input was fed into the CNN. The N prediction results were combined to obtain the final prediction based on majority voting.

2.3   Uncertainty Estimation

Both model-based (*epistemic*) uncertainty and image-based (*aleatoric*) uncertainty have been investigated for deep CNNs in recent years [18]. The *epistemic* uncertainty is often obtained by Bayesian approximation-based methods such as test-time dropout [10]. In [27], test-time augmentation was used to estimate the *aleatoric* uncertainty of segmentation results in a consistent mathematical framework. In this paper, we use test-time augmentation to obtain segmentation results as well as the associated *aleatoric* uncertainty according to [27].

The uncertainty estimation is obtained by measuring the diversity of the predictions for a given image. Both the variance and entropy of the distribution can be used to estimate uncertainty. Since variance is not sufficiently representative in the context of multi-modal distributions, we use entropy for the pixel-wise uncertainty estimation desired for segmentation tasks. Let X denote the input image and Y denote the output segmentation. We use Y^i to denote the predicted label for the i-th pixel. With the Monte Carlo simulation described in Sect. 2.2, a set of values for Y^i are obtained $\mathcal{Y}^i = \{y_1^i, y_2^i, \ldots, y_N^i\}$. The entropy of the distribution of Y^i is therefore approximated as:

$$H(Y^i|X) \approx -\sum_{m=1}^{M} \hat{p}_m^i \ln(\hat{p}_m^i) \tag{1}$$

where $\hat{p}_m^i$ is the frequency of the m-th unique value in $\mathcal{Y}^i$.

3   Experiments and Results

Data and Implementation Details. We used the BraTS 2018[1] [3–6,24] dataset for experiments. The training set contains images from 285 patients, including 210 cases of HGG and 75 cases of LGG. The BraTS 2018 validation and testing set contain images from 66 and 191 patients with brain tumors of unknown grade, respectively. Each patient was scanned with four sequences: T1, T1ce, T2 and FLAIR. As a pre-processing performed by the organizers, all the images were skull-striped and re-sampled to an isotropic $1\,\mathrm{mm}^3$ resolution, and the four modalities of the same patient had been co-registered. The ground truth were provided by the BraTS organizers. We uploaded the segmentation results obtained by our method to the BraTS 2018 server, and the server provided quantitative evaluations including Dice score and Hausdorff distance compared with the ground truth.

We implemented the 3D UNet [2], multi-class WNet and cascaded networks [28] in Tensorflow[2] [1] using NiftyNet[3][4] [11]. The Adaptive Moment

---

[1] http://www.med.upenn.edu/sbia/brats2018.html.
[2] https://www.tensorflow.org.
[3] http://niftynet.io.
[4] https://github.com/taigw/brats18.

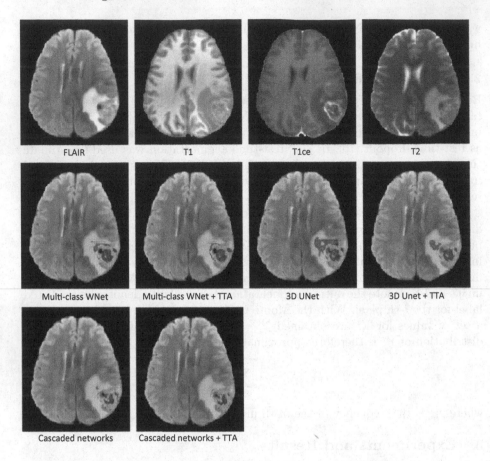

Fig. 1. An example of brain tumor segmentation results obtained by different networks and test-time augmentation (TTA). The first row shows the four modalities of the same patient. The second and third rows show segmentation results. Green: edema; Red: non-enhancing tumor core; Yellow: enhancing tumor core. (Color figure online)

Estimation (Adam) [19] strategy was used for training, with initial learning rate 10^{-3}, weight decay 10^{-7}, and maximal iteration 20k. The training patch size was $96 \times 96 \times 96$ for 3D UNet and $96 \times 96 \times 19$ for multi-class WNet. The batch size was 2 and 4 for these two networks respectively. For the cascaded networks, we followed the configurations in [28]. The training process was implemented on an NVIDIA TITAN X GPU. As a pre-processing, each image was normalized by the mean value and standard deviation. The Dice loss function [9,25] was used for training.

At test time, the augmented prediction number was set to $N = 20$ for all the network structures. The multi-class WNet and cascaded networks were trained in axial, sagittal and coronal views respectively, and the predictions in these three views were fused by averaging at test time.

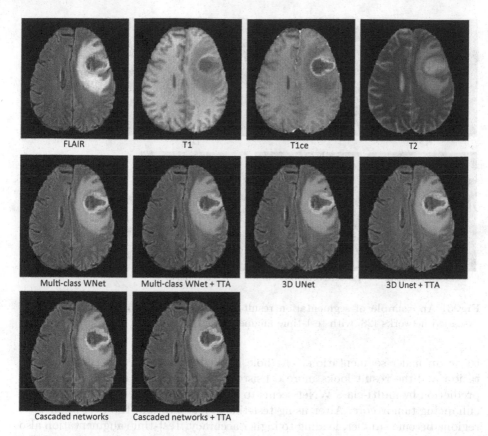

Fig. 2. Another example of brain tumor segmentation results obtained by different networks and test-time augmentation (TTA). The first row shows the four modalities of the same patient. The second and third rows show segmentation results. Green: edema; Red: non-enhancing tumor core; Yellow: enhancing tumor core. (Color figure online)

Segmentation Results. Figure 1 shows an example from the BraTS 2018 validation set. The first row shows the input images of four modalities: FLAIR, T1, T1ce and T2. The second and third rows present the segmentation results of 3D UNet, multi-class WNet, cascaded networks and their corresponding results with test-time augmentation. It can be observed that the initial output of the 3D UNet seems to be noisy with some false positives of edema and non-enhancing tumor core. After using test-time augmentation, the result becomes more spatially consistent. The output of multi-class WNet also seems to be noisy for the non-enhancing tumor core. A smoother segmentation is obtained by multi-class WNet with test-time augmentation. For the cascaded networks, test-time augmentation also leads to visually better results of the tumor core.

Figure 2 shows another example from the BraTS 2018 validation set. It can be observed that the 3D UNet obtains a hole in the tumor core, which seems

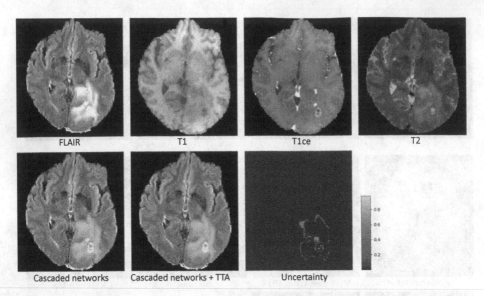

Fig. 3. An example of segmentation result and uncertainty estimation obtained by cascaded networks [28] with test-time augmentation.

to be an under-segmentation. The hole is filled after using test-time augmentation and the result looks more consistent with the input images. The initial prediction by multi-class WNet seems to have an over segmentation of the non-enhancing tumor core. After using test-time augmentation, the over-segmented regions become smaller, leading to higher accuracy. Test-time augmentation also helps to improve the result of cascaded networks. Figure 3 shows a case from the BraTS 2018 testing set, where test-time augmentation obtains a better spatial consistency for the tumor core. In addition, it leads to an uncertainty estimation of the segmentation output. It can be observed that most uncertain results focus on the border of the tumor and some potentially mis-segmented regions.

A quantitative evaluation of our different methods on the BraTS 2018 validation set is shown in Table 1. The initial output of 3D UNet achieved Dice scores of 73.44%, 86.38% and 76.58% for enhancing tumor core, whole tumor and tumor core respectively. 3D UNet with test-time augmentation achieved a better performance than the baseline of 3D UNet, leading to Dice scores of 75.43%, 87.31% and 78.32% respectively. For the initial output of multi-class WNet, the Dice score was 75.70%, 88.98% and 72.53% for these three structures respectively. After using test-time augmentation, an improvement was achieved, and the Dice score was 77.70%, 89.56% and 73.04% for these three structures respectively. For the cascaded networks, test-time augmentation leads to higher accuracy for the enhancing tumor core and tumor core. Table 2 presents the performance of our cascaded networks with test-time augmentation on BraTS 2018 testing set. The average Dice scores for enhancing tumor core, whole tumor and tumor core are 74.66%, 87.78% and 79.64%, respectively. The corresponding values of Hausdorff distance are 4.16 mm, 5.97 mm and 6.71 mm, respectively.

Table 1. Mean values of Dice and Hausdorff measurements of different methods on BraTS 2018 validation set. ET, WT, TC denote enhancing tumor core, whole tumor and tumor core, respectively. TTA: test-time augmentation.

	Dice (%)			Hausdorff (mm)		
	ET	WT	TC	ET	WT	TC
3D UNet	73.44	86.38	76.58	9.37	12.00	10.37
3D UNet + TTA	75.43	87.31	78.32	4.53	5.90	8.03
Multi-class WNet	75.70	88.98	72.53	4.24	4.99	12.13
Multi-class WNet + TTA	77.07	89.56	73.04	4.44	4.92	11.13
Cascaded networks	79.19	90.31	85.40	3.34	5.38	6.61
Cascaded networks + TTA	79.72	90.21	85.83	3.13	6.18	6.37

Table 2. Dice and Hausdorff measurements of our cascaded networks with test-time augmentation on BraTS 2018 testing set. ET, WT, TC denote enhancing tumor core, whole tumor and tumor core, respectively.

	Dice (%)			Hausdorff (mm)		
	ET	WT	TC	ET	WT	TC
Mean	74.66	87.78	79.64	4.16	5.97	6.71
Standard deviation	25.85	11.92	24.97	7.07	8.56	10.27
Median	83.38	91.33	89.68	2.00	3.32	3.16
25 Quantile	72.87	86.69	78.24	1.41	2.24	2.00
75 Quantile	88.64	94.09	93.58	3.00	5.48	6.40

4   Discussion and Conclusion

For test-time augmentation, we only used flipping, rotation and scaling for spatial transformations. It is also possible to employ more complex transformations such as elastic deformations used in [2]. However, such deformations take longer time for testing and have a lower efficiency. The results show that test-time augmentation leads to an improvement of segmentation accuracy for different CNNs including 3D UNet [2], multi-class WNet and cascaded networks [28]. Test-time augmentation can be applied to other CNN models as well. The uncertainty estimation obtained by our method can be used for downstream analysis such as uncertainty-aware volume measurement [8] and guiding user interactions [29]. It would be of interest to assess the impact of test-time augmentation on CNNs trained with state-of-the-art policies such as in [14]. By using test-time augmentation, we investigated the test image-based (*aleatoic*) uncertainty for brain tumor segmentation. It is of interest to investigate how ensemble of CNNs [16] can produce *epistemic* uncertainty for this task. For a comprehensive study of uncertainty, it is promising to combine ensemble of models or test-time dropout with test-time augmentation. This will be left for future work.

In conclusion, we explored the effect of test-time augmentation on CNN-based brain tumor segmentation. We used 3D U-Net, 2.5D multi-class WNet and cascaded networks as the underpinning network structures. For training and testing, we augmented the image by 3D rotation, flipping, scaling and adding random noise. Experiments with BraTS 2018 training and validation set show that test-time augmentation helps to improve the brain tumor segmentation accuracy for different CNN structures and obtain uncertainty estimation of the segmentation results.

Acknowledgements. We would like to thank the NiftyNet team. This work was supported through an Innovative Engineering for Health award by the Wellcome Trust [WT101957, WT97914, 203145Z/16/Z, 203148/Z/16/Z], Engineering and Physical Sciences Research Council (EPSRC) [NS/A000027/1, NS/A000049/1, NS/A000050/1], the National Institute for Health Research University College London Hospitals Biomedical Research Centre (NIHR BRC UCLH/UCL High Impact Initiative), hardware donated by NVIDIA, and the Health Innovation Challenge Fund [HICF-T4-275].

References

1. Abadi, M., et al.: TensorFlow: a system for large-scale machine learning. In: USENIX Symposium on Operating Systems Design and Implementation, pp. 265–284 (2016)
2. Çiçek, Ö., Abdulkadir, A., Lienkamp, S.S., Brox, T., Ronneberger, O.: 3D U-Net: learning dense volumetric segmentation from sparse annotation. In: Ourselin, S., Joskowicz, L., Sabuncu, M.R., Unal, G., Wells, W. (eds.) MICCAI 2016. LNCS, vol. 9901, pp. 424–432. Springer, Cham (2016). https://doi.org/10.1007/978-3-319-46723-8_49
3. Bakas, S., et al.: Segmentation labels and radiomic features for the pre-operative scans of the TCGA-LGG collection. The Cancer Imaging Archive (2017)
4. Bakas, S., et al.: Segmentation labels and radiomic features for the pre-operative scans of the TCGA-GBM collection. The Cancer Imaging Archive (2017)
5. Bakas, S., et al.: Advancing the cancer genome atlas glioma MRI collections with expert segmentation labels and radiomic features. Nat. Sci. Data **4**, 170117 (2017)
6. Bakas, S., Reyes, M., et al.: Identifying the best machine learning algorithms for brain tumor segmentation, progression assessment, and overall survival prediction in the BRATS challenge (2018). https://arxiv.org/abs/1811.02629
7. Chen, H., Dou, Q., Yu, L., Qin, J., Heng, P.A.: VoxResNet: deep voxelwise residual networks for brain segmentation from 3D MR images. NeuroImage **170**, 446–455 (2018)
8. Eaton-Rosen, Z., Bragman, F., Bisdas, S., Ourselin, S., Cardoso, M.J.: Towards safe deep learning: accurately quantifying biomarker uncertainty in neural network predictions. In: International Conference on Medical Image Computing and Computer-Assisted Intervention, pp. 691–699 (2018)
9. Fidon, L., Li, W., Garcia-Peraza-Herrera, L.C.: Generalised Wasserstein Dice score for imbalanced multi-class segmentation using holistic convolutional networks. arXiv preprint arXiv:1707.00478 (2017)
10. Gal, Y., Ghahramani, Z.: Dropout as a Bayesian approximation: representing model uncertainty in deep learning. In: International Conference on Machine Learning, pp. 1050–1059 (2016)

11. Gibson, E., et al.: NiftyNet: a deep-learning platform for medical imaging. Comput. Methods Programs Biomed. **158**, 113–122 (2018)
12. Grosgeorge, D., Petitjean, C., Dacher, J.N., Ruan, S.: Graph cut segmentation with a statistical shape model in cardiac MRI. Comput. Vis. Image Underst. **117**(9), 1027–1035 (2013)
13. Havaei, M., et al.: Brain tumor segmentation with deep neural networks. Med. Image Anal. **35**, 18–31 (2016)
14. Isensee, F., Kickingereder, P., Wick, W., Bendszus, M., Maier-Hein, K.H.: No new-net. arXiv preprint arXiv:1809.10483 (2018)
15. Jin, H., Li, Z., Tong, R., Lin, L.: A deep 3D residual CNN for false positive reduction in pulmonary nodule detection. Med. Phys. **45**(5), 2097–2107 (2018)
16. Kamnitsas, K., et al.: Ensembles of multiple models and architectures for robust brain tumour segmentation. In: Crimi, A., Bakas, S., Kuijf, H., Menze, B., Reyes, M. (eds.) BrainLes 2017. LNCS, vol. 10670, pp. 450–462. Springer, Cham (2018). https://doi.org/10.1007/978-3-319-75238-9_38
17. Kamnitsas, K., et al.: Efficient multi-scale 3D CNN with fully connected CRF for accurate brain lesion segmentation. Med. Image Anal. **36**, 61–78 (2017)
18. Kendall, A., Gal, Y.: What uncertainties do we need in Bayesian deep learning for computer vision? In: Advances in Neural Information Processing Systems, pp. 5580–5590 (2017)
19. Kingma, D.P., Ba, J.L.: Adam: a method for stochastic optimization. In: International Conference on Learning Representations (2015)
20. Krähenbühl, P., Koltun, V.: Efficient inference in fully connected CRFs with Gaussian edge potentials. In: Proceedings of the 24th International Conference on Neural Information Processing Systems, NIPS 2011, pp. 109–117. Curran Associates Inc., USA (2011)
21. Long, J., Shelhamer, E., Darrell, T.: Fully convolutional networks for semantic segmentation. In: IEEE Conference on Computer Vision and Pattern Recognition, pp. 3431–3440 (2015)
22. Louis, D.N., et al.: The 2016 world health organization classification of tumors of the central nervous system: a summary. Acta Neuropathologica **131**(6), 803–820 (2016)
23. Matsunaga, K., Hamada, A., Minagawa, A., Koga, H.: Image classification of melanoma, nevus and seborrheic keratosis by deep neural network ensemble. arXiv preprint arXiv:1703.03108 (2017)
24. Menze, B.H., et al.: The multimodal brain tumor image segmentation benchmark (BRATS). IEEE Trans. Med. Imaging **34**(10), 1993–2024 (2015)
25. Milletari, F., Navab, N., Ahmadi, S.A.: V-Net: fully convolutional neural networks for volumetric medical image segmentation. In: International Conference on 3D Vision, pp. 565–571 (2016)
26. Ronneberger, O., Fischer, P., Brox, T.: U-Net: convolutional networks for biomedical image segmentation. In: Navab, N., Hornegger, J., Wells, W.M., Frangi, A.F. (eds.) MICCAI 2015. LNCS, vol. 9351, pp. 234–241. Springer, Cham (2015). https://doi.org/10.1007/978-3-319-24574-4_28
27. Wang, G., Li, W., Aertsen, M., Deprest, J., Ourselin, S., Vercauteren, T.: Aleatoric uncertainty estimation with test-time augmentation for medical image segmentation with convolutional neural networks. arXiv preprint arXiv:1807.07356 (2018)
28. Wang, G., Li, W., Ourselin, S., Vercauteren, T.: Automatic brain tumor segmentation using cascaded anisotropic convolutional neural networks. In: Crimi, A., Bakas, S., Kuijf, H., Menze, B., Reyes, M. (eds.) BrainLes 2017. LNCS, vol. 10670, pp. 178–190. Springer, Cham (2018). https://doi.org/10.1007/978-3-319-75238-9_16

29. Wang, G., et al.: Interactive medical image segmentation using deep learning with image-specific fine-tuning. IEEE Trans. Med. Imaging **37**(7), 1562–1573 (2018)
30. Wang, G., Zhang, S., Xie, H., Metaxas, D.N., Gu, L.: A homotopy-based sparse representation for fast and accurate shape prior modeling in liver surgical planning. Med. Image Anal. **19**(1), 176–186 (2015)
31. Yu, F., Koltun, V.: Multi-scale context aggregation by dilated convolutions. CoRR abs/1511.07122 (2015)
32. Zhang, H., Cisse, M., Dauphin, Y.N., Lopez-Paz, D.: Mixup: beyond empirical risk minimization. arXiv preprint arXiv:1710.09412, pp. 1–11 (2017)

Extending 2D Deep Learning Architectures to 3D Image Segmentation Problems

Alberto Albiol[1]([✉])[ID], Antonio Albiol[1][ID], and Francisco Albiol[2][ID]

[1] iTeam, Universitat Politècnica de València, Valencia, Spain
{alalbiol,aalbiol}@iteam.upv.es
[2] Instituto de Física Corpuscular (IFIC), Universitat de València,
Consejo Superior de Investigaciones Científicas, Valencia, Spain
kiko.albiol@ific.uv.es

Abstract. Several deep learning architectures are combined for brain tumor segmentation. All the architectures are inspired on recent 2D models where 2D convolution have been replaced by 3D convolutions. The key differences between the architectures are the size of the receptive field and the number of feature maps on the final layers. The obtained results are comparable to the top methods of previous Brats Challenges when median is use to average the results. Further investigation is still needed to analyze the outlier patients.

Keywords: Brain segmentation · Brats · 3D inception · 3D VGG · 3D densely connected · 3D Xception

1 Introduction

Brain tumor segmentation is an important problem which has received a considerable attention by the research community and particularly since the advent of deep learning.

Glial cells are the cause of gliomas that are the most common brain tumors. Gliomas are usually classified into low-grade gliomas (LGG) and high grade gliomas (HGG) which are malignant and more aggressive.

Brain tumors are usually imaged using several Magnetic Resonance (MR) sequences, such as T1-weighted, contrast enhanced T1-weighted (T1c), T2-weighted and Fluid Attenuation Inversion Recovery (FLAIR) images. From a pure pattern recognition point of view, these modalities provide complimentary information and can be used as different feature input maps. In other words, image modalities play a role similar to color planes of RGB natural images.

The Multimodal Brain Tumor Segmentation Challenge 2018 provided a set of MR sequences for training and evaluation of brain tumor segmentation algorithms. Ground truth for all the scans have been manually provided by expert

© Springer Nature Switzerland AG 2019
A. Crimi et al. (Eds.): BrainLes 2018, LNCS 11384, pp. 73–82, 2019.
https://doi.org/10.1007/978-3-030-11726-9_7

board-certified neuroradiologists, so that every voxel is categorized into these classes [11]:

- Label 0: background.
- Label 1: necrotic and non-enhancing tumor.
- Label 2: edema.
- Label 4: enhancing tumor.

We did not pay much attention on the medical details of this problem. Our main contribution was to extend some of the recent approaches used for 2D image classification: VGG, inception, Xception, densely connected models to be used with 3D data in a real segmentation problem.

2   Methods

Our approach uses an ensemble of deep neural networks with different architectures. The idea is that the ensemble provides a more robust solution with less variance compared to individual methods. Also, some architectures may compensate for other architectures weaknesses and thus improve the global performance. The idea of using an ensemble with multiple architectures was also used by the winning method of the last Brats competition [9].

This section describes the different architectures used in our approach. All the architectures have in common that every voxel is independently labeled using a deep neural network architecture. We are aware that better results could had been obtanied if some post processing that considered the spatial constraints had been used, similar to the CRF proposed in [10].

The key differences between the architectures are the number of parameters, the number of feature planes and the size of the receptive field associated to each voxel. These hyper-parameters were chosen as a trade-off usually limited by the memory of the GPU. More specifically, we mixed four different architectures in our final ensemble: VGG-Like, inception-2, inception-3 and densely connected. These models are described in detail in the following subsections.

2.1   VGG-like Model

This model is inspired on the well known VGG model proposed by [12]. The differences between our approach and the original VGG are:

- 2-D convolutions are replaced by 3-D convolutions.
- Maxpool layers are not used.
- The network is replicated in a convolutional way so that every pixel is labeled independently.

Table 1 describes in detail the layers used in this model. Note that all convolutional layers are preceded by batch normalization and followed by a ReLU activation function, except the last layer which is followed by a softmax activation function.

Table 1. Description of our VGG-like architecture.

Layer name	Kernel size	Num filters
conv_1_1	$3 \times 3 \times 3$	30
conv_1_2	$3 \times 3 \times 3$	30
conv_2_1	$3 \times 3 \times 3$	60
conv_2_2	$3 \times 3 \times 3$	60
conv_3_1	$3 \times 3 \times 3$	120
conv_3_2	$3 \times 3 \times 3$	120
conv_4_1	$3 \times 3 \times 3$	240
conv_4_2	$3 \times 3 \times 3$	240
fc_1	$1 \times 1 \times 1$	400
fc_2	$1 \times 1 \times 1$	200
logits	$1 \times 1 \times 1$	4

2.2 Dense-Like Model

This architecture is inspired by the recent work [8]. The key difference between the original method and the one used in this paper, is that 2D convolutions are replaced by 3D convolutions. The advantage of densely connected networks (compared to VGG like models) is that features are reused on subsequent layers and each layer adds a few new features only. This allows to increase the number of layers and therefore the size of the receptive field associated to each voxel. This architecture also allows to combine features with relatively small receptive fields (first layers) with features with large receptive fields (last layers). This is particularly useful in segmentation problems, where large receptive fields provide context information and small receptive fields provide fine-grained information that helps to increase the precision of the segmentation.

Table 2 summarizes the architecture of our densely connected network. Note that each layer concatenates all the output features from the previous layers, for this reason the number of input feature grows steadily until layer conv_20. Then two fully connected layers similar to the VGG architecture are used.

2.3 Inception-Like Model

This architecture is inspired by some of the ideas proposed in [14] and [13]. The key idea proposed by the inception model is to replace convolutional layers by several parallel structures with different kernel shapes. This reduces the number of parameters (regularization) and forces diversity on the output features of each layer.

We took these ideas and adapted them to the problem of brain segmentation. The main limitation of inception layers is that they require much GPU memory because each layer is composed of several simpler sub-layers, for instance some

Table 2. Description of our Dense-like architecture.

Layer name	Kernel size	Num inputs	Num output filters
conv_1	$3 \times 3 \times 3$	4	8
conv_2	$3 \times 3 \times 3$	12	8
conv_3	$3 \times 3 \times 3$	20	8
conv_4	$3 \times 3 \times 3$	28	8
conv_5	$3 \times 3 \times 3$	36	8
conv_6	$3 \times 3 \times 3$	44	8
conv_7	$3 \times 3 \times 3$	52	8
conv_8	$3 \times 3 \times 3$	60	8
conv_9	$3 \times 3 \times 3$	68	8
conv_10	$3 \times 3 \times 3$	76	8
conv_11	$3 \times 3 \times 3$	84	8
conv_12	$3 \times 3 \times 3$	92	8
conv_13	$3 \times 3 \times 3$	100	8
conv_14	$3 \times 3 \times 3$	108	8
conv_15	$3 \times 3 \times 3$	106	8
conv_16	$3 \times 3 \times 3$	114	8
conv_17	$3 \times 3 \times 3$	122	8
conv_18	$3 \times 3 \times 3$	130	8
conv_19	$3 \times 3 \times 3$	138	8
conv_20	$3 \times 3 \times 3$	146	8
fc_1	$1 \times 1 \times 1$	154	400
fc_2	$1 \times 1 \times 1$	400	200
logits	$1 \times 1 \times 1$	200	4

inception layers use 1-D convolutions along each spatial dimension. In the case of 2D convolutions, this option doubles the number of layers and the required memory used to store intermediate results and gradients. In the case of 3D segmentation, this problem is even worse because the use of 1-D convolutions implies to use three times more memory.

For this reason, we created two simplified GoogLenet-like models with a few inception layers before the fully connected layers as detailed in Table 3.

Figure 1 shows the internal structure of the inception layers. As it can be seen, four different branches are used. The first layer extracts new features and reduces the dimensionality. The second and third branches introduce spatial convolution; the fourth brach is an average layer without pooling. This structure is similar to the structure of Fig. 5 in [13].

Table 3. Description of the inception architectures used in the final ensemble.

Layer name	kernel size	num filters
conv_1_1	$3 \times 3 \times 3$	30
conv_1_2	$3 \times 3 \times 3$	30
conv_2_1	$3 \times 3 \times 3$	60
conv_2_2	$3 \times 3 \times 3$	60
conv_3_1	$3 \times 3 \times 3$	120
conv_3_2	$3 \times 3 \times 3$	120
inception	see. Fig. 1	240
inception	see. Fig. 1	240
fc_1	$1 \times 1 \times 1$	400
fc_2	$1 \times 1 \times 1$	200
logits	$1 \times 1 \times 1$	4

Inception2

Layer name	kernel size	num filters
conv_1_1	$3 \times 3 \times 3$	30
conv_1_2	$3 \times 3 \times 3$	30
conv_2_1	$3 \times 3 \times 3$	60
conv_2_2	$3 \times 3 \times 3$	60
conv_3_1	$3 \times 3 \times 3$	120
conv_3_2	$3 \times 3 \times 3$	120
inception	see. Fig. 1	240
inception	see. Fig. 1	240
inception	see. Fig. 1	240
fc_1	$1 \times 1 \times 1$	400
fc_2	$1 \times 1 \times 1$	200
logits	$1 \times 1 \times 1$	4

Inception 3

2.4   Other Architectures Not in the Final Ensemble

We also made experiments with other architectures not included in the final ensemble for their lower performance on our training data using cross-validation.

The most innovative structure in this group was based in the Xception architecture presented in [6]. This architecture assumes that correlation in feature planes can be decoupled from spatial correlation, and therefore separability is applied. We implemented this separable 3D spatial filters from scratch in Tensorflow (the library only provides this feature for 2D images).

We also made experiments with other inception architectures similar to those presented in Figs. 6 and 7 of [13]. However, the results on our cross-validated training set were not good enough.

The main limitation of these other inception architectures and also the Xception layers is that they require more GPU memory compared to the simpler VGG architecture, for this reason total number of layers needs to be reduced so that the model fits into memory. The main advantage of these architectures in 2D images is that they require a smaller number of parameters which help to regularize the model. However, we found that overfitting was not the problem for any of our models (the training cost and training error was not negligible), and therefore models with many parameters (as the VGG) could be trained without overfitting.

Finally, we also made experiments with field bias correction of the input data [15]. In these experiments, we corrected the bias of the T1 and T1ce input modalities and compared the performance without the field bias correction and the same neural network architecture. The results with the bias correction were always worse compared to using the original raw data with the same model architecture, and for this reason we omitted field bias correction.

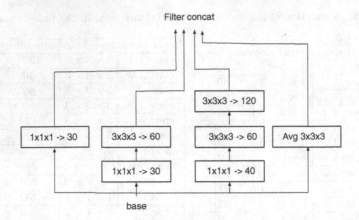

Fig. 1. Structure of the inception layers used in the models of Table 3. Each box shows the kernel size and the number of output features.

2.5   Number of Parameters and Receptive Field Size

Table 4 shows the number of parameters and receptive field size for the models included in our final ensemble. The model that requires more parameters is the VGG-like. This constraint limits the number of layers of the VGG model to avoid GPU memory problems. This is the reason why the VGG-like model has the smallest receptive field size.

The inception models halve the number of parameters (the latter layers are the ones with more parameters) and have a larger receptive field.

Finally, the densely connected model is the model with less parameters and largest receptive field.

The idea of our ensemble is to be able to combine models with large receptive field (more context), as the densely connected model, with very expressive models, i.e. models with many deep features (VGG-like) so that each model compensate for the weaknesses of the others.

Table 4. Number of parameters and receptive field size for the models used in our ensemble

Model	#parameters	Receptive field size
VGG-like	3270252	$17 \times 17 \times 17$
Inception2	1375872	$21 \times 21 \times 21$
Inception3	1611882	$25 \times 25 \times 25$
Densely connected	494220	$41 \times 41 \times 41$

3   Experiments and Results

3.1   Data

Our system was evaluated on the data from the Brain Tumor Segmentation Challenge 2018 (BRATS) [2–4,11]. As in previous editions, the training set consists of 210 cases with high grade glioma (HGG) and 75 cases with low grade glioma (LGG), for which manual segmentations are provided. The segmentations include the following tumor tissue labels: (1) necrotic core and non enhancing tumor, (2) edema, (4) enhancing core. Label 3 is not used. The validation set consists of 66 cases, both HGG and LGG but the grade is not revealed. For each subject, four MRI sequences are available, FLAIR, T1, T1 contrast enhanced (T1ce) and T2. The datasets are pre-processed by the organisers and provided as skull-stripped, registered to a common space and resampled to isotropic 1mm3 resolution. Dimensions of each volume are $240 \times 240 \times 155$.

3.2   Implementation Detais

We implemented everything in python. Input/output data for MRI scans was handled with the nibabel library [7] and neural networks were implemented using tensorflow [1]. The code used in this work has been dockerized and released to the challenge organizers so it will be available to the community.

We did not try any bias field correction of the input scans. The only intensity normalization that we used was z-score normalization of the input scans using the mean and standard deviation of the brain volume only (so the mean and std deviation are not dependent of the brain size).

Models were trained using crops of the original MRI scans. As in [10], the size of each crop was larger than the size of the receptive field. More specifically, the size of the crop is set $(9 + r_f) \times (9 + r_f) \times (9 + r_f)$, where r_f is the size of the receptive field. Thus, each crop contributes to the cost function with $9 \times 9 \times 9$ voxels. This approach increases the computational efficiency (reuses many computations) and we think that it also acts as a regularizer, forces the model to be smooth during labeling. For each mini batch, we increased the number of crops to fill the GPU memory (12Gb in our machine). These crops were randomly sampled using a uniform distribution among the four classes: healthy, oedema, core and enhancing core. During evaluation the size of the crops were increased and consecutive crops had some overlap to handle the reduced size of the network output (we used convolutions with only valid support).

Training was done using gradient descent with the Adam optimizer using a constant learning rate of 0.0001 for about 40k steps. We did not observed any overfitting during training, and for this reason we did not investigate into adding any L2, L1 regularization, learning rate decay.... Perhaps one of the reason why we did not observed overfitting is because we implemented a strong data augmentation that generated affine 3D transformations of the MRI scans on the fly.

3.3   Training Results

We split the training data in two random sets, so that one of the sets that contained 20% of the patients was used to evaluate the training progress. For each model architecture we generated two different training partitions using different random seeds, so that all training data was used by the models in the ensemble.

We ranked the model architectures using the Dice scores on our validation subset. Table 5 shows the Dice scores for our best models on our validation split. As it can be seen the differences among models are very small, however since the receptive field size and number of parameters is very different we think that the models might have captured complimentary information.

The last row in Table 6 shows the results on the Brats test set that we obtained on the challenge. The results on this set are clearly worse than those obtained for the validation set, this fact could be a clear symptom of some overfitting on the training and validation sets. However, we suspect that there could be also some differences due to other factors, such as different acquisition conditions because we did not made any model selection on the validation set and in that case we did not observed any difference with the results on our cross validation partition.

Table 5. Results of the selected model architectures on our validation split

Model name	Dice_WT	Dice_TC	Dice_ET
VGG-like	0.880	0.771	0.689
Inception2	0.882	0.792	0.685
Inception3	0.880	0.789	0.695
Densely connected	0.883	0.787	0.683

3.4   Results on the Validation and Test Sets

We submitted the predicted labels for each of the described models and also for the ensemble model for the validation set. There ensemble model averages the probabilities of 8 trained models (one for each architecture, and two random partitions of the training set).

Table 6 shows the results provided by the Brats evaluation platform on the blind validation dataset. The results are quite consistent with the results shown on Table 5, and hence we can conclude that we did not overfit the training dataset and the models generalize quite well on new data. However, the evaluation on the Brats platform shows an interesting point, median values of the Dice scores are much larger than the mean values. This confirms the existence of image outliers. The last row in Table 6 shows the results on the contest test set, the results for all other contest participants can be found in [5].

Table 6. Results of the selected model architectures on the validation set

Model name	Set	Mean Dice_WT	Mean Dice_TC	Mean Dice_ET	Median Dice_WT	Median Dice_TC	Median Dice_ET
VGG-like	Validation	0.872	0.760	0.751	0.900	0.837	0.844
Inception2	Validation	0.877	0.773	0.7533	0.909	0.866	0.858
Inception3	Validation	0.873	0.776	0.781	0.907	0.852	0.858
Densely connected	Validation	0.874	0.755	0.729	0.903	0.837	0.846
Ensemble	Validation	0.881	0.777	0.773	0.912	0.873	0.860
Ensemble	Test	0.850	0.740	0.723	0.894	0.856	0.828

4   Discussion and Conclusion

In this paper, we have extended some well known architectures for 2D image classification to the problem of 3D image segmentation. This can be easily done by replacing 2D convolutions by their 3D counterparts and adjusting the number of layers and number of feature maps to more appropriate ranges so that models can be fitted in memory.

We selected four model architectures so that we had models with large/small receptive fields, many/less parameters. The idea is that different configurations can capture complimentary information and an ensemble model can outperform each separate model.

The results on the validation set, show that there no exist many performance differences between the different model architectures, however the ensemble model outperforms each model. These results confirms our hypothesis and are also consistent with the results that we had previously obtained on the training data. The results on the Brats test set are clearly worse, we think that the cause of this behaviour is that there are some differences in the image acquisition and our method is not robust enough to deal with these variations.

We also tried other models, not included in the final ensemble, such as the 3D Xception that assumes independence between spatial and feature dimensions. We also tried to use bias field correction however our results showed that this was not useful for our models.

Finally, it is worth to highlight that the obtained results shows the existence of image outliers that are not well segmented. This issue severely drops our global performance as shown by the huge difference of using the mean or median metrics. We need to make further research on the causes of these outliers.

References

1. Abadi, M., et al.: Tensorflow: large-scale machine learning on heterogeneous systems (2015). https://www.tensorflow.org/, software available from tensorflow.org
2. Bakas, S., et al.: Segmentation labels and radiomic features for the pre-operative scans of the TCGA-LGG collection. The Cancer Imaging Archive (2017)
3. Bakas, S., et al.: Advancing the cancer genome atlas glioma MRI collections with expert segmentation labels and radiomic features. Nat. Sci. Data **4**, 170117 (2017)
4. Bakas, S., et al.: Segmentation labels and radiomic features for the pre-operative scans of the TCGA-GBM collection. The Cancer Imaging Archive (2017)
5. Bakas, S., Reyes, M., Menze, B.: Identifying the best machine learning algorithms for brain tumor segmentation, progression assessment, and overall survival prediction in the brats challenge. In: arXiv preprint arXiv:1811.02629 (2018)
6. Chollet, F.: Xception: deep learning with depthwise separable convolutions (2016), cite arxiv:1610.02357
7. NIPY developers: nibabel 1.0.2, August 2016. https://doi.org/10.5281/zenodo.60861
8. Huang, G., Liu, Z., van der Maaten, L., Weinberger, K.Q.: Densely connected convolutional networks. In: Proceedings of the IEEE Conference on Computer Vision and Pattern Recognition (2017)
9. Kamnitsas, K., et al.: Ensembles of multiple models and architectures for robust brain tumour segmentation. In: International MICCAI Brainlesion Workshop (2017)
10. Kamnitsas, K., et al.: Efficient multi-scale 3D CNN with fully connected CRF for accurate brain lesion segmentation. Med. Image Anal. **36**, 61–78 (2017)
11. Menze, B.H., et al.: The multimodal brain tumor image segmentation benchmark (brats). IEEE Trans. Med. Imaging **34**(10), 1993–2024 (2015). https://doi.org/10.1109/TMI.2014.2377694
12. Simonyan, K., Zisserman, A.: Very deep convolutional networks for large-scale image recognition. In: arXiv: 1409.1556 (2014)
13. Szegedy, C., Vanhoucke, V., Ioffe, S., Shlens, J., Wojna, Z.: Rethinking the inception architecture for computer vision. In: 2016 IEEE Conference on Computer Vision and Pattern Recognition (CVPR), pp. 2818–2826, June 2016. https://doi.org/10.1109/CVPR.2016.308
14. Szegedy, C., et al.: Going deeper with convolutions. In: Computer Vision and Pattern Recognition (CVPR) (2015), http://arxiv.org/abs/1409.4842
15. Tustison, N.J., et al.: N4ITK: improved N3 bias correction. IEEE Trans. Med. Imaging **29**(6), 1310–1320 (2010). https://doi.org/10.1109/tmi.2010.2046908

Tumor Segmentation and Survival Prediction in Glioma with Deep Learning

Li Sun[1], Songtao Zhang[1], and Lin Luo[2](✉)

[1] Southern University of Science and Technology, Shenzhen 518055, China
[2] Peking University, Beijing 100871, China
luol@pku.edu.cn

Abstract. Every year, about 238,000 patients are diagnosed with brain tumor in the world. Accurate and robust tumor segmentation and prediction of patients' overall survival are important for diagnosis, treatment planning and risk factor characterization. Here we present a deep learning-based framework for brain tumor segmentation and survival prediction in glioma using multimodal MRI scans. For tumor segmentation, we use ensembles of three different 3D CNN architectures for robust performance through majority rule. This approach can effectively reduce model bias and boost performance. For survival prediction, we extract 4524 radiomic features from segmented tumor region. Then decision tree and cross validation are used to select potent features. Finally, a random forest model is trained to predict the overall survival of patients. On 2018 MICCAI Multimodal Brain Tumor Segmentation Challenge (BraTS), our method ranks at second place and 5th place out of 60+ participating teams on survival prediction task and segmentation task respectively, achieving a promising 61.0% accuracy on classification of long-survivors, mid-survivors and short-survivors.

Keywords: Survival prediction · Brain tumor segmentation · 3D CNN · Multimodal MRI

1 Introduction

Brain tumor is cancerous or noncancerous mass or growth of abnormal cells in the brain, malignant brain tumor is one of the most aggressive and fatal tumors. Originated in the glial cells, gliomas are the most common brain tumor. [6] Depending on the pathologic evaluation of the tumor, gliomas can be categorized into glioblastoma (GBM/HGG) and lower grade glioma (LGG). Gliomas contain various heterogeneous histological sub-regions, including peritumoral edema, necrotic core, enhancing and non-enhancing tumor core. Magnetic resonance imaging (MRI) is commonly used in radiology to portray the phenotype and intrinsic heterogeneity of gliomas, since multimodal MRI scans, such as T1-weighted, contrast enhanced T1-weighted (T1c), T2-weighted and Fluid Attenuation Inversion Recovery (FLAIR) images, provide complementary profiles for

© Springer Nature Switzerland AG 2019
A. Crimi et al. (Eds.): BrainLes 2018, LNCS 11384, pp. 83–93, 2019.
https://doi.org/10.1007/978-3-030-11726-9_8

different sub-regions of gliomas. For example, the enhancing tumor sub-region is described by areas that show hyper-intensity in T1Gd scan when compared to T1 scan.

Accurate and robust prediction of overall survival through automated algorithms for patients diagnosed with gliomas can provide valuable guidance for diagnosis, treatment planning and outcome prediction. However, the selection of reliable and potent prognostic is difficult. Medical imaging (e.g. MRI, CT) can provide radiographic phenotype of tumor, and it has been exploited increasingly to extract and analyze quantitative imaging features. [7] Clinical data, including patient age, resection status and others, also provide important information about patients' outcome.

Segmentation of gliomas in pre-operative MRI scans, conventionally done by expert board-certified neuroradiologists, can provide quantitative morphological characterization and measurement of gliomas sub-regions. It is also pre-requisite for survival prediction since most potent features are derived from the tumor region. This quantitative analysis has great potential for diagnosis and research, as it can be used for grade assessment of gliomas and planning of treatment strategies. But this task is challenging due to the high variance in appearance and shape, ambiguous boundaries and imaging artifacts. Until now, automatic segmentation of brain tumors in multimodal MRI scans is still one of the most difficult tasks in medical image analysis. In recent years, deep convolutional neural networks (CNNs) have achieved great success in the field of computer vision. Inspired by the biological structure of visual cortex, CNNs are artificial neural networks with multiple hidden convolutional layers between the input and output layers. They have non-linear property and are capable of extracting higher level representative features. CNNs have been applied into a wide range of fields and achieved state-of-the-art performance on tasks such as image recognition, instance detection, and semantic segmentation.

In this paper, we present a novel deep learning based framework to segment brain tumor and its subregion from MRI scans, then perform survival prediction based on radiomic features extracted from segmented tumor sub-regions as well as clinical feature. Our automatic framework for brain tumor segmentation and survival prediction ranks at second place and 5th place out of 60+ participating teams on survival prediction task and segmentation task on 2018 MICCAI BraTS Challenge respectively, achieving a promising 61.0% accuracy on classification of long-survivors, mid-survivors and short-survivors.

2   Methodology

2.1   Overview

Our proposed framework for survival prediction using MRI scans consists of the following steps, as illustrated in the figure below. First, tumor subregions are segmented using an ensemble model comprising of three different convolutional neural network architectures for robust performance through voting/majority rule. Then radiomics features are extracted from tumor sub-regions and total tumor volume. Next, decision tree regressor with gradient boosting is used to fit the training data and rank the importance of each feature based on variance reduction, and cross validation is used to choose the optimal number of top-ranking features to use. Finally, a random forest model is used to fit the training data and predict the overall survival of patient (Fig. 1).

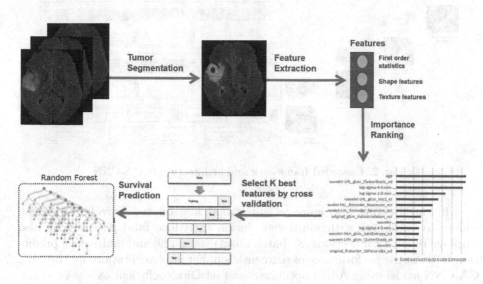

Fig. 1. Framework overview

2.2   Data Preprocessing

Since the intensity value of MRI is dependent on the imaging protocol and scanner used, we applied intensity normalization to reduce the bias in imaging. More specifically, the intensity value of each MRI is subtracted the mean and divided by the standard deviation of the brain region. In order to reduce overfitting, we applied random flipping and random gaussian noise to augment the training set.

2.3   Network Architecture

In order to perform accurate and robust brain tumor segmentation, we use an ensemble model comprising of three different convolutional neural network architectures. A variety of models have been proposed for tumor segmentation. Generally, they differ in model depth, filter number, connection way and others. Different model architectures can lead to different model performance and behavior. By training different kinds of model separately and merge the result, the model variance can be decreased and the overall performance can be improved. [11] We use three different CNN models and fuse the result by voting/majority rule. The detailed description of each model will be discussed as follows.

CA-CNN. The first network we employ is Cascaded Anisotropic Convolutional Neural Network (CA-CNN) proposed by Wang et al. [17]. The cascade is used to convert multi-class segmentation problem into a sequence of three hierarchical binary segmentation problems. The network is illustrated as follows (Fig. 2):

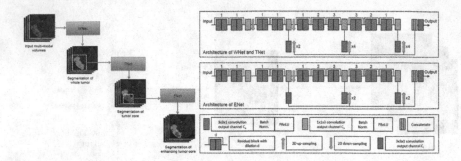

Fig. 2. Cascaded framework and architecture of CA-CNN

This architecture also employs anisotropic and dilated convolution filters, which are combined with multi-view fusion to reduce false positives. It also employs residual connections [8], batch normalization [9] and multi-scale prediction to boost the performance of segmentation. For implementation, we train the CA-CNN model using Adam optimizer, and set Dice coefficient as loss function. We set initial learning rate to 1×10^{-3}, weight decay 1×10^{-7}, batch size 5, and maximal iteration $30k$.

DFKZ Net. The second network we employ is DFKZ Net, which was proposed by Isensee et al. [10] from German Cancer Research Center (DFKZ). This network is inspired by U-Net. It employs a context encoding pathway that extracts increasingly abstract representations of the input, and a decoding pathway used to recombine these representations with shallower features to precisely segment the structure of interest. The context encoding pathway consists of three content modules, each has two $3 \times 3 \times 3$ convolutional layers and a dropout layer with residual connection. The decoding pathway consists of three localization modules, each contains a $3 \times 3 \times 3$ convolutional layer followed by a $1 \times 1 \times 1$ convolutional layer. For the decoding pathway, the output of layers of different depth is integrated by elementwise summation, thus the supervision can be injected deep in the network (Fig. 3).

For implementation, we train the network using Adam optimizer. To address the problem of class imbalance, we utilize the multi-class Dice loss function [10]:

$$L = -\frac{2}{|K|} \sum_{k \in K} \frac{\sum_i u_{i(k)} v_{i(k)}}{\sum_i u_{i(k)} + \sum_i v_{i(k)}} \tag{1}$$

where u denotes output possibility, v denotes one-hot encoding of ground truth, k denotes the class, K denotes the total number of classes and $i(k)$ denotes the

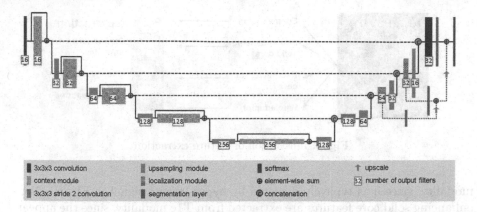

Fig. 3. Architecture of DFKZ Net

number of voxels for class k in patch. We set initial learning rate 5×10^{-4} and use instance normalization. We train the model for 90 epochs.

3D U-Net. U-Net [5,14] is a classical network for biomedical image segmentation. It consists of a contracting path to capture context and a symmetric expanding path that enables precise localization with extension. Each pathway has three convolutional layers with dropout and pooling. And the contracting pathway and expanding pathway are linked by skip-connections. Each layer contains $3 \times 3 \times 3$ convolutional kernels. The first convolutional layer has 32 filters, while deeper layers contains twice filters than previous shallower layer.

For implementation, we use Adam optimizer [12], and we use instance normalization [15]. In addition, we utilize cross entropy as loss function. The initial learning rate is 0.001, the model is trained for 4 epochs.

Ensemble of Models. In order to enhance segmentation performance and reduce model variance. We use voting/majority rule to build an ensemble model. During training process, different models are trained separately. In the testing stage, each model independently predicts the class for each voxel, the final class is determined by majority rule.

2.4 Feature Extraction

Quantitative phenotypic features from MRI scans can reveal the characteristics of brain tumors. Based on the segmentation result, we extract radiomics features from edema, non-enhancing solid core and necrotic/cystic core and the whole tumor region respectively using *Pyradiomics* toolbox [16] (Fig. 4).

The modality used for feature extraction is depended on the intrinsic property of tumor subregion. For example, edema features are extracted from FLAIR

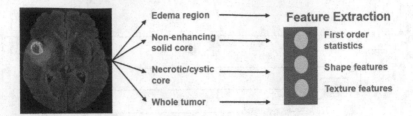

Fig. 4. Illustration of feature extraction

modality, since it is typically depicted by hyper-intense signal in FLAIR. Non-enhancing solid core features are extracted from T1c modality, since the appearance of the necrotic (NCR) and the non-enhancing (NET) tumor core is typically hypo-intense in T1-Gd when compared to T1. Necrotic/cystic core tumor features are extracted from T1c modality, since it is described by areas that show hyper-intensity in T1Gd when compared to T1.

The features we extracted can be grouped into three categories. The first category is first order statistics, which includes maximum intensity, minimum intensity, mean, median, 10th percentile, 90th percentile, standard deviation, variance of intensity value, energy, entropy and others. These features characterize the grey level intensity of tumor region.

The second category is shape features, which include volume, surface area, surface area to volume ratio, maximum 3D diameter, maximum 2D diameter for axial, coronal and sagittal plane respectively, major axis length, minor axis length and least axis length, sphericity, elongation and other features. These features characterize the shape of tumor region.

The third category is texture features, which include 22 grey level co-occurrence matrix (GLCM) features, 16 gray level run length matrix (GLRLM) features, 16 Grey level size zone matrix (GLSZM) features, five neighboring gray tone difference matrix (NGTDM) features and 14 gray level dependence matrix (GLDM) Features. These features characterize the texture of tumor region.

Not only do we extract features from original images, but we also extract features from Laplacian of Gaussian (LoG) filtered images and images generated by wavelet decomposition. Because LoG filtering can enhance the edge of images, possibly enhance the boundary of tumor, and wavelet decomposition can separate images into multiple levels of detail components (finer or coarser). More specifically, from each region, 1131 features are extracted, including 99 features extracted from the original image, and 344 features extracted from Laplacian of Gaussian filtered images, since we use 4 filters with sigma value 2.0, 3.0, 4.0, 5.0 respectively, and 688 features extracted from 8 wavelet decomposed images (all possible combinations of applying either a High or a Low pass filter in each of the three dimensions). In total, for each patient, we extract $1131 \times 4 = 4524$ radiomic features, these features are combined with clinical data (age and resection state) for survival prediction. The values of these features are normalized by subtracting the mean and scaling to unit variance.

2.5   Feature Selection

A portion of features we extracted are redundant or irrelevant to survival prediction. In order to enhance performance and reduce overfitting, we applied feature selection to select a subset of features that have the most predictive power. Feature selection is divided into two steps: importance ranking and cross validation. We rank the importance of features by fitting a decision tree regressor with gradient boosting using training data, then the importance of features can be determined by how effectively the feature can reduce intra-node standard deviation in leaf nodes. The second step is to select the optimal number of best features for prediction by cross validation. In the end, we select 14 features and their importance are listed as follows: (Abbreviations: wt = edema, tc = tumor core, et = enhancing tumor, full = whole tumor; The detailed feature definition can be found at https://pyradiomics.readthedocs.io/en/latest/features.html, last accessed on 30 June 2018) (Table 1).

Not surprisingly, age has the most predictive power among all features. The rest of features selected come from both original images and derived images. And we found that most features selected are come from images generated by wavelet decomposition.

2.6   Survival Prediction

Based on the 14 features selected, we trained a random forest regressor for final survival prediction. We set the number of base regressor as 100, and bootstrap samples when building trees.

Table 1. Selected most predicative features

Extracted from	Name	Subregion	Score
clinical	age	NA	0.037375134
wavelet-LHL	glcm_ClusterShade	wt	0.036912293
log-sigma-4.0mm-3D	glcm_Correlation	tc	0.035558309
log-sigma-2.0mm-3D	gldm_LargeDependenceHighGrayLevelEmphasis	tc	0.026591038
wavelet-LHL	glcm_Informational Measure of Correlation	et	0.022911978
wavelet-HLL	firstorder_Maximum	et	0.020121927
wavelet-LHL	firstorder_Skewness	et	0.019402119
original image	glcm_Autocorrelation	et	0.014204463
wavelet-HHH	gldm_LargeDependenceLowGrayLevelEmphasis	full	0.014085406
log-sigma-4.0mm-3D	firstorder_Mwtian	wt	0.013031814
wavelet-HLH	glcm_JointEntropy	wt	0.013023534
wavelet-LHH	glcm_ClusterShade	tc	0.012335471
wavelet-HLL	glszm_LargeAreaHighGrayLevelEmphasis	full	0.011980896
original image	firstorder_10Percentile	wt	0.011803132

3   Experiments

3.1   Dataset

We utilize the BraTS 2018 dataset [1–4,13] to evaluate the performance of our methods. The training set contains images from 285 patients, including 210 HGG and 75 LGG. The validation set contains MRI scans from 66 patients with brain tumors of unknown grade. The test set contains images from 191 patients with brain tumor, in which 77 patients have resection state of Gross Total Resection (GTR) and are evaluated for survival prediction. Each patient was scanned with four sequences: T1, T1c, T2 and FLAIR. All the images were skull-striped and re-sampled to an isotropic $1\,\mathrm{mm}^3$ resolution, and the four sequences of the same patient had been co-registered. The ground truth was obtained by manual segmentation results given by experts. Segmentation annotations comprise of the following tumor subtypes: Necrotic/non-enhancing tumor (NCR), peritumoral edema (ED), and Gd-enhancing tumor (ET). Resection status and patient age are also provided. The overall survival (OS) data, defined in days is also included in training set (Fig. 5).

3.2   Segmentation Result

We train the model using the 2018 MICCAI BraTS training set with methods described above. Then we applied the trained model for prediction on validation set and test set. We compared the segmentation result of ensemble model with individual model on validation set, the result demonstrates that the ensemble model performs better than individual models on enhancing tumor and whole tumor, while CA-CNN performs marginally better on tumor core (Table 2).

Table 2. Evaluation result of ensemble model and individual model

Model	Enhancing tumor	Whole tumor	Tumor core
CA-CNN	0.77682	0.90282	**0.85392**
DFKZ Net	0.76759	0.89306	0.82459
3D U-Net	0.78088	0.88762	0.82567
Ensemble model	**0.80522**	**0.90944**	0.84943

The predicted segmentation labels are uploaded to the CBICA's Image Processing Portal (IPP) for evaluation. BraTS Challenge uses two schemes for evaluation: Dice score and the Hausdorff distance (95%). In test phase, we rank at 5th place out of 60+ teams. The evaluation result of segmentation on validation set and test set are listed as follows (Table 3).

Table 3. Evaluation result of ensemble model for segmentation

Stage	Metric	Enhancing tumor	Whole tumor	Tumor core
Validation	Mean Dice	0.80522	0.90444	0.84943
	Mean Hausdorff95 (mm)	2.77719	6.32753	6.37318
Test	Mean Dice	0.71712	0.87615	0.79773
	Mean Hausdorff95 (mm)	4.97823	7.20086	6.47348

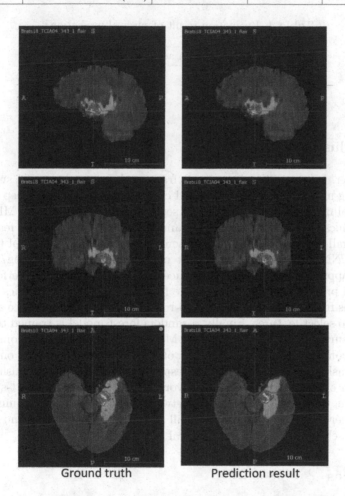

Ground truth   Prediction result

Fig. 5. Examples of segmentation result compared with ground truth Green: edema, Yellow: non-enhancing solid core, Red: enhancing core (Color figure online)

3.3 Survival Prediction Result

Based on the segmentation result of brain tumor subregions, we extract features from brain tumor sub-regions segmented from MRI scans and trained the survival prediction model as described above. Then we use the model to predict patient's overall survival on validation set and test set. The predicted overall survival is uploaded to the IPP for evaluation. We use two schemes for evaluation: classification of subjects as long-survivors (>15 months), short-survivors (<10 months), and mid-survivors (between 10 and 15 months) and median error (in days). In test phase, we rank at second place out of 60+ teams. The evaluation result is listed as follows (Table 4).

Table 4. Evaluation result of survival prediction

Stage	Classification accuracy	Median error
Validation	46.4%	217.92
Test	61.0%	181.37

4 Conclusion

In this paper, we present an automatic framework for prediction of survival in glioma using multimodal MRI scans and clinical features. Firstly deep convolutional neural network (CNN) is used to segment tumor region from MRI scans, then radiomics features are extracted and combined with clinical features to predict overall survival. For tumor segmentation, we use ensembles of three different 3D CNN architectures for robust performance through voting/majority rule. This approach can effectively reduce model bias and boost performance. For survival prediction, we extract shape features, first order statistics and texture features from segmented tumor sub-region, then use decision tree and cross validation to select features. Finally, a random forest model is trained to predict the overall survival of patients. On 2018 MICCAI BraTS Challenge, our method ranks at second place and 5th place out of 60+ participating teams on survival prediction task and segmentation task respectively, achieving a promising 61.0% accuracy on classification of long-survivors, mid-survivors and short-survivors. In the future, we will explore different network architectures and training strategies to further improve our result. We will also design new features and optimize our feature selection methods for survival prediction.

References

1. Bakas, S., et al.: Segmentation labels and radiomic features for the pre-operative scans of the TCGA-GBM collection. Cancer Imaging Arch. **286** (2017)
2. Bakas, S., et al.: Segmentation labels and radiomic features for the pre-operative scans of the TCGA-LGG collection. Cancer Imaging Arch. **286** (2017)
3. Bakas, S., et al.: Advancing the cancer genome atlas glioma MRI collections with expert segmentation labels and radiomic features. Sci. Data **4**, 170117 (2017)
4. Bakas, S., Reyes, M., et al.: Identifying the best machine learning algorithms for brain tumor segmentation, progression assessment, and overall survival prediction in the BRATS challenge. arXiv preprint arXiv:1811.02629 (2018)

5. Çiçek, Ö., Abdulkadir, A., Lienkamp, S.S., Brox, T., Ronneberger, O.: 3D U-Net: learning dense volumetric segmentation from sparse annotation. In: Ourselin, S., Joskowicz, L., Sabuncu, M.R., Unal, G., Wells, W. (eds.) MICCAI 2016. LNCS, vol. 9901, pp. 424–432. Springer, Cham (2016). https://doi.org/10.1007/978-3-319-46723-8_49

6. Ferlay, J., Shin, H.R., Bray, F., Forman, D., Mathers, C., Parkin, D.M.: Estimates of worldwide burden of cancer in 2008: GLOBOCAN 2008. Int. J. Cancer **127**(12), 2893–2917 (2010)

7. Gillies, R.J., Kinahan, P.E., Hricak, H.: Radiomics: images are more than pictures, they are data. Radiology **278**(2), 563–577 (2016)

8. He, K., Zhang, X., Ren, S., Sun, J.: Deep residual learning for image recognition. In: Proceedings of the IEEE Conference on Computer Vision and Pattern Recognition, pp. 770–778 (2016)

9. Ioffe, S., Szegedy, C.: Batch normalization: accelerating deep network training by reducing internal covariate shift. arXiv preprint arXiv:1502.03167 (2015)

10. Isensee, F., Kickingereder, P., Wick, W., Bendszus, M., Maier-Hein, K.H.: Brain tumor segmentation and radiomics survival prediction: contribution to the BRATS 2017 challenge. In: Crimi, A., Bakas, S., Kuijf, H., Menze, B., Reyes, M. (eds.) BrainLes 2017. LNCS, vol. 10670, pp. 287–297. Springer, Cham (2018). https://doi.org/10.1007/978-3-319-75238-9_25

11. Kamnitsas, K., et al.: Ensembles of multiple models and architectures for robust brain tumour segmentation. In: Crimi, A., Bakas, S., Kuijf, H., Menze, B., Reyes, M. (eds.) BrainLes 2017. LNCS, vol. 10670, pp. 450–462. Springer, Cham (2018). https://doi.org/10.1007/978-3-319-75238-9_38

12. Kingma, D.P., Ba, J.: Adam: a method for stochastic optimization. In: International Conference on Learning Representations (2015)

13. Menze, B.H., et al.: The multimodal brain tumor image segmentation benchmark (BRATS). IEEE Trans. Med. Imaging **34**(10), 1993 (2015)

14. Ronneberger, O., Fischer, P., Brox, T.: U-Net: convolutional networks for biomedical image segmentation. In: Navab, N., Hornegger, J., Wells, W.M., Frangi, A.F. (eds.) MICCAI 2015. LNCS, vol. 9351, pp. 234–241. Springer, Cham (2015). https://doi.org/10.1007/978-3-319-24574-4_28

15. Ulyanov, D., Vedaldi, A., Lempitsky, V.: Instance normalization: the missing ingredient for fast stylization. arxiv 2016. arXiv preprint arXiv:1607.08022

16. Van Griethuysen, J.J.M., et al.: Computational radiomics system to decode the radiographic phenotype. Cancer Res. **77**(21), e104–e107 (2017)

17. Wang, G., Li, W., Ourselin, S., Vercauteren, T.: Automatic brain tumor segmentation using cascaded anisotropic convolutional neural networks. In: Crimi, A., Bakas, S., Kuijf, H., Menze, B., Reyes, M. (eds.) BrainLes 2017. LNCS, vol. 10670, pp. 178–190. Springer, Cham (2018). https://doi.org/10.1007/978-3-319-75238-9_16

Multi-planar Spatial-ConvNet for Segmentation and Survival Prediction in Brain Cancer

Subhashis Banerjee[1,2]([⊠]), Sushmita Mitra[1], and B. Uma Shankar[1]

[1] Machine Intelligence Unit, Indian Statistical Institute, Kolkata, India
mail.sb88@gmail.com, {sushmita,uma}@isical.ac.in
[2] Department of CSE, University of Calcutta, Kolkata, India

Abstract. A new deep learning method is introduced for the automatic delineation/segmentation of brain tumors from multi-sequence MR images. A Radiomic model for predicting the Overall Survival (OS) is designed, based on the features extracted from the segmented Volume of Interest (VOI). An encoder-decoder type ConvNet model is designed for pixel-wise segmentation of the tumor along three anatomical planes (axial, sagittal and coronal) at the slice level. These are then combined, using a consensus fusion strategy, to produce the final volumetric segmentation of the tumor and its sub-regions. Novel concepts such as spatial-pooling and unpooling are introduced to preserve the spatial locations of the edge pixels for reducing segmentation error around the boundaries. We also incorporate shortcut connections to copy and concatenate the receptive fields from the encoder to the decoder part, for helping the decoder network localize and recover the object details more effectively. These connections allow the network to simultaneously incorporate high-level features along with pixel-level details. A new aggregated loss function helps in effectively handling data imbalance. The integrated segmentation and OS prediction system is trained and validated on the BraTS 2018 dataset.

Keywords: Deep learning · Convolutional neural network ·
Spatial-pooling · Brain tumor segmentation · Survival prediction ·
Radiomics · Class imbalance handling

1 Introduction

Gliomas are the most common and aggressive malignant brain tumors originating in the glial cells of the central nervous system. Based on their aggressiveness in infiltration, they are broadly classified into two categories, viz. High-Grade Glioma or GlioBlastoma Multiforme (HGG/GBM) and Low-Grade Glioma (LGG). Magnetic Resonance Imaging (MRI) has been extensively employed over the last few decades, in diagnosing brain and nervous system abnormalities; mainly due to its improved soft tissue contrast. Typically the MR sequences

© Springer Nature Switzerland AG 2019
A. Crimi et al. (Eds.): BrainLes 2018, LNCS 11384, pp. 94–104, 2019.
https://doi.org/10.1007/978-3-030-11726-9_9

include $T1$-weighted, $T2$-weighted, $T1$-weighted Contrast enhanced ($T1C$), and $T2$-weighed with FLuid-Attenuated Inversion Recovery (FLAIR). The rationale behind using all four sequences lies in the fact that different tumor regions become more visible in different sequences; thereby enabling more accurate demarcation of the tumor [5,6].

Accurate delineation of tumor regions in MRI sequences is of great importance since it allows: *(i)* volumetric measurement of the tumor, *(ii)* monitoring of tumor growth in patients between multiple MRI scans, over treatment span and *(iii)* treatment planning with follow-up evaluation, including the prediction of overall survival (OS). Manual segmentation of tumors from MRI is a highly tedious, time-consuming and error-prone task, mainly due to factors such as human fatigue, overabundance of MRI slices per patient, and an increasing number of patients. Such manual operations often lead to inaccurate delineation. The need for an automated or semi-automated Computer Aided Diagnosis thus becomes apparent [7,8,15]. The large spatial and structural variability among brain tumors makes automatic segmentation a challenging problem. The distinctive segmentation of both HGG and LGG by the same model is also a difficult proposition.

Inspired by the success of Convolutional Neural Networks (ConvNets) [9,12], we develop a novel ConvNet model with spatial-pooling called Spatial-ConvNet. This can preserve the edge information during automated segmentation of gliomas from multi-sequence MRI data. The segmented Volume of Interest (VOI) or tumor is used to extract two categories of Radiomic features [10,11,18], viz. "semantic" and "agnostic", for predicting the OS of patients. A new loss function helps in class imbalance handling.

The rest of the paper is organized as follows. Section 2 provides details about the data, preparation of patch database for the ConvNet training, the proposed multi-planar ConvNet model with spatial-pooling layer, the aggregated loss function for imbalanced segmentation, and radiomic analysis of the segmented VOI for OS prediction. Section 3 describes the experimental results of the segmentation and OS prediction, demonstrating their effectiveness both qualitatively and quantitatively. Finally, conclusions are provided in Sect. 4.

2   Materials and Methods

In this section we discuss the BRATS 2018 data, and the steps of tumor segmentation and survival rate prediction. The proposed segmentation method comprises of extraction of patches, training and testing of the segmentation model, post-processing, radiomic feature extraction for overall survival prediction, followed by training and testing of the classifier for OS prediction.

2.1   Dataset

Brain tumor MRI scan datasets and patient Overall Survival (OS) data, used in this research, were provided by BraTS 18 Challenge [1–4,13]. It consists of

210 HGG/GBM and 75 LGG glioma cases as training dataset and 66 combined cases of HGG/GBM and LGG as validation dataset. The OS data was included with correspondences to the pseudo-identifiers of the GBM/HGG imaging data having 163 and 53 validation data points respectively. Each patient MRI scan set consist of four MRI sequences or channels, encompassing native $(T1)$ and post-contrast enhanced $T1$-weighted $(T1C)$, $T2$-weighted $(T2)$, and $T2$ FLuid-Attenuated Inversion Recovery $(FLAIR)$ volumes, having 155 slices of 240×240 resolution images. The data is already aligned to the same anatomical template, skull-stripped, and interpolated to $1\,\mathrm{mm}^3$ voxel resolution. The manual segmentation of volume structures have been performed by experts following the same annotation protocol, and their annotations revised and approved by board-certified neuro-radiologists. Annotation labels included are the gadolinium enhancing tumor (ET), the peritumoral edema (ED), and the necrotic and non-enhancing tumor (NCR/NET). The predicted labels are evaluated by merging three regions, *viz.* whole tumor (WT: all the three labels), tumor core (TC: ET and NCR/NET) and enhancing tumor (ET).

The OS data is defined in terms of days, and also includes the age of patients along with their resection status. Only these subjects with resection status GTR (Gross Total Resection) are considered for evaluating OS prediction. Based on the number of survival days, the subjects are grouped into three classes viz. long-survivors (>15 months), short-survivors (<10 months), and mid-survivors (between 10 to 15 months).

2.2   Multi-planar ConvNet with Spatial-Pooling for Segmentation

MRI scans are volumetric and can be represented in three-dimensions using multi-planar representation along axial (X-Z axes), coronal (Y-X axes), and sagittal (Y-Z axes) planes. Taking advantage of this multi-view property, we propose a deep learning based segmentation model that uses three separate ConvNets for segmenting the tumor along the three individual planes at slice level. These are then combined using a consensus fusion strategy to produce the final volumetric segmentation of the tumor and its sub regions. It is observed that the integrated prediction from multiple planes is superior, in terms of accuracy and robustness of decision, with respect to the estimation based on any single plane. This is perhaps because of utilizing more information, while minimizing the loss.

The ConvNet architecture, used for slice wise segmentation along each plane, is an encoder-decoder type of network. The encoder or the contracting path uses pooling layers to down sample an image into a set of high-level features, followed by a decoder or an expanding part which uses the feature information to construct a pixel-wise segmentation mask. The main problem with this type of networks is that, during the down sampling or the pooling operation the network loses spatial information. Up sampling in the decoder network then tries to approximate this through interpolation. This produces segmentation error around the boundary of the region-of-interest (ROI) or volume-of-interest

(VOI). It is a major drawback in medical image segmentation, where accurate delineation is of utmost importance.

In order to circumvent this problem we introduce an elitist spatial-max-pooling layer, which can retain the maximum locations to be subsequently used during unpooling through the spatial-max-unpooling layer. The procedure is illustrated in Fig. 1. We also incorporate shortcut connections to copy and concatenate the receptive fields (after convolution block) from the encoder to the decoder part, in order to help the decoder network localize and recover the object details more effectively. These connections allow the network to simultaneously incorporate high-level features with the pixel-level details. The entire segmentation model architecture is depicted in Fig. 2.

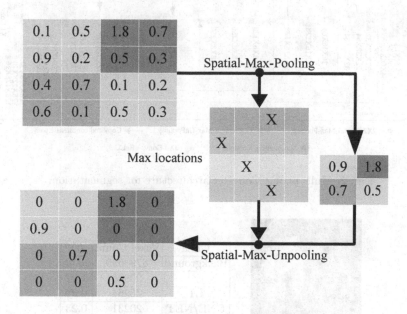

Fig. 1. Spatial-pooling and unpooling operations.

Tumors are typically heterogeneous, depending on cancer subtypes, and contain a mixture of structural and patch-level variability. Applying a ConvNet directly to the entire slice has its inherent drawbacks. Since the size of each slice is 240×240, therefore if we train the ConvNet on the whole image/slice then the number of parameters to train will be huge. Moreover, very little difference is observeable in adjacent MRI slices at the global level; whereas patches generated from the same slice often exhibit significant dissimilarity. Besides, the segmentation classes are highly imbalanced. Approximately 98% of the voxels belong to either the healthy tissue or to the black surrounding area. The NCR/NET volumes are of the lowest size amongst all the three classes, as depicted in Fig. 3.

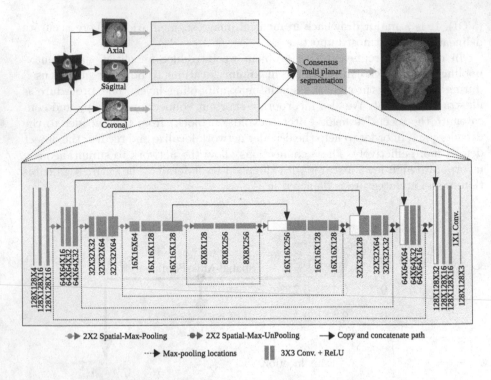

Fig. 2. Multi-planar ConvNet architecture for segmentation.

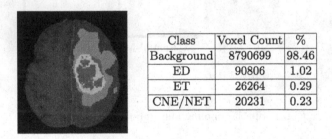

Class	Voxel Count	%
Background	8790699	98.46
ED	90806	1.02
ET	26264	0.29
CNE/NET	20231	0.23

Fig. 3. Tumor sub-class distribution for a sample MRI slice.

Each ConvNet is trained on patches of size $128 \times 128 \times 4$, extracted from all four MRI sequences corresponding to a particular plane. A randomized patch extraction algorithm, developed by us, is employed. The patch selection is done using an entropy based criterion. The three ConvNets (along the three planes) are trained end-to-end/pixel-to-pixel, based on the patches extracted from the corresponding ground truth images. During testing the stack of slices are fed to the model, to produce pixel-wise segmentation of the tumor along the three planes. The training performance is evaluated using Dice overlap score [14], for the three segmented sub-regions WT, ET and TC. Since the dataset is highly imbalanced therefore standard loss functions used in literature are not suitable for training and optimizing the ConvNet. This is because most classifiers focus on learning the larger classes, thereby resulting in poor classification accuracy for the smaller classes. Hence we propose a new loss function, which is an aggregation of two loss components; *viz.* – Generalized Dice loss [17] and Weighted Cross-entropy [16].

2.3 Overall Survival Prediction Based on Radiomic Features

For the OS prediction task we extract two types of Radiomic features, viz. "semantic" and "agnostic" [5]. The former includes attributes like size, shape, location, vascularity, spiculation, necrosis; and the latter attempts to capture lesion heterogeneity through quantitative descriptors like histogram, texture, etc. We extracted 33 semantic and 50 agnostic features from each segmented VOI. These are provided as input to a Multilayer Perceptron (MLP), having two hidden layers, to predict the number of survival days; which is further used to determine the survival class (short, mid or long).

3   Preliminary Experimental Results

The ConvNet models were developed using TensorFlow, with Keras in Python. The experiments were performed on the Intel AI DevCloud platform having cluster of Intel Xeon Scalable processors. Codes developed for our experiments will soon be made available. The proposed segmentation model is trained and validated on the corresponding training and validation datasets provided by the BraTS 2018 [1–3] organizers.

The preliminary quantitative evaluation results obtained by our segmentation model on the BraTS 2018 training and validation datasets are displayed in Table 1. The box-and-whisker plots in Fig. 4 reports the detailed quantitative segmentation results generated on 66 patients from the BraTS 2018 validation dataset. Quantitative metrics used for evaluating the segmentation results w.r.t. the gold standard (in case of training) and through the Leaderboard/blind testing (in case of validation) are (i) Dice score, (ii) sensitivity, (ii) specificity and (iii) Hausdorff distance computed for WT, TC and ET. The box-and-whisker plots report the minimum, lower quartile, median upper quartile and maximum. Points which fall outside 1.5 times the interquartile range are considered as outliers. It is evident from the box-and-whisker plots, that in most cases our algorithm produces significantly good segmentation accuracy w.r.t the manual segmentation by the radiologists for most of the cases. Qualitative segmentation results, obtained by our method for sample HGG and LGG patients from the BraTS 2018 training dataset and for a sample patient from the BraTS 2018 validation dataset, are shown in Figs. 5 and 6.

Table 1. Performance evaluation of proposed method on the BraTS 2018 training and validation datasets.

Evaluation metrics		Dice			Sensitivity			Specificity			Hausdorff95		
Training		ET	WT	TC	ET	WT	TC	ET	WT	TC	ET	WT	TC
	Mean	0.83	0.91	0.89	0.86	0.89	0.92	1.00	1.0	1.00	3.14	4.24	5.84
	StdDev	0.16	0.08	0.10	0.17	0.11	0.09	0.00	0.00	0.00	5.00	7.94	12.31
	Median	0.87	0.92	0.92	0.91	0.92	0.95	1.00	1.00	1.00	1.41	3	3
	25quantile	0.80	0.90	0.86	0.84	0.87	0.92	1.00	1.00	1.00	1.41	2	2
	75quantile	0.91	0.95	0.94	0.96	0.96	0.97	1.00	1.00	1.00	2.24	4.24	4.47
Validation	Mean	0.77	0.88	0.80	0.84	0.86	0.79	1.00	1.00	1.00	4.29	4.90	6.59
	StdDev	0.24	0.13	0.24	0.25	0.17	0.26	0.00	0.00	0.00	3.90	4.71	6.10
	Median	0.86	0.91	0.90	0.90	0.91	0.91	1.00	1.00	1.00	2.00	3.16	4.12
	25quantile	0.80	0.88	0.78	0.80	0.86	0.72	1.00	1.00	1.00	1.41	2.24	2.00
	75quantile	0.90	0.93	0.94	0.97	0.95	0.96	1.00	1.00	1.00	2.87	5.10	9.11

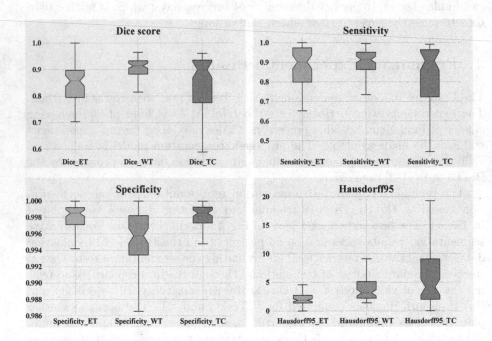

Fig. 4. Box-and-whisker plot of segmentation accuracy of the three sub-regions ET, WT and TC observed with Dice score, Sensitivity, Specificity and Hausdorff95.

Preliminary results of the proposed OS prediction method is reported in Table 2. We used 80% of the training data (130 patients) for training, and the remaining 20% (33 patients) for validation. The model was finally tested on 28 patients, having resection status GTR from the validation set, through the Leaderboard blind testing.

Table 2. OS prediction result on the BraTS 2018 validation dataset.

Accuracy	MSE	medianSE	stdSE	SpearmanR
0.54	180959.429	44665.0	340939.903	0.273

Proposed   Ground-truth

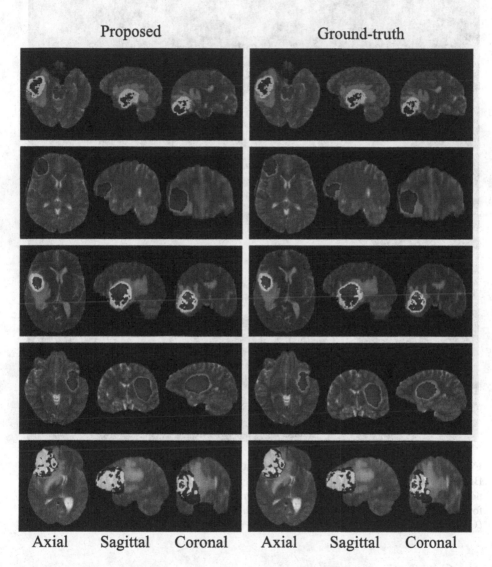

Axial   Sagittal   Coronal   Axial   Sagittal   Coronal

Fig. 5. Example segmentation result for five patients from the BraTS 2018 training dataset. The green label is edema, the red label is nonenhancing or necrotic tumor core, and the yellow label is enhancing tumor core. (Color figure online)

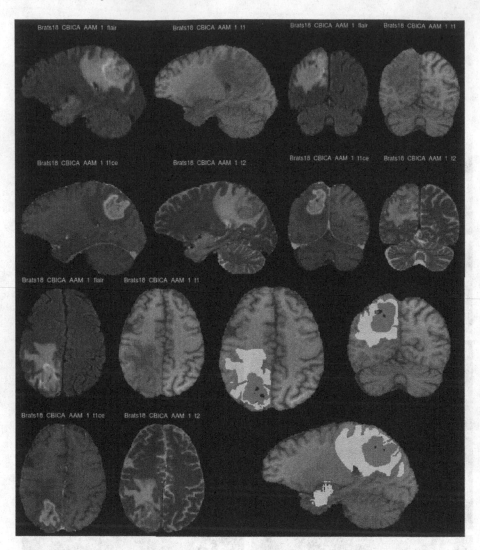

Fig. 6. Segmentation result for a sample patient (ID: BraTS18_CBICA_AAM_1) from BraTS 2018 validation data. The green label is edema, the red label is nonenhancing or necrotic tumor core, and the yellow label is enhancing tumor core. The Dice coefficients for the ET, WT and TC segmentation for this patient are 0.92, 0.90, 0.93 respectively (Color figure online)

4   Conclusion

We have designed a new deep learning based method for the automatic delineation/segmentation of brain tumors from multi-sequence MR images. The encoder-decoder type ConvNet model for pixel-wise segmentation performed better than other patch-based models. Integrated prediction from multiple anatomical planes (axial, sagittal and coronal) was superior, in terms of accuracy and robustness of decision, with respect to the estimation based on any single plane. Novel concepts such as spatial-pooling and unpooling reduced segmentation error around the boundary of the VOI. We also incorporated shortcut connections to copy and concatenate the receptive fields, from the encoder to the decoder parts, to help the decoder network localize and recover the object details more effectively. Very good validation accuracy was obtained for the segmentation task. We are currently exploring some other new features and feature selection methods, in order to improve the accuracy of predicting OS.

Acknowledgment. We gratefully acknowledge the support of Intel Corporation for providing access to the Intel AI DevCloud platform used in this work.

S. Banerjee acknowledges the support provided to him by the Intel Corporation, through the Intel AI Student Ambassador Program.

This publication is an outcome of the R&D work undertaken project under the Visvesvaraya PhD Scheme of Ministry of Electronics & Information Technology, Government of India, being implemented by Digital India Corporation.

S. Mitra acknowledges the support provided to her by the Indian National Academy of Engineering, through the INAE Chair Professorship.

References

1. Bakas, S., et al.: Advancing the cancer genome atlas glioma MRI collections with expert segmentation labels and radiomic features. Sci. Data **4**, 170117 (2017)
2. Bakas, S., et al.: Segmentation labels and radiomic features for the pre-operative scans of the TCGA-GBM collection. The Cancer Imaging Archive (2017). https://doi.org/10.7937/K9/TCIA.2017.KLXWJJ1Q
3. Bakas, S., et al.: Segmentation labels and radiomic features for the pre-operative scans of the TCGA-LGG collection. The Cancer Imaging Archive (2017). https://doi.org/10.7937/K9/TCIA.2017.GJQ7R0EF
4. Bakas, S., Reyes, M., et al.: Identifying the best machine learning algorithms for brain tumor segmentation, progression assessment, and overall survival prediction in the brats challenge. arXiv preprint arXiv:1811.02629 (2018)
5. Banerjee, S., Mitra, S., Shankar, B.U.: Synergetic neuro-fuzzy feature selection and classification of brain tumors. In: Proceedings of IEEE International Conference on Fuzzy Systems (FUZZ-IEEE), pp. 1–6 (2017)
6. Banerjee, S., Mitra, S., Uma Shankar, B.: Single seed delineation of brain tumor using multi-thresholding. Inf. Sci. **330**, 88–103 (2016)
7. Banerjee, S., Mitra, S., Uma Shankar, B.: Automated 3D segmentation of brain tumor using visual saliency. Inf. Sci. **424**, 337–353 (2018)
8. Banerjee, S., Mitra, S., Uma Shankar, B., Hayashi, Y.: A novel GBM saliency detection model using multi-channel MRI. PLOS ONE **11**(1), e0146388 (2016)
9. Banerjee, S., Mitra, S., Sharma, A., Shankar, B.U.: A CADe system for gliomas in brain MRI using convolutional neural networks. arXiv preprint arXiv:1806.07589 (2018)
10. Coroller, T., et al.: Early grade classification in meningioma patients combining radiomics and semantics data. Med. Phys. **43**, 3348–3349 (2016)
11. Gillies, R.J., Kinahan, P.E., Hricak, H.: Radiomics: images are more than pictures, they are data. Radiology **278**, 563–577 (2015)

12. LeCun, Y., Bengio, Y., Hinton, G.: Deep learning. Nature **521**(7553), 436–444 (2015)
13. Menze, B.H., et al.: The multimodal Brain Tumor image Segmentation benchmark (BraTS). IEEE Trans. Med. Imaging **34**, 1993–2024 (2015)
14. Milletari, F., Navab, N., Ahmadi, S.A.: V-net: fully convolutional neural networks for volumetric medical image segmentation. In: 2016 Fourth International Conference on 3D Vision (3DV), pp. 565–571. IEEE (2016)
15. Mitra, S., Banerjee, S., Hayashi, Y.: Volumetric brain tumour detection from MRI using visual saliency. PLOS ONE **12**, 1–14 (2017)
16. Ronneberger, O., Fischer, P., Brox, T.: U-net: convolutional networks for biomedical image segmentation. In: Navab, N., Hornegger, J., Wells, W.M., Frangi, A.F. (eds.) MICCAI 2015. LNCS, vol. 9351, pp. 234–241. Springer, Cham (2015). https://doi.org/10.1007/978-3-319-24574-4_28
17. Sudre, C.H., Li, W., Vercauteren, T., Ourselin, S., Jorge Cardoso, M.: Generalised dice overlap as a deep learning loss function for highly unbalanced segmentations. In: Cardoso, M.J., et al. (eds.) DLMIA/ML-CDS -2017. LNCS, vol. 10553, pp. 240–248. Springer, Cham (2017). https://doi.org/10.1007/978-3-319-67558-9_28
18. Zhou, M., et al.: Radiomics in brain tumor: image assessment, quantitative feature descriptors, and machine-learning approaches. Am. J. Neuroradiol. **39**, 208–216 (2018)

A Pretrained DenseNet Encoder
for Brain Tumor Segmentation

Jean Stawiaski[✉]

Stryker Corporation, Navigation, Freiburg im Breisgau, Germany
`jean.stawiaski@stryker.com`

Abstract. This article presents a convolutional neural network for the automatic segmentation of brain tumors in multimodal 3D MR images based on a U-net architecture. We evaluate the use of a densely connected convolutional network encoder (DenseNet) which was pretrained on the ImageNet data set. We detail two network architectures that can take into account multiple 3D images as inputs. This work aims to identify if a generic pretrained network can be used for very specific medical applications where the target data differ both in the number of spatial dimensions as well as in the number of inputs channels. Moreover in order to regularize this transfer learning task we only train the decoder part of the U-net architecture. We evaluate the effectiveness of the proposed approach on the BRATS 2018 segmentation challenge [1–5] where we obtained dice scores of 0.79, 0.90, 0.85 and 95% Hausdorff distance of 2.9 mm, 3.95 mm, and 6.48 mm for enhanced tumor core, whole tumor and tumor core respectively on the validation set. This scores degrades to 0.77, 0.88, 0.78 and 95% Hausdorff distance of 3.6 mm, 5.72 mm, and 5.83 mm on the testing set [1].

Keywords: Brain tumor · Convolutional neural network · Densely connected network · Image segmentation

1 Introduction

Automatic segmentation of brain tumor structures has a great potential for surgical planning and intraoperative surgical resection guidance. Automatic segmentation still poses many challenges because of the variability of appearances and sizes of the tumors. Moreover the differences in the image acquisition protocols, the inhomogeneity of the magnetic field and partial volume effects have also a great impact on the image quality obtained from routinely acquired 3D MR images. However brain gliomas can be well detected using modern magnetic resonance imaging. The whole tumor is particularly visible in T2-FLAIR, the tumor core is visible in T2 and the enhancing tumor structures as well as the necrotic parts can be visualized using contrast enhanced T1 scans. An example is illustrated in Fig. 1.

In the recent years, deep neural networks have shown to provide state-of-the-art performance for various challenging image segmentation and classification

ⓒ Springer Nature Switzerland AG 2019
A. Crimi et al. (Eds.): BrainLes 2018, LNCS 11384, pp. 105–115, 2019.
https://doi.org/10.1007/978-3-030-11726-9_10

Fig. 1. Example of images from the BRATS 2018 dataset. From left to right: T1 image, T2 image: the whole tumor and its core are visible, T2 FLAIR image: discarding the cerebrospinal fluid signal from the T2 image highlights the tumor region only, T1ce: contrast injection permits to visualize the enhancing part of the tumor as well as the necrotic part. Finally the expected segmentation result is overlaid on the T1ce image. The edema is shown in red, the enhancing part in white and the necrotic part of the tumor is shown in blue. (Color figure online)

problems [6–10]. Medical image segmentation problems have also been successfully tackled by such approaches [11,12,14,15,19]. However training deep neural networks can still be challenging in the case of a limited number of training data. In such situations it is often necessary to limit the complexity and the expressivity of the network. It has been observed that initializing weights of a convolutional network that has been pretrained on a large data set improves its accuracy on specific tasks where a limited number of training data is available [16]. We evaluate in this work the accuracy of a U-net architecture [11,12] where the encoder is a densely connected convolutional network [17] which has been pretrained on the ImageNet data set [18]. We study an extreme case of transfer learning where we fix the weights of the pretrained DenseNet encoder. Moreover we consider a segmentation problem where the input data dimensionality does not match the native input dimensions of the pretrained network. We will thus make use of a fixed pretrained network trained on 2D color images in order to segment 3D multimodal medical images. We will see that fixing the weights of the encoder is a simple but effective way to regularize the segmentation results.

2   Method

This section details the proposed network architectures, the loss function used to train the network as well as the training data preparation.

2.1   Convolutional Neural Network Architectures

The network processes 2D images of size (224, 224) pixels containing three channels. An input image is composed of three successive slices of the input volume along one of the three anatomical orientations: either along the coronal, the sagittal or the transverse plane. We use a pretrained network that has been designed to take a single 2D color image as input. In order to be able to process multi modal inputs, we have designed two distinct architectures:

- the first solution (M1) consists in removing the stem of the original DenseNet and only make use of the following convolutional layers which input is a tensor of size (64, 112, 112). This architecture is illustrated in Fig. 2. The proposed network is composed of a "precoder" which produces an adequate high dimensional input tensor for the pretrained network. This architecture is illustrated in Fig. 3. It processes independently each input images and concatenates the resulting tensors. This approach is very flexible and could take as input an image of any dimensions.

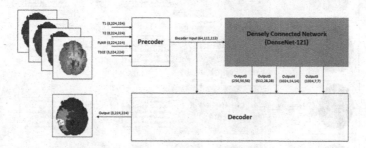

Fig. 2. Network architecture (M1). The network is composed of a "precoder" producing a high order tensor which is fed to a pretrained densely connected convolutional network. Several intermediate layers are then used to reconstruct a high resolution segmentation map.

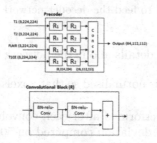

Fig. 3. Precoder architecture (M1). The precoder architecture process independently the input images by a sequence of multiple residual blocks (R1, R2) and concatenates the resulting output tensors. A residual block (R) is also illustrated. All convolution operations are computed with (3×3) kernels.

- the second solution (M2) consists in evaluating the different input modality separately through the original DenseNet encoder. Each input image modality is processed with the same encoder which shares its weights across the different modalities. Outputs at different scales are then concatenated and fed to the decoder. This architecture is illustrated in Figs. 4 and 6. This architecture does not permit to vary the number of input slices but has the advantage to fully leverage the original DenseNet weights.

For both architectures, the decoder consists in upsampling a low resolution layer, concatenate it with a higher resolution layer before applying a sequence of convolution operations. The first convolutional layer reduces the number of input channels by applying a (1×1) convolution. Following layers are composed of spatial (3×3) convolutions with residual connections.

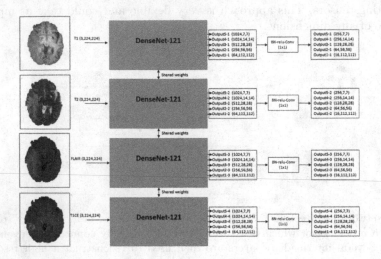

Fig. 4. Encoder architecture (M2). The network processes the different input image modality with the same encoder, a DenseNet composed of 121 layers. Intermediate layers of the encoder are used to feed the decoder network.

We give here additional details about the network architectures:

- each sample 3D image y is normalised so that voxels values falls in the interval $[0, 1]$.
- batch normalisation is performed after each convolutional layer using a running mean and standard deviation computed on 5000 samples:
- each layer is composed of residual connections as illustrated in Fig. 6,
- the activation function used in the network is a rectified linear unit,
- convolutions are computed using reflective border padding,
- upsampling is performed by nearest neighbor interpolation (Fig. 5).

2.2 Training

We used the BRATS 2018 training and validation sets for our experiments [2–5]. The training set contains 285 patients (210 high grade gliomas and 75 low grade gliomas). The BRATS 2018 validation set contains 66 patients with brain tumors of unknown grade with unknown ground truth segmentations. Each patient contains four modalities: T1, T1 with contrast enhancement, T2 and T2 FLAIR. The aim of this experiment is to segment automatically the whole tumor, the tumor

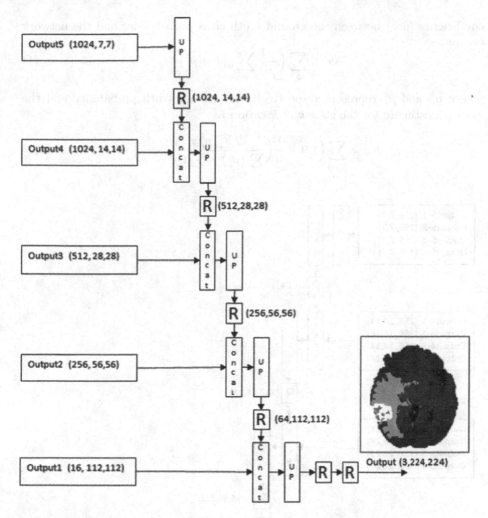

Fig. 5. Decoder architecture of the first model (M1). The decoder consists in a sequence of upsampling and residual convolution operations in order to produce a high resolution segmentation map.

core and the tumor enhancing parts. Note that the outputs of our neural network corresponds directly to the probability that a pixel belongs to a tumor, the core of a tumor and the enhancing part of the tumor. The last layer of the proposed architecture is thus composed of three independent (1×1) convolutional layers because we directly model the problem as a multi-label segmentation problem where a pixel can be assigned to multiple classes. Note that only weights of the "precoder" and the decoder are learned. Original weights of the pretrained DenseNet-121 stay fixed during the training procedure.

The network produces a segmentation maps by minimizing a loss function defined as the combination of the mean cross entropy (mce) and the mean Dice coefficients (dce) between the ground truth class probabilities and the network estimates:

$$ce = \sum_k \left(\frac{-1}{n} \sum_i y_i^k log(p_i^k) \right) \tag{1}$$

where y_i^k and p_i^k represent respectively the ground truth probability and the network estimate for the class k at location i.

$$dce = \sum_{k \neq 0} \left(1.0 - \frac{1}{n} \left(\frac{2 \sum_i p_i^k y_i^k + \epsilon}{\sum_i (p_i^k) + \sum_i (y_i^k) + \epsilon} \right) \right). \tag{2}$$

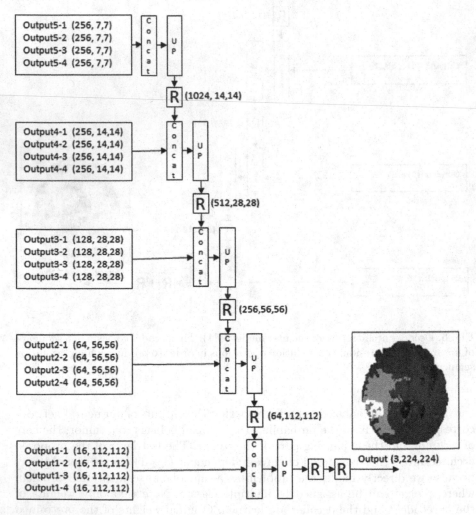

Fig. 6. Decoder architecture of the second model (M2). The decoder concatenates the encoding layers of each modalities. The segmentation is produced with a sequence of upsampling and convolution operations.

Note that we exclude the background class for the computation of the dice coefficient. The network is implemented using Microsoft CNTK[1]. We use stochastic gradient descent with momentum to train the network and L2 weights regularization. We use a cyclic learning rate where the learning rate varies from 0.0002 to 0.00005. An example of the evolution of the accuracy and the learning rate is illustrated in Fig. 7. We train the network for 160 epochs on a Nvidia GeForce GTX 1080 GPU. A full epoch consists in analyzing all images of the BRATS training data set and extracting 20 2D random samples from the 3D MR volumes.

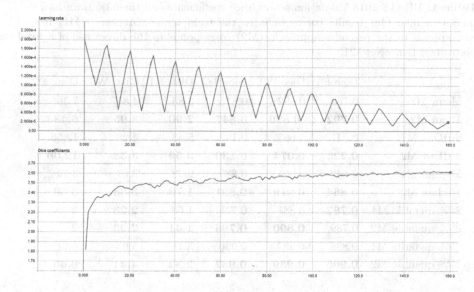

Fig. 7. Network training. Illustration of the cyclic learning rate schedule (top). Evolution of the sum of the dice coefficients of the three classes during training (bottom).

2.3   Testing

Segmentation results are obtained by evaluating the network along slices extracted from the three anatomical orientations and averaging the results. A segmentation map is then obtained by assigning to each voxel the label having the maximum probability among the three classes: tumor, tumor core or enhancing tumor. Finally connected components composed of less than 100 voxels are removed. We are not making use of test time image augmentation or ensembling methods.

---

[1] https://www.microsoft.com/en-us/cognitive-toolkit/.

3   Results

We uploaded our segmentation results to the BRATS 2018 server[2] which evaluates the segmentation and provides quantitative measurements in terms of Dice scores, sensitivity, specificity and Hausdorff distances of enhanced tumor core, whole tumor, and tumor core. Results of the BRATS 2018 validation phase are presented in Table 1. The validation phase is composed of 66 datasets with unknown ground truth segmentations.

Table 1. BRATS 2018 Validation scores, dice coefficients and the 95% Hausdorff distances in mm. Our results corresponds to the team name "Stryker". (M1) results corresponds to the precoder approach, (M2) corresponds to the direct use of a fixed pretrained DenseNet-121.

	Dice ET	Dice WT	Dice TC	Dist. ET	Dist. WT	Dist. TC
Mean M1	0.768	0.892	0.815	3.85	4.85	7.56
Mean M2	**0.792**	**0.899**	**0.847**	**2.90**	**3.95**	**6.48**
StdDev M1	0.241	0.065	0.187	5.43	4.28	12.56
StdDev M2	**0.223**	**0.074**	**0.130**	**3.59**	**3.38**	**12.06**
Median M1	0.849	0.905	0.889	2.23	3.67	3.74
Median M2	**0.864**	**0.919**	**0.891**	**1.73**	**3.08**	**3.30**
25% quantile M1	**0.792**	0.881	0.758	1.68	**2.23**	2
25% quantile M2	0.789	**0.890**	**0.796**	**1.41**	**2.23**	2
75% quantile M1	0.888	0.933	0.930	3.16	5.65	8.71
75% quantile M2	**0.906**	**0.939**	**0.932**	**2.82**	**4.41**	**6.65**

Results of the BRATS 2018 testing phase are presented in Table 2. The testing phase is composed of 191 datasets with unknown ground truth segmentations.

Table 2. BRATS 2018 Testing scores, dice coefficients and the 95% Hausdorff distances in mm.

	Dice ET	Dice WT	Dice TC	Dist. ET	Dist. WT	Dist. TC
Mean M2	**0.776**	**0.878**	**0.786**	**3.63**	**5.72**	**5.83**
StdDev M2	0.223	0.104	0.257	5.29	7.31	7.93
Median M2	0.828	0.908	0.891	2.23	3.60	3.46
25% quantile M2	0.749	0.857	0.796	1.41	2.23	2.1
75% quantile M2	0.895	0.935	0.924	3.0	6.08	6.13

---

[2] https://www.cbica.upenn.edu/BraTS18/lboardValidation.html.

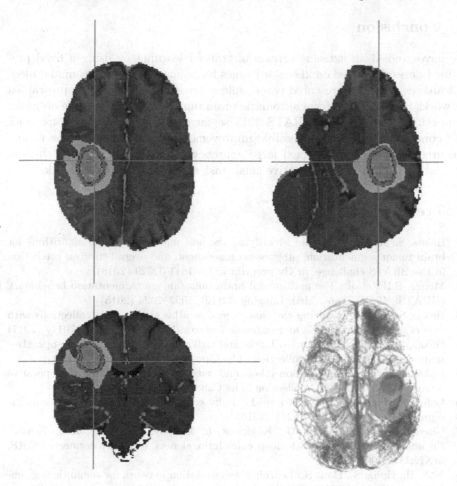

Fig. 8. Segmentation result obtained on a image of the testing data.

4   Discussion

The validation and testing results obtained on the BRATS segmentation challenge show that the proposed approaches are indeed efficient. Despite the fact that the used encoder has been trained on natural color images, it turns out that the learned features can be leveraged for a large class of applications including segmentation of medical images. Using a fixed encoder is thus an effective way to regularize the neural network. Note that we did not make use of advanced image augmentations or ensembling methods. The two approaches produce comparable results and have both advantages and drawbacks. The model (M1) is more versatile since it can use any number of input modalities (channels) and any number of spatial dimensions. However current experiments shows that the model (M2), despite its simplicity, produces slightly better results. A major limitation of the proposed approach is the lack of 3D spatial consistency (Fig. 8).

5 Conclusion

We have studied an extreme version of transfer learning by using a fixed pre-trained network trained on 2D color images for segmenting 3D multi modal medical images. We have presented two simple approaches for leveraging pretrained networks in order to perform automatic brain tumor segmentation. We obtained competitive scores on the BRATS 2018 segmentation challenge[3]. Future work will concentrate on several possible improvements by additionally fine tuning the pretrained encoder. A fixed large expressive 2D neural network is thus an interesting alternative to a relative small task specific 3D neural networks.

References

1. Bakas, S., Reyes, M., et al.: Identifying the best machine learning algorithms for brain tumor segmentation, progression assessment, and overall survival prediction in the BRATS challenge. arXiv preprint arXiv:1811.02629 (2018)
2. Menze, B.H., et al.: The multimodal brain tumor image segmentation benchmark (BRATS). IEEE Trans. Med. Imaging **34**(10), 1993–2024 (2015)
3. Bakas, S., et al.: Advancing the cancer genome atlas glioma MRI collections with expert segmentation labels and radiomic features. Nat. Sci. Data **4**, 170117 (2017)
4. Bakas, S., et al.: Segmentation labels and radiomic features for the pre-operative scans of the TCGA-GBM collection. The Cancer Imaging Archive (2017)
5. Bakas, S., et al.: Segmentation labels and radiomic features for the pre-operative scans of the TCGA-LGG collection. The Cancer Imaging Archive (2017)
6. Long, J., Shelhamer, E., Darrell, T.: Fully convolutional networks for semantic segmentation. arXiv:1605.06211 (2016)
7. Chen, L.-C., Papandreou, G., Kokkinos, I., Murphy, K., Yuille, A.L.: Semantic image segmentation with deep convolutional nets and fully connected CRF. arXiv:1412.7062 (2014)
8. Noh, H., Hong, S., Han, B.: Learning deconvolution network for semantic segmentation. arXiv:1505.04366 (2015)
9. Badrinarayanan, V., Kendall, A., Cipolla, R.: SegNet: a deep convolutional encoder-decoder architecture for image segmentation. arXiv:1511.00561 (2015)
10. Yu, F., Koltun, V.: Multi-scale context aggregation by dilated convolutions. arXiv:1511.07122 (2016)
11. Ronneberger, O., Fischer, P., Brox, T.: U-Net: convolutional networks for biomedical image segmentation. arXiv:1505.04597 (2015)
12. Milletari, F., Navab, N., Ahmadi, S.-A.: V-Net: fully convolutional neural networks for volumetric medical image segmentation. arXiv:1606.04797 (2016)
13. Lee, C.-Y., Xie, S., Gallagher, P., Zhang, Z., Tu, Z.: Deeply-supervised nets. arXiv:1409.5185 (2014)
14. Zhu, Q., Du, B., Turkbey, B., Choyke, P.L., Yan, P.: Deeply-supervised CNN for prostate segmentation. arXiv:1703.07523 (2017)
15. Yu, L., Yang, X., Chen, H., Qin, J., Heng, P.-A.: Volumetric ConvNets with mixed residual connections for automated prostate segmentation from 3D MR images. In: Proceedings of the Thirty-First AAAI Conference on Artificial Intelligence (AAAI-2017) (2017)

---

[3] https://www.cbica.upenn.edu/BraTS17/lboardValidation.html.

16. Iglovikov, V., Shvets, A.: TernausNet: U-Net with VGG11 encoder pre-trained on ImageNet for image segmentation. arXiv:1801.05746 (2018)
17. Huang, G., Liu, Z., van der Maaten, L., Weinberger, K.Q.: Densely connected convolutional networks. arXiv:1608.06993 (2017)
18. Russakovsky, O., et al.: ImageNet large scale visual recognition challenge. arXiv:1409.0575 (2014)
19. Christ P.F., et al.: Automatic liver and lesion segmentation in ct using cascaded fully convolutional neural networks and 3D conditional random fields. arXiv:1610.02177 (2016)

Hierarchical Multi-class Segmentation of Glioma Images Using Networks with Multi-level Activation Function

Xiaobin Hu[✉], Hongwei Li, Yu Zhao, Chao Dong, Bjoern H. Menze, and Marie Piraud

Department of Computer Science, Technische Universität München, Munich, Germany
xiaobin.hu@tum.de

Abstract. For many segmentation tasks, especially for the biomedical image, the topological prior is vital information which is useful to exploit. The containment/nesting is a typical inter-class geometric relationship. In the MICCAI Brain tumor segmentation challenge, with its three hierarchically nested classes 'whole tumor', 'tumor core', 'active tumor', the nested classes relationship is introduced into the 3D-residual-Unet architecture. The network comprises a context aggregation pathway and a localization pathway, which encodes increasingly abstract representation of the input as going deeper into the network, and then recombines these representations with shallower features to precisely localize the interest domain via a localization path. The nested-class-prior is combined by proposing the multi-class activation function and its corresponding loss function. The model is trained on the training dataset of Brats2018, and 20% of the dataset is regarded as the validation dataset to determine parameters. When the parameters are fixed, we retrain the model on the whole training dataset. The performance achieved on the validation leaderboard is 86%, 77% and 72% Dice scores for the whole tumor, enhancing tumor and tumor core classes without relying on ensembles or complicated post-processing steps. Based on the same start-of-the-art network architecture, the accuracy of nested-class (enhancing tumor) is reasonably improved from 69% to 72% compared with the traditional Softmax-based method which blind to topological prior.

Keywords: Topological prior · Nested classes · 3D-residual-Unet · Multi-class activation function

1 Introduction

Glioma are the most common family of brain tumors, and forms some of highest-mortality and economically costly diseases of brain cancer [1–3]. The diagnosed method is highly relayed on manual segmentation and analysis of multi-modal MRI scans by bio-medical experts. Nevertheless, this diagnosed way is severely

© Springer Nature Switzerland AG 2019
A. Crimi et al. (Eds.): BrainLes 2018, LNCS 11384, pp. 116–127, 2019.
https://doi.org/10.1007/978-3-030-11726-9_11

limited by the labor-intensive character of the manual segmentation process and disagreement or mistakes between manual segmentation. Consequently, there exists a great need for a fast and robust automated segmentation algorithm. Convolutional neural networks (CNNs) have been verified to be extremely effective for a variety of semantic segmentation tasks [4].

While CNN segmentation algorithms are abundant in biomedical imaging, only very few make use of nested-topological prior information. Among the few that do [5–11], we find three different approaches. First, the use of cascaded algorithms where the network consists of successive segmentation networks. Second, the information on the nested-classes is incorporated into the loss function, imposing penalties on solutions that do not respect the nested geometry relations. Third, Markov random fields are used to formalizing class relationship in the post-processing of the network output. Here, we make use of a new activation function [12] that is directly implementing class hierarchy in the network training and generalize it to 3 nested classes. For the glioma labels we assume that active tumor regions are always contained in the tumor core which is surrounded by the tumor edema, resulting in a hierarchical three-class model. In sharp contrast with nested-class method, the softmax-based method of multi-class ignores the geometric prior between different classes, and assumes the classes are mutually-exclusive, meaning one pixel cannot belong to different classes at the same time, which absolutely discards the topological information and sometimes leads the unreasonable segmentation results. The comparison of Dice score criteria between two different methods is implemented and it obviously indicates the nested-class method achieves higher accuracy than the softmax-based method, especially for the internal-classes.

In the following, we introduce a brief overview of start-of-the-art 3D-residual U-net architecture and multi-class-nested activation and loss function. We then propose and evaluate our model architectures for Brats tumor segmentation. Finally, we implement the comparison between two main avenues and illustrate the multi-level activation performs better especially in the inter-class.

2   Methodology

2.1   Network Architecture

The nested-classes relationship between different labels are shown in Fig. 2. The general network structure shown in Fig. 1 is stemming from the previously used glioma segmentation network by Isensee [13] to process large 3D input blocks of $144 \times 144 \times 144$ voxels. The original network is inspired by the U-net [14] which allows the network to intrinsically recombine different scales throughout the entire network. This vertical depth is set as 5, which balances between the spatial resolution and feature representations. The context module is a pre-activation residual block, and is connected by $3 \times 3 \times 3$ convolutions with input stride 2. The purpose of the localization pathway is to extract features from the lower levels of the network and transform them to a high spatial resolution by means of a simple upscale technology. The upsampled features and its corresponding level of the context aggregation feature are recombined via concatenation. Furthermore, the localization module, consisting of a $3 \times 3 \times 3$ convolution followed by a $1 \times 1 \times 1$ convolution, is designed to gather these features.

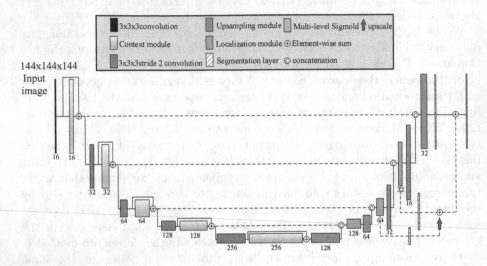

Fig. 1. Network architecture from [13]: Context pathway (left) aggregates high level information; Localization pathway (right) localizes precisely

The deep supervision is introduced in the localization pathway by integrating segmentation layers at different levels of the network and combining them via elementwise summation to form the final network output. The output activation layer is multi-level Sigmoid layer instead of softmax layer in the Isensee's network which converting the multi-class problem to binary ones. Intrinsically, the multi-level activation is the assemble of multi-sigmoid function and then straightforwardly maps to multi-class segmentation incorporating the topological prior. Consequently, it overcomes the softmax-based method's shortcoming which is blind to the geometric prior.

2.2 Crop Preprocessing

For 3D network architecture, the larger patch size of training dataset contains more continuous context knowledge and localization information which are beneficial to improve the segmentation accuracy. In order to acquire to the larger cube size patch of 3D image, the valuable knowledge in the MRI is extracted as much as possible while the meaningless information is cropped. Then the crop processing is implemented, and the maximum size of cube patch is selected as $[144, 144, 144]$.

The crop preprocessing equation is defined as:

$$array = [a_{min} - (b_{size} - a)/2 : a_{min} + (b_{size} + a)/2]$$
$$a = a_{max} - a_{min}$$

(1)

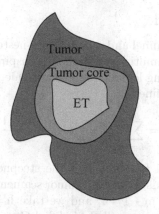

Fig. 2. Schematic description of the nesting of classes in the BRATS challenge, which respects the following hierarchy: Enhancing Tumor (ET) $\in$ Tumor core $\in$ Tumor

where a_{min} and a_{max} are the min and max non-zero information index of MRI image, and a represents the length of non-zero information. b_{size} is the cube patch size and selected as 144.

The index is recorded and used in the image post-processing stage to recovery back to the original shape $[155, 240, 240]$. However, a little of meaningful information which exceeds the cube patch size 144 is unavoidably ignored and have little effect on the segmentation result. In order to equally compare the softmax-based with the multi-level method, no data augmentation operation is used in the stage of image pre-processing.

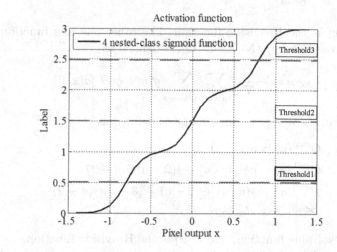

Fig. 3. Multi-class activation function, Eq. (1) with m + 1 = 4, h = 0.8 and k = 10

2.3   Multi-level Method

Here, we use one output channel and a multi-class-nested activation function, as first proposed in [12]. The multi-level method is inspired by continuous regression, and thereby generalizing logistic regression to hierarchically-nested classes. It is shown in Fig. 3 and defined as

$$a(x) = \sum_{n=1}^{m} \sigma(k[x + h(n - \frac{m+1}{2})])$$ (2)

Where σ is the sigmoid function, k is the steepness and h is the spacing between consecutive Sigmoids. For Brain tumor segmentation challenge 4-classes nested label case, we have $m + 1 = 4$, and we take $h = 0.5$ and steepness $= 10$. The corresponding loss function, called Modified Cross-Entropy (MCE) in [12], is defined as

$$L_{MCE} = -\frac{1}{N_{tot}} \sum_{pixel\,i} \sum_{classes\,c} y_i^c w^c log(P^c[a(x_i)])$$ (3)

where w^c is the weight of corresponding label, which we take as $w^{c\alpha}(w^{c\alpha} = (\frac{N_{tot}}{N_c})^\alpha)$, where N_{tot} is the sum number of pixels, N_c the number of pixels in each class, and where $y^c = 1$ for the ground-truth label c of pixel i and $y^c = 0$ otherwise. Furthermore, the mapping function P^c is defined as

$$\begin{aligned}
P^{c=0}(a) &= 1 - a/3 \\
P^{c=1}(a) &= a\Theta(1-a) + (3-a)/2\Theta(a-1) \\
P^{c=2}(a) &= a/2\Theta(2-a) + (3-a)\Theta(a-2) \\
P^{c=3}(a) &= a/3.
\end{aligned}$$ (4)

Where $\Theta(x)$ is the Heaviside function. The other one loss function, called Normalized Cross-Entropy (NCE) in [12], is defined as

$$L_{NCE} = -\frac{1}{N_{tot}} \sum_{pixel\,i} \sum_{classes} y_i^c w^c log(\Theta^c[a(x_i)])$$ (5)

Furthermore, the mapping function Q^c is defined as

$$\begin{aligned}
Q^{c=0}(a) &= s(1-a) \\
Q^{c=1}(a) &= a\Theta(1-a) + s(2-a)\Theta(a-2) \\
P^{c=2}(a) &= s(a-1)\Theta(2-a) + (3-a)\Theta(a-2) \\
P^{c=3}(a) &= s(a-2).
\end{aligned}$$ (6)

where s is the softplus function, and $\Theta(x)$ is the Heaviside function.

Weighted modified and Normalized cross-entropy losses are naturally combined with standard cross-entropy loss and mitigate the class unbalance problem. They also have the ability to encode of any hierarchical and mutually-exclusive topological relationship of classes in a network architecture.

2.4   Evaluation Metrics

In the task for BRATS, the number of positives and negatives are highly unbalanced. Consequently, four typical different metrics are used by the organizers to evaluate the performance of the algorithm and then rank the different teams.

Give a ground-truth segmentation map G and a segmentation map corresponding one class generated by the algorithm. The four evaluation criteria are defined as following.

Dice similarity coefficient (DSG):

$$DSC = \frac{2(G \cap P)}{|G| + |P|} \tag{7}$$

The Dice similarity coefficient measures the overlap in percentage between G and P.

Hausdorff distance (95th percentile) is defined as:

$$H(G, P) = max(supinf_{x \in G, y \in P} d(x, y), supinf_{y \in P, x \in G} d(x', y)) \tag{8}$$

where $d(x, y)$ denotes the distance of x and y, sup denotes the supremum and inf for the infimum. This measures how far two subsets of a metric space are from each other. As used in this challenge, it is modified to obtain a robustified version by using the 95th percentile instead of the maximum (100 percentile) distance.

Sensitivity (also called the true positive rate) measures the proportion of actual positives that are correctly identified. Specificity (also called the true negative rate) measures the proportion of actual negatives that are correctly identified. Assume P is the number of real positive prediction pixel of lesion and N is the number of real negative prediction pixel of lesion. Condition positive P consists with true positive TP and false negative FN. Besides, the condition negative N is also divided into TN true negative and FP false positive.

Then, the metrics of Sensitivity and Specificity are illustrated as:

$$Sensitivity = \frac{TR}{P} = \frac{TP}{TP + FN} \tag{9}$$

$$Specificity = \frac{TN}{N} = \frac{TN}{TN + FP} \tag{10}$$

Then the values of those four metrics were computed by the organizers independently and made available in the validation leaderboard.

3   Experiment Results

In BRATS 2018 dataset [15–19], there are four types, Necrotic core, Edema, Non-enhancing core and Enhancing core that form the three tumor classes in Fig. 2. The dataset contains 4 different modalities for MRI, native (T1), post-contrast T1-weighted (T1Gd), T2-weighted (T2) and T2 Fluid Attenuated

(a) T1image   (b) Flair image   (c) Prediction   (d) Ground-truth

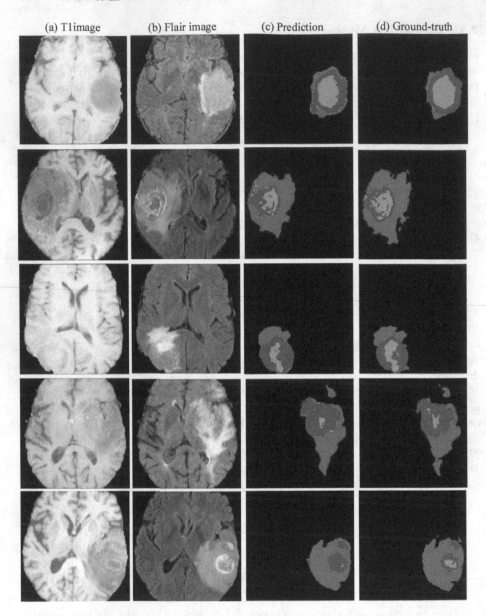

Fig. 4. Segmentation results, for five different validation cases. The tumor class is depicted in red, tumor core in green and enhancing tumor in blue. (Color figure online)

Inversion Recovery (FLAIR) which are all used as different input channels. We train the networks using ADAM optimizer with an initial learning rate of 0.0005, and to regularize the network, we use early stopping when the precision on the 20% of the training dataset reserved for validation is no longer improved, and dropout (with rate 0.3) in all residual block before the multi-class sigmoid function. Some slices of segmentation results containing the tumor, tumor core and enhancing core are shown in Fig. 4. We observe that the topology geometry between different labels is constrained to the nested-classes relationship, consequently avoiding errors stemming from the lack of topological prior.

Table 1. Validation results presented on the leaderboard

	Dice score			
	Enhancing core	Whole tumor	Tumor core	Weight scheme
Multi-level (MCE)	**0.719**	0.857	**0.769**	0.4
Multi-level (NCE)	0.676	0.857	0.755	0.4
Multi-level (NCE)	0.633	0.837	0.736	0.5
Multi-level (NCE)	0.655	0.856	0.758	0.3
Softmax-based method	0.691	**0.861**	0.763	-

Table 2. Quantitative evaluation of Dice score

Dice score	Enhancing core	Whole tumor	Tumor core
Mean	0.71965	0.85685	0.76906
StdDev	0.28526	0.09802	0.21962
Median	0.84268	0.87823	0.84325
25quantile	0.6889	0.83379	0.70743
75quantile	0.8876	0.90895	0.91292

The segmentation result is severely affected by highly unbalanced problems existing in the Brats dataset. As class imbalance in a data set increases, the performance of a neural net trained on that data has been shown to decrease dramatically [20]. In order to mitigate this issue, many methods [21–23] were proposed to modify the loss function to alleviate this problems. Here, the weighted cross entropy incorporating the nested-class information is proposed and investigated. We experimented with different weighting schemes ($\alpha = 1, 0.5, 0.4, 0.3$) and with the different losses (MCE and NCE) proposed in [12]. The best performing combination turned out to be $\alpha = 0.4$ and MCE loss function. The segmentation thresholds to determine the boundaries between classes, were set to $[0.95, 1.65, 2.2]$ on the validation process. For this final configuration, we reached Dice scores of 86% for the complete tumor, 77% for the tumor core and 72% for the enhancing core as presented in Table 1. The weighted-modified-cross-entropy performs much better than the result achieved by normalized cross-entropy, and weight scheme affects the segmentation result severely since the extraordinary unbalance problem. The different weight schemes $[0.5, 0.4, 0.3]$ are compared and the optimal weight scheme is taken as 0.4. In comparison with the softmax-based method based on the same network architecture proposed by Isensee without ensembles operation, any complicated image pre-processing and post-processing steps and extra training dataset, it indicates that the Dice score of nested-class (enhancing core) drastically improved from 0.691 to 0.719 while the Dice core of whole tumor and tumor core almost remains at same extent. The quantitative evaluation (Mean, std, Median, 25%, 75% quantile) of Dice score of enhancing core and whole tumor and tumor core are showed in Table 2. And other evaluation metrics (the proportion of actual positives correctly identified—Sensitivity, the proportion of actual negatives correctly identified—Specificity and Hausdorff95) are listed in Table 3.

Table 3. Sensitivity, Specificity and Hausdorff95 results presented on the leaderboard

Mean	Enhancing core	Whole tumor	Tumor core
Sensitivity	0.74119	0.93916	0.78743
Specificity	0.9974	0.98715	0.99591
Hausdorff95	5.50007	10.84397	9.98557

3.1 Threshold Scheme Definition and Analysis

Setting the optimal threshold is an important component of the multi-class segmentation task, and it is straightforwardly linked to segmentation boundary. From the activation function (4 nested-class sigmoid function) Fig. 3, the 4 classes segmentation problem is corresponding with the threshold scheme with 3 parameters [Threshold-1, Threshold-2, Threshold-3]. The threshold scheme is optimally chosen during the validation procedure, and then fixed and applied into test dataset.

In order to analyze how the threshold affects the segmentation accuracy, the relationship between boundary threshold and Dice score is illustrated in Fig. 5. The target threshold is changed to the value taken from a specific interval which is considered to be possible to achieve optimal segmentation result when other thresholds are fixed at the optimal value. The criteria Dice score of three classes is very sensitive to the threshold-3 value compared with other two threshold indexes, that it may drop into Dice score valley within interval [2.2, 2.4]. The threshold-2 index has little impact on the Dice score of whole classes except for threshold greater than 1.8. Consequently, it is easier to make an optimal threshold scheme after determining indexes of threshold-3 and threshold-2. After experiment and optimization, the suitable threshold scheme in the Brats challenge is selected as [0.95, 1.65, 2.2].

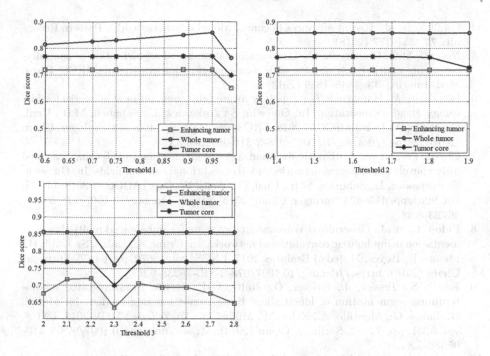

Fig. 5. Boundary division of threshold scheme

4   Conclusions

In this paper we applied the technique of multi-level activation to the nested classes segmentation of glioma. The results of our experiments indicate that the multi-level activation function and its corresponding loss function are efficient compared to Softmax output layer based on the same network framework. Using the MCE loss function and a reweighting scheme with power-law $= 0.4$, we obtain Dice scores 86% for complete tumor, 77% for tumor core and 72% for enhancing core on the validation leaderboard of the 2018 BRATS challenge, proving the applicability of the multi-level activation scheme. Finally, this activation could be combined with other network architectures. Using it with the best performing architecture of the BRATS challenge could even lead to further improved results.

References

1. Davis, M.E.: Glioblastoma: overview of disease and treatment. Clin. J. Oncol. Nurs. **20**(5), S2–S8 (2016). https://doi.org/10.1188/16.CJON.S1.2-8
2. Hanif, F., Muzaffar, K., Perveen, K., Malhi, S.M., Simjee, S.U.: Glioblastoma multiforme: a review of its epidemiology and pathogenesis through clinical presentation and treatment. Asian Pac. J. Cancer Prev. **18**, 3–9 (2017)
3. Birbrair, A., et al.: Novel peripherally derived neural-like stem cells as therapeutic carriers for treating glioblastomas. STEM CELLS Transl. Med. **6**, 471–481 (2017)

4. Gu, J.X., et al.: Recent advances in convolutional neural networks. Pattern Recognit. **77**, 354–377 (2018)
5. Nosrati, M.S., Hamarneh, G.: Local optimization based segmentation of spatially-recurring, multi-region objects with part configuration constraints. IEEE Trans. Med. Imaging **33**, 1845–1859 (2014)
6. BenTaieb, A., Hamarneh, G.: Topology aware fully convolutional networks for histology gland segmentation. In: Ourselin, S., Joskowicz, L., Sabuncu, M.R., Unal, G., Wells, W. (eds.) MICCAI 2016. LNCS, vol. 9901, pp. 460–468. Springer, Cham (2016). https://doi.org/10.1007/978-3-319-46723-8_53
7. Christ, P.F., et al.: Automatic liver and lesion segmentation in CT using cascaded fully convolutional neural networks and 3D conditional random fields. In: Ourselin, S., Joskowicz, L., Sabuncu, M.R., Unal, G., Wells, W. (eds.) MICCAI 2016. LNCS, vol. 9901, pp. 415–423. Springer, Cham (2016). https://doi.org/10.1007/978-3-319-46723-8_48
8. Fidon, L., et al.: Generalised Wasserstein dice score for imbalanced multi-class segmentation using holistic convolutional networks. In: Crimi, A., Bakas, S., Kuijf, H., Menze, B., Reyes, M. (eds.) BrainLes 2017. LNCS, vol. 10670, pp. 64–76. Springer, Cham (2018). https://doi.org/10.1007/978-3-319-75238-9_6
9. Bauer, S., Tessier, J., Krieter, O., Nolte, L.-P., Reyes, M.: Integrated spatio-temporal segmentation of longitudinal brain tumor imaging studies. In: Menze, B., Langs, G., Montillo, A., Kelm, M., Müller, H., Tu, Z. (eds.) MCV 2013. LNCS, vol. 8331, pp. 74–83. Springer, Cham (2014). https://doi.org/10.1007/978-3-319-05530-5_8
10. Alberts, E., et al.: A nonparametric growth model for brain tumor segmentation in longitudinal MR sequences. In: Crimi, A., Menze, B., Maier, O., Reyes, M., Handels, H. (eds.) BrainLes 2015. LNCS, vol. 9556, pp. 69–79. Springer, Cham (2016). https://doi.org/10.1007/978-3-319-30858-6_7
11. Liu, Z.W., Li, X.X., Luo, P., Loy, C.C., Tang, X.O.: Deep learning Markov random field for semantic segmentation. IEEE Trans. Pattern Anal. Mach. Intell. **1–1**, 8828 (2017)
12. Piraud, M., Sekuboyina, A., Menze, B.H.: Multi-level activation for segmentation of hierarchically-nested classes. In: Computer Vision and Pattern Recognition Workshop (2018)
13. Isensee, F., Kickingereder, P., Wick, W., Bendszus, M., Maier-Hein, K.H.: Brain tumor segmentation and radiomics survival prediction: contribution to the BRATS 2017 challenge. In: Crimi, A., Bakas, S., Kuijf, H., Menze, B., Reyes, M. (eds.) BrainLes 2017. LNCS, vol. 10670, pp. 287–297. Springer, Cham (2018). https://doi.org/10.1007/978-3-319-75238-9_25
14. Ronneberger, O., Fischer, P., Brox, T.: U-Net: convolutional networks for biomedical image segmentation. In: Navab, N., Hornegger, J., Wells, W.M., Frangi, A.F. (eds.) MICCAI 2015. LNCS, vol. 9351, pp. 234–241. Springer, Cham (2015). https://doi.org/10.1007/978-3-319-24574-4_28
15. Menze, B.H., Jakab, A., Bauer, S., et al.: The multimodal brain tumor image segmentation benchmark (BRATS). IEEE Trans. Med. Imaging **34**(10), 1993–2024 (2015)
16. Bakas, S., et al.: Advancing The Cancer Genome Atlas glioma MRI collections with expert segmentation labels and radiomic features. Nat. Sci. Data **4**, 170117 (2017)
17. Bakas, S., et al.: Segmentation labels and radiomic features for the pre-operative scans of the TCGA-GBM collection. The Cancer Imaging Archive (2017). https://doi.org/10.7937/K9/TCIA.2017.KLXWJJ1Q

18. Bakas, S., et al.: Segmentation labels and radiomic features for the pre-operative scans of the TCGA-LGG collection. The Cancer Imaging Archive (2014). https://doi.org/10.7937/K9/TCIA.2017.GJQ7R0EF

19. Bakas, S., Reyes, M., et al.: Identifying the best machine learning algorithms for brain tumor segmentation, progression assessment, and overall survival prediction in the BRATS challenge, arXiv preprint arXiv:1811.02629 (2018)

20. Mazurowski, M.A., Habas, P.A., Zurada, J.M., Lo, J.Y., Baker, J.A., Tourassi, G.D.: Training neural network classifiers for medical decision making: the effects of imbalanced datasets on classification performance. Neural Netw. **21**(2), 427–436 (2017)

21. Milletari, F., Navab, N., Ahmadi, S.A.: V-Net: fully convolutional neural networks for volumetric medical image segmentation. In: Fourth International Conference on 3D Vision, pp. 565–571 (2016)

22. Sudre, C.H., Li, W., Vercauteren, T., Ourselin, S., Jorge Cardoso, M.: Generalised dice overlap as a deep learning loss function for highly unbalanced segmentations. In: Cardoso, M.J., et al. (eds.) DLMIA/ML-CDS 2017. LNCS, vol. 10553, pp. 240–248. Springer, Cham (2017). https://doi.org/10.1007/978-3-319-67558-9_28

23. Crum, W.R., Camara, O., Hill, D.L.G.: Generalized overlap measures for evaluation and validation in medical image analysis. IEEE Trans. Med. Imaging **25**(11), 1451–1461 (2006)

Brain Tumor Segmentation and Tractographic Feature Extraction from Structural MR Images for Overall Survival Prediction

Po-Yu Kao[1(✉)], Thuyen Ngo[1], Angela Zhang[1], Jefferson W. Chen[2], and B. S. Manjunath[1(✉)]

[1] Vision Research Lab, University of California, Santa Barbara, CA, USA
{poyu_kao,manj}@ece.ucsb.edu
[2] UC Irvine Health, University of California, Irvine, CA, USA

Abstract. This paper introduces a novel methodology to integrate human brain connectomics and parcellation for brain tumor segmentation and survival prediction. For segmentation, we utilize an existing brain parcellation atlas in the MNI152 1 mm space and map this parcellation to each individual subject data. We use deep neural network architectures together with hard negative mining to achieve the final voxel level classification. For survival prediction, we present a new method for combining features from connectomics data, brain parcellation information, and the brain tumor mask. We leverage the average connectome information from the Human Connectome Project and map each subject brain volume onto this common connectome space. From this, we compute tractographic features that describe potential neural disruptions due to the brain tumor. These features are then used to predict the overall survival of the subjects. The main novelty in the proposed methods is the use of normalized brain parcellation data and tractography data from the human connectome project for analyzing MR images for segmentation and survival prediction. Experimental results are reported on the BraTS2018 dataset.

Keywords: Brain tumor segmentation · Brain parcellation ·
Group normalization · Hard negative mining · Ensemble modeling ·
Overall survival prediction · Tractographic feature

1 Introduction

Glioblastomas, or Gliomas, are one of the most common types of brain tumor. They have a highly heterogeneous appearance and shape and may happen at any location in the brain. High-grade glioma (HGG) is one of the most aggressive types of brain tumor with median survival of 15 months [17]. There is a significant amount of recent work on brain tumor segmentation and survival prediction. Kamnitsas et al. [11] integrate seven different 3D neural network

© Springer Nature Switzerland AG 2019
A. Crimi et al. (Eds.): BrainLes 2018, LNCS 11384, pp. 128–141, 2019.
https://doi.org/10.1007/978-3-030-11726-9_12

models with different parameters and average the output probability maps from each model to obtain the final brain tumor mask. Wang et al. [20] design a hierarchical pipeline to segment the different types of tumor compartments using anisotropic convolutional neural networks. The network architecture of Isensee et al. [8] is derived from a 3D U-Net with additional residual connections on context pathway and additional multi-scale aggregation on localization pathways, using the Dice loss in the training phase to circumvent class imbalance. For the brain tumor segmentation task, we propose a methodology to integrate multiple DeepMedics [12] and patch-based 3D U-Nets adjusted from [5] with different parameters and different training strategies in order to get a robust brain tumor segmentation from multi-modal structural MR images. We also utilize the existing brain parcellation to bring location information to the patch-based neural networks. In order to increase the diversity of our ensemble, 3D U-Nets with dice loss and cross-entropy loss are included. The final segmentation mask of the brain tumor is calculated by taking the average of the output probability maps from each model in our ensemble.

For the overall survival (OS) prediction task, Shboul et al. [16] extract 40 features from the predicted brain tumor mask and use a random forest regression to predict the glioma patient's OS. Jungo et al. [10] extract four features from each subject and use a support vector machine (SVM) with radial basis function (RBF) kernel to classify glioma patients into three different OS groups. In this paper, we propose a novel method to extract the tractographic features from the lesion regions on structural MR images via an average diffusion MR image which is from a total of 1021 IICP subjects [19] (Q1-Q4, 2017). We then use these tractographic features to predict the patient's OS with a SVM classifier with linear kernel.

2   Glioma Segmentation

2.1   Materials

The Brain Tumor Segmentation (BraTS) 2018 dataset [1–3,14] provides 285 training subjects with four different types of MR images (MR-T1, MR-T1ce, MR-T2 and MR-FLAIR) and expert-labeled ground-truth of lesions, including necrosis & non-enhancing tumor, edema, and enhancing tumor. The dataset consists of 66 validation subjects and 191 test subjects with four different types of MR images. These MR images are co-registered to the same anatomical template, interpolated to the same resolution ($1 \, mm^3$) and skull-stripped. For each subject, a standard z-score normalization is applied within the brain region as our pre-processing step for brain tumor segmentaion.

2.2   Brain Parcellation Atlas as a Prior for Tumor Segmentation

Current state-of-the-art deep network architectures [8,11,20] for brain tumor segmentation do not consider location information. However, from Fig. 1, it is clear that the lesions are not uniformly distributed in different brain regions. This distribution is computed by dividing the total volume of the lesions by the total volume of the corresponding brain parcellation region. Our proposed method (Fig. 2) explicitly includes the location information as input into a patch-based neural network. First, we register the brain parcellation atlas to the subject space using FLIRT [9] from FSL. This registration enables associating each subject voxel with a structure label indicating the voxel location normalized across all subjects. Thus, the input to the neural network will include both the image data and the corresponding parcellation labels.

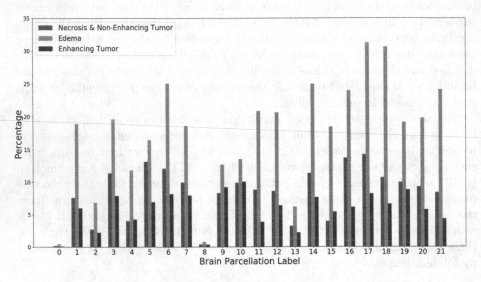

Fig. 1. The percent of brain lesion types observed in different parcellation regions of the Harvard-Oxford subcortical atlas [6]. The x-axis indicates the parcellation label. Regions not covered by the Harvard-Oxford subcortical atlas are in label 0.

2.3   Network Architecture and Training

We integrate multiple state-of-the-art neural networks in our ensemble[1] for robustness. Our ensemble combines 26 neural networks adapted from [5,12]. The detailed network architecture and training method for each model is shown in Table 1. Each 3D U-Net uses group normalization [21] and each DeepMedic uses batch normalization in our ensemble. We utilize a hard negative mining strategy to solve the class imbalance problem while we train a 3D U-Net with cross-entropy loss. Finally, we take the average of the output probability maps from each neural network and get the final brain tumor segmentation. The average training time for each DeepMedic is approximately 3 h and for each 3D U-Net is approximately 12 h, and the average testing time for a subject is approximately 20 min on a NVIDIA GTX Titan X and a Intel Xeon CPU E5-2696 v4 @ 2.20 GHz.

---

[1] The ensemble is publicly available at https://hub.docker.com/r/pykao/brats2018/.

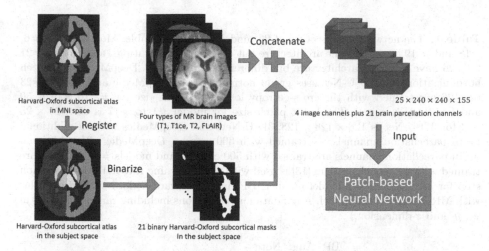

Fig. 2. Incorporating brain parcellation atlas into a patch-based neural network. First, Harvard-Oxford subcortical atlas is registered to the subject space, and the parcellation label is binarized into a 21-dimension vector. This vector is concatenated with the original MR images as input to a patch-based neural network.

Group Normalization. The deep network architectures used for segmentation are computationally demanding. For the 3D U-Nets, our GPU resources enable us to use only 2 samples (of dimensions $128 \times 128 \times 128$ voxels) per iteration. With this small batch size of 2 samples, batch statistics collected during conventional batch normalization method [7] are unstable and thus not suitable for training. In batch normalization, statistics are computed for each feature dimension. Recently Wu et al. [21] propose to group several feature dimensions together while computing batch statistics. This so-called group normalization helps to stabilize the computed statistics. In our implementation, the number of groups is set to 4.

Hard Negative Mining. We train a 3D U-Net with $128 \times 128 \times 128$ patches randomly cropped from the original data. With such large dimensions, the majority of voxels are not classified as lesion and the standard cross-entropy loss would encourage the model to favor the background class. To cope with this problem, we only select negative voxels with the largest losses (hard negative) to backpropagate the gradients. In our implementation, the number of selected negative voxels is at most three times the number of positive voxels. Hard negative mining not only improves the tumor segmentation performance of our model but also decreases its false positive rate.

2.4   Experimental Results

We first examine the brain tumor segmentation performance using MR images and the Harvard-Oxford subcortical brain parcellation masks as input to

Table 1. The network architecture of 26 models in our ensemble. Models #1 to #6, #18 and # 19 have the same architecture but different initializations, and models #21 to #26 have the same architecture but different initializations. DeepMedic uses batch normalization and 3D U-Net uses group normalization. DeepMedic and models #23 to #26 are trained with the cross-entropy loss. The batch size for #3 to #19 is 50 and for 3D U-Net is 2. The input patch size for model #1 to #17 is $25 \times 25 \times 25$ and for 3D U-Net is $128 \times 128 \times 128$. 3D U-Nets and DeepMedics without additional brain parcellation channels are trained with 300 epochs, DeepMedic with additional brain parcellation channels are trained with 500 epochs, and models #18 and #19 are trained with 600 epochs. Adam [13] is used with 0.001 learning rate in the optimization step for all models. (# : model number, BP: input Harvard-Oxford subcortical atlas with MR images to the model, Aug.: data augmentations including random flipping in x-, y- and z-dimension.)

	#	BP	Aug.	Note
DeepMedic	1			Batch size: 36
	2		√	
	3			
	4	√		
	5		√	
	6	√	√	
	7		√	1.5 times 3D convolutional kernels
	8	√	√	
	9			Double 3D convolutional kernels
	10		√	
	11	√	√	
	12			2.5 times 3D convolutional kernels
	13		√	
	14	√	√	
	15			Triple 3D convolutional kernels
	16		√	
	17	√	√	
	18		√	Input patch size: $22 \times 22 \times 22$
	19		√	Input patch size: $28 \times 28 \times 28$
3D U-Net	20			From [8] with Dice loss
	21			Dice loss
	22	√		
	23			Hard negative mining within one batch
	24	√		
	25			Hard negative mining within one image
	26	√		

DeepMedic and 3D U-Net. The quantitative results are shown in Table 2. This table demonstrates that adding brain parcellation masks as additional inputs to a patch-based neural network improves its performance. For segmentation of the enhancing tumor, whole tumor and tumor core, the average *Hausdorff 95* scores for DeepMedic-based models improve from 5.205 to 3.922, from 11.536 to 8.507 and from 11.215 to 8.957, respectively. The average *Dice* scores for models based on 3D U-Net also improve from 0.753 to 0.764, from 0.889 to 0.894 and from 0.766 to 0.775, respectively, for each of the three tumor compartments.

Table 2. Quantitative results of the performance of adding additional brain parcellation masks with MR images to DeepMedic and 3D U-Net on the BraTS2018 validation dataset. Bold numbers highlight the improved results with additional brain parcellation masks. Models with BP use binary brain parcellation masks and MR images as input, while models without BP use only MR images as input. For comparison, each model without brain parcellation (BP) is paired with the same model using BP, the pair having the same parameters and weights initially. The scores for DeepMedic without BP is the average scores from model #3, #5, #7, #10, #13 and #16, and the scores for DeepMedic with BP is the average scores from model #4, #6, #8, #11, #14 and #17. The scores for 3D U-Net without BP is the average scores from model #21, #23 and #25, and the scores for 3D U-Net with BP is the average scores from model #22, #24 and #26. Tumor core (TC) is the union of necrosis & non-enhancing tumor and enhancing tumor (ET). Whole tumor (WT) is the union of necrosis & non-enhancing tumor, edema and enhancing tumor. Results are reported as mean.

	Description	ET	WT	TC
Dice	DeepMedic without BP	0.758	0.892	0.804
	DeepMedic with BP	**0.766**	**0.894**	0.804
	3D U-Net without BP	0.753	0.889	0.766
	3D U-Net with BP	**0.764**	**0.894**	**0.775**
Hausdorff 95 (in *mm*)	DeepMedic without BP	5.205	11.536	11.215
	DeepMedic with BP	**3.992**	**8.507**	**8.957**
	3D U-Net without BP	4.851	5.337	10.550
	3D U-Net with BP	5.216	5.544	**10.442**

We then evaluate the brain tumor segmentation performance of our proposed ensemble on the BraTS2018 training, validation and test datasets. The quantitative results are shown in Table 3. This table shows the robustness of our ensemble on the brain tumor segmentation task. Our ensemble has consistent brain tumor segmentation performance on the BraTS2018 training, validation and test datasets.

Table 3. Quantitative results of the tumor segmentation performance of our ensemble on BraTS2018 training dataset with 5-fold cross-validation, validation dataset and test dataset. Tumor core (TC) is the union of necrosis & non-enhancing tumor and enhancing tumor (ET). Whole tumor (WT) is the union of necrosis & non-enhancing tumor, edema and enhancing tumor. Results are reported as mean.

	Dataset	ET	WT	TC
Dice	BraTS2018 training	0.735	0.902	0.813
	BraTS2018 validation	0.788	0.905	0.813
	BraTS2018 test	0.749	0.875	0.793
Hausdorff 95 (in *mm*)	BraTS2018 training	5.433	5.398	6.932
	BraTS2018 validation	3.812	4.323	7.555
	BraTS2018 test	4.219	6.479	6.522

3   Overall Survival Prediction for Brain Tumor Patients

3.1   Material

The BraTS2018 dataset also includes the age (in years), survival (in days) and resection status for each of 163 subjects in the training dataset, and 59 of them have the resection status of Gross Total Resection (GTR). The validation dataset has 53 subjects with the age (in years) and resection status, and 28 of them have the resection status of GTR. The test dataset has 131 subjects with the age (in years) and resection status, and 77 of them have the resection status of GTR. For this task, we only predict the overall survival (OS) for glioma patients with resection status of GTR.

3.2   Methodology

Our proposed training pipeline, shown in Fig. 3, includes three stages: In the first stage, we use the proposed ensemble from the Sect. 2 to obtain the predicted tumor mask for each subject. In the second stage, We extract the tractographic features explained in section below from each subject. We then perform feature normalization and selection. In the final stage, we train a SVM classifier with linear kernel using the tractographic features extracted from the training subjects. We evaluate the overall survival classification performance of tractographic features on the BraTS2018 training dataset with the 1000-time repeated stratified 5-fold cross-validation, valdiation datset and test dataset.

Glioma Segmentation: To segment the glioma, we use the proposed ensemble in the previous section to obtain the prediction of three different types of tissue including necrosis & non-enhancing tumor, edema, and enhancing tumor.

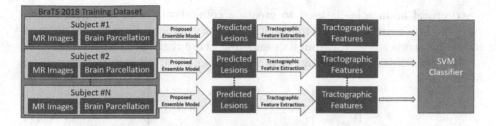

Fig. 3. Training pipeline for overall survival prediction.

Tractographic Feature Extraction from the Glioma Segmentation:
After we obtain the predicted lesion mask, we extract the tractographic features from the whole tumor region which is the union of all different lesions for each subject.

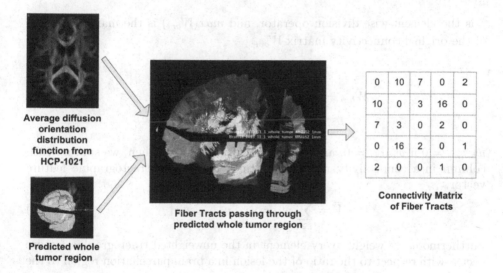

Fig. 4. Workflow for building a connectivity matrix for each subject. The fiber tracts are created by DSI Studio (http://dsi-studio.labsolver.org/), and ITK-SNAP [24] is used for visualizing the 3D MR images and 3D labels.

Tractographic Features: Tractographic features describe the potentially damaged parcellation regions impacted by the brain tumor through fiber tracking. Figure 4 shows the workflow for building a connectivity matrix for each subject. First, the predicted whole tumor mask and the average diffusion orientation distribution function from HCP-1021, created by QSDR [22], are obtained for each subject. FLIRT is used to map the whole tumor mask from subject space to MNI152 1 mm space. Second, we use a deterministic diffusion fiber tracking method [23] to create approximately 1,000,000 tracts from the whole tumor region. Finally, a structural brain atlas is used to create a connectivity matrix $\vec{W}_{ori}$ for each subject. This matrix contains information about whether a fiber connecting one region to another passed through or ended at those regions, as shown:

$\vec{W}_{ori}$ is a $N \times N$ matrix, and N is the number of parcellation in a structural brain atlas.

$$\vec{W}_{ori} = \begin{bmatrix} w_{ori,11} & w_{ori,12} & \cdots & w_{ori,1N} \\ w_{ori,21} & w_{ori,22} & \cdots & w_{ori,2N} \\ \vdots & \vdots & \ddots & \vdots \\ w_{ori,N1} & w_{ori,N2} & \cdots & w_{ori,NN} \end{bmatrix} \tag{1}$$

If w_{ij} is pass-type, it shows the number of tracts passing through region j and region i. if w_{ij} is end-type, it shows the number of tracts starting from a region i and ending in a region j. From the original connectivity matrix $\vec{W}_{ori}$, we create a normalized version $\vec{W}_{nrm}$ and a binarized version $\vec{W}_{bin}$.

$$\vec{W}_{nrm} = \vec{W}_{ori}/max(\vec{W}_{ori}) \tag{2}$$

/ is the element-wise division operator, and $max(\vec{W}_{ori})$ is the maximum value of the original connectivity matrix $\vec{W}_{ori}$.

$$\vec{W}_{bin} = \begin{bmatrix} w_{bin,11} & w_{bin,12} & \cdots & w_{bin,1N} \\ w_{bin,21} & w_{bin,22} & \cdots & w_{bin,2N} \\ \vdots & \vdots & \ddots & \vdots \\ w_{bin,N1} & w_{bin,N2} & \cdots & w_{bin,NN} \end{bmatrix} \tag{3}$$

$w_{bin,ij} = 0$ if $w_{ori,ij} = 0$, and $w_{bin,ij} = 1$ if $w_{ori,ij} > 0$. Then, we sum up each column in a connectivity matrix to form a unweighted tractographic feature vector.

$$\vec{V} = \sum_{i=1}^{N} w_{ij} = [v_1, v_2, \ldots, v_N] \tag{4}$$

Furthermore, we weight every element in the unweighted tractographic feature vector with respect to the ratio of the lesion in a brain parcellation region to the volume of this brain parcellation region.

$$\vec{V}_{wei} = \vec{\alpha} \odot \vec{V}, \vec{\alpha} = [t_1/b_1, t_2/b_2, \ldots, t_N/b_N] \tag{5}$$

$\odot$ is the element-wise multiplication operator, t_i is the volume of the whole brain tumor in the i-th brain parcellation, and b_i is the volume of the i-th brain parcellation. This vector $\vec{V}_{wei}$ is the tractographic feature extracted from brain tumor.

In this paper, automated anatomical labeling (AAL) [18] is used for building the connectivity matrix. AAL has 116 brain parcellation regions, so the dimension of the connectivity matrix $\vec{W}$ is 116×116 and the dimension of each tractographic feature $\vec{V}_{wei}$ is 1×116. In the end, we extract six types of tractographic features for each subject. Six types of tractographic features are computed from: (1) the pass-type of the original connectivity matrix, (2) the pass-type of the normalized connectivity matrix, (3) the pass-type of the binarized connectivity matrix, (4) the end-type of the original connectivity matrix, (5) the end-type of the normalized connectivity matrix and (6) the end-type of the binarized connectivity matrix.

Feature Normalization and Selection: First, we remove features with low variance between subjects, and then apply a standard z-score normalization on the remaining features. In the feature selection step, we combine recursive feature elimination with the 1000-time repeated stratified 5-fold cross-validation and a SVM classifier with linear kernel. These feature processing steps are implemented by using scikit-learn [15].

Overall Survival Prediction: We first divide all 59 training subjects into three groups: long-survivors (e.g., >15 months), short-survivors (e.g., <10 months), and mid-survivors (e.g., between 10 and 15 months). Then, we train a SVM classifier with linear kernel on all training subjects with 1000-time repeated stratified 5-fold cross-validation in order to evaluate the performance of the proposed tractographic feature on overall survival prediction for brain tumor patients. We also evaluate the OS prediction performance of tractographic features on the BraTS2018 validation and test dataset.

3.3   Experimental Results

In this task, we first examine the overall survival classification performance of our proposed tractographic feature compared to other types of features including age, volumetric features, spatial features, volumetric spatial features and morphological features.

Volumetric Features: The volumetric features include the volume and the ratio of brain to the different types of lesions, as well as the tumor compartments. 19 volumetric features are extracted from each subject.

Spatial Features: The spatial features describe the location of the tumor in the brain. The lesions are first registered to the MNI152 1 mm space by using FLIRT, and then the centroids of whole tumor, tumor core and enhancing tumor are extracted as our spatial features. For each subject, we extract 9 spatial features.

Volumetric Spatial Features: The volumetric spatial features describe the volume of different tumor lesions in different brain regions. First, the Harvard-Oxford subcortical structural atlas brain parcellation regions are registered to the subject space by using FLIRT. The volumes of different types of tumor lesions in each of parcellation regions, left brain region, middle brain region, right brain region and other brain region are extracted as volumetric spatial features. For each subject, we extract 78 volumetric spatial features.

Morphological Features: The morphological features include the length of the major axis of the lesion, the length of the minor axis of the lesion and the surface irregularity of the lesions. We extract 19 morphological features from each subject.

In the first experiment, the ground-truth lesion is used to extract different types of features, and the pass-type of the binarized connectivity matrix is built to compute the tractographic feature. Recursive feature elimination with cross-validation (RFECV) is used in the feature selection step to shrink the feature. A SVM classifier with linear kernel is trained with each feature type, and stratified 5-fold cross-validation is conducted 1000 times in order to achieve a reliable metric. The average and standard deviation of overall survival classification accuracy for different types of features on the BraTS2018 training dataset is shown in Fig. 5. This figure demonstrates that the proposed tractographic features have the best overall survival classification performance compared to age, volumetric features, spatial features, volumetric spatial features and morphological features. Initial analysis based on feature selection indicate that 12 out of 116 AAL regions are more influential in affecting overall survival of the brain tumor patient.

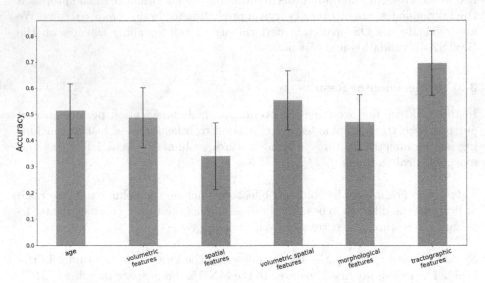

Fig. 5. Overall survival classification accuracy between different types of features on BraTS2018 training dataset. 1000-time repeated stratified 5-fold cross-validation is used to obtain the average classification accuracy.

Next, the pass-type of the binarized connectivity matrix is built from the predicted lesion and the tractographic feature is computed from this connectivity matrix. The overall survival classification performance of this tractographic feature is compared with the tractographic feature from our first experiment. In this experiment, we follow the same feature selection method and training strategy, using the same SVM classifier with linear kernel. The average and standard deviation of overall survival classification accuracy on the BraTS2018 training dataset is reported in Table 4. From this table, the average classification accuracy drops to 63% when we use predicted lesions instead of ground-truth lesions to generate the tractographic features. This drop is likely caused by the imperfection of our tumor segmentation tool.

Table 4. The overall survival classification performance of the proposed tractographic features from the ground-truth lesions and from the predicted lesions on the BraTS2018 training dataset with 1000-time repeated stratified 5-fold cross-validation.

The source of tractographic features	Classification accuracy (mean±std)
Ground-truth lesions	0.70 ± 0.12
Predicted lesions	0.63 ± 0.13

For the training data, the tractographic features are computed using the ground-truth whole tumor, and a linear SVM classifier trained on these features. We used stratified 5-fold cross validation on the training dataset, averaged over 1000 independent trials. The average OS classification accuracy using the tractographic features was 0.892 on the training set and 0.697 on the cross-validation set. However, when applied to the BraTS2018 validation and test datasets, the accuracy dropped to 0.357 and 0.416, respectively [4]. Note that for the validation and test data, there is no ground-truth segmentation available. So we first predicted the whole tumor and then the tractography features are extracted from these predicted tumors, followed by the OS classification using the previously trained linear SVM. We speculate that the automated segmentation to predict the whole tumor is one possible reason for the significant variation in performance between the training and validation/test data, in addition any data specific variations.

4   Discussion

For brain tumor segmentation, our proposed method, which combines the lesion occurrence probabilities in structural regions with MR images as inputs to a patch-based neural network, improves the patch-based neural network's performance. The proposed ensemble results in a more robust tumor segmentation. For overall survival prediction, the novel use of tractographic features appears to be promising for aiding brain tumor patients. To the best of our knowledge, this is the first paper to integrate brain parcellation and human brain connectomics for brain tumor segmentation and overall survival prediction.

Acknowledgements. This research was partially supported by a National Institutes of Health (NIH) award # 5R01NS103774-02.

References

1. Bakas, S., et al.: Segmentation labels and radiomic features for the pre-operative scans of the TCGA-LGG collection. The Cancer Imaging Archive (2017). https://doi.org/10.7937/K9/TCIA.2017.GJQ7R0EF
2. Bakas, S., et al.: Segmentation labels and radiomic features for the pre-operative scans of the TCGA-GBM collection. The Cancer Imaging Archive (2017). https://doi.org/10.7937/K9/TCIA.2017.KLXWJJ1Q
3. Bakas, S., et al.: Advancing the cancer genome atlas glioma mri collections with expert segmentation labels and radiomic features. Sci. Data **4**, 170117 (2017)
4. Bakas, S., Reyes, M., et al.: Identifying the best machine learning algorithms for brain tumor segmentation, progression assessment, and overall survival prediction in the brats challenge. arXiv preprint arXiv:1811.02629 (2018)
5. Çiçek, Ö., Abdulkadir, A., Lienkamp, S.S., Brox, T., Ronneberger, O.: 3D U-Net: learning dense volumetric segmentation from sparse annotation. In: Ourselin, S., Joskowicz, L., Sabuncu, M.R., Unal, G., Wells, W. (eds.) MICCAI 2016. LNCS, vol. 9901, pp. 424–432. Springer, Cham (2016). https://doi.org/10.1007/978-3-319-46723-8_49
6. Desikan, R.S., et al.: An automated labeling system for subdividing the human cerebral cortex on mri scans into gyral based regions of interest. Neuroimage **31**(3), 968–980 (2006)
7. Ioffe, S., Szegedy, C.: Batch normalization: Accelerating deep network training by reducing internal covariate shift. arXiv preprint arXiv:1502.03167 (2015)
8. Isensee, F., Kickingereder, P., Wick, W., Bendszus, M., Maier-Hein, K.H.: Brain tumor segmentation and radiomics survival prediction: contribution to the BRATS 2017 challenge. In: Crimi, A., Bakas, S., Kuijf, H., Menze, B., Reyes, M. (eds.) BrainLes 2017. LNCS, vol. 10670, pp. 287–297. Springer, Cham (2018). https://doi.org/10.1007/978-3-319-75238-9_25
9. Jenkinson, M., Smith, S.: A global optimisation method for robust affine registration of brain images. Med. Image Anal. **5**(2), 143–156 (2001)
10. Jungo, A., et al.: Towards uncertainty-assisted brain tumor segmentation and survival prediction. In: Crimi, A., Bakas, S., Kuijf, H., Menze, B., Reyes, M. (eds.) BrainLes 2017. LNCS, vol. 10670, pp. 474–485. Springer, Cham (2018). https://doi.org/10.1007/978-3-319-75238-9_40
11. Kamnitsas, K., et al.: Ensembles of multiple models and architectures for robust brain tumour segmentation. In: Crimi, A., Bakas, S., Kuijf, H., Menze, B., Reyes, M. (eds.) BrainLes 2017. LNCS, vol. 10670, pp. 450–462. Springer, Cham (2018). https://doi.org/10.1007/978-3-319-75238-9_38
12. Kamnitsas, K., et al.: Efficient multi-scale 3D CNN with fully connected CRF for accurate brain lesion segmentation. Med. Image Anal. **36**, 61–78 (2017)
13. Kingma, D.P., Ba, J.: Adam: A method for stochastic optimization. arXiv preprint arXiv:1412.6980 (2014)
14. Menze, B.H., et al.: The multimodal brain tumor image segmentation benchmark (brats). IEEE Trans. Med. Imaging **34**(10), 1993 (2015)
15. Pedregosa, F., et al.: Scikit-learn: machine learning in python. J. Mach. Learn. Res. **12**, 2825–2830 (2011)
16. Shboul, Z.A., Vidyaratne, L., Alam, M., Iftekharuddin, K.M.: Glioblastoma and survival prediction. In: Crimi, A., Bakas, S., Kuijf, H., Menze, B., Reyes, M. (eds.) BrainLes 2017. LNCS, vol. 10670, pp. 358–368. Springer, Cham (2018). https://doi.org/10.1007/978-3-319-75238-9_31

17. Thakkar, J.P., et al.: Epidemiologic and molecular prognostic review of glioblastoma. Cancer Epidemiol. Prev. Biomark. **23**(10), 1985–1996 (2014)
18. Tzourio-Mazoyer, N., et al.: Automated anatomical labeling of activations in spm using a macroscopic anatomical parcellation of the mni mri single-subject brain. Neuroimage **15**(1), 273–289 (2002)
19. Van Essen, D.C., et al.: The wu-minn human connectome project: an overview. Neuroimage **80**, 62–79 (2013)
20. Wang, G., Li, W., Ourselin, S., Vercauteren, T.: Automatic brain tumor segmentation using cascaded anisotropic convolutional neural networks. In: Crimi, A., Bakas, S., Kuijf, H., Menze, B., Reyes, M. (eds.) BrainLes 2017. LNCS, vol. 10670, pp. 178–190. Springer, Cham (2018). https://doi.org/10.1007/978-3-319-75238-9_16
21. Wu, Y., He, K.: Group normalization. arXiv preprint arXiv:1803.08494 (2018)
22. Yeh, F.C., Tseng, W.Y.I.: Ntu-90: a high angular resolution brain atlas constructed by q-space diffeomorphic reconstruction. Neuroimage **58**(1), 91–99 (2011)
23. Yeh, F.C., Verstynen, T.D., Wang, Y., Fernández-Miranda, J.C., Tseng, W.Y.I.: Deterministic diffusion fiber tracking improved by quantitative anisotropy. PloS one **8**(11), e80713 (2013)
24. Yushkevich, P.A., et al.: User-guided 3D active contour segmentation of anatomical structures: significantly improved efficiency and reliability. Neuroimage **31**(3), 1116–1128 (2006)

Glioma Prognosis: Segmentation of the Tumor and Survival Prediction Using Shape, Geometric and Clinical Information

Mobarakol Islam[1,2], V. Jeya Maria Jose[2,3], and Hongliang Ren[2(✉)]

[1] NUS Graduate School for Integrative Sciences and Engineering (NGS),
National University of Singapore, Singapore, Singapore
[2] Department of Biomedical Engineering,
National University of Singapore, Singapore, Singapore
mobarakol@u.nus.edu, jeyamariajose7@gmail.com, ren@nus.edu.sg
[3] Department of Instrumentation and Control Engineering,
National Institute of Technology, Tiruchirappalli, India

Abstract. Segmentation of brain tumor from magnetic resonance imaging (MRI) is a vital process to improve diagnosis, treatment planning and to study the difference between subjects with tumor and healthy subjects. In this paper, we exploit a convolutional neural network (CNN) with hypercolumn technique to segment tumor from healthy brain tissue. Hypercolumn is the concatenation of a set of vectors which form by extracting convolutional features from multiple layers. Proposed model integrates batch normalization (BN) approach with hypercolumn. BN layers help to alleviate the internal covariate shift during stochastic gradient descent (SGD) training by zero-mean and unit variance of each mini-batch. Survival Prediction is done by first extracting features (Geometric, Fractal, and Histogram) from the segmented brain tumor data. Then, the number of days of overall survival is predicted by implementing regression on the extracted features using an artificial neural network (ANN). Our model achieves a mean dice score of 89.78%, 82.53% and 76.54% for the whole tumor, tumor core and enhancing tumor respectively in segmentation task and 67.9% in overall survival prediction task with the validation set of BraTS 2018 challenge. It obtains a mean dice accuracy of 87.315%, 77.04% and 70.22% for the whole tumor, tumor core and enhancing tumor respectively in the segmentation task and a 46.8% in overall survival prediction task in the BraTS 2018 test data set.

Keywords: Brain tumor segmentation · Glioma ·
Convolutional neural network · Hypercolumn · PixelNet ·
Magnetic resonance imaging · Survival prediction

ⓒ Springer Nature Switzerland AG 2019
A. Crimi et al. (Eds.): BrainLes 2018, LNCS 11384, pp. 142–153, 2019.
https://doi.org/10.1007/978-3-030-11726-9_13

1   Introduction

Gliomas are the most frequent brain tumor with the highest mortality rate which develops from glial cell [7]. Early detection, accurate segmentation and estimation of the relative volume are very crucial for overall survival (OS) prediction, treatment and surgical planning. In addition, manual segmentation of tumor tissue is tedious, time consuming and requires strong supervision by a human expert. It is also prone to inter and intra-rater variability. So it is highly necessary to develop an automatic segmentation system to diagnose and estimate the volume, size, shape and location of the tumor. Overall survival prediction along with automatic segmentation would be a very useful tool that would help in better clinical diagnosis.

In recent years, the success of deep learning in this field is huge as it shows state of art performance in the applications of segmentation, classification, regression and detection. Iftekharuddin et al. [9] exploits convolutional neural network (CNN) for glioma segmentation and extracts handcrafted features like histogram, co-occurrence matrix, neighbourhood gray tone difference, run length, volume and areas to predict OS using random forest regression model. Jungo et al. [15] uses the residual convolutional neural network with Bayesian dropout to segment tumor and calculates the geometric features (e.g. volume, heterogeneity, rim width, surface irregularity etc.) from the segmented tumor. Later, a simple artificial neural network (ANN) is utilized to predict the exact days of OS. 3D U-net and linear regression approaches are also exploited to segment and predict OS [1].

In this paper, we propose a batch normalized CNN architecture with hypercolumn features inspired by multi-modal PixelNet [6,11,12] where a modest number of features are extracted from multiple convolution layers and trained with a multi-layer perceptron (MLP) to predict segmentation classes. We discuss about the various features (Geometric, Fractal, Image and Clinical) extracted from the segmentation output. We propose a new method of overall survival prediction by combining all the meaningful features that contribute to the number of days of survival left for the patient. Details about how the features are selected and various experiments that are run for finding the best regression technique has been discussed. The problems of generalizing a network over its performance on a validation data set is also discussed.

2   Methods

In this section, we discuss about the dataset provided, the model and methodologies to process data for segmentation and about the extraction of features, feature selection and regression techniques that were exploited for overall survival prediction.

2.1   Dataset

BraTS 2018 (Brain Tumor Image Segmentation Benchmark) training database [2–4,19] consists in total 285 cases of patients out of which the overall survival prediction data was provided for 163 cases. BraTS 2018 Validation dataset [2–4,19] consists of 53 cases. All the data are a multi-modal MRI scan of 210 high-grade glioma (HGG) and 75 low-grade glioma (LGG) and 4 different modalities including T1 (spin-lattice relaxation), T1c (T1-contrasted), T2 (spin-spin relaxation) and FLAIR (fluid attenuation inversion recovery). Each scan is a continuous 3D volume of 155 2D slices of size 240 × 240. The volume of the various modalities is already skull-stripped, aligned with T1c and interpolated to 1 mm voxel resolution.

The provided ground truth with manual segmentation includes three labels: GD-enhancing tumor (ET—label 4), the peritumoral edema (ED—label 2), and the necrotic and non-enhancing tumor (NCR/NET—label 1). The predicted labels are evaluated by merging three regions: whole tumor (WT: all four labels), tumor core (TC: 1, 2) and enhancing tumor (ET: 4). Figure 1 illustrates all the four modalities for one of the samples of the training data set of BraTS 2018. Figure 2 illustrates the provided ground truth for the same sample. The green label corresponds to GD-enhancing tumor, yellow label corresponds to peritumoral edema and the red label corresponds to necrotic and non-enhancing tumor. ITK-SNAP [24] is the tool that was used to visualize the data.

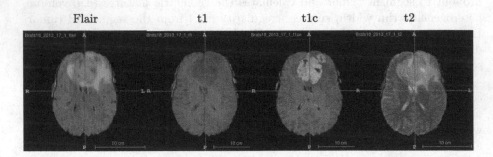

Fig. 1. Visualization of the different modalities in the BraTS 2018 Training data set.

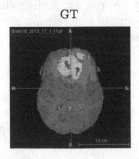

Fig. 2. Ground Truth with different segmentation labels as given in BraTS 2018 Training data set. Red, Yellow and Green colors represent Necrotic, Enhancing and Edema respectively. (Color figure online)

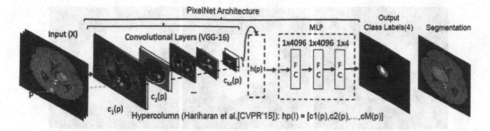

Fig. 3. Batch Normalized PixelNet architecture.

2.2 Segmentation

Proposed Model. Our proposed model is inspired from PixelNet [6, 11] where we integrate additional 3 convolution layers and batch normalization after all the convolution layers. The model consists of 18 pixel-block and a hypercolumn layer followed by a multi-layer perceptron (MLP) of 3 fully connected layers as in Fig. 3. A pixel-block contains convolution, batch normalization (BN) [10] and ReLu layers sequentially. Hypercolumn layer extracts the features from multiple convolution layers and concatenates them into a feature vector which propagates to the MLP for pixel-wise classification.

2.3 Survival Prediction

Following the extraction of tumor from the MRI scans, the segmented tumor along with certain other parameters are used for survival prediction. The following paragraphs elucidate the features those are extracted along with the regression model that is built for predicting survival.

Feature Extraction. The tumor geometry and its location hold a very important role in deciding the number of days of survival [21]. Figure 4 visually illustrates how the features such as location or centroid of tumor, size and shape of tumor affect the overall survival of the patient. It is evident that more the proximity of tumor to the centre of brain, the lower is the overall survival of patient. Also, lesser the size or smaller the shape of tumor, higher is the overall survival of patient. So, we extract geometrical features which include First axis coordinates, Second axis coordinates, Third axis coordinates, Eigen Values, First axis length, Second axis length, Third axis Length, Centroid coordinates, meridional eccentricity and equatorial eccentricity for individual tumor types as well as whole tumor. Figure 5 illustrates the tumor extracted from the brain and gives an intuition of how the geometric features like centroid, eccentricity and axis lengths are calculated. Also, the volume of tumor and its ratio with respect to the total volume was calculated. Features of the tumor image including mean, variance, standard deviation, entropy, skewness [18], Kurtosis [18], entropy and histogram feature intensities were extracted. Fractal dimension of the necrotic

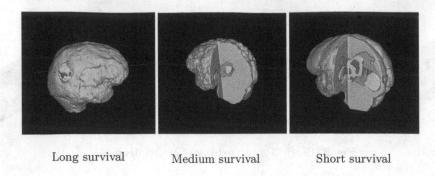

Long survival   Medium survival   Short survival

Fig. 4. Visualization of tumor with different survival rate.

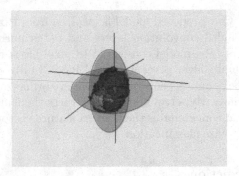

Fig. 5. Visualization of the extracted tumor from brain.

tumor has been found to play a pivotal role in the survival prediction, according to [22]. So, fractal dimension and fractal ratio were extracted for the necrotic part of the segmented tumor core. In addition, the age of the subject provided by the BraTS 2018 training data set was also included.

Feature Selection. Several experiments were conducted using different combinations of extracted features. After analyzing the cross-validation errors of the experiments, the most informative features are alone retained and others are neglected. Features like eigen values, eccentricity, skewness, mean and variance were not found to have an important role in survival prediction. Also, geometric features of the GD-enhancing tumor (ET—label 4) was found to be only increasing the cross-validation errors and hence was removed. So finally, a total of 50 features that were found to be the most informative are used in the regression model. These features are First axis coordinates, Second axis coordinates, Third axis coordinates, First axis length, Second axis length, Third axis Length, Centroid coordinates for part wise non-enhancing tumor core (NCR/NET—label 1), peritumoral edema (ED—label 2) as well as for the whole tumor without including GD-enhancing tumor (ET—label 4) in addition to Kurtosis, Entropy, Histogramic intensity, Fractal dimension and age.

Regression Model. With the selected features, we train a fully connected artificial neural network (ANN) with one hidden layer [23] and ReLu activation function. Figure 6 shows the ANN model that takes 50 features as input and gives the number of days of survival as the output. We run an experiment to find which configuration of the hidden layers gives the lowest mean squared error (MSE). The best configuration was found to be 100 neurons in the hidden layer. Then, we find the best epoch by finding the epoch for which the MSE is minimum by using a cross-validation data set while training. However, after analyzing many experiments, we find that when MSE is minimum, the accuracy is low and vice-versa. So, we find an epoch where the accuracy and the MSE of the model are balanced. Adam optimizer [16] with MSE loss function has been used to conduct all experiments with ANN. We also tune the hyper parameters like learning rate and batch size. The best result for BraTS 2018 Validation Data was acquired for the 900th epoch with a batch size of 10.

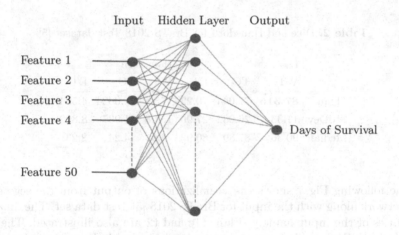

Fig. 6. Regression model using ANN.

3   Experiments and Results

3.1   Segmentation

We convert the 3D voxel of $240 \times 240 \times 155$ into 2D slices of 240×240 by ignoring blank slices of scan and ground-truth. We choose a sample of 2000 pixels per image and batch size of 10 in training time. We normalize data to zero mean and standard deviation and augment by flipping left-right. In testing phase, hyper-column has been formed with all the pixels inside brain region and predict slice by slice to form MRI. Finally, we adopt largest component analysis to remove false positive as a post-processing technique. We utilize Caffe [13] framework with a single Nvidia GPU 1080Ti GPU to perform all the experiments. Table 1 represents the Dice and Hausdorff performance of our model. It obtains dice accuracy of 89.78%, 82.53% and 76.54% of whole tumor (WT), tumor core (TC) and enhance tumor (ET) respectively. Table 2 shows the performance metrics that the segmentation method has achieved on the test data set. It obtains dice accuracy of 87.315%, 77.04% and 70.22% of whole tumor (WT), tumor core (TC) and enhance tumor (ET) respectively.

Table 1. Dice and Hausdorff for BraTS 2018 validation dataset

	Dice			Hausdorff		
	WT	TC	ET	WT	TC	ET
Mean	**89.78**	82.53	76.54	5.09	7.11	3.60
StdDev	8.35	17.80	23.23	7.04	8.04	5.58
Median	90.51	86.87	83.15	3.08	4.53	2.23

Table 2. Dice and Hausdorff for BraTS 2018 Test dataset [5]

	Dice			Hausdorff		
	WT	TC	ET	WT	TC	ET
Mean	**87.315**	77.044	70.22	7.15	8.0024	4.594
StdDev	11.33	26.082	27.96	13.08	11.937	8.802
Median	90.36	87.89	80.93	3.6	4.24	2.23

The following Fig. 7 shows the visualizations of output from the segmentation network along with the input for BraTS 2018 [5] Test data set. The different modalities of the input namely- Flair, t1c and t2 are also illustrated. The prediction shows the labelling of tumor done by the network. It consists of 3 labels, green corresponds to GD-enhancing tumor, yellow corresponds to peritumoral edema and red corresponds to necrotic and non-enhancing tumor core.

3.2 Survival Prediction

Evaluation of the survival prediction model is performed on BraTS 2018 survival validation dataset (which is a subset (28 out of 53 subjects) of the segmentation validation dataset) [2–4,19]. Quantitative details for the ANN that gave the best accuracy of 67.9% has been listed in Table 3. This is the best accuracy that was achieved in the Validation Leader board till the time this paper is written. However, the same ANN was found to give relatively less accuracy for the test data set. Table 4 shows the quantitative results for BraTS 2018 [5] Test data set.

Flair   t1c   t2   Prediction

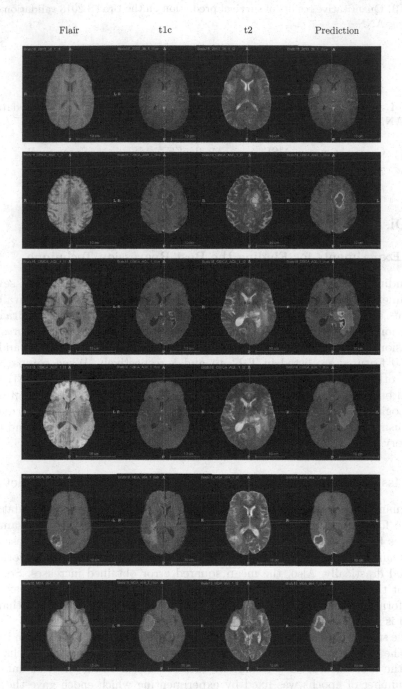

Fig. 7. Visualization of the predicted segmentation labels for BraTS 2018 Test data set. (Color figure online)

Table 3. Quantitative results of survival prediction on the BraTS 2018 validation data set using ANN

Cases	Accuracy	MSE	MedianSE	stdSE	SpearmanR
28	**67.9%**	96161.713	59473.481	117207.189	0.496

Table 4. Quantitative results of survival prediction on the BraTS 2018 Test data set using ANN

Cases	Accuracy	MSE	MedianSE	stdSE	SpearmanR
77	**46.8%**	341387.439	92892.2261	788491.731	0.148

4 Discussion

4.1 Experiments for Finding the Best Regression Technique

For finding out the best regression model and the meaningful features, several experiments are conducted. Experiments with different combinations of available features and regression models are conducted on the 33 subjects of BraTS17 validation dataset [2,3,19] to find the best possible combination of features and regression model. The details of the experiments on the entire features and the best 50 features are listed in Tables 4 and 5 respectively. When compared to ANN, other regression models such as Support Vector Machine (SVM) with Radial basis function (RBF) kernel [14], Random Forest [17], Linear Regression [20], Logistic Regression [8] are investigated but resulted in inferior performance. Also, using only the best 50 features showed improvement in accuracy and MSE for every models.

4.2 Issues with Generalizing a Network on Validation Data Set

The quantitative results of survival prediction on the BraTS 2018 test data set can be found in Table 4. When we use the same algorithm with the same 50 features that gave the best results in the validation data set for the test data set of survival prediction task, we find that the accuracy of the network has reduced drastically. Also, the mean squared error obtained increases. So, it is evident that the network does not perform well with the test data set as good it performs for the validation data set. The reason for this behaviour is that the model is not generalized. It is over-fitted to the train data.

The same architecture, with the same number of neurons in the hidden layers, with the same hyper parameters; trained for a certain number of epochs that gave the best accuracy for validation data set was used for the test data set. The number of epochs was fixed by experimenting which epoch gave the least mean squared error for the validation data set. This is where the network was over-fitted. Generalizing a network architecture and the number of epochs till which it should be trained should not be done by comparing the results with a

Table 5. Performance comparison with all the features in different machine learning models for BraTS 2017 validation data set.

Models	Accuracy	MSE	MedianSE	stdSE	SpearmanR
Linear Regression	50.5%	252353.061	95419.102	**429879.191**	0.263
SVM	33.3%	242147.277	62044.450	563941.194	0.142
Random Forest	42.4%	**208660.63**	**33367.111**	502312.762	0.213
Logistic Regression	39.4%	286470	40401	540201.363	**0.479**
ANN	**54.5%**	211967.681	53967.305	540221.112	0.206

Table 6. Performance comparison with the best 50 features in different machine learning models for BraTS 2017 validation data set.

Models	Accuracy	MSE	MedianSE	stdSE	SpearmanR
Linear Regression	50.5%	237148.501	78915.034	362705.604	0.221
SVM	42.4%	233367.604	127524.752	374037.279	0.218
Random Forest	39.4%	262224.703	**42507.088**	506754.007	**0.324**
Logistic Regression	36.4%	**181509.182**	58564	**249213.006**	0.13
ANN	**60.6%**	214207.487	60832.523	354332.371	0.293

few data. The data with which a network is generalized can be of a very meagre amount and might contain similar characteristics. So generalizing a network over the validation data set here has caused the network to perform badly in a fresh test data set.

Looking at Table 6, models like Logistic regression or Random Forest which would not have over fitted the data could have been used. Thus, selection of validation data is very crucial for any neural network as well as deep learning model where the data set is really small.

5   Conclusion

Batch Normalized pixelnet is found to give quality segmentation results for the BraTS 2018 Validation data set. The main advantage of the pixelnet is that it has freedom of sampling pixel during training phase. The background of the scan is removed during training and this helps the network to converge faster. For the survival prediction task, a lot of features were studied and extracted. The features that were found to increase the error in the regression problem of survival prediction problem were removed. Various regression models were experimented and ANN was found to give the best results for the BraTS 2018 Validation data set. However, due to overfitting the same network was not able to give good results for the BraTS 2018 test data set.

Acknowledgement. This work is supported by the Singapore Academic Research Fund under Grant R-397-000-227-112, NUSRI China Jiangsu Provincial Grant BK20150386 and BE2016077 and NMRC Bedside & Bench under grant R-397-000-245-511 awarded to Dr. Hongliang Ren.

References

1. Amorim, P.H.A.: 3D U-Nets for brain tumor segmentation in MICCAI 2017 BraTS challenge. In: International MICCAI Brainlesion Workshop, pp. 9–14. Springer (2017)
2. Bakas, S., et al.: Advancing the cancer genome atlas glioma MRI collections with expert segmentation labels and radiomic features. Sci. Data **4**, 170117 (2017)
3. Bakas, S., et al.: Segmentation labels and radiomic features for the pre-operative scans of the TCGA-GBM collection. Cancer Imaging Arch. **286** (2017). https://doi.org/10.7937/K9/TCIA.2017.KLXWJJ1Q
4. Bakas, S., et al.: Segmentation labels and radiomic features for the pre-operative scans of the TCGA-LGG collection. Cancer Imaging Arch. (2017). https://doi.org/10.7937/K9/TCIA.2017.GJQ7R0EF
5. Bakas, S., Reyes, M., et al.: Identifying the best machine learning algorithms for brain tumor segmentation, progression assessment, and overall survival prediction in the brats challenge. arXiv preprint arXiv:1811.02629 (2018)
6. Bansal, A., Chen, X., Russell, B., Gupta, A., Ramanan, D.: PixelNet: representation of the pixels, by the pixels, and for the pixels. arXiv preprint arXiv:1702.06506 (2017)
7. Holland, E.C.: Progenitor cells and glioma formation. Curr. Opin. Neurol. **14**(6), 683–688 (2001)
8. Hosmer Jr., D.W., Lemeshow, S., Sturdivant, R.X.: Applied Logistic Regression, vol. 398. Wiley, New York (2013)
9. Shboul, Z.A., Vidyaratne, L., Alam, M., Iftekharuddin, K.M.: Glioblastoma and survival prediction. In: Crimi, A., Bakas, S., Kuijf, H., Menze, B., Reyes, M. (eds.) BrainLes 2017. LNCS, vol. 10670, pp. 358–368. Springer, Cham (2018). https://doi.org/10.1007/978-3-319-75238-9_31
10. Ioffe, S., Szegedy, C.: Batch normalization: accelerating deep network training by reducing internal covariate shift. arXiv preprint arXiv:1502.03167 (2015)
11. Islam, M., Ren, H.: Multi-modal PixelNet for brain tumor segmentation. In: Crimi, A., Bakas, S., Kuijf, H., Menze, B., Reyes, M. (eds.) BrainLes 2017. LNCS, vol. 10670, pp. 298–308. Springer, Cham (2018). https://doi.org/10.1007/978-3-319-75238-9_26
12. Islam, M., Ren, H.: Class balanced PixelNet for neurological image segmentation. In: Proceedings of the 2018 6th International Conference on Bioinformatics and Computational Biology, pp. 83–87. ACM (2018)
13. Jia, Y., et al.: Caffe: convolutional architecture for fast feature embedding. In: Proceedings of the 22nd ACM International Conference on Multimedia, pp. 675–678. ACM (2014)
14. Joachims, T.: Making large-scale SVM learning practical. Technical report, SFB 475: Komplexitätsreduktion in Multivariaten Datenstrukturen, Universität Dortmund (1998)

15. Jungo, A., et al.: Towards uncertainty-assisted brain tumor segmentation and survival prediction. In: Crimi, A., Bakas, S., Kuijf, H., Menze, B., Reyes, M. (eds.) BrainLes 2017. LNCS, vol. 10670, pp. 474–485. Springer, Cham (2018). https://doi.org/10.1007/978-3-319-75238-9_40
16. Kingma, D.P., Ba, J.: Adam: a method for stochastic optimization. arXiv preprint arXiv:1412.6980 (2014)
17. Liaw, A., Wiener, M., et al.: Classification and regression by randomforest. R News 2(3), 18–22 (2002)
18. Mardia, K.V.: Measures of multivariate skewness and kurtosis with applications. Biometrika 57(3), 519–530 (1970)
19. Menze, B.H., et al.: The multimodal brain tumor image segmentation benchmark (BRATS). IEEE Trans. Med. Imaging 34(10), 1993 (2015)
20. Neter, J., Wasserman, W., Kutner, M.H.: Applied linear regression models (1989)
21. Pérez-Beteta, J., et al.: Glioblastoma: does the pre-treatment geometry matter? A postcontrast T1 MRI-based study. Eur. Radiol. 27(3), 1096–1104 (2017)
22. Reishofer, G., Studencnik, F., Koschutnig, K., Deutschmann, H., Ahammer, H., Wood, G.: Age is reflected in the fractal dimensionality of MRI diffusion based tractography. Sci. Rep. 8(1), 5431 (2018)
23. Schalkoff, R.J.: Artificial Neural Networks, vol. 1. McGraw-Hill, New York (1997)
24. Yushkevich, P.A., Gao, Y., Gerig, G.: ITK-SNAP: an interactive tool for semi-automatic segmentation of multi-modality biomedical images. In: 2016 IEEE 38th Annual International Conference of the Engineering in Medicine and Biology Society (EMBC), pp. 3342–3345. IEEE (2016)

Segmentation of Brain Tumors
Using DeepLabv3+

Ahana Roy Choudhury[1]([⊠]), Rami Vanguri[2,3], Sachin R. Jambawalikar[2],
and Piyush Kumar[1]

[1] Department of Computer Science, Florida State University, Tallahassee, FL, USA
{roychoud,piyush}@cs.fsu.edu
[2] Department of Radiology, Columbia University Medical Center, New York, USA
{rv2368,sj2532}@cumc.columbia.edu
[3] Data Science Institute, Columbia University, New York, NY, USA

Abstract. Multi-modal MRI scans are commonly used to grade brain
tumors based on size and imaging appearance. As a result, imaging
plays an important role in the diagnosis and treatment administered to
patients. Deep learning based approaches in general, and convolutional
neural networks (CNNs) in particular, have been utilized to achieve superior performance in the fields of object detection and image segmentation. In this paper, we propose to utilize the DeepLabv3+ network for
the task of brain tumor segmentation. For this task, we build 18 different models using various combinations of the T1CE, FLAIR, T1 and
T2 images to identify the whole tumor, the tumor core and the enhancing core of the brain tumor for the testing and validation data sets. We
use the MICCAI BraTS training data, which consists of 285 cases, to
train our network. Our method involves the segmentation of individual
slices in three orientations using 18 different combinations of slices and
a majority voting-based combination of the results of some of the classifiers that use the same combination of slices, but in different orientations.
Finally, for each of the three regions, we train a separate model, which
uses the results from the 18 classifiers as its inputs. The outputs of the
18 models are combined using bit packing to prepare the inputs to the
final classifiers for the three regions. We achieve mean Dice coefficients
of 0.7086, 0.7897 and 0.8755 for the enhancing tumor, the tumor core
and the whole tumor regions respectively.

Keywords: Image segmentation · Convolutional neural networks ·
DeepLab · MRI · Tumor · Enhancing tumor · Tumor core

1 Introduction

Gliomas are a type of tumor that start in glial cells and occur in the brain
and spinal cord. They are fatal and may be classified as either high or low grade
gliomas [19] based on their aggressiveness. While high grade gliomas (HGGs) are
malignant, grow rapidly and result in high mortality, low grade gliomas (LGGs)

© Springer Nature Switzerland AG 2019
A. Crimi et al. (Eds.): BrainLes 2018, LNCS 11384, pp. 154–167, 2019.
https://doi.org/10.1007/978-3-030-11726-9_14

develop more slowly, may be benign or malignant, but can develop into HGGs if no treatment is provided to the patient [18]. Thus, it is essential that both HGGs and LGGs should be treated sooner rather than later as the survival rate of the patient depends on the characteristics of the tumor. Based on the characteristics of the tumor, different methods of treatment, such as surgery, chemotherapy, or radiation therapy [10], are used. In order to diagnose, treat and predict the survival rate after surgery; various modes of MRI scans are used. The MRI scans are usually segmented into tumor regions by experts. However, manual delineation of the tumor is complex and time consuming due to the variation in tumor structure. As a result, it is important to develop algorithms and methods that can accurately perform automatic image segmentation in order to delineate the different regions of a brain tumor. However, due to the heterogeneity in the size, location, and shape of gliomas, developing algorithms for automatic segmentation is challenging.

With the advent of deep learning, it has become possible to develop networks that can achieve a moderate level of accuracy in tasks that involve automatic image segmentation. Moreover, the introduction of convolutional neural network (CNN) architectures [16] have further improved the performance and reduced the complexity of performing image segmentation. This is because CNNs do not require the use of hand-crafted features and involve the use of fewer number of parameters than the number of parameters that are used by DNNs in order to perform image segmentation. CNNs have been used for segmenting sections of the brain [23] and for brain tumor segmentation [14,15,21] by several researchers.

To accomplish automatic segmentation, we propose to use the DeepLabv3+ framework [8], which has been shown to successfully identify objects in natural images. DeepLabv3+ achieves a mIoU of 89.0% on the test data for the PASCAL VOC 2012 challenge [11]. To do this, we utilize the pre-trained Xception network of DeepLabv3+ and further train it on multimodal MRI scans, treating the individual modes as image channels. DeepLabv3+ involves the use of atrous parallel convolutions at different strides to capture information at different scales and encoder-decoder pathways to achieve sharp delineation of object boundaries.

In this paper, we present our technique for performing brain tumor segmentation in order to participate in the MICCAI BraTS 2018 challenge. The data consists of MRI scans in four modes: FLAIR, T1CE, T1, and T2. The goal of the segmentation challenge is to segment the brain scans and identify the whole tumor, tumor core, and enhancing tumor regions. Our method involves the use of 21 models, where 18 models are trained directly on the input MRI scans and the other 3 models use the outputs from the first 18 models as their inputs. All our models are trained and validated using slices of the brain images and we considered the 3 possible orientations in each case. Thus, our initial 18 models can be considered to be made up of 6 sets of 3 models each. We use different combinations of inputs for each of the initial 18 models and then use bit packing to combine these results and prepare the inputs for the final 3 models. Each of the final 3 models, is used to segment one region of the tumor, and the results are combined to get the final segmentation. During this final combination, we consider the fact that the tumor core is a sub-region of the whole tumor and the enhancing tumor is a sub-region of the tumor core.

Our contributions are as follows:

1. Build a pipeline to perform brain tumor segmentation and identify the whole tumor, tumor core and enhancing tumor regions.
2. Combine several DeepLabv3+ models to achieve higher accuracy than can be achieved by using a single instance of DeepLabv3+.
3. Design and implement a bit packing-based algorithm to combine the results of several models and prepare the inputs for the final 3 models in the pipeline.
4. On the BraTS 2018 testing data, we are able to achieve Dice coefficients of 0.7086, 0.7897 and 0.8755 for the enhancing tumor, tumor core and whole tumor regions.

2   Background: DeepLabv3+ Architecture

Convolutional neural networks typically apply convolution with different strides as well as pooling. These methods cause a reduction in the sizes of the feature

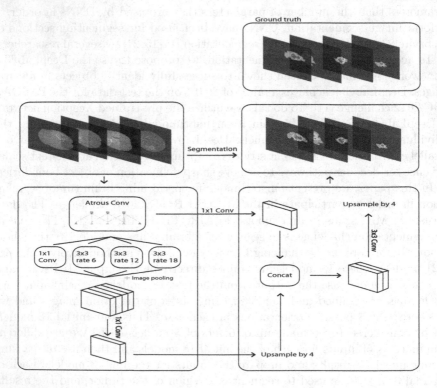

Fig. 1. DeepLabv3+ involves the use of atrous convolution at multiple scales in the encoder network as well as a decoder network

maps that are operated on by the succeeding layers, and as a result, the last feature map lacks comprehensive information associated with object boundaries. In order to overcome this challenge, an encoder-decoder pathway is used in several architectures [1,17,22]. While the U-Net [22] decoder concatenates the feature maps at corresponding scales in the encoder and decoder pathways, SegNet [1] saves and uses the max pooling indices from the encoder pathway. On the other hand, DeepLab [6] and Deeplabv3 [7] use parallel atrous convolutions at different rates in order to capture the information at different scales. Effectively, this technique, called the Atrous Spatial Pyramid Pooling (ASPP) allows the encoder network to use different fields of view. Figure 1 shows how DeepLabv3+ [8] combines these two techniques and uses an encoder-decoder pathway as well as ASPP in order to achieve precise delineation of object boundaries.

DeepLabv3+ utilizes ASPP, but, due to computation cost and complexity, it is not possible to extract features that have a resolution greater than $\frac{1}{8}^{th}$ of the input image. The last feature map generated by the encoder network, which has 256 channels, is used as the input to the decoder network. The decoder network first uses bilinear upsampling, then a 1×1 convolution is performed followed by concatenation with the corresponding features from the encoder network, 3×3 convolutions and upsampling.

DeepLabv3+ modifies the Xception model [9] by making the network deeper, using depth-wise separable atrous convolution instead of max pooling, and introducing additional batch normalization and non-linear activation layers. At present, the resulting architecture [12], which is implemented using Tensorflow, leads the Pascal VOC leaderboard for image segmentation, and hence, we decided to use this network in order to develop our automatic brain tumor segmentation software.

3 Brain Tumor Segmentation Using 21 Models

3.1 Data

The training dataset for the MICCAI BraTS 2018 challenge consists of 285 sets of NIfTI image files, with 210 sets of image files for HGG type of tumor and 75 sets for LGG. Each set of image files consists of MRI scans using four different modes: T1, T2, T1CE and FLAIR. Also, the corresponding segmentation map is provided for each of the 285 patients. The segmentation labels are assigned as follows:

- Label 4 is assigned to the enhancing tumor region.
- The edema is denoted by label 2.
- The NCR+NET region is identified using label 1.

Our task is to identify the enhancing tumor, the tumor core, which consists of the enhancing tumor and the NCR+NET regions, and the whole tumor, which consists of the tumor core and the edema. Details about the BraTS challenge and the annotated dataset are provided in [3,20]. The relevant data sources are [2,4].

3.2 Data Preprocessing

In order to perform segmentation of the different regions of a brain tumor, we slice the 3-D NIfTI brain images along 3 orientations: sagittal, coronal, and axial. By doing so, we get 155 axial slices and 240 slices each in the sagittal and coronal directions. In the original image, the intensity of each voxel is stored as a 16 bit integer. In order to use DeepLabv3+ for segmenting the slices, we store each slice as a separate .png file and the maximum possible intensity of a pixel in a .png file is 255. Thus, we scale the intensities in the original NIfTI image to a value between 0 and 255. Each NIfTI image file stores the intensities of $240 \times 240 \times 155$ voxels. Thus, each axial slice consists of 240×240 pixels, while the coronal and sagittal slices have 240×155 pixels each. We use zero padding to pad each of the coronal and sagittal slices to increase their size to 240×240 pixels.

We use the following 6 combinations of images in the three separate orientations to create the inputs to the initial 18 DeepLabv3+ models. Since DeepLabv3+ uses only 3 channels, we use various combinations of image slices in order to obtain RGB images that contain useful data.

1. To generate the inputs for the first set, called the RGB set, we combine the T1CE, T2 and FLAIR images for the corresponding slices in our data and create RGB images, where the FLAIR image occupies channel 1 (R), T1CE occupies channel 2 (G) and T2 occupies channel 3 (B). The FLAIR images contain information that helps to identify the whole tumor and the T2 and T1CE images help to identify the tumor core and enhancing tumor sections respectively.
2. For the second set, we use the FLAIR image as the first channel, the T2 image as the second channel and, for the third channel, we use slices created by taking the pixel-wise differences of the corresponding T1CE and the T1 slices. We use the pixel-wise difference of T1CE and T1 images because it contains information that identifies the enhancing tumor and the NCR regions. We refer to this as the T1CE-T1 set.
3. The third set of images, called FLAIRs3, consists of 3 FLAIR slices. Thus, to segment slice n, we use slices n-1, n, n+1 as the 3 slices for the 3 channels. This combination contains information from the two neighboring slices and should increase the accuracy of the prediction.
4. The fourth set of images, called T1CEs3, consists of 3 consecutive T1CE slices, prepared in the same way as the FLAIRs3 slices.
5. The fifth set of images, called T2s3, consists of 3 consecutive T2 slices.
6. The sixth set of images, called T1s3, consists of 3 consecutive T1 slices.

For the training data, we use the labels provided in the training data set, while assigning label 3 to all pixels that are labeled with label 4. This is because, DeepLabv3+ uses only consecutive labels and the label 3 is not used in the data provided.

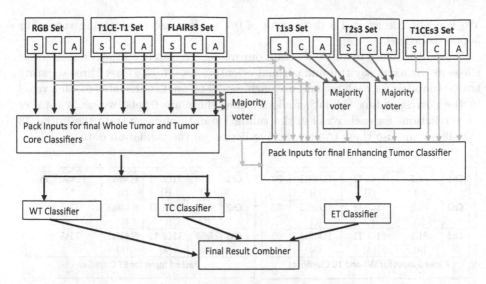

Fig. 2. Brain Tumor Segmentation Architecture: There are 18 initial classifiers that belong to 6 sets, each containing classifiers for slices in the sagittal (S), coronal (C) and axial (A) directions. The results of these are packed to form packed inputs for final three classifiers for the 3 different target classes.

3.3 Brain Tumor Segmentation Using DeepLabv3+

As shown in Fig. 2, we use 18 initial classifiers and use bit packing to pack the pixel-wise predictions from these models. Then, we use these packed results as the inputs to our final three models, one for each of the three target classes. Finally, a combiner is used to combine the results of the three final models to get the combined prediction.

Training and Prediction Using the Initial 18 Classifiers: We use 18 initial models, which consist of 6 sets of 3 models each (one model for each of the three orientations). The input data for each of these 18 models have been described in Sect. 3.2.

During training, we only use the slices in our training set which actually have tumors. On an average, in the training data provided, about 1.07% of the brain is occupied by tumor in case of HGG and about 1.24% in case of LGG. As a result, the portion of the brain that contains tumors is significantly smaller than the healthy part. Thus, to minimize the class imbalance, we only use the slices that have tumors during the training phase. However, the validation and testing sets consist of all the slices in a specific direction (axial, sagittal or coronal).

We train our models using an existing DeepLabv3+ pre-trained model. Specifically, we use a checkpoint of the Xception model [13], which is pre-trained on the augmented Pascal VOC dataset. We use a learning rate of 0.005, a batch size of 8 (due to GPU memory constraints) and train for a total of 75000 steps.

We use poly decay and a decay rate of 0.1 to decrease the learning rate after 30000 iterations.

Using the above method, we train our model separately on the 6 sets of input slices in the axial, sagittal and coronal orientations. Thus, we get three separate trained models for each set and use each model to predict the label of each voxel in the corresponding validation slices. Since there are 6 sets, we get a total of 18 predictions for each voxel in the validation data after we train the initial 18 classifiers and use them to perform prediction on the validation data.

Ch1	RGB (S)	T1CE-T1 (S)	FLAIRs3 (S)	00
Ch2	RGB (C)	T1CE-T1 (C)	FLAIRs3 (C)	00
Ch3	RGB (A)	T1CE-T1 (A)	FLAIRs3 (A)	00
Packed Input for WT and TC Classifiers				

Ch1	RGB (S)	T1CE-T1 (S)	FLAIRs3 (S)	FLAIRs3_m
Ch2	RGB (C)	T1CE-T1 (C)	FLAIRs3 (C)	T2s3_m
Ch3	RGB (A)	T1CE-T1 (A)	FLAIRs3 (A)	T1s3_m
Packed Input for ET Classifier				

Fig. 3. A diagrammatic representation to show the exact bits occupied by the predictions from different initial classifiers indicating the channels occupied by sagittal (S), coronal (C) and axial (A) predictions from the same set. The most significant bits are the left most ones. For example, sagittal RGB predictions occupy the MSBs of channel 1. $FLAIRs3\_m$, $T2s3\_m$ and $T1s3\_m$ are generated after using our majority voting-based combiner.

Bit Packing of the Results from the Initial 18 Classifiers to Prepare the Inputs to the Final 3 Classifiers: Since the results of the initial 18 classifiers assign a label from 0 to 3 to each pixel, the number of bits required to represent or store the labels is 2. However, DeepLabv3+ takes images that use 8 bits for each pixel as the input. So the value of each pixel can have a value between 0 and 255. Thus, for each pixel, we can use bit packing to pack the results of 4 initial classifiers to get the value for one channel of the input pixel to the final classifiers. Also, since DeepLabv3+ uses 3 channels, the total number of initial classifiers whose results we can directly pack to get a valid input slice for the final classifiers is 12. In Fig. 3, for each pixel, each channel stores 8 bits, which is depicted by dividing the 8 bits into 4 blocks which store 2 bits each. The 2 bits in each block correspond to the 2 bit predictions from each of the 18 initial classifiers.

Preparation of Packed Input for the Final Classifiers to Segment the Tumor Core and Whole Tumor: For the input to the final classifiers that detect the whole tumor and the tumor core sections, we use the results from 9 of the initial classifiers. The least significant two bits of the input to the final classifiers for each channel are set to zero. For the whole tumor region, we rank the 6 sets of initial classifiers in accordance with the accuracy achieved for the whole tumor region, select the top three sets and use their results in order to prepare the packed input for the final classifier. Also, after packing, the bits occupied by the results of a specific set depends on the rank of the set, such that, the results from the set that gives most accurate results occupy the most significant bits. We tried to do the same for the tumor core, but found that though the T1CE images using 3 consecutive slices have better performance than 3 consecutive slices of FLAIR images for the tumor core, the packed model that uses the results from the FLAIR slices shows better performance. Thus, for the tumor core and whole tumor classifiers, we use the following 3 sets in 3 different orientations:

- The RGB set, which consists of the FLAIR, T1CE and T2 slices and gives most accurate results.
- The T1CE-T1 set, which consists of the FLAIR and T2 slices for two channels and the difference of T1CE and T1 as the third channel, and achieves rank 2 based on the accuracy of the result.
- The FLAIRs3 set constructed using 3 consecutive FLAIR slices, which is ranked third among the three selected sets.

The results for the slices belonging to the same set but in different orientations occupy the same bits for all the three channels. Also, the predictions from the classifiers that use the sagittal and coronal slices are recombined and then sliced axially to create slices in the axial direction. This ensures that the predictions from the three different orientations can be combined to form RGB images after the packing is done.

Preparation of Packed Input for the Final Classifiers to Segment the Enhancing Tumor: In case of the enhancing tumor region, we use the results from all of the initial 18 classifiers for creating the packed input for the final classifier. For the most significant 6 bits of each channel for the packed input, we use the results from the RGB, T1CE-T1, and T1CEs3 classifiers. This decision is also made based on the accuracy of the results of each of the 18 initial classifiers for the enhancing tumor region. Based on the rank achieved by each set, the RGB set achieves maximum accuracy and its results occupy the most significant 2 bits, while the results of T1CEs3 have a rank of 3 among the 3 top-ranked classifier sets, and thus occupy bit 5 and bit 6 of the packed input to the final classifier. In all these cases the results obtained from the 3 different orientations in the same classifier set are used to create the inputs for the 3 different channels. Finally, we use majority voting in order to combine the results of each of the 9 remaining classifiers. We use intra-set majority voting to combine the results from the 3 models (one each for the slices in each orientation) in a set and get a single 2 bit result. Thus, we get 2 bits each after combining the results from the 3 orientations of T1s3, T2s3, and FLAIRs3. Then, we create the packed input to the final classifier by using the results from FLAIRs3, T2s3, and T1s3 as the least significant 2 bits of channel 1, 2, and 3 respectively.

Our simplistic majority voting scheme works by first recombining all the predicted slices in the axial direction to get the 3-D brain segmentation. We also perform the same operation along different axes to reconstruct the segmented images using the sagittal and coronal slices. Here we refer to the prediction for voxel at position i, j, k considering the sagittal direction slices as $p_{i,j,k}^{sagittal}$, the axial direction slices as $p_{i,j,k}^{axial}$ and coronal direction slices as $p_{i,j,k}^{coronal}$. If, for any voxel, any two of these predictions are the same, that is either $p_{i,j,k}^{sagittal} = p_{i,j,k}^{coronal}$ or $p_{i,j,k}^{sagittal} = p_{i,j,k}^{axial}$ or $p_{i,j,k}^{axial} = p_{i,j,k}^{coronal}$, then we assign the label predicted by the majority to that voxel. However, if all the three predictions differ, we give priority to the prediction that uses the slices in the orientation that achieves the best performance. From our experiments, we found that, other than the classifiers that use T1CE-T1 slices, the highest average Dice coefficient is achieved by the model that uses the cross-sectional slices.

Training and Generation of Final Prediction Using the 3 Final Classifiers: Using the above inputs, we trained DeepLabv3+ models for 150000 iterations and checked the performance on the validation set after every 5000 iterations. For the model that is trained for identifying the enhancing tumor section, we get the best performance after training for 115000 iterations. While the inputs for the models for the whole tumor and the tumor core were the same, we achieve the best performance for the whole tumor section after training for 100000 iterations, and for the tumor core after training for 130000 iterations. All the other training parameters are the same as the ones used for the initial 18 classifiers. For the testing phase, we save and use the best performing checkpoints to perform our predictions.

Finally, we combine the outputs of the final 3 classifiers by using the fact that the tumor core is a subset of the whole tumor and the enhancing tumor is a subset of the tumor core. Thus, we combine the results and ensure that if there are any pixels that are predicted as tumor core but lie outside the mask for the whole tumor, then it is not considered to belong to tumor core. We use a similar logic to allocate the labels for the enhancing tumor region.

Our code can be found at: https://ar16m@bitbucket.org/ar16m/brats.

4   Results

We evaluate the performance of our network by training the 18 initial classifiers consisting of 6 sets of 3 classifiers each and then providing their performance on the validation set, which consists of 66 cases. The models are used to perform predictions on the validation slices of the corresponding orientations. We report the Dice coefficient of each of the trained models by creating a separate table for each of the 3 orientations. We also report the Dice coefficient for the results of each set after using majority voting to combine the results obtained from the 3 orientations.

In Tables 1, 2, 3 and 4, we provide the mean Dice coefficient achieved for the three classes: whole tumor, tumor core, and enhancing tumor.

Table 1. Dice coefficient achieved by the 6 models trained on the sagittal slices

Model	Enhancing tumor	Whole tumor	Tumor core
RGB	0.68459	0.8939	0.80119
T1CE-T1	0.6562	0.8655	0.77151
FLAIRs3	0.33235	0.84988	0.64296
T1CEs3	0.65405	0.74965	0.74233
T1s3	0.2589	0.72237	0.57025
T2s3	0.39554	0.82681	0.65191

Table 2. Dice coefficient achieved by the 6 models trained on the coronal slices

Model	Enhancing Tumor	Whole Tumor	Tumor Core
RGB	0.66206	0.87746	0.77098
T1CE-T1	0.61373	0.87475	0.74592
FLAIRs3	0.27256	0.87242	0.63564
T1CEs3	0.64578	0.75146	0.73137
T1s3	0.21524	0.7272	0.51093
T2s3	0.37821	0.83565	0.652361

Table 3. Dice coefficient achieved by the 6 models trained on the axial slices

Model	Enhancing Tumor	Whole Tumor	Tumor Core
RGB	0.69409	0.88508	0.80733
T1CE-T1	0.63185	0.88176	0.77198
FLAIRs3	0.38685	0.87763	0.69402
T1CEs3	0.69508	0.75459	0.75409
T1s3	0.33677	0.74698	0.61879
T2s3	0.3873	0.82306	0.66986

Table 4. Dice coefficient achieved by the 6 majority-voted predictions (one corresponding to each set)

Model	Enhancing Tumor	Whole Tumor	Tumor Core
RGB	0.72705	0.89152	0.82154
T1CE-T1	0.68443	0.88669	0.79011
FLAIRs3	0.38325	0.88484	0.69455
T1CEs3	0.70896	0.77544	0.77949
T1s3	0.3184	0.75872	0.60877
T2s3	0.43856	0.84126	0.69651

From the above results, we see that the use of the basic majority voting-based combiner improves the Dice coefficient. Thus, in the packed input used by the final classifier for the enhancing tumor region, we use the results obtained after performing majority voting for the 3 lowest ranked models. In this way, we reduce the number of bits occupied by the results from the classifiers that perform poorly, ensure that we can pack the results of the 18 initial classifiers into the 8×3 bits available for each pixel, and utilize more accurate results for the final classifier inputs.

Finally, we provide some measures of the accuracy achieved by combining the results from our final 3 classifiers. In Table 5, we first provide the mean Dice coefficient achieved by the 3 final classifiers for the corresponding regions and then we provide the Dice coefficient, sensitivity, specificity and Hausdroff distance achieved by the combined prediction on the validation data as well as the Dice coefficient and Hausdroff distance for testing data.

Table 5. Performance of our combined model using four common metrics.

Metric	Enhancing tumor	Whole tumor	Tumor core
Individual final models			
Dice coefficient	0.75055	0.896	0.82327
Final combined model			
Dice coefficient	0.75158	0.896	0.82313
Sensitivity	0.79062	0.9013	0.81701
Specificity	0.99704	0.99389	0.99766
Hausdroff distance	3.87866	4.47801	6.24132
Test set results using final combined model			
Dice coefficient	0.70859	0.87549	0.78969
Hausdroff distance	4.48324	6.45026	6.91055

We also provide here a sample image in order to visualize our image segmentation results after combining the results from the three final models. Figure 4 depicts the T1, FLAIR, T1CE and T2 images as well as our predicted segmentation and the ground truth segmentation. The figure shows the level of accuracy that we have achieved after combining the results from the three final models. However, in some cases the uneven boundaries are not identified accurately.

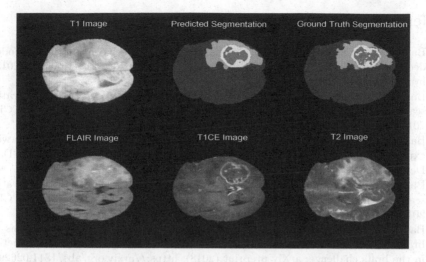

Fig. 4. Visualization of segmentation: The figure depicts the T1, FLAIR, T1CE and T2 images as well as the ground truth and predicted segmentations.

5 Conclusion and Future Work

We used the DeepLabv3+ deep learning framework to perform brain tumor segmentation. We trained our network separately on the coronal, axial and sagittal slices and developed a network consisting of 18 initial classifiers, which are divided into 6 sets of 3 classifiers each (one for each orientation). Every set has a different combination of input slices as the input, and the outputs of these 18 classifiers are combined using bit packing to prepare the inputs for the final 3 classifiers, one for each of the three target regions. The outputs of these 3 classifiers are combined to produce the final segmented brain images. On the testing data, we achieve Dice coefficients of 0.7086, 0.7897 and 0.8755 for the enhancing tumor, tumor core and whole tumor regions respectively. A summary of the BraTS 2018 challenge can be found in [5].

In future, we intend to extend our model so that we can use patches of at least 5 slices to perform the segmentation of the central slice. This neighborhood information should enable our model to achieve higher accuracy because, in general, if there is a tumor in the neighboring slices, the slice under consideration has a higher probability of having a tumor. However, this information is not considered in cases where the segmentation is done for individual slices or the information may be insufficient in cases where we use only two neighboring slices.

References

1. Badrinarayanan, V., Kendall, A., Cipolla, R.: Segnet: a deep convolutional encoder-decoder architecture for image segmentation. CoRR abs/1511.00561 (2015). http://arxiv.org/abs/1511.00561
2. Bakas, S., et al.: Segmentation labels and radiomic features for the pre-operative scans of the TCGA-LGG collection (2017). https://doi.org/10.7937/K9/TCIA.2017.GJQ7R0EF
3. Bakas, S., et al.: Advancing the cancer genome atlas glioma MRI collections with expert segmentation labels and radiomic features. Sci. Data **4**, 170117 (2017). Data Descriptor http://dx.doi.org/10.1038/sdata.2017.117
4. Bakas, S., et al.: Segmentation labels and radiomic features for the pre-operative scans of the TCGA-GBM collection (2017). https://doi.org/10.7937/K9/TCIA.2017.KLXWJJ1Q
5. Bakas, S., Reyes, M., et al.: Identifying the best machine learning algorithms for brain tumor segmentation, progression assessment, and overall survival prediction in the brats challenge. arXiv preprint (2018). https://arxiv.org/abs/1811.02629
6. Chen, L., Papandreou, G., Kokkinos, I., Murphy, K., Yuille, A.L.: Deeplab: Semantic image segmentation with deep convolutional nets, atrous convolution, and fully connected CRFs. CoRR abs/1606.00915 (2016). http://arxiv.org/abs/1606.00915
7. Chen, L., Papandreou, G., Schroff, F., Adam, H.: Rethinking atrous convolution for semantic image segmentation. CoRR abs/1706.05587 (2017). http://arxiv.org/abs/1706.05587
8. Chen, L., Zhu, Y., Papandreou, G., Schroff, F., Adam, H.: Encoder-decoder with atrous separable convolution for semantic image segmentation. CoRR abs/1802.02611 (2018). http://arxiv.org/abs/1802.02611
9. Chollet, F.: Xception: deep learning with depthwise separable convolutions. CoRR abs/1610.02357 (2016). http://arxiv.org/abs/1610.02357
10. DeAngelis, L.M.: Brain tumors. N. Engl. J. Med. **344**(2), 114–123 (2001). https://doi.org/10.1056/NEJM200101113440207. pMID: 11150363
11. Everingham, M., Van Gool, L., Williams, C.K.I., Winn, J., Zisserman, A.: The PASCAL Visual Object Classes Challenge 2012 (VOC2012) Results. http://www.pascal-network.org/challenges/VOC/voc2012/workshop/index.html
12. Google: Deeplabv3+ (2018). https://github.com/tensorflow/models/tree/master/research/deeplab
13. Google: deeplabv3_pascal_train_aug (2018). http://download.tensorflow.org/models/deeplabv3_pascal_train_aug_2018_01_04.tar.gz
14. Kamnitsas, K., et al.: Ensembles of multiple models and architectures for robust brain tumour segmentation. In: Crimi, A., Bakas, S., Kuijf, H., Menze, B., Reyes, M. (eds.) Brainlesion: Glioma, Multiple Sclerosis, Stroke and Traumatic Brain Injuries, pp. 450–462. Springer International Publishing, Cham (2018)
15. Kamnitsas, K., et al.: Efficient multi-scale 3d CNN with fully connected CRF for accurate brain lesion segmentation. CoRR abs/1603.05959 (2016). http://arxiv.org/abs/1603.05959
16. Krizhevsky, A., Sutskever, I., Hinton, G.E.: Imagenet classification with deep convolutional neural networks. In: Pereira, F., Burges, C.J.C., Bottou, L., Weinberger, K.Q. (eds.) Advances in Neural Information Processing Systems 25, pp. 1097–1105. Curran Associates, Inc. (2012). http://papers.nips.cc/paper/4824-imagenet-classification-with-deep-convolutional-neural-networks.pdf

17. Li, R., et al.: Deepunet: a deep fully convolutional network for pixel-level sea-land segmentation. CoRR abs/1709.00201 (2017). http://arxiv.org/abs/1709.00201

18. Liu, J., Li, M., Wang, J., Wu, F., Liu, T., Pan, Y.: A survey of mri-based brain tumor segmentation methods. Tsinghua Sci. Technol. **19**(6), 578–595 (2014)

19. Louis, D.N., et al.: The 2016 world health organization classification of tumors of the central nervous system: a summary. Acta Neuropathol. **131**(6), 803–820 (2016). https://doi.org/10.1007/s00401-016-1545-1

20. Menze, B.H., et al.: The multimodal brain tumor image segmentation benchmark (brats). IEEE Trans. Med. Imaging **34**(10), 1993–2024 (2015). https://doi.org/10.1109/TMI.2014.2377694

21. Pereira, S., Pinto, A., Alves, V., Silva, C.A.: Brain tumor segmentation using convolutional neural networks in MRI images. IEEE Trans. Med. Imaging **35**(5), 1240–1251 (2016). https://doi.org/10.1109/TMI.2016.2538465

22. Ronneberger, O., Fischer, P., Brox, T.: U-net: convolutional networks for biomedical image segmentation. CoRR abs/1505.04597 (2015). http://arxiv.org/abs/1505.04597

23. Wachinger, C., Reuter, M., Klein, T.: Deepnat: deep convolutional neural network for segmenting neuroanatomy. NeuroImage **170**, 434–445 (2018)

Brain Tumor Segmentation on Multimodal MR Imaging Using Multi-level Upsampling in Decoder

Yan Hu[1], Xiang Liu[2], Xin Wen[2], Chen Niu[2], and Yong Xia[1(✉)]

[1] National Engineering Laboratory for Integrated Aero-Space-Ground-Ocean Big Data Application Technology, School of Computer Science and Engineering, Northwestern Polytechnical University, Xi'an 710072, People's Republic of China
yxia@nwpu.edu.cn

[2] The First Affiliated Hospital of Xi'an Jiao Tong University, Xi'an 710061, People's Republic of China

Abstract. Accurate brain tumor segmentation plays a pivotal role in clinical practice and research settings. In this paper, we propose the multi-level up-sampling network (MU-Net) to learn the image presentations of transverse, sagittal and coronal view and fuse them to automatically segment brain tumors, including necrosis, edema, non-enhancing, and enhancing tumor, in multimodal magnetic resonance (MR) sequences. The MU-Net model has an encoder–decoder structure, in which low level feature maps obtained by the encoder and high level feature maps obtained by the decoder are combined by using a newly designed global attention (GA) module. The proposed model has been evaluated on the BraTS 2018 Challenge validation dataset and achieved an average Dice similarity coefficient of 0.88, 0.74, 0.69 and 0.85, 0.72, 0.66 for the whole tumor, core tumor and enhancing tumor on the validation dataset and testing dataset, respectively. Our results indicate that the proposed model has a promising performance in automated brain tumor segmentation.

Keywords: Magnetic resonance imaging · Brain tumor segmentation · Encoder–decoder · Multi-level upsampling · Global attention

1 Introduction

Glioma is a type of tumors that starts in the glial cells of the brain or the spin, comprising about 30% of all brain tumors and central nervous system tumors, and 80% of all malignant brain tumors [1]. Shape and localization of tumors are crucial for diagnosis, treatment planning and follow-up observation in clinical, while the manual segmentation of brain tumor in magnetic resonance (MR) images requires a high degree of skills and concentration, and is time-consuming, expensive and prone to operator bias. Thus, a fully automated and reliable segmentation algorithm is of great significance. However, despite considerable research efforts being devoted to this task [2], automated segmentation of brain tumors remains a challenge, largely due to the variable shapes and locations, diffusion and poor contrast of brain tissues in MR images.

© Springer Nature Switzerland AG 2019
A. Crimi et al. (Eds.): BrainLes 2018, LNCS 11384, pp. 168–177, 2019.
https://doi.org/10.1007/978-3-030-11726-9_15

In recent years, deep learning techniques, especially deep convolutional neural networks (DCNNs), have led to significant breakthroughs in computer vision, since they provide an 'end-to-end' framework for simultaneous presentation learning and image segmentation and thus free users from the troublesome extraction of handcrafted features. Such breakthroughs have prompted many researchers to use DCNNs for brain tumor segmentation. The solutions published in the literature can be roughly divided into two groups. One group of solutions are based on the classification of image patches. Pereira et al. [3] designed an 11-layer CNN and a 9-layer CNN to classify the patches extracted from high grade gliomas (HGG) and low grade gliomas (LGG), respectively. To simultaneously learn the presentation of both fine details and coarse structures from input images, Zhao et al. [4] proposed a three-convolutional-pathway network, in which the input patches for three pathways have a size of 48×48, 28×28 and 12×12, respectively, and concatenated these three outputs for classification. Kamnitsas et al. [5] adopted a 3D CNN architecture, i.e. DeepMedic, with multiple input image resolutions, residual connections and fully connected conditional random field. Castillo et al. [6] developed a neural network with four contracting pathways and residual connections that receive patches centered on the same voxel, but with different spatial resolutions. Lopez et al. [7] removed max pooling layers in dilated residual network [8] to avoid loss of upsampling the prediction by interpolation, but at the same time enlarge the receptive field through dilated convolutional operations. McKinley et al. [9] also replaced max pooling layers by dilated convolutions without influencing the receptive field of the classifier in Densenet. The other group of solutions are based on fully convolutional networks (FCNs). Pereira et al. [3] employed two U-Nets, one for the localization of tumors and the other for the segmentation of intra-tumor structures. Li et al. [10] used three parallel end-to-end networks for three views and generated the segmentation results using majority voting. Kamnitsas et al. [11] trained seven end-to-end networks and used ensemble learning to produce robust segmentation results. Wang et al. [12] proposed a cascade of fully convolutional neural networks to decompose the multi-class segmentation problem into a sequence of three binary segmentation problems according to the subregion hierarchy. In our previous work [13], we used a cascaded U-Net model and a patch-wise CNN to detect and segment brain tumors.

In this paper, we propose a FCN called the multi-level upsampling network (MU-Net) to segment brain tumor structures, including necrosis, edema and enhancing tumor from multimodality MR. Our main contributes are: (a) we designed a global attention (GA) module to combine the low level feature from encoder and high level feature from decoder; (b) we designed a multi-level decoding architecture. The proposed algorithm has been evaluated on the BraTS 2018 Challenge validation dataset and achieved a promising result.

2   Dataset

The proposed MU-Net model was evaluated on the Brain Tumor Segmentation 2018 (BraTS 2018) Challenge dataset [14–16]. There are 285 cases for training, including 210 HGG and 75 LGG cases. Each case has four multimodal MR scans, including the

T1, T1c, T2, and FLAIR. All these scans were co-registered to the same anatomical template, interpolated to the same dimension of $240 \times 240 \times 155$ and the same voxel size of $1.0 \times 1.0 \times 1.0$ mm^3 and skull-stripped. Each case has been segmented manually, by up to four raters, following the same annotation protocol, and their annotations were approved by experienced neuro-radiologists. Annotations of tumor tissues comprise the enhancing tumor (ET-label 4), the peritumoral edema (ED-label 2), and the necrotic and non-enhancing tumor core (NCR/NET-label 1). The validation and testing datasets consist of 66 and 191 cases, respectively, but their grade and ground truth are unseen.

3   Methods

The 3D brain MR sequences are resliced from three views, transverse, sagittal and coronal respectively. Three probability maps of these three views are learned by three identical MU-Nets, respectively, and concatenated together as the input of a multi-view fusion network. The pipeline of proposed algorithm is shown in Fig. 1.

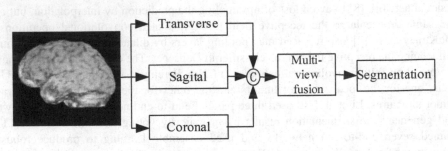

Fig. 1. Pipeline of proposed algorithm.

3.1   MU-Net

The proposed MU-Net model adopts the encoder-decoder structure, consisting of five convolutional blocks, a spatial pyramid pooling (SPP) module [17], five global attention (GA) modules, and nine upsampling feature (UF) modules. The architecture of this model is shown in Fig. 2.

The encoder branch is a variants of ResNet-101. The convolutional layer with 64 7×7 kernels and a stride of 2 in the root block (i.e. Block 1) is replaced with five convolutional layers, each consisting 64 3×3 kernels. The stride of the third convolutional layer is 2, and the stride of other convolutional layers is 1. Other blocks in this branch is the same as those in ResNet-101 [18].

Between the encoder and decoder, we add a SPP module, in which there are five parallel operators, including three 3×3 dilated convolution with a dilation rate of 6, 12, and 18, respectively, a 1×1 convolution and a global pooling (see Fig. 3(a)). The input of the SPP module is processed by these operators simultaneously, and the feature maps generated by these operators are concatenated as the output of the SPP module.

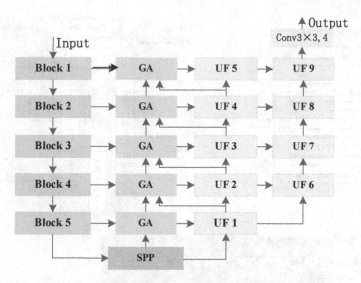

Fig. 2. Architecture of the proposed MU-Net model

The major part of the decoder branch contains five decode modules (i.e. UF 1 – UF 5), which are designed to recover the size of feature maps. Usually, there are two 3 × 3 convolutions and a bilinear interpolation between them in each UF module (see Fig. 3 (c)). However, since there is no down-sampling operation in the encoder block 3–5, the interpolation operation is omitted in UF 1, UF 2, and UF 5 modules such that the output feature maps have the same size as the input of MU-Net. Meanwhile, to combine low-level feature maps and high-level feature maps in the decoding process, we add five GA modules to the MU-Net model. Each GA module takes two groups of inputs - low-level feature maps from the corresponding encoder block and high-level feature maps from the UF module at the previous level. Two 3 × 3 convolutions are applied to low-level feature maps, respectively. High-level feature maps are also processed by two operations – one is the global average pooling followed by a 1 × 1 convolution as, and the other is a 3 × 3 convolution. The processed high-level feature maps are then used as the element-wise weighting mask of the processed low-level feature maps (see Fig. 3(b)). In addition, the output of each of UF 2 – UF 5 are fed simultaneously to the UF module (UF 6 – UF 9) at the next level. Eventually, the output of the UF 6 and the output of UF1 are concatenated and fed to a 3 × 3 convolution another UF module to produce the segmentation results.

3.2 Multi-view Fusion

Three views are fused by a shallow encoder-decoder network. The encoder consists of three convolutional layers with 64, 128 and 256 3 × 3 kernels, followed by three max pooling layers respectively. The decoder comprises three deconvolutional layers with 256, 128 and 64 kernels of size 3 × 3. Then, we convolve the output of the decoder by four 3 × 3 kernels and predict by max possibility.

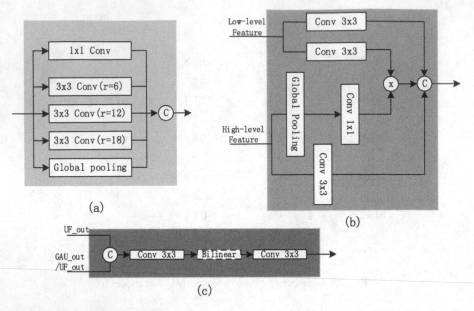

Fig. 3. Architecture of modules used in segmentation model. (a) shows the SPP module;(b) shows GA module; (c) shows the UF module.

3.3 Implementation

With the proposed MU-Net model, brain tumor segmentation can be performed on a slice-by-slice basis. The slices in each training dataset were cropped and padded to 224×224, 224×160, 224×160 for transverse, sagittal, and coronal view, respectively, and the voxel values of each modality were normalized by the min-max normalization. The encoding branch was initialized by the pre-trained ResNet-101 [19]. The positive slices (with tumor) and negative slices (without tumor) were randomly selected at a rate of 5:1. The cross entropy was used as the loss function, and the adaptive moment estimator (Adam) with an exponentially descending learning rate of 0.001–0.00001 was adopted as the optimizer. It took about twenty hours to train each MU-Net model with a batch size of 8 and epochs of 30 on two GPUs (NVIDIA 1080 Ti, 12 GB RAM) four hours to train the fusion network with a batch size of 16 and epochs of 20.

4   Experiments and Results

Following the request of the challenge, four intra-tumor structures have been grouped into three mutually inclusive tumor regions: (a) whole tumor (WT) that consists of all tumor tissues, (b) tumor core (TC) that consists of the enhancing tumor and necrotic and non-enhancing tumor core, and (c) enhancing tumor (ET). The performance of segmenting each tumor region was quantitatively evaluated through an online system by using three metrics, including the average Dice similarity coefficient, sensitivity and Hausdorff distance.

Preliminary results for the BraTS 2018 Training dataset have been obtained by hold-out using 80% of the data (228 cases) for training and the remaining 20% for validation (57 cases). Table 1 shows the quantitative evaluation and Fig. 4 presents some examples of the predictions against the ground truth on predicted cases from BraTS 2018 training data. It appears that this proposed model works well when the edge is relatively smooth, as the first three examples shown in Fig. 4. However, similarly to other semantic image segmentation task, our deep model works weakly on pixels distributed near the edge as th last two examples shown in Fig. 4. Tables 2 and 3 give the quantitative evaluation of our algorithm on 66 validation and 191 testing unseen subjects. We can observe that performance on training data, validation data and testing data are consistent, which indicates that this model generalizes well to unseen examples. Figure 5 shows the visualization of segmentation result from validation dataset.

Table 1. Quantitative result of validation on BraTS 2018 training set.

	Dice			Sensitivity			Hausdorf-95		
	ET	WT	TC	ET	WT	TC	ET	WT	TC
Mean	0.61	0.83	0.73	0.83	0.89	0.75	41.48	47.23	41.14
StdDev	0.27	0.11	0.17	0.17	0.09	0.21	37.49	23.49	28.90
Median	0.72	0.86	0.77	0.88	0.91	0.82	41.69	46.70	44.77
25quantile	0.50	0.78	0.64	0.80	0.85	0.62	5.12	30.36	12.37
75quantile	0.80	0.90	0.86	0.95	0.96	0.93	61.25	59.84	59.67

5 Discussion

5.1 Multi-level Upsampling

To demonstrate the performance improvement resulted from using the GA module, we trained a similar network but without using multi-level upsampling on the BraTS 2018 training dataset and tested it on the validation dataset. Table 4 gives the performance of both models measured by the average Dice similarity coefficient, sensitivity, specificity and Hausdorf-95. It reveals that multi-level upsampling connection is able to improve the performance.

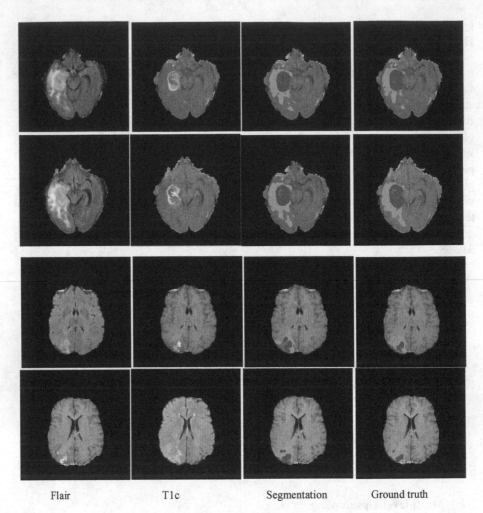

Flair   T1c   Segmentation   Ground truth

Fig. 4. Segmentation examples from the validation set. From top to bottom: the 55th, 57th, slices from the subject Brats18_TCIA01_147_1 and the 55th, 57th slices from the subject Brats18_TCIA10_629_1. Red - NCR&NET, Blue - ET, Green – ED. (Color figure online)

Table 2. Quantitative result on BraTS 2018 validation set.

	Dice			Sensitivity			Hausdorf-95		
	ET	WT	TC	ET	WT	TC	ET	WT	TC
Mean	0.69	0.88	0.74	0.71	0.87	0.77	6.69	4.76	10.67
StdDev	0.27	0.10	0.24	0.28	0.15	0.25	12.43	4.04	9.87
Median	0.80	0.91	0.84	0.82	0.93	0.88	2.83	3.00	6.78
25quantile	0.66	0.88	0.69	0.63	0.84	0.68	1.73	2.24	4.36
75quantile	0.86	0.94	0.90	0.90	0.95	0.93	5.39	5.74	14.73

Table 3. Quantitative result on BraTS 2018 testing set.

	Dice			Hausdorf-95		
	ET	WT	TC	ET	WT	TC
Mean	0.66	0.85	0.72	5.94	6.29	9.04
StdDev	0.29	0.15	0.27	9.04	9.25	11.47
Median	0.77	0.90	0.82	2.83	4.00	5.39
25quantile	0.60	0.84	0.65	2.00	2.34	3.16
75quantile	0.85	0.93	0.90	5.15	6.28	11.32

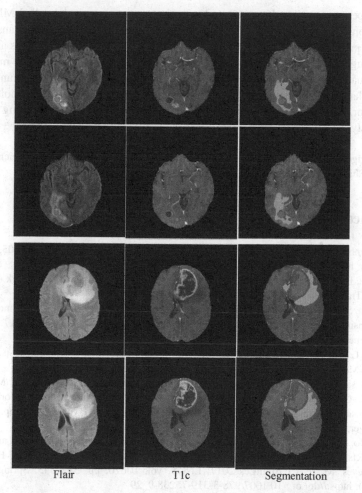

Flair   T1c   Segmentation

Fig. 5. Segmentation examples from the validation set. From top to bottom: the 54th and 57th slices from the subject Brats18_CBICA_ANK_1 and 87th and 91th, slices from the subject Brats18_CBICA_ANK_1. *Red* - NCR&NET, *Blue* - ET, *Green* – ED (Color figure online)

Table 4. Comparison of model with multi-level upsampling and without multi-level upsampling on BraTS 2018 validation dataset

	Dice			Sensitivity			Hausdorf-95		
	ET	WT	TC	ET	WT	TC	ET	WT	TC
With MU	0.69	0.88	0.74	0.71	0.87	0.77	6.69	4.76	10.67
Without MU	0.65	0.84	0.69	0.74	0.84	0.70	29.82	34.51	38.73

6 Conclusion

In this paper, we proposed a novel end-to-end segmentation model called MU-Net to segment brain tumors and their intra structures from multimodal MR scans, which learns the presentation of MR scans in transverse, sagittal and coronal views and fused them through a convolutional neural network for image segmentation. This model has been evaluated on the BraTS 2018 Challenge online system and achieved an average Dice similarity coefficient of 0.88, 0.74, 0.69 and 0.85, 0.72, 0.66 for whole tumor, core tumor, and enhancing tumor on the validation dataset and testing dataset, respectively.

Acknowledgement. This work was supported in part by the National Natural Science Foundation of China under Grants 61471297 and 61771397.

References

1. Goodenberger, M.L., Jenkins, R.B.: Genetics of adult glioma. Cancer Genet. **205**, 613–621 (2012). https://doi.org/10.1016/j.cancergen.2012.10.009
2. Menze, B.H., et al.: The multimodal brain tumor image segmentation benchmark (BRATS). IEEE Trans. Med. Imaging **34**, 1993–2024 (2015)
3. Pereira, S., Oliveira, A., Alves, V., Silva, C.A.: On hierarchical brain tumor segmentation in MRI using fully convolutional neural networks: a preliminary study. In: 2017 IEEE 5th Portuguese Meeting on Bioengineering (ENBENG), pp. 1–4. IEEE (2017)
4. Zhao, L., Jia, K.: Multiscale CNNs for brain tumor segmentation and diagnosis. Comput. Math. Methods Med. **2016** (2016)
5. Kamnitsas, K., et al.: Deepmedic for brain tumor segmentation. In: Crimi, A., Menze, B., Maier, O., Reyes, M., Winzeck, S., Handels, H. (eds.) International Workshop on Brainlesion: Glioma, Multiple Sclerosis, Stroke and Traumatic Brain Injuries, pp. 138–149. Springer, Cham (2016). https://doi.org/10.1007/978-3-319-55524-9_14
6. Castillo, L.S., Daza, L.A., Rivera, L.C., Arbeláez, P.: Brain Tumor segmentation and parsing on MRIs using multiresolution neural networks. In: Crimi, A., Bakas, S., Kuijf, H., Menze, B., Reyes, M. (eds.) BrainLes 2017. LNCS, vol. 10670, pp. 332–343. Springer, Cham (2018). https://doi.org/10.1007/978-3-319-75238-9_29
7. Moreno Lopez, M., Ventura, J.: Dilated convolutions for brain tumor segmentation in MRI scans. In: Crimi, A., Bakas, S., Kuijf, H., Menze, B., Reyes, M. (eds.) BrainLes 2017. LNCS, vol. 10670, pp. 253–262. Springer, Cham (2018). https://doi.org/10.1007/978-3-319-75238-9_22

8. Yu, F., Koltun, V., Funkhouser, T.A.: Dilated residual networks. In: Computer Vision and Pattern Recognition, pp. 636–644 (2017)
9. McKinley, R., Jungo, A., Wiest, R., Reyes, M.: Pooling-free fully convolutional networks with dense skip connections for semantic segmentation, with application to brain tumor segmentation. In: Crimi, A., Bakas, S., Kuijf, H., Menze, B., Reyes, M. (eds.) BrainLes 2017. LNCS, vol. 10670, pp. 169–177. Springer, Cham (2018). https://doi.org/10.1007/978-3-319-75238-9_15
10. Li, Y., Shen, L.: Deep learning based multimodal brain tumor diagnosis. In: Crimi, A., Bakas, S., Kuijf, H., Menze, B., Reyes, M. (eds.) BrainLes 2017. LNCS, vol. 10670, pp. 149–158. Springer, Cham (2018). https://doi.org/10.1007/978-3-319-75238-9_13
11. Kamnitsas, K., et al.: Ensembles of multiple models and architectures for robust brain tumour segmentation. In: Crimi, A., Bakas, S., Kuijf, H., Menze, B., Reyes, M. (eds.) BrainLes 2017. LNCS, vol. 10670, pp. 450–462. Springer, Cham (2018). https://doi.org/10.1007/978-3-319-75238-9_38
12. Wang, G., Li, W., Ourselin, S., Vercauteren, T.: Automatic brain tumor segmentation using cascaded anisotropic convolutional neural networks. In: Crimi, A., Bakas, S., Kuijf, H., Menze, B., Reyes, M. (eds.) BrainLes 2017. LNCS, vol. 10670, pp. 178–190. Springer, Cham (2018). https://doi.org/10.1007/978-3-319-75238-9_16
13. Hu, Y., Xia, Y.: 3D deep neural network-based brain tumor segmentation using multimodality magnetic resonance sequences. In: Crimi, A., Bakas, S., Kuijf, H., Menze, B., Reyes, M. (eds.) BrainLes 2017. LNCS, vol. 10670, pp. 423–434. Springer, Cham (2018). https://doi.org/10.1007/978-3-319-75238-9_36
14. Bakas, S.: Advancing the cancer genome atlas glioma MRI collections with expert segmentation labels and radiomic features. Sci. Data **4**, 170117 (2017)
15. Bakas, S., et al.: Segmentation labels and radiomic features for the pre-operative scans of the TCGA-GBM collection. The Cancer Imaging Archive (2017). https://doi.org/10.7937/K9/TCIA.2017.KLXWJJ1Q
16. Bakas, S., et al.: Segmentation labels and radiomic features for the preoperative scans of the TCGA-LGG collection. The Cancer Imaging Archive (2017). https://doi.org/10.7937/K9/TCIA.2017.GJQ7R0EF
17. Chen, L.C., Zhu, Y., Papandreou, G., Schroff, F., Adam, H.: Encoder-decoder with atrous separable convolution for semantic image segmentation. arXiv preprint arXiv:1802.02611 (2018)
18. He, K., Zhang, X., Ren, S., Sun, J.: Identity mappings in deep residual networks. In: Leibe, B., Matas, J., Sebe, N., Welling, M. (eds.) ECCV 2016. LNCS, vol. 9908, pp. 630–645. Springer, Cham (2016). https://doi.org/10.1007/978-3-319-46493-0_38
19. Pre-trained Resnet_v2_101 model. http://download.tensorflow.org/models/resnet_v2_101_2017_04_14.tar.gz
20. Bakas, S., Reyes, M., Jakab, A, Bauer et al.: Identifying the best machine learning algorithms for brain tumor segmentation, progression assessment, and overall survival prediction in the BRATS challenge. arXiv preprint arXiv:1811.02629 (2018)

Neuromorphic Neural Network for Multimodal Brain Image Segmentation and Overall Survival Analysis

Woo-Sup Han and Il Song Han[✉]

ODIGA Ltd, London, UK
{phil.han,ishan}@odiga.co.uk

Abstract. Image analysis of brain tumors is one of key elements for clinical decision, while manual segmentation is time consuming and known to be subjective to clinicians or radiologists. In this paper, we examined the neuromorphic convolutional neural network on this task of multimodal images, using a down-up resizing network structure. The controlled rectifier neuron function was incorporated in neuromorphic neural network, for introducing the efficiency of segmentation and saliency map generation used in noisy image processing of X-ray CT data and dark road video data. The neuromorphic neural network is proposed to the brain imaging analytic, based on the visual cortex-inspired deep neural network developed for 3 dimensional tooth segmentation and robust visual object detection. Experiment results illustrated the effectiveness and feasibility of our proposed method with flexible requirements of clinical diagnostic decision data, from segmentation to overall survival analysis. The survival prediction was 71% accuracy for the data with true result and 50.6% accuracy of predicting survival days for the individual challenge data without any clinical diagnostic data.

Keywords: Convolutional neural network · Neuromorphic processing · Brain tumor · Image segmentation · Survival analysis · Visual cortex

1 Introduction

The assessment of brain tumors delivers valuable information and becomes one of key elements of clinical diagnosis. Therefore, the automatic brain image segmentation emerges as a critical technology, as there are advantages of faster, more objective and potentially desirable accuracy. Due to the irregular nature of tumors as well as noisy 3D MRI images, the development of practical solution is still challenging throughout the BRATS Challenge [1–5]. Particularly, the accuracy of categorical estimates ranged from 23% up to 78% for the survival prediction among the expert clinicians [6].

Overall survival (OS) analysis of patients has been also the subject of interests, which is evaluated from the baseline to the time of last available follow-up. There observed a great variability in survival prediction, and even experienced physicians are relatively poor at predicting the individual's survival period [7–9]. A time threshold of 18 months can be defined to differentiate the patients into 2 groups, those with short- or long-term survival [9]. In this paper, the segmentation algorithm is proposed and

© Springer Nature Switzerland AG 2019
A. Crimi et al. (Eds.): BrainLes 2018, LNCS 11384, pp. 178–188, 2019.
https://doi.org/10.1007/978-3-030-11726-9_16

applied to evaluate OC based on BRAT 2018 high grade glioblastomas (HGG) data set and the survival data.

Since the convolutional network like U-net have been widely in use [10, 11], the convolutional neural network (CNN) and its derivatives have attracted more attentions on segmentation tasks. Recently, the neuromorphic neural network introduced its feasibility of segmentation of 3D CT images, by the successful 3D dental tooth segmentation including roots in the gum [12].

In this paper, we will give an experimental study of the effectiveness of neuromorphic neural network on multimodal brain tumor segmentation. This paper is intended for Tumor Segmentation Challenge 2018 [1–5]. Since the available dataset is limited in size, we utilized the convolutional filters developed for other applications of segmenting objects in medical images and other noisy images. Experiment results using multimodal brain MRI images show the bio-inspired convolutional filters and controlled linear rectifier neurons can boost the performance of the segmentation tasks.

2   Methods

The neuromorphic convolutional neural network in Fig. 1 is inspired by a neuromorphic neuron of simple cell neuron in visual cortex experimentation by Hubel and Wiesel, with the various orientation selective features. The system has the process of orientation feature extraction using neuromorphic processing mimicking the simple cell neuron, based on the convolution with filter banks. The introduction of down-up resizing in Fig. 1 enables the abstract feature extraction, with robust saliency map generation combined with the controlled rectifier neuron.

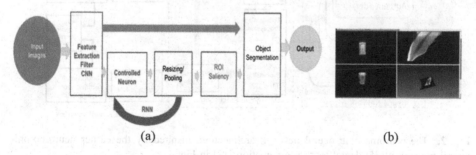

(a)   (b)

Fig. 1. (a) The earlier neuromorphic convolutional recurrent neural networks of object segmentation, inspired by the robust visual processing of visual cortex, (b) 3D tooth segmentation of dental X-ray CBCT images [12].

The neuromorphic convolutional recurrent neural network in Fig. 1 demonstrated the feasibility of segmenting object in unclear images. In principle, the front-end convolutional neural network acts as the feature extractor for the first round of RoI (Region of Interest) processing, where RoI represents the segmented area of teeth. Extracted features of orientation components improve the RoI processing substantially by removing the noisy image components of scattered non-tooth bone objects or tissues in the gum. The neuromorphic neural network in Fig. 1(a) demonstrated the successful tooth segmentation of dental X-ray Cone Beam CT images, as shown in Fig. 1(b) [12]. In this paper, the concept of neuromorphic convolution neural network and resizing in Fig. 1(a) is adopted for processing the saliency map of brain tumors, for the tumor segmentation and overall survival diagnosis.

The architecture of neuromorphic neural network in Fig. 2 is designed for brain imaging analytic, where the neuromorphic convolutional neural network is based on the pre-trained filters for 3D tooth segmentation of X-ray CT [12]. The differential processing module is introduced to evaluate the multimodal images of four different modes. The differential processing is based on image data manipulation, which is unsigned 8-bit óperation The recurrent structure of Fig. 1 is skipped for the effective configuration of differential processing stage, considering the brain matter of less solid object than tooth. Down-Up resizing neural networks provide the function saliency map generation, which is applied to the controlled rectifier neuron for brain tumor segmentation. (0.05–20) was determined for Down-Up resizing for the neuromorphic neural network for brain imaging analytic. The tumor segmentation in Fig. 2 is based on two hypotheses, when one is the transfer learning of trained neural network for deep learning and the other is the alternate brain-mimicking model to the inaccurate clinical diagnosis reference data of 23%–78%. The fully connected neural network with two segmentation data is used for predicting the overall survival (OS), utilizing the patients' record for training the network.

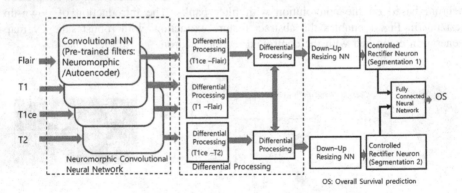

Fig. 2. The neuromorphic neural network architecture, inspired by the earlier neuromorphic neural network of 3D dental tooth segmentation [12] in Fig. 1.

The fully connected neural network in Fig. 3 is the feedforward neural network with two inputs and three hidden layers, which is 2(inputs) × 20 × 20 × 5 × 1 (output). Two input variables are total pixel numbers of each segmented images. The output represents the OS prediction of the shorter survival period ('0': less than 18 months) or the longer survival period ('1': more than 18 months). The input data of segmentation 1 and segmentation 2 are normalized for training.

The training process of neural network in Fig. 3 was based on 'MICCAI_BraTS _2018_Data_Training', where 80% of data was used for training, 5% for test, and 15% for validation.

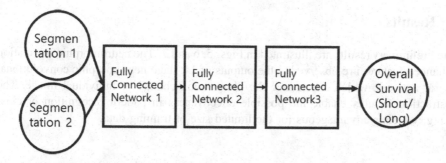

Fig. 3. The fully connected neural network for the diagnosis of overall survival – short term (less than 18 months) and long term (over 18 months).

The fully connected neural network in Fig. 4 was designed for evaluating days of overall survival period, as required by individually allocated dataset of BraTS 2018. The patients' statistics were assumed to be similar to those found in 'MICCAI_BraTS_2018_Data_Training'. The 8-bit resolution was aimed with the multiplying scale of 50. The feedforward network in Fig. 4 has 4 hidden layers and 8 output neurons, which is 2(inputs) $\times$ 50 $\times$ 50 $\times$ 30 $\times$ 30 $\times$ 8(output). The output represents the OS prediction in days after multiplying with the designated scale. The survival period for training was encoded to 8-bit after divided by 50. The input data of segmentation 1 and segmentation 2 are normalized for training, with the tuned threshold of controlled rectifier neuron. The rationale of tuning the threshold is based on the assumed consistency in patient statistics, regarding the severity of tumor.

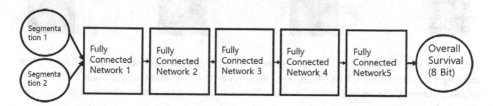

Fig. 4. The fully connected neural network for the diagnosis of overall survival period in days.

The training process of neural network in Fig. 4 was based on 'MICCAI_BraTS_ 2018_Data_Training', where 75% of data was used for training, 15% for test, and 10% for validation. The accuracy of trained network in Fig. 4 is evaluated as 71.9% based on the target-output confusion matrix.

3 Results

The preliminary results are illustrated in Figs. 5, 6 and 7. The neuromorphic orientation enhanced features are observed at the outputs of 1st stage neuromorphic convolutional neural network, which can reduce the illumination change of individual image. The abstraction features enable the possible effectiveness in pattern recognition or clustering, which is advantageous for the limited size of training data.

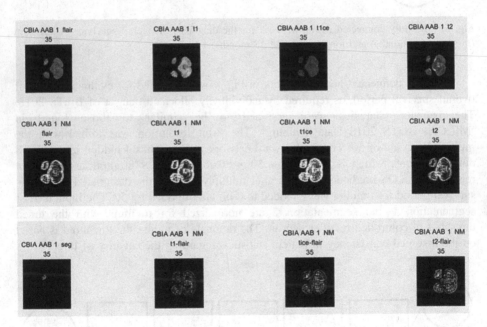

Fig. 5. Example outcomes of proposed neuromorphic neural network using BRATS 2018 training data set, (HGG, CBIA_AAB_1 A, layer no = 35). Top: multimodal brain MRI images, middle: intermediate outputs of 1st stage CNN, bottom: segmentation results (left end) and intermediate results.

The segmentation procedure utilizes the averaging and threshold process during Down-Up resizing neural network operation. The bottom left object illustrates the segmentation 1 in Fig. 5, which is some way close to the enhancing tumor in the provided ground truth. The similar functions were observed in dental tooth segmentation [12], which illustrated the automatic segmentation function by neuromorphic convolution filters and Down-Up resizing neural network. The tumor saliency maps of segmentation are automatically produced and examples are shown as 'NM seg Detected RoI' and 'NM seg Detected RoI 2' in Figs. 6 and 7.

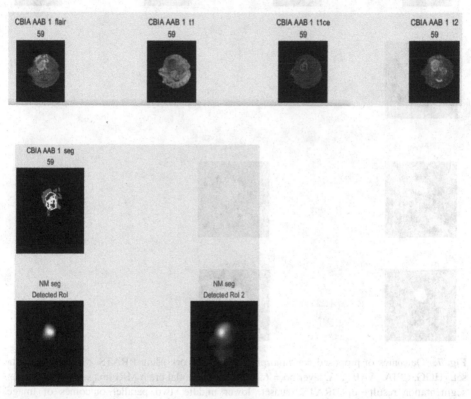

Fig. 6. Outcomes of proposed neuromorphic neural network using BRATS 2018 training data set, (HGG, CBIA_AAB_1 A, layer no = 59). Top: Multimodal brain MRI images, middle: segmentation results of BRATS dataset, bottom: two parallel outcomes of image segmentation intermediate results.

The controlled linear rectifier neurons are employed to produce the data for OS prediction, with both the tuned threshold value and the fixed threshold value. The converted saliency maps are illustrated in the bottom images in Fig. 7.

The fully connected network of Fig. 3 was trained by the limited number of 161 data sets, and the result of Table 1 is summarized. Since there is a substantial difference among human experts of tumor segmentation, it would be challenging to implement the

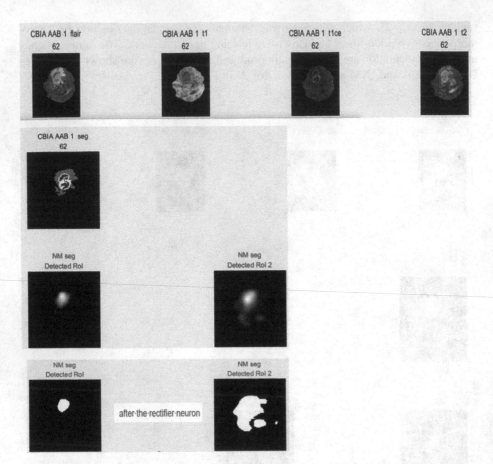

Fig. 7. Outcomes of proposed neuromorphic neural network using BRATS 2018 training data set, (HGG, CBIA_AAB_1 A, layer no = 62). Top: multimodal brain MRI images, upper middle: segmentation results of BRATS dataset, lower middle: two parallel outcomes of image segmentation intermediate results, bottom: segmented images converted by controlled rectifier neurons.

Table 1. OS of BRATS 2018 HGG dataset by neuromorphic neural network

Accuracy and sensitivity	Cases
Correct prediction	71% (115 cases among 161 cases)
Positive failure (mistaken as the long OS period: more than 18 months)	11%
Negative failure (fail to predict the short OS period: less than 18 months)	18%

system with the accurate result much better than the human expert with 23%–78% [6]. The estimated performance is around 71% of accuracy. The hypothesis other than the saliency map is kept minimum for evaluating the OS, as there exists the large variance among clinical experts.

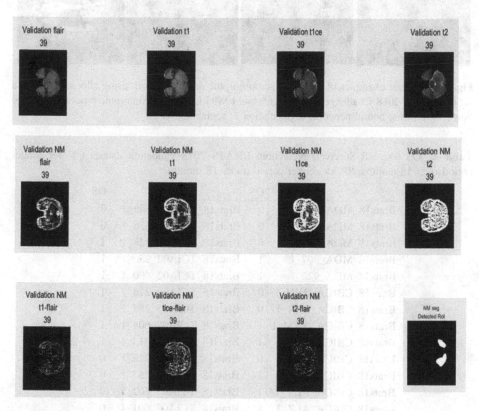

Fig. 8. Outcomes of proposed neuromorphic neural network using BRATS 2018 validation data set, Top: multimodal brain MRI images. Middle: intermediate outputs of 1st stage CNN, bottom: intermediate results and segmentation result (right end: segmentation 1 of Fig. 2)

The contents of Table 2 illustrate the OS prediction of validation data released for BRATS 2018, without the ground truth of tumor segmentation. Images of Fig. 8 represent the result using validation dataset.

The contents of Table 3 illustrate the OS prediction of allocated test data provided for BRATS 2018. Images of Fig. 9 represent the outcome examples using allocated dataset, where survival days were evaluated. The 50.6% accuracy of OS diagnosis was reported for patients' survival days prediction, according to the result from BRATS 2018 committee.

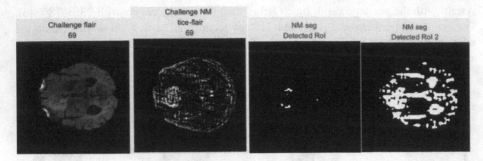

Fig. 9. Outcome examples of proposed neuromorphic neural network using allocated test data set of BRATS 2018 Challenge, left to right: brain MRI image, neuromorphic processing before down-up resizing neural network, segmentation 1, segmentation 2.

Table 2. OS (Overall Survival) prediction: BRATS 2018 validation dataset ('1' for longer period over 18 months, '0' for shorter period under 18 months

	OS		OS
Brats18_MDA_1012_1	0	Brats18_TCIA13_646_1	0
Brats18_MDA_1015_1	0	Brats18_TCIA13_636_1	0
Brats18_MDA_1081_1	0	Brats18_TCIA09_248_1	1
Brats18_MDA_907_1	1	Brats18_TCIA02_230_1	1
Brats18_MDA_922_1	1	Brats18_TCIA02_400_1	1
Brats18_CBICA_BHN_1	0	Brats18_TCIA03_216_1	1
Brats18_CBICA_BLK_1	0	Brats18_TCIA03_288_1	1
Brats18_CBICA_AAM_1	1	Brats18_TCIA03_604_1	1
Brats18_CBICA_ABT_1	1	Brats18_TCIA03_313_1	1
Brats18_CBICA_ALA_1	0	Brats18_TCIA04_212_1	0
Brats18_CBICA_ALT_1	1	Brats18_TCIA04_253_1	0
Brats18_CBICA_ALV_1	1	Brats18_TCIA07_602_1	0
Brats18_CBICA_ALZ_1	1	Brats18_TCIA07_601_1	0
Brats18_CBICA_AMF_1	0	Brats18_TCIA07_600_1	0
Brats18_CBICA_AMU_1	0	Brats18_UAB_3446_1	1
Brats18_CBICA_ANK_1	1	Brats18_UAB_3448_1	1
Brats18_CBICA_APM_1	1	Brats18_UAB_3449_1	1
Brats18_CBICA_AQE_1	1	Brats18_UAB_3454_1	1
Brats18_CBICA_ARR_1	0	Brats18_UAB_3455_1	1
Brats18_CBICA_ATW_1	0	Brats18_UAB_3456_1	1
Brats18_CBICA_AUC_1	1	Brats18_UAB_3490_1	1
Brats18_CBICA_AUE_1	0	Brats18_UAB_3498_1	1
Brats18_CBICA_AZA_1	1	Brats18_UAB_3499_1	1
Brats18_TCIA13_652_1	0	Brats18_WashU_S037_1	0
Brats18_TCIA13_638_1	0	Brats18_WashU_W033_1	1
Brats18_TCIA13_617_1	0	Brats18_WashU_W038_1	1
		Brats18_WashU_W047_1	1

Table 3. OS (Overall Survival) days prediction examples: BRATS 2018 allocated test dataset

ID	Survival days	ID	Survival days
Brats18_CBICA_AAA_1	571	Brats18_CBICA_ANJ_1	381
Brats18_CBICA_AAC_1	566	Brats18_CBICA_ANR_1	397
Brats18_CBICA_AAE_1	568	Brats18_CBICA_ANW_1	471
Brats18_CBICA_AAF_1	568	Brats18_CBICA_AOB_1	339
Brats18_CBICA_AAK_1	568	Brats18_CBICA_AOG_1	386
Brats18_CBICA_AAN_1	573	Brats18_CBICA_AOQ_1	429
Brats18_CBICA_ABH_1	572	Brats18_CBICA_APS_1	457
Brats18_CBICA_ABP_1	565	Brats18_CBICA_AQB_1	311
Brats18_CBICA_AKQ_1	571	Brats18_CBICA_AQC_1	325
Brats18_CBICA_AKY_1	327	Brats18_CBICA_AQK_1	375
Brats18_CBICA_ALP_1	460	Brats18_CBICA_ATH_1	574
Brats18_CBICA_ALW_1	468	Brats18_CBICA_AUD_1	464
Brats18_CBICA_AMA_1	459	Brats18_CBICA_AUF_1	338
Brats18_CBICA_AMB_1	457	Brats18_CBICA_AUK_1	461
Brats18_CBICA_AMD_1	392	Brats18_CBICA_AVC_1	437
Brats18_CBICA_AMG_1	425	Brats18_CBICA_AVS_1	339
Brats18_CBICA_AMI_1	462	Brats18_CBICA_AVZ_1	461
Brats18_CBICA_AMK_1	458	Brats18_CBICA_AWA_1	405
Brats18_CBICA_AMN_1	456	Brats18_CBICA_AWB_1	330
Brats18_CBICA_AMO_1	574	Brats18_CBICA_AWK_1	339
Brats18_CBICA_AMP_1	482	Brats18_CBICA_AWP_1	572
Brats18_CBICA_AMQ_1	456	Brats18_CBICA_BGF_1	433
Brats18_CBICA_AMS_1	460	Brats18_CBICA_BHI_1	462
Brats18_CBICA_AMY_1	451	Brats18_CBICA_BKX_1	423
Brats18_CBICA_ANB_1	338	Brats18_CBICA_BLE_1	348

4   Discussion

Our proposed algorithm has the feature of mimicking human visual recognition process, where the neuromorphic convolutional neural network unlikely incurs the heavy computing resources for learning. The decent size of neural network is more favorable to the fast or real-time operation for enhanced medical imaging analytics. The algorithm can be implemented by integer based computing, where early demonstrators were implemented for the real-time applications using a mobile GPU. The accuracy of overall survival (OS) prediction was evaluated as 50%–70%, by neuromorphic saliency processing of neural network.

The major finding is that the pre-trained neuromorphic neural network can diagnosis the OS without the ground truth data of tumor segmentation, which should be provided by the clinicians or radiologists. The proposed neuromorphic neural network only requires the medical care records of patients' survival days and MRI images, with the reasonable accuracy of OS diagnosis. It demonstrated the feasibility of saving the clinicians' or radiologists' time and resource for the medical care of patients.

We expect to develop the neuromorphic neural network system of improved accuracy via further training and optimization of overall network in addition to the current trained fully connected network and the pre-trained convolutional filters, delivering further features of the patient-specific and progressive response to treatment providing a longer survival.

Acknowledgement. Authors appreciate the comments of reviewers for their advice and constructive feedback to our article for the improvement.

References

1. Menze, B., et al.: The multimodal brain tumor image segmentation benchmark (BRATS). IEEE Trans. Med. Imaging **34**(10), 1993–2024 (2015)
2. Bakas, S., et al.: Advancing The Cancer Genome Atlas glioma MRI collections with expert segmentation labels and radiomic features. Nat. Sci. Data **4**, 170117 (2017)
3. Bakas, S., et al.: Segmentation labels and radiomic features for the pre-operative scans of the TCGA-GBM collection. The Cancer Imaging Archive. https://doi.org/10.7937/k9/tcia.2017. klxwjj1q (2017)
4. Bakas, S., et al: Segmentation labels and radiomic features for the pre-operative scans of the TCGA-LGG collection. The Cancer Imaging Archive. https://doi.org/10.7937/k9/tcia.2017. gjq7r0ef (2017)
5. Bakas, S., Reyes, M., et al.: Identifying the best machine learning algorithms for brain tumor segmentation, progression assessment, and overall survival prediction in the BRATS challenge, arXiv preprint: arXiv:1811.02629 (2018)
6. White, N., Reid, F., Harris, A., Harries, P., Stone, P.: A systematic review of predictions of survival in palliative care: how accurate are clinicians and who are the experts? PLOS One (2016). https://doi.org/10.1371/journal.pone.0161407
7. Cheon, S., et al.: The accuracy of clinicians' predictions of survival in advanced cancer: a review. Ann. Palliat. Med. **5**(1), 22–29 (2016). https://doi.org/10.3978/j.issn.2224-5920. 2015.08.04
8. Kondziolka, D., et al.: The accuracy of predicting survival in individual patients with cancer. AANS J. Neurosurg. **120**, 24–30 (2014)
9. Zacharaki, E., Morita, N., Bhatt, P., O'Rourke, D., Melhem, E., Davatzikos, C.: Survival analysis of patients with high-grade gliomas based on data mining of image variables. AJNR Am. J. Neuroradiol. **33**(6), 1065–1071 (2012)
10. Cao, S., et al.: 3D U-Net for multimodal brain tumor segmentation. In: 2017 International MICCAI BraTS Challenge Proceedings, pp. 30–33, Quebec City (2017)
11. Isensee, F., Kickingereder, P., Wick, W., Bendszus, M., Maier-Hein, K.H.: Brain tumor segmentation and radiomics survival prediction: contribution to the BRATS 2017 challenge. In: Crimi, A., Bakas, S., Kuijf, H., Menze, B., Reyes, M. (eds.) BrainLes 2017. LNCS, vol. 10670, pp. 287–297. Springer, Cham (2018). https://doi.org/10.1007/978-3-319-75238-9_25
12. Han, W., Han, I.: Object segmentation for vehicle video and dental CBCT by neuromorphic convolutional recurrent neural network. In: Bi, Y., Kapoor, S., Bhatia, R. (eds.) Intelligent Systems and Application, pp. 264–284. Springer, Cham (2018). https://doi.org/10.1007/978-3-319-33386-1. ISBN 978-3-319-69266-1

Glioma Segmentation
with Cascaded UNet

Dmitry Lachinov[1,2]([envelope]) [iD], Evgeny Vasiliev[1] [iD], and Vadim Turlapov[1] [iD]

[1] Lobachevsky State University, Gagarina ave. 23, 603950
Nizhny Novgorod, Russian Federation
dlachinov@gmail.com, eugene.unn@gmail.com, vadim.turlapov@gmail.com
[2] Intel, Nizhny Novgorod, Russian Federation
dmitry.lachinov@intel.com

Abstract. MRI analysis takes central position in brain tumor diagnosis and treatment, thus its precise evaluation is crucially important. However, its 3D nature imposes several challenges, so the analysis is often performed on 2D projections that reduces the complexity, but increases bias. On the other hand, time consuming 3D evaluation, like segmentation, is able to provide precise estimation of a number of valuable spatial characteristics, giving us understanding about the course of the disease.

Recent studies focusing on the segmentation task, report superior performance of Deep Learning methods compared to classical computer vision algorithms. But still, it remains a challenging problem. In this paper we present deep cascaded approach for automatic brain tumor segmentation. Similar to recent methods for object detection, our implementation is based on neural networks; we propose modifications to the 3D UNet architecture and augmentation strategy to efficiently handle multimodal MRI input, besides this we introduce approach to enhance segmentation quality with context obtained from models of the same topology operating on downscaled data. We evaluate presented approach on BraTS 2018 dataset and achieve promising results on test dataset with 14th place and Dice score of 0.720/0.878/0.785 for enhancing tumor, whole tumor and tumor core segmentation respectively.

Keywords: Segmentation · BraTS · UNet · Cascaded UNet · Multiple encoders

1 Introduction

Multimodal magnetic resonance imaging (MRI) is a powerful tool for studying human brain. Among it's different applications, it is mainly used for disease diagnosis and treatment planning. Accurate assessment of MRI results is critical throughout all these steps. Since MRI scans are the set of multiple three dimensional arrays, it's manual analysis and evaluation is a non-trivial procedure and requires time, attention and expertise. Lack of these resources can lead to unsatisfying results. Typically, these scans are analyzed by clinical experts using two

© Springer Nature Switzerland AG 2019
A. Crimi et al. (Eds.): BrainLes 2018, LNCS 11384, pp. 189–198, 2019.
https://doi.org/10.1007/978-3-030-11726-9_17

dimensional cut and projection planes. It limits the amount of data taken into account for decision making, thus it adds bias to the resulting evaluation. On the other hand, accurate segmentation and 3D reconstruction is able to provide more insights on disease progression and help a therapist to plan the treatment better. However these methods are not widely used due to unreasonable amount of time needed for manual labeling.

Denoting the problem of automatic glioma segmentation Brain Tumor Segmentation (BraTS) challenge [1,11] was created and became an annual competition allowing participants to evaluate and compare their state of the art methods using unified framework. Participants are called to develop their algorithms and produce segmentation labels of the different glioma sub-regions: "enhancing tumor" (ET), "tumor core" (TC) and "whole tumor" (WT). The training data [2,3] consists of 210 high grade and 75 low grade glioma MRIs manually labeled by experts in the field. Testing data is split into two parts: **validation set** that can be used for evaluation throughout the challenge and **test set** for final evaluation. Performance of the methods is measured using Dice coefficient, Sensitivity, Specificity and Hausdorff distance.

Above-named challenge made a significant impact on the evolution of computational approaches for tumor segmentation. In the last few years, a variety of algorithms were proposed to solve this problem. Compared with other methods, convolutional neural networks have been showing the best state of the art performance for computer vision tasks in general and for biomedical image processing tasks in particular.

In this paper we present cascaded variant of the popular UNet network [6,12] that iteratively refines segmentation results of it's previous stages. We employ this approach for brain tumor segmentation task in the scope of BRATS 2018 challenge and evaluate it's performance. We also compare regular 3D UNet [6] with it's cascaded counterpart.

2   Method

In this study we propose neural networks based approach for brain tumor segmentation. Our method can be represented as a chain of multiple classifiers C_i of the same topology F refining segmentation output of previous iterations. Every classifier C_i shares the same topology but has it's own set of parameters W_i that is subject to optimization during training. Y_i - the result of the i-th step can be represented as $Y_i = F(X_i, Y_{i-1}, Y_{i-2}, W_i)$, where X_i is the i-th input.

Described approach is illustrated in Fig. 1. Each of the basic blocks C_i is a UNet network modified with respect to the task of glioma segmentation. Compared to the original UNet architecture described in [12] and extended for 3D case in [6], we employ multiple encoders separately handling input modalities and introduce the way to merge their output. In this paper we describe UNet modification with multiple encoders first. Then we propose ensembling strategy to efficiently merge segmentation results obtained on different scales.

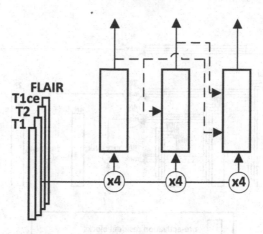

Fig. 1. Schematic representation of approach employed in this paper. T1, T2, T1ce, FLAIR stands for input MRI modalities. x4, x2 indicate downsampling factor for the network input. Dotted arrows indicate connections between networks C_i that are illustrated as basic blocks.

2.1 Multiple Encoders UNet

Traditional UNet architecture [12] extended for handling volumetric input [6] has two stages: encoder part where network learns feature representations on different scales and aggregates contextual information, and decoder part where network extracts information from observed context and previously learned features. Skip connections employed between corresponding encoder and decoder layers enable efficient training of the deep parts of the network and comparison of identically scaled features with different receptive fields.

This method allows to handle multimodal MRI input, however, it mixes and processes signals of different types identically. In contrast, we propose approach that learns feature representations for every modality separately and combines them at later stages. This is achieved by employing grouped convolutions in the encoder path with number of groups equals to the number of input modalities. Resulting features are calculated as a maximum of the feature maps produced by encoders. In order to preserve feature maps' sizes we employ point-wise convolution right after max operation. Similar to the original UNet, the number of filters is doubled with every downsampling operation and reduced by half with every upsampling operation, ReLU is used as activation function after every convolution layer. Described architecture is illustrated in Fig. 2.

The network is built of basic pre-activation residual blocks [7] that consist of two instance normalization layers, two relu activation layers and two convolutions with kernel size 3. This basic building block is illustrated in Fig. 3.

The motivation behind this architecture is to encourage model to extract features separately for every modality. In combination with feature maps merging strategy and channel-out augmentation it allows to build more robust model that can process data with one or more corrupted modalities (Fig. 4).

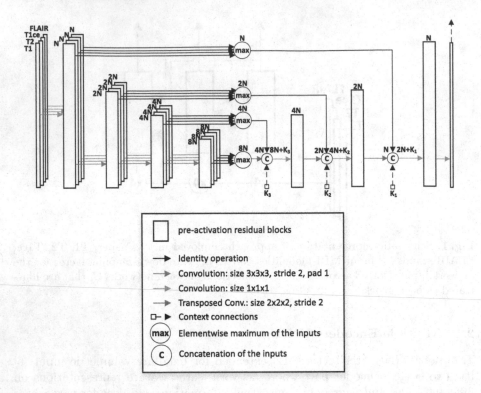

Fig. 2. Architecture of multiple encoders UNet. T1, T2, T1CE, FLAIR stand for input modalities. N is a base number of filters, K is a number of filters in context feature map obtained from lower scale models.

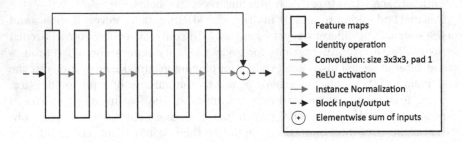

Fig. 3. Design of the residual block

Cascaded UNet. Proposed network is illustrated in Fig. 1 and consists of three basic blocks. Each block by itself is a modified UNet network with it's own loss function at the end. Every next block takes downsampled volume as an input and produces segmentation of the corresponding size. Similar to DeepMedic [10], this architecture simultaneously processes the input image at multiple scales and extracts scale-specific features. The feature map before the last convolutional layer in every block is concatenated to the corresponding feature map of higher-scale block. It enables the context information flow between networks with different scales.

In UNet architecture decoder output at each scale i depends on encoder output at the same scale (skip connections) and decoder output of the previous scale: $d_i^t = f(e_i^t, d_{i-1}^t)$, where d_i^t is decoder output, and e_i^t is encoder output at scale i, and t is the index of the network. Expanding the first convolution of f we get $d_i^t = g(W_{i,e}^t e_i^t + W_{i,d}^t d_{i-1}^t)$, where W are trainable parameters. Here we propose to incorporate context of the lower scale networks by concatenating corresponding network output y^t (see Fig. 2, illustrated as dotted arrows) so d_i^t becomes $d_i^t = g(W_{i,e}^t e_i^t + W_{i,d}^t d_{i-1}^t + W_{i,y}^t y^{t-i})$. This approach fuses multiple networks operating at different scales together and encourages model to iteratevily refine results of previous iterations.

The connections between networks are illustrated as dotted arrows in Fig. 1. Each basic UNet network produces two outputs: feature map (dotted arrows) and softmax operation over this feature map (straight arrows). The resulting probability tensor can be further used for ensembling, yet, we are interested in a final feature map. Since it has the most meaningful information about segmentation on the given scale, we want to propagate this feature map to higher resolution networks. To achieve the flow of the context between classifiers of different scale we propose to concatenate their output feature map to corresponding feature map of the higher scale network (see Fig. 2, illustrated as dotted arrows).

By employing following ensembling strategy we are building quite deep convolutional neural network. Compared to standard approach of doubling the number of feature channels after each pooling operator, out method takes less parameters and introduces bottlenecks between networks. Having same number of parameters, presented approach performs better than models with the same depth or the same number of parameters.

2.2   Data

In this paper we are focusing at brain tumor segmentation with deep neural networks. For training and evaluation purposes we are using BraTS 2018 [1–3, 11] dataset. It contains clinically acquired preoperative multimodal MRI scans of glioblastoma and lower grade glioma obtained in different institutions with different protocols. These multimodal scans contain native T1, post-contrast T1-weighted, T2-weighted, and T2 Fluid Attenuated Inversion Recovery (FLAIR) volumes, and co-registered to the same anatomical template, interpolated to the same resolution $(1mm^3)$ and skull-stripped. These MRI scans were manually annotated by one to four raters, and approved by experienced radiologist. Segmentation labels describe different glioma sub-regions: "enhancing tumor" (ET), "tumor core" (TC) and "whole tumor" (WT). In total, dataset has 285 MRIs for training (210 high grade and 75 low grade glioma images), 67 validation and 192 testing MRIs.

2.3   Preprocessing and Data Augmentation

We have found data preprocessing employed in [8] to be especially effective. Like in [8], we perform z-score normalization on non-zero (brain) voxels. After that we are eliminating outliers and noise by clamping all values to the range from –5 to 5. At the final step we shift brain voxels to the range [0;10] and assign zeros to background.

For offline data augmentation we artificially increase number of samples by employing b-spline transformation to the original data. It has been done with ITK implementation [9].

During training we randomly flip input image along sagittal plane and "mute" input modalities with predefined probability. Without this augmentation the network was only considering one of the input modalities while making a prediction and not taking others into account. To deal with this issue we are randomly filling input channels with Gaussian noise. We introduce probability to apply this augmentation for every channel and set it to 0.1, so there is 34% chance to mute at least one out of four modalities. This also helps to aggregate information allover input data and to deal with noisy or corrupted input images like illustrated in the Fig. 4.

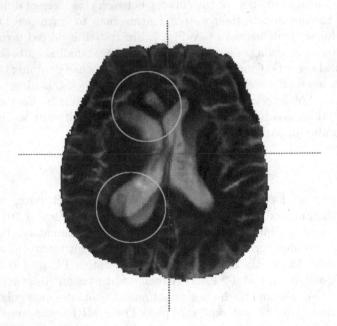

Fig. 4. Example of the registration artifacts found in the training dataset. This series contain one corrupted modality (shown) and three correct ones. Overlapping structures of the brain are marked with red circles. Visualization is done with ITK-SNAP [13]. (Color figure online)

2.4   Training

The training procedure is conducted on brain regions resampled to $128 \times 128 \times 128$ voxels. We are operating with downsampled data to preserve the context since we believe it plays important role for robust segmentation of multimodal MRI scans obtained from different institutions and scanners. We use Mean Dice loss $L_{mean\_dice}(g, p)$ where g is a ground truth, p is a model's prediction. We trained our network with stochastic gradient descent with initial learning rate of 0.1, exponential learning rate decay with rate 0.99 for every epoch, weight decay of 0.9 and minibatch size equal to 4 samples.

$$L_{mean\_dice}(g, p) = 1 - \frac{1}{|C|} \sum_{c \in C} \frac{\sum_{i_c} p_c^i g_c^i}{\sum_{i_c} p_c^i + g_c^i},$$

where C is a set of different classes.

This CNN was implemented in MXNet framework [5] and trained using four GTX 1080TI with batch size 4 to enable data parallelism. Training was performed for 500 epoches.

3   Results

In this section we report evaluation results obtained with online validation system provided by organizers. With intention to penalize model for relying on the one single modality we apply channel-out augmentation to the input data by randomly filling input modalities with Gaussian noise in addition to standard augmentations like mirroring and elastic transformations. Then we compare results obtained with this augmentation disabled (Table 1) and enabled (Table 2). The challenge validation data [2,3] contains 66 MRI scans obtained with different scanners and from different institutions. Results of evaluation on validation dataset are reported in Table 3; and on test dataset in Table 4.

Table 1. Evaluation of glioma segmentation without channel-out augmentation; Dice index is reported, WT stands for whole tumor, ET stands for enhancing tumor, TC stands for tumor core, ME UNet stands for Multiple Encoders UNet and C ME UNet stands for Cascaded Multiple Encoders UNet. Tested networks has the same number of parameters.

Method	WT	ET	TC
UNet	0.901	0.767	0.797
ME UNet	0.904	0.763	0.823
C ME UNet	0.906	0.772	0.836

Table 2. Evaluation of glioma segmentation with channel-out augmentation; Dice index is reported, WT stands for whole tumor, ET stands for enhancing tumor, TC stands for tumor core, ME UNet stands for Multiple Encoders UNet and C ME UNet stands for Cascaded Multiple Encoders UNet. Tested networks has the same number of parameters.

Method	WT	ET	TC
UNet	0.901	0.779	0.837
ME UNet	0.907	0.784	0.827
C ME UNet	0.908	0.784	0.844

Table 3. Performance of proposed method on BraTS 2018 validation data, Dice index is reported.

	WT	ET	TC
Mean	0.908	0.784	0.844
StdDev	0.065	0.237	0.161
Median	0.926	0.858	0.906
25quantile	0.900	0.805	0.791
75quantile	0.943	0.897	0.947

Table 4. Performance of proposed method on BraTS 2018 test data, Dice index is reported.

	WT	ET	TC
Mean	0.878	0.720	0.795
StdDev	0.119	0.278	0.251
Median	0.913	0.818	0.901
25quantile	0.870	0.711	0.804
75quantile	0.940	0.877	0.936

Ground truth   Prediction

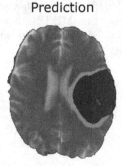

Fig. 5. Example of segmentation labels produces by proposed method in comparison with ground truth annotation.

4   Discussion and Conclusion

Analyzing the segmentation results provided by out model (Fig. 5) we noticed that it produces more smooth results compared to ground truth. According to BraTS 2018 challenge summarizing manuscript [4], out method took 14th place in final ranking. Analyzing the results we found out model to produce high number of inaccurate enhancing tumor segmentation labels (24th rank by DICE ET). This issue could be potentially overcame with learning ET, TC, WT labels instead of labels provided by annotation. However our model showed relatively high score for segmentation of Tumor Core (11th place by DICE TC) and Whole Tumor (10th place by DICE WT). Furthermore, it achieved ranks as high as 9th, 5th, 12th for segmentation of ET, TC, WT w.r.t. Hausdorff distance.

To sum it up, in this paper we presented automatic segmentation algorithm solving two main problem arising during brain tumor segmentation with multi-modal scans: complex input consisting of multiple modalities and overconfidence of the classifier. Solving the problem of heterogeneous input we proposed to use multiple encoders, so that every individual input modality produces corresponding feature maps independently from others; and we introduced the way to merge encoded feature maps. Also we explored influence of channel-out augmentation on model's output quality and we showed that proposed architecture benefits from this aggressive augmentation. It encourages model to take into account whole input by implicitly penalizing classifiers that rely only on one single modality. As a result model becomes robust to the presence of noise and corrupted data that could be encountered in the training and validation datasets. Moreover we introduced the way to efficiently fuse multiple models operating on the different resolution that forms a cascade of classifiers. Every next classifiers takes results of previous ones and refines the segmentation for it's specific scale. It enables iterative result refinement with less parameters than in corresponding deep models. As a part of BraTS 2018 challenge [1,11] we implemented and evaluated our approach with online validation tools. As a result we achieved high mean score and notably high median score. The mean Dice score of 0.878/0.72/0.795 was reported on testing dataset for the Whole tumor, Enhancing tumor and Tumor core correspondingly

References

1. Bakas, S., et al.: Advancing the cancer genome atlas glioma MRI collections with expert segmentation labels and radiomic features. Sci Data 4, 170117 (2017). https://doi.org/10.1038/sdata.2017.117, http://www.ncbi.nlm.nih.gov/pmc/articles/PMC5685212/, 28872634[pmid]
2. Bakas, S., et al.: Segmentation labels and radiomic features for the pre-operative scans of the TCGA-GBM collection. The Cancer Imaging Archive (2017). https://doi.org/10.7937/K9/TCIA.2017.KLXWJJ1Q
3. Bakas, S., et al.: Segmentation labels and radiomic features for the pre-operative scans of the TCGA-LGG collection. The Cancer Imaging Archive (2017). https://doi.org/10.7937/K9/TCIA.2017.GJQ7R0EF

4. Bakas, S., Reyes, M., et al.: Identifying the best machine learning algorithms for brain tumor segmentation, progression assessment, and overall survival prediction in the brats challenge. CoRR abs/1811.02629 (2018). http://arxiv.org/abs/1811. 02629

5. Chen, T., et al.: Mxnet: a flexible and efficient machine learning library for heterogeneous distributed systems. CoRR abs/1512.01274 (2015). http://arxiv.org/abs/ 1512.01274

6. Çiçek, Ö., Abdulkadir, A., Lienkamp, S.S., Brox, T., Ronneberger, O.: 3D U-Net: learning dense volumetric segmentation from sparse annotation. In: Ourselin, S., Joskowicz, L., Sabuncu, M.R., Unal, G., Wells, W. (eds.) MICCAI 2016. LNCS, vol. 9901, pp. 424–432. Springer, Cham (2016). https://doi.org/10.1007/978-3-319-46723-8_49

7. He, K., Zhang, X., Ren, S., Sun, J.: Deep residual learning for image recognition. In: 2016 IEEE Conference on Computer Vision and Pattern Recognition (CVPR), pp. 770–778, June 2016. https://doi.org/10.1109/CVPR.2016.90

8. Isensee, F., Kickingereder, P., Wick, W., Bendszus, M., Maier-Hein, K.H.: Brain tumor segmentation and radiomics survival prediction: contribution to the BRATS 2017 challenge. In: Crimi, A., Bakas, S., Kuijf, H., Menze, B., Reyes, M. (eds.) BrainLes 2017. LNCS, vol. 10670, pp. 287–297. Springer, Cham (2018). https:// doi.org/10.1007/978-3-319-75238-9_25

9. Johnson, H.J., McCormick, M., Ibáñez, L., Consortium, T.I.S.: The ITK Software Guide. Kitware Inc, third edn. (2013, In press). http://www.itk.org/ ItkSoftwareGuide.pdf

10. Kamnitsas, K., et al.: Efficient multi-scale 3D CNN with fully connected CRF for accurate brain lesion segmentation. CoRR abs/1603.05959 (2016). http://arxiv. org/abs/1603.05959

11. Menze, B.H., et al.: The multimodal brain tumor image segmentation benchmark (brats). IEEE Trans. Med. Imaging 34(10), 1993–2024 (2015). https://doi.org/10. 1109/TMI.2014.2377694

12. Ronneberger, O., Fischer, P., Brox, T.: U-Net: convolutional networks for biomedical image segmentation. In: Navab, N., Hornegger, J., Wells, W.M., Frangi, A.F. (eds.) MICCAI 2015. LNCS, vol. 9351, pp. 234–241. Springer, Cham (2015). https://doi.org/10.1007/978-3-319-24574-4_28

13. Yushkevich, P.A., et al.: User-guided 3D active contour segmentation of anatomical structures: significantly improved efficiency and reliability. Neuroimage 31(3), 1116–1128 (2006)

Segmentation of Gliomas and Prediction of Patient Overall Survival: A Simple and Fast Procedure

Elodie Puybareau$^{(\boxtimes)}$, Guillaume Tochon, Joseph Chazalon,
and Jonathan Fabrizio

EPITA Research and Development Laboratory (LRDE), Le Kremlin-Bicêtre, France
elodie.puybareau@lrde.epita.fr

Abstract. This paper proposes, in the context of brain tumor study, a fast automatic method that segments tumors and predicts patient overall survival. The segmentation stage is implemented using a fully convolutional network based on VGG-16, pre-trained on ImageNet for natural image classification, and fine tuned with the training dataset of the MICCAI 2018 BraTS Challenge. It relies on the "pseudo-3D" method published at ICIP 2017, which allows for segmenting objects from 2D color-like images which contain 3D information of MRI volumes. With such a technique, the segmentation of a 3D volume takes only a few seconds. The prediction stage is implemented using Random Forests. It only requires a predicted segmentation of the tumor and a homemade atlas. Its simplicity allows to train it with very few examples and it can be used after any segmentation process. The presented method won the second place of the MICCAI 2018 BraTS Challenge for the overall survival prediction task. A Docker image is publicly available on https://www.lrde.epita.fr/wiki/NeoBrainSeg.

Keywords: Glioma · Tumor segmentation · Survival prediction ·
Fully convolutional network · Random forest

1 Introduction

Gliomas are the most common brain tumors in adults, growing from glial cells and invading the surrounding tissues [9]. Two classes of tumors are observed. The patients with the more aggressive ones, classified as high-grade gliomas (HGG), have a median overall survival of two years or less and imply immediate treatment [13,16]. The less aggressive ones, the low-grade gliomas (LGG), allow an overall survival of several years, with no need of immediate treatment. Multimodal magnetic resonance imaging (MRI) helps practitioners to evaluate the degree of the disease, its evolution and the response to treatment. Images are analyzed based on qualitative or quantitative measures of the lesion [8,21]. Developing automated brain tumor segmentation techniques that are able to

© Springer Nature Switzerland AG 2019
A. Crimi et al. (Eds.): BrainLes 2018, LNCS 11384, pp. 199–209, 2019.
https://doi.org/10.1007/978-3-030-11726-9_18

analyze these tumors is challenging, because of the highly heterogeneous appearance and shapes of these lesions. Manual segmentations by experts can also be a challenging task, as they show significant variations in some cases. During the past 20 years, different algorithms for segmentation of tumor structures has been developed and reviewed [1,6,7]. A fair comparison of those implies a benchmark based on the same dataset, and MICCAI BraTS Challenges [15] serve this purpose.

The work we present here has been done in the context of the MICCAI 2018 Multimodal Brain Tumor Segmentation Challenge (BraTS)[1]. The goal of the challenge was to provide a fully automated pipeline for the segmentation of the glioma from multi modal MRI scans without any manual assistance and to predict the patient overall survival. Despite the relevance of glioma segmentation, this task is challenging due to the high heterogeneity of tumors. The development of an algorithm that can perform fully automatic glioma segmentation and overall prediction of survival would be an important improvement for patients and practitioners. A review and results of the 2018 Challenge can be found in [5].

During the challenge, multiple datasets were provided with different volumes (T1, T1ce, T2 and FLAIR):

- a training dataset of 285 patients preprocessed and with ground truth annotated [2–4],
- a dataset without public ground truth but with the possibility to evaluate online our method and obtain preliminary results,
- a final dataset without ground truth, used to rank the participants.

Our contribution is composed of two independent modules: one for tumor segmentation and one for survival time prediction. The tumor segmentation module (Sect. 2.1) blends ideas from two previous publications. It first builds on a work published in the IEEE Intl. Conf. on Image Processing (ICIP) in 2017 [22], which proposed to segment 3D brain MR volumes using fully convolutional network (FCN). It leveraged transfer learning thanks to a VGG network [18] pre-trained on the ImageNet dataset and later fine-tuned on the training set of the challenge. Its input were 2D color-like images composed of 3 consecutive slices of the 3D volume (see Fig. 1). This method used only one modality, and reached good results for brain segmentation with a decent speed. Based on this architecture, we incorporated the ideas of [24] which reused this architecture to take slices from several modalities as input. Our final segmentation solution provides specially designed pre- and post-processing dedicated to the challenge and makes use of both local 3D and multi-modal information. The survival prediction module (Sect. 2.2) we introduce here is based on Random Forests and relies on a very light training to cope with the limited amount of examples available in the challenge. Despite its apparent simplicity, it provides a reasonable survival time estimate as reported in the results (Sect. 3).

---

[1] https://www.med.upenn.edu/sbia/brats2018.html.

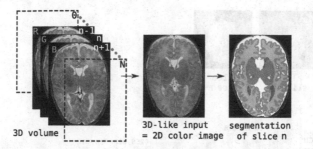

Fig. 1. Illustration of 3D-like color image and associated segmentation used in [22]. (Color figure online)

2   Method

This section describes the method we submitted to the MICCAI 2018 Multimodal Brain Tumor Segmentation Challenge (BraTS). As it was previously mentioned, it is composed of a tumor segmentation module and a survival prediction module, mapping the two tasks of the challenge.

2.1   Tumor Segmentation

An overview of the proposed segmentation method is given in Fig. 2. The method is fully automatic, and takes pseudo-3D images as input. It is really fast as about 10 s are needed to process a complete volume with a GPU-equipped machine. It consists in three sub-stages: a data pre-processing, a deep network inference, and a segmentation post-processing.

Pre-processing. We first normalize the input data to fit in the range imposed by the original network (before fine tuning). Let n, m be respectively the minimum non-null and maximum gray-level value of the histogram. We requantize all voxel values using a linear function so that the gray-level range $[n, m]$ is mapped to $[-127, 127]$.

Then, as our inference network processes 2D color-like images (3 channels of 2D slices), the question amounts to how to prepare appropriate inputs given that a brain MR image is a 3D volume. Our second step is therefore to stack successive 2D slices: for each nth slice of the volume to segment, we consider three images corresponding to the $(n-1)^{\text{th}}$, n^{th}, and $(n+1)^{\text{th}}$ slices of the original volume. These three gray-level 2D images are assembled to form a 2D color-like image (one image per channel). Each 2D color-like image thus forms a representation of a part (a small volume) of the MR volume. This image is the input of the FCN to obtain a 2D segmentation of the nth slice. This process is depicted in Fig. 2 (left).

To combine information from different modalities, we complete this process. The nth slice is taken from one modality and its $(n-1)^{\text{th}}$ and $(n+1)^{\text{th}}$ from another one. This combination brings not only 3D information but also multimodality information. Figure 3 illustrates this variant.

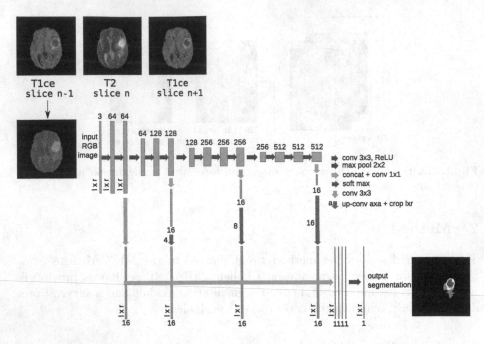

Fig. 2. Architecture of the proposed network. We fine tune it and combine linearly fine to coarse feature maps of the pre-trained VGG network [18]. Note that each input color image is built from the slice n and its neighboring slices $n-1$ and $n+1$. (Color figure online)

Deep FCN for Tumor Segmentation. Fully convolutional network (FCN) and transfer learning has proved their efficiency for natural image segmentation [12]. In a previous work, Xu et al. [22] proposed to rely on a FCN and transfer learning to segment 3D brain MR images, although those images are very different from natural images. As it was a success, we adapted it to glioma segmentation. We rely on the 16-layer VGG network [18], which was pre-trained on millions of natural images from ImageNet for image classification [11]. For our application, we keep only the 4 stages of convolutional parts called "*base network*", and we discard the fully connected layers at the end of VGG network. This base network is mainly composed of convolutional layers, Rectified Linear Unit (ReLU) layers for non-linear activation function, and max-pooling layers between two successive stages. The three max-pooling layers divide the base network into four stages of fine to coarse feature maps. Inspired by the works in [12,14], we add specialized convolutional layers (with a 3×3 kernel size) with K (e.g. $K = 16$) feature maps after the convolutional layers at the end of each stage. All the specialized layers are then rescaled to the original image size, and concatenated together. We add a last convolutional layer with kernel size 1×1 at the end. This last layer combine linearly the fine to coarse feature maps in the concatenated specialized layers, and provide the final segmentation result. The proposed network architecture is illustrated in Fig. 2. This architecture is

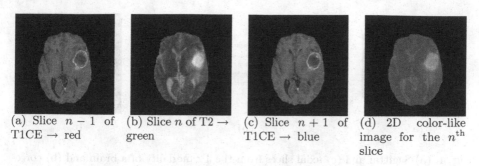

(a) Slice $n - 1$ of T1CE → red

(b) Slice n of T2 → green

(c) Slice $n + 1$ of T1CE → blue

(d) 2D color-like image for the nth slice

Fig. 3. Three successive slices (a–c) are used to build a 2D color-like image (d) from for example T1CE and T2 images. (Color figure online)

also very similar with the one used in [14] for retinal image analysis, where the retinal images are already 2D color-like images. Using such a 2D representation avoids the expensive computational and memory requirements of fully 3D FCN.

For the training phase, we use the multinomial logistic loss function for a one-of-many classification task, passing real-valued predictions through a softmax to get a probability distribution over classes. During training, we use the classical data augmentation strategy by scaling and rotating. We rely on the ADAM optimization procedure [10] (AMSGrad variant [17]) to minimize the loss of the network. The relevant parameters of the methods are the following: the learning rate is set to 0.002 (we did not use learning rate decay), the beta_1 and beta_2 are respectively set to 0.9 and 0.999, and we use a fuzz factor (epsilon) of 0.001.

At test time, after having pre-processed the 3D volume, we prepare the set of 2D color-like images and pass every image through the network. We run the train and test phase on an NVIDIA GPU. The testing one lasts less than 10 s for a complete volume.

Post-processing. The output of the network for one slice during the inference phase is a 2D segmented slice. After treating all the slices of the volume, all the segmented slices are stacked to recover a 3D volume with the same shape as the initial volume, and containing only the segmented lesions.

This segmentation procedure is repeated three times as we slice the initial volume three times (along the three axis). We get three different segmentations and we merge them to get the final segmentation by a majority voting procedure.

Then, as a final step, we regularize the segmented volumes using a morphological closing to fill small holes lying within tumor regions.

2.2 Patient Survival Prediction

The second task of the MICCAI 2018 BraTS challenge is concerned with the prediction of patient overall survival from pre-operative scans (only for subjects with gross total resection (GTR) status). Note that, to comply with the evaluation framework, the classification procedure is conducted by labeling subjects

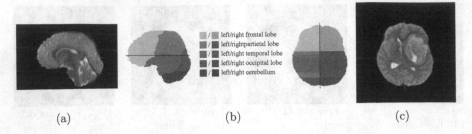

Fig. 4. (a) Sagittal and (c) axial slices from the T2 modality of a brain and (b) corresponding rescaled brain atlas.

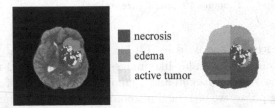

Fig. 5. T2 slice (left) and corresponding atlas slice (right) with segmented tumor overlaid. (Color figure online)

into three classes: short-survivors (less than 10 months), mid-survivors (between 10 and 15 months) and long-survivors (greater than 15 months).

Definition and Extraction of Relevant Features. The first step of the prediction task is the definition and extraction of relevant features impacting the survival of patients. Beside the patient age, we decided to focus on the tumor size and its localization within the brain. More specifically, we denote by S_i the segmented volume predicted by our Deep FCN architecture, as described in Sect. 2.1 for the ith patient. Voxels in S_i are labeled by 1, 2 and 4 (corresponding to ■, ■ and ▢ in Fig. 5, respectively), depending whether they were classified as necrosis, edema or active tumor, respectively.

Thus, we define the relative size of each class in S_i with respect to the total brain size (the number of non-zero voxels in the patient T2 modality) as the features related to the tumor size.

In order to describe the tumor position, we created a crude brain atlas divided in 10 regions accounting for the frontal, parietal, temporal and occipital lobes and the cerebellum for each hemisphere, as displayed by Fig. 4(b). The 3D atlas was first shaped to the average bounding box dimensions of all patients with GTR status, *i.e.* $170 \times 140 \times 140$ pixels. It is then adjusted to each patient bounding box dimensions by nearest-neighbors interpolation, and finally masked by all non-zero voxels in the patient T2 modality. Finally, we retrieve the centroid coordinates of the region within the atlas that is affected the most by the necrosis

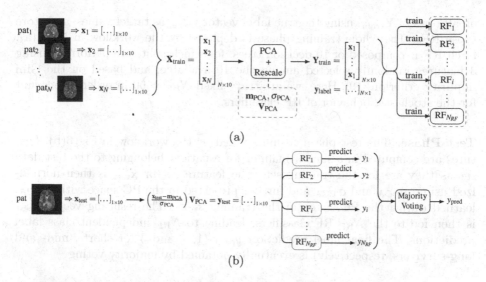

(a)

(b)

Fig. 6. (a) Training and (b) test procedures. The stored information after the training phase is encircled in dashed red in the training workflow (a). (Color figure online)

(*i.e.*, the region that has the most voxels labeled as necrosis in S_i with respect to its own size) relatively to the brain bounding box as well as the relative centroid coordinates of the necrosis + active tumor and defined those as the relevant features accounting for the tumor position.

In summary, each patient is defined by the following 6 criteria:

1. the patient age.
2. the relative size of necrosis with respect to brain size.
3. the relative size of edema with respect to brain size.
4. the relative size of active tumor with respect to brain size.
5. the relative centroid coordinates of the region in the atlas that is the most affected by necrosis with respect to the brain bounding box.
6. the relative centroid coordinates of the binarized tumor (only considering necrosis and active tumor) with respect to the brain bounding box.

This leads to a total of **10 features per patient** (since both centroids coordinates are 3-dimensional).

Training Phase. For the training phase, we first extract the feature vector $\mathbf{x}_i \in \mathbb{R}^{10}$ of each of the N patients in the training set (with $N = 59$), as described in Sect. 2.2 above. All those feature vectors are stacked in a $N \times 10$ feature matrix $\mathbf{X}_{\text{train}}$ on which a principal component analysis (PCA) is performed. The feature-wise mean $\mathbf{m}_{\text{PCA}}$ and standard deviation $\boldsymbol{\sigma}_{\text{PCA}}$ vectors computed during the scaling phase of the PCA, as well as the projection matrix $\mathbf{V}_{\text{PCA}}$ are stored for further use. Finally, the PCA output is normalized again, yielding the $N \times 10$ matrix $\mathbf{Y}_{train}$. Finally, we train N_{RF} random forest (RF) classifiers [20] on all rows of $\mathbf{Y}_{\text{train}}$, using the true label vector $\mathbf{y}_{\text{label}}$ as target values, and store those RFs. The whole training phase is depicted by the workflow in Fig. 6(a). Each RF is composed of 10 decision trees, for which splits are performed using 3 features randomly selected among the 10 available, and based on the Gini impurity criterion [19]. Here, we arbitrary fixed $N_{\text{RF}} = 50$ in order to account for the stochastic behavior of RF classifiers.

Test Phase. The test phase is summarized by the workflow in Fig. 6(b). Features are computed in a similar manner for a patient belonging to the test data set as they are for the training set. The feature vector $\mathbf{x}_{\text{test}}$ is then normalized using $\mathbf{m}_{\text{PCA}}$ and $\boldsymbol{\sigma}_{\text{PCA}}$ and further projected in the PC space with $\mathbf{V}_{\text{PCA}}$, learned during the PCA step of the training stage. The resulting vector $\mathbf{y}_{\text{test}}$ is then fed to the N_{RF} RF classifiers, leading to N_{RF} independent class label predictions. The final label prediction y_{pred} (1, 2 and 3 for short-, mid- and long-survivors, respectively) is eventually obtained by majority voting.

3   Setup and Results

This section presents the setup of the experiments and results obtained during the development of our method (using the training dataset), and the final ranking during the challenge.

3.1   Setup and Experiments for Tumor Segmentation

In this part, we used only the training scans provided during the challenge.

Modalities. Instead of using only one modality to form the pseudo-3D color images (the input of the network), we formed multi-modality pseudo-3D images using T1ce and T2 modalities: for each slice n, we combined the slice n of T2 with the slices $n-1$ and $n+1$ from T1ce.

Axis and Combination. Our method deals with 2D color-like images that are pseudo-3D. To take advantage of the entire volume, we associated three networks, each network being trained on a particular axis (axial, sagittal and coronal), and combined their results to obtain the final segmentation. We trained 3 networks, one for each axis. The inference was done according to the axis, so for one volume we obtained 3 segmentation. These segmentations are then combined: for each voxel, the final segmentation is the result of the majority voting procedure.

Training and Testing. To train our model we select randomly 90% of scans from Brats challenge training dataset. The model was trained using the parameters described in the Sect. 2.1. We tested on the 10% remaining scans.

3.2   Results

Tumor Segmentation. The results on the training dataset evaluate to a dice of 0.82 for the whole tumor segmentation (evaluation on 10% of the training set). More precisely, we obtained for the 3 classes: 0.6 for the GD-enhancing tumor, 0.63 for the peritumoral edema, and 0.56 for the necrotic and non-enhancing tumor core. We did not achieved a good ranking during the challenge for this task. Precise results can be found in [5].

Survival Time Prediction. For the survival prediction task we obtained an accuracy of 0.54 on the training dataset. During the challenge, we obtained an accuracy of 0.61. This allowed us to reach the **2nd place of the challenge for the survival prediction task.**

3.3   Discussion

The prediction task is the final aim of the entire pipeline. The segmentation task is a basis for the prediction task but is not a finality. We developed the prediction procedure using the ground truth segmentations, so that our prediction method is independent from our segmentation method. We can notice that the prediction procedure can deal with precise segmentations (i.e. ground truth segmentations) and coarser ones (such as our segmentation results).

This is the main advantage of our prediction method as it does not require a lot of data: it relies on a coarse segmentation and a brain atlas. Simple descriptors are extracted to perform the prediction. A strong point which differentiates our method from others is that our method does not rely on a specific modality to work: the method relies on a segmentation result, regardless of how it has been obtained, and not directly on a modality. It can be used without the constraints of working on one modality or an other. Furthermore the segmentation does not need to be precise to permits the prediction as illustrated by our results during the challenge.

4   Conclusion

We proposed in this article a method to first segment glioma in few seconds based on transfer learning from VGG-16, a pre-trained network used to classify natural images, and then to predict the survival time of the patients. This segmentation method takes the advantage of keeping 3D information of the MRI volume and the speed of processing only 2D images, thanks to the pseudo-3D concept while the prediction method uses only a segmentation result and a homemade brain atlas.

This method can also deal with multi-modality, and can be applied to other segmentation problems, such as in [24], where a similar method is proposed to segment white matter hyperintensities, but pseudo-3D has been replaced by an association of multimodality and mathematical morphology pre-processing to improve the detection of small lesions. Hence, we might also try to modify our inputs thanks to some highly non-linear filtering to help the network segment tumors, precisely some mathematical morphology operators [23].

The strength of this method is its modularity and its simplicity. It is easy to implement, fast, and does not need a huge amount of annotated data for training (in the work on brain segmentation [22], there is only 2 images for training for some cases).

From a segmentation result, we introduced a simple and efficient method to predict the patient overall survival, based on Random Forests. This method only needs as input a segmentation, a brain atlas and a brain volume for atlas registration. This method is not only fast; it is also easy to train with few samples and can be used after any tumor segmentation module.

Finally, we made a Docker image of the overall method publicly available at https://www.lrde.epita.fr/wiki/NeoBrainSeg.

Acknowledgments. The authors would like to thank the organizers of the BraTS 2018 Challenge and the MICCAI Brainles Workshop, and Dr. Marie Donzel from Claude Bernard University Lyon 1 medical school for the useful discussions regarding the definition of relevant brain features for the survival prediction. The GPU card "Quadro P6000" used for the work presented in this paper was donated by NVIDIA Corporation.

References

1. Angelini, E.D., Clatz, O., Mandonnet, E., Konukoglu, E., Capelle, L., Duffau, H.: Glioma dynamics and computational models: a review of segmentation, registration, and in silico growth algorithms and their clinical applications. Curr. Med. Imaging Rev. **3**(4), 262–276 (2007)
2. Bakas, S., et al.: Segmentation labels and radiomic features for the pre-operative scans of the TCGA-GBM collection. Cancer Imaging Arch. (2017). https://doi.org/10.7937/K9/TCIA.2017.KLXWJJ1Q
3. Bakas, S., et al.: Segmentation labels and radiomic features for the pre-operative scans of the TCGA-LGG collection. Cancer Imaging Arch. (2017). https://doi.org/10.7937/K9/TCIA.2017.GJQ7R0EF
4. Bakas, S., et al.: Advancing the cancer genome atlas glioma MRI collections with expert segmentation labels and radiomic features. Sci. Data **4**, 170117 (2017)
5. Bakas, S., Reyes, M., et al.: Identifying the best machine learning algorithms for brain tumor segmentation, progression assessment, and overall survival prediction in the brats challenge. arXiv preprint arXiv:1811.02629 (2018)
6. Bauer, S., Wiest, R., Nolte, L.P., Reyes, M.: A survey of MRI-based medical image analysis for brain tumor studies. Phys. Med. Biol. **58**(13), R97 (2013)
7. Bonnín Rosselló, C.: Brain lesion segmentation using Convolutional Neuronal Networks. B.S. thesis, Universitat Politècnica de Catalunya (2018)
8. Eisenhauer, E.A., et al.: New response evaluation criteria in solid tumours: revised RECIST guideline (version 1.1). Eur. J. Cancer **45**(2), 228–247 (2009)
9. Holland, E.C.: Progenitor cells and glioma formation. Curr. Opin. Neurol. **14**(6), 683–688 (2001)

10. Kingma, D.P., Ba, J.: Adam: a method for stochastic optimization. CoRR abs/1412.6980 (2014)
11. Krizhevsky, A., Sutskever, I., Hinton, G.E.: ImageNet classification with deep convolutional neural networks. In: Advances in Neural Information Processing Systems, pp. 1097–1105 (2012)
12. Long, J., Shelhamer, E., Darrell, T.: Fully convolutional networks for semantic segmentation. In: Proceedings of IEEE International Conference on Computer Vision and Pattern Recognition, pp. 3431–3440 (2015)
13. Louis, D.N., et al.: The 2007 who classification of tumours of the central nervous system. Acta Neuropathol. 114(2), 97–109 (2007)
14. Maninis, K.-K., Pont-Tuset, J., Arbeláez, P., Van Gool, L.: Deep retinal image understanding. In: Ourselin, S., Joskowicz, L., Sabuncu, M.R., Unal, G., Wells, W. (eds.) MICCAI 2016. LNCS, vol. 9901, pp. 140–148. Springer, Cham (2016). https://doi.org/10.1007/978-3-319-46723-8_17
15. Menze, B.H., et al.: The multimodal brain tumor image segmentation benchmark (BRATS). IEEE Trans. Med. Imaging 34(10), 1993 (2015)
16. Ohgaki, H., Kleihues, P.: Population-based studies on incidence, survival rates, and genetic alterations in astrocytic and oligodendroglial gliomas. J. Neuropathol. Exp. Neurol. 64(6), 479–489 (2005)
17. Reddi, S.J., Kale, S., Kumar, S.: On the convergence of Adam and beyond. In: International Conference on Learning Representations (2018)
18. Simonyan, K., Zisserman, A.: Very deep convolutional networks for large-scale image recognition. CoRR abs/1409.1556 (2014)
19. Strobl, C., Boulesteix, A.L., Zeileis, A., Hothorn, T.: Bias in random forest variable importance measures: illustrations, sources and a solution. BMC Bioinform. 8(1), 25 (2007)
20. Svetnik, V., Liaw, A., Tong, C., Culberson, J.C., Sheridan, R.P., Feuston, B.P.: Random forest: a classification and regression tool for compound classification and QSAR modeling. J. Chem. Inf. Comput. Sci. 43(6), 1947–1958 (2003)
21. Wen, P.Y., et al.: Updated response assessment criteria for high-grade gliomas: response assessment in neuro-oncology working group. J. Clin. Oncol. 28(11), 1963–1972 (2010)
22. Xu, Y., Géraud, T., Bloch, I.: From neonatal to adult brain MR image segmentation in a few seconds using 3D-like fully convolutional network and transfer learning. In: Proceedings of the 23rd IEEE International Conference on Image Processing (ICIP), Beijing, China, pp. 4417–4421, September 2017
23. Xu, Y., Géraud, T., Najman, L.: Connected filtering on tree-based shape-spaces. IEEE Trans. Pattern Anal. Mach. Intell. 38(6), 1126–1140 (2016)
24. Xu, Y., Géraud, T., Puybareau, É., Bloch, I., Chazalon, J.: White matter hyperintensities segmentation in a few seconds using fully convolutional network and transfer learning. In: Crimi, A., Bakas, S., Kuijf, H., Menze, B., Reyes, M. (eds.) BrainLes 2017. LNCS, vol. 10670, pp. 501–514. Springer, Cham (2018). https://doi.org/10.1007/978-3-319-75238-9_42

Brain Tumour Segmentation Method Based on Supervoxels and Sparse Dictionaries

J. P. Serrano-Rubio[1(✉)] and Richard Everson[2]

[1] Information Technologies Laboratory, Technological Institute of Irapuato, Irapuato, Guanajuato, Mexico
juserrano@itesi.edu.mx
[2] College of Engineering, Mathematics and Physical Sciences, University of Exeter, Exeter, UK
R.M.Everson@exeter.ac.uk

Abstract. This paper presents an automatic method for brain tumour segmentation from magnetic resonance images. The method uses the feature vectors obtained by an efficient feature encoding approach which combines the advantages of the supervoxels and sparse coding techniques. Extremely Randomized Trees (ERT) algorithm is trained using these feature vectors to detect the whole tumour and for multi-label classification of abnormal tissues. A Conditional Random Field (CRF) algorithm is implemented to delimit the region where the brain tumour is located. The obtained predictions of the ERT are used to estimate probability maps. The probability maps of the images and the Euclidean distance between the feature vectors of neighbour supervoxels define the conditional random field energy function. The minimization of the energy function is performed via graph cuts. The proposed methods are evaluated on real patient data obtained from BraTS 2018 challenge. Results demonstrate that proposed method achieves a competitive performance on the validation dataset using Dice score is: 0.5719, 0.7992 and 0.6285 for enhancing tumuor, whole tumour and tumour core respectively. The achieved performance of this method on testing set using Dice score is: 0.5081, 0.7278 and 0.5778 for enhancing tumuor, whole tumour and tumour core respectively.

Keywords: Brain tumour segmentation · Sparse coding techniques · Supervoxels

1 Introduction

Gliomas are a type of brain tumours with the highest mortality rate in adults. The gliomas are classified according to their histopathological appearances into Low Grade Gliomas (LGG) and High Grade Gliomas (HGG) for determining the best treatment for the patient [1]. Furthermore, the treatment of gliomas

© Springer Nature Switzerland AG 2019
A. Crimi et al. (Eds.): BrainLes 2018, LNCS 11384, pp. 210–221, 2019.
https://doi.org/10.1007/978-3-030-11726-9_19

depend on its size and location within the brain, since the tumours can grow and infiltrate over the surrounding normal brain tissue what sometimes it complicates its surgical removal.

One of the most important and critical procedures for the diagnosis of brain tumours is the brain segmentation. Brain segmentation has been useful for the analysis and visualization of brain structures [2,3] with the purpose of monitoring, guidance and surgery planning [4]. The goal of the brain tumour segmentation is to delineate the pathological regions such as the peritumoral edema, non-enhancing tumour and enhancing tumour. Nowadays, brain tumour segmentation methods are divided into two categories: (a) semi-automatic and (b) automatic methods. Semi-automatic methods are extremely costly due to they require an expert to detect the tumour. Automatic methods automatically detect and segment the brain tumours by using machine learning algorithms which can assign each tissue to its respectively class. Brain tumour segmentation becomes a challenging task for segmentation automatic methods, because the tumours can appear anywhere in the brain and they can present a wide number of shapes and sizes.

In this paper we propose a fully automatic method for segmentation of brain tumours. This method incorporates an efficient feature encoding approach which is based on three algorithms: (1) sparse coding technique, (2) a Gaussian pyramid and (3) supervoxels. 3D image patches of images are projected into sparse dictionary spaces in order to obtain feature vectors which compose the pattern vectors of each supervoxel. An Extremely Randomly Trees (ERT) algorithm is trained using the feature vectors to detect the whole tumour and for multi-label classification of abnormal tissues. The tumour segmentation method is performed in two phases:

- The whole tumour is detected by identifying the region where the tumour is located. A Conditional Random Field (CRF) is implemented to complement the segmentation task. The obtained predictions of ERT are used to estimate probability maps for each MRI. The energy function of the CRF is defined by the probability maps and the euclidean distance between the sparse feature vectors of neighbour supervoxels. The minimization of the energy function is performed via graph cuts.
- A multi-label classification is performed on the delimited region where the tumour is located in order to recognize the enhancing tumour, peritumoral edema and enhancing tumour.

The method described in this paper has been adapted from an automatic method for segmentation of vertebrae using magnetic resonance images which has been proposed in [5]. The rest of the paper is organized as follows: Sect. 2 presents the details of the Magnetic Resonance Image (MRI) database which is employed to evaluated the proposal method and Sect. 3 presents the feature learning approach to obtain the sparse feature vectors for normal and abnormal tissue and the brain tumour segmentation model. Section 4 presents the experimental details and results. Finally, Sect. 5 presents the conclusions.

2   BraTS 2018 Database

The database of magnetic resonance images is adopted from BraTS 2018 challenge as part of the International Conference on Medical Image Computing and Computer-Assisted Interventions (MICCAI) conference [2,6–8]. The magnetic resonance images are 3D volumes whose dimension is $240 \times 240 \times 155$. The training data set is composed by volumes of 285 subjects. Each subject has four modalities of MRI volumes (Flair, T1, T1 contrast enhanced (T1C) and T2). The training data set is divided into 210 cases of High Grade Glioma (HGG) and 75 cases with Low-Grade Glioma (LGG). In addition, each subject presents a ground truth labels which has been manually revised by expert board-certified neurologists. The labels include the following tissue labels: for normal tissue and background regions (label 0), non-enhancing tumour (label 1), peritumoral edema (label 2) and enhancing tumour (label 4). Figure 1 presents one frame for all employed MRI modalities and its ground truth.

The validation data set and testing set consist of 66 and 191 subjects respectively. Both data sets contain HGG and LGG gliomas but the grade is not revealed.

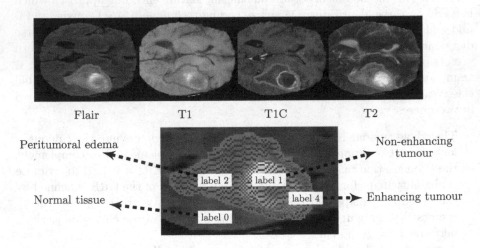

Fig. 1. Flair, T1, T1C and T2 MRI modalities and ground truth segmentation.

3   Method

Each MRI is normalized by setting the mean to 0 and standard deviation to 1 of the voxel intensities. Our automatic segmentation method is based on two stages. The first stage consist of the recognition of the whole tumour with the purpose of delimiting the region where the brain tumour is located. The second stage uses the delimited region to perform the multi-label classification of brain tumour tissues. The automatic tumour segmentation method is based on five algorithms:

1. Sparse Coding Technique [9],
2. Gaussian pyramid method [10],
3. An extension to supervoxels of the Superpixels Extracted via Energy-Driven Sampling (SEEDS) [11],
4. Extremely Randomized Trees [1],
5. Conditional Random Fields [12].

3.1 Sparse Coding Technique

The brain tumour segmentation method incorporates a sparse coding technique to capture structures and patterns of the magnetic resonance images as well for defining an efficient representation of the input magnetic resonance images. We call to this efficient representation as sparse dictionaries. Below is a description of how is obtained a sparse dictionary.

Let $\mathbf{I}$ be a set of images. 3D image patches $p_1, p_2, ..., p_{\psi-1}, p_\psi \in \mathbb{R}^{kxkxk}$ are randomly selected from $\mathbf{I}$. All 3D image patches are reshaped into the vector $w_i \in \mathbb{R}^{k^3}$. The sparse dictionary $\mathbf{B} \in \mathbb{R}^{k^3 \times n}$ is obtained by using the sparse coding technique over n linear filters using Eq. 1.

$$\text{minimize}_{\{\mathbf{B}, a_i\}} \sum_{i=1}^{\psi} \underbrace{|| w_i - \mathbf{B}a_i ||_2^2}_{\text{residue function}} + \underbrace{\beta \, || a_i ||_1}_{\text{sparsity function}} \tag{1}$$

$$\text{subject to} \;\; || \mathbf{B}_{(:,j)} ||_2 = 1 \; \forall_j \tag{2}$$

where a_i is a coefficient vector, $\mathbf{B}_{(:,j)}$ is the j-th column of the sparse dictionary and β is the parameter which controls the sparsity of the solution [9]. Figure 2 illustrates the linear combination for representing the image patch p_i and one component of the sparse dictionary. In summary, the sparse coding technique finds a set of basis vector (sparse dictionary) such that an input vector can be represented as a linear combination of these basis vectors.

3.2 Gaussian Pyramid

A Gaussian pyramid is implemented as a structured data for multi-scale magnetic resonance image representation [10]. The Gaussian pyramid allows to generate a sequence of images over different scales for each magnetic resonance image. Each scale is assigned as a level of the pyramid. For experimental purposes we use three levels of the pyramid $l = 0, 1, 2$. Figure 3 presents and example of Gaussian pyramid using only one frame of a magnetic resonance image. The scale factor for reducing the image volume is 2^l. Sparse dictionaries are calculated by using the set of images which are obtained over the three levels of Gaussian pyramid. Therefore, we obtained twelve sparse dictionaries that corresponds to each MRI modality over the three levels of the Gaussian pyramid.

Reshape 3D patch:

One component of the sparse dictionary

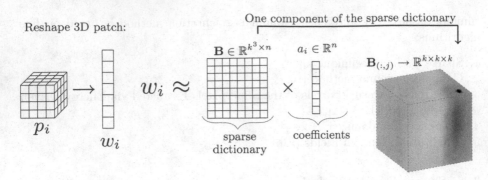

Fig. 2. Linear combination for representing the image patch p_i as a linear combination of a basis vectors (sparse dictionary).

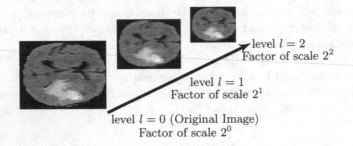

level $l = 2$
Factor of scale 2^2

level $l = 1$
Factor of scale 2^1

level $l = 0$ (Original Image)
Factor of scale 2^0

Fig. 3. Example of Gaussian Pyramid for only one MRI frame.

3.3 Supervoxels

The MRIs are partitioned into regions of perceptually similar voxels called supervoxels (set of voxels delimited by a boundary region) [13]. The supervoxels are calculated by decreasing the computational complexity of image volumes as well as obtaining meaningful structures of brain regions. Supervoxels present an irregular shape which provides a better alignment with the tissue boundaries than cubic image patches [14].

We use an extension of the Superpixels Extracted via Energy-Driven Sampling (SEEDS) algorithm [15]. The four modalities of MRI volumes are incorporated to calculate the supervoxels. The set of supervoxels for only one volume is denoted as $\mathbf{S} = \{s_1, s_2, ..., s_{\eta-1}, s_\eta\}$. The label set c for each supervoxel is assigned by majority vote.

Connected-Component Label algorithm is implemented to validate that all supervoxels are spatially connected as an only set. Each supervoxel is considered as an atomic unit for obtaining a feature vector for normal and abnormal brain tissues classification. Figure 4 shows six supervoxels which partition the structure of one brain tumour. All modalities of MRI volumes are used to estimated the supervoxels. Note that all supervoxels are estimated as volume of irregular shapes, however for illustration purposes, Fig. 4 shows the supervoxels for only one slide of the image volume.

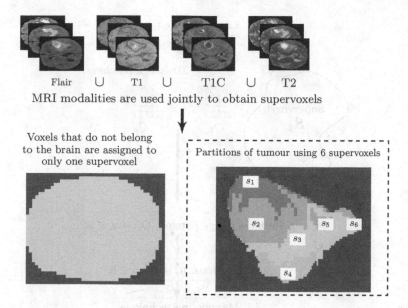

Fig. 4. Example of supervoxel segmentation by using MRI modalities (Flair, T1, T1C, T2).

3.4 Sparse Feature Vectors

Figure 5 shows the method for obtaining the feature vector for supervoxel s_i. The feature vector is denoted as v_{s_i}. This feature vector is estimated using the sparse dictionary that corresponds to a level of the Gaussian pyramid as well as for a specific modality of MRI volume. In the following list the calculation of the feature vector is enumerated:

1. Obtain a set of 3D image patches $\mathbf{Q}_1, \mathbf{Q}_2, .., \mathbf{Q}_{\kappa-1}, \mathbf{Q}_\kappa \in \mathbb{R}^{k \times k \times k}$ located into the boundary region delimited by the supervoxel.
2. 3D image patches are reshaped into vectors $q_1, q_2, .., q_\kappa, q_{\kappa-1} \in \mathbb{R}^{k^3}$ and projected into the sparse dictionary space. The projected vectors are $u_1, u_2, ..., u_\kappa$.
3. Feature vector v_{s_i} can be obtained using a maxpoling technique which uses the highest value of the positive and negative components of the projected vectors as shown in Fig. 5.

A pattern vector $\mathbf{D}_i$ is associated to each supervoxel i. The pattern vector $\mathbf{D}_i$ is formed by the concatenation of several features vectors as is shown in Fig. 6. Vector $\{d_{s_i}\}^l$ denotes the concatenation of features vectors for all MRI modalities over the level l of the Gaussian pyramid for supervoxel s_i. Finally the pattern vector for supervoxel s_i is formed by the concatenation of the vectors $\{d_{s_i}\}^{l=0}, \{d_{s_i}\}^{l=1}$ and $\{d_{s_i}\}^{l=2}$.

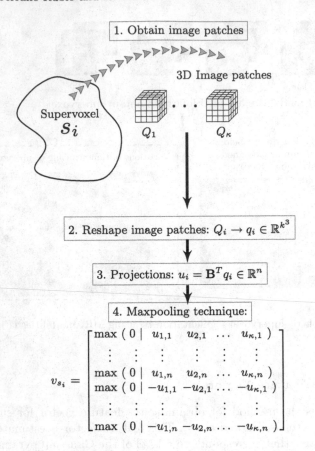

Fig. 5. Method for obtaining a feature vector for supervoxel s_i.

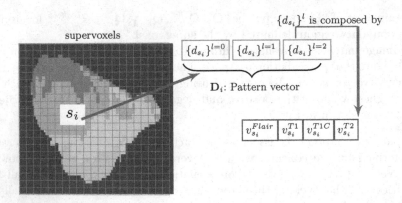

Fig. 6. Composition of one pattern vector from supervoxel s_i

3.5   Segmentation of Gliomas

Figure 7 shows the method to build the training set. For each subject in the training database, it is randomly selected a subset of 50 feature vectors of tumour tissue and normal tissue. The training set is composed by the concatenation of all subsets. An ERT classifier is trained to automatically assign each supervoxel to its respectively class. ERT combines the predictions of several classifiers using random splits to generate different trees and then calculate an output final classification [16]. Each tree learns a weak predictor for each pattern vector $\theta(D_i)$. Equation 3 presents the calculation of the final classification for the class c which is given by the most frequent estimated class and where T is the number of trees.

$$c = mode[\theta_1(\mathbf{D}_i), \theta_2(\mathbf{D}_i), ..., \theta_T(\mathbf{D}_i)] \tag{3}$$

Equation 4 is used to calculated the probability map for the images of each patient. The probability map assigns all supervoxels probability values of belonging to each tissue class c.

$$P(c|\theta(\mathbf{D}_i)) = \frac{1}{T}\sum_{t=1}^{T} \mathbb{1}_A(\theta_t(\mathbf{D}_i)), \text{ where: } \mathbb{1}_A(\theta_t(\mathbf{D}_i)) = \begin{cases} 1, & \theta_t(\mathbf{D}_i) = c \\ 0, & \theta_t(\mathbf{D}_i) \neq c \end{cases} \tag{4}$$

3.6   Recognition of the Whole Tumour

To complement the segmentation task for the first stage of the brain tumour segmentation a Conditional Random Field (CRF) is implemented. The CRF operates over the supervoxels of the volumes of images. The goal of the implementation of CRF is to delimit the region where the brain tumour is located. The CRF's aim is to obtain the labelling of the supervoxels such that minimize the following equation:

$$E(\mathbf{X}, \mathbf{D}) = \underbrace{\sum_{i \in S} f_i(\mathbf{X}_i|\mathbf{D}_i)}_{\text{Unary potential}} + \beta \underbrace{\sum_{i \in S} \sum_{j \in N_i} f_{i,j}(\mathbf{X}_i, \mathbf{X}_j|\mathbf{D}_i, \mathbf{D}_j)}_{\text{Pairwise potential}} \tag{5}$$

where S denotes the set of supervoxels, N_i is the set of neighbours of supervoxel i, β controls the relative importance of the smoothing term. The energy function of the conditional random field defines a posterior probability distribution $P(\mathbf{X} \mid \mathbf{D})$ given a set of class label $\mathbf{X}$ and a set of features for each supervoxel $\mathbf{D}$. The tumour segmentation problem is described as the partitioning of vertices of a graph into disjoint subsets. Graph cuts algorithm is performed to partition the graph and obtain the maximum a posteriori inference of the labels $\mathbf{X}$ [12]. Equation 6 presents the unary potential which is defined by the probability map of the supervoxels. Equation 7 presents the pairwise potential which models the relationship among neighbouring of supervoxels.

$$f_i(\mathbf{X}_i|\mathbf{D}_i) = -log(P(\mathbf{X}_i|\mathbf{D}_i)) \tag{6}$$

$$f_{i,j}(\mathbf{X}_i, \mathbf{X}_j|\mathbf{D}_i, \mathbf{D}_j) = \begin{cases} \exp(-\parallel \mathbf{D}_i - \mathbf{D}_j \parallel), & \text{if } \mathbf{X}_i \neq \mathbf{X}_j \\ 0 & , \text{ otherwise} \end{cases} \tag{7}$$

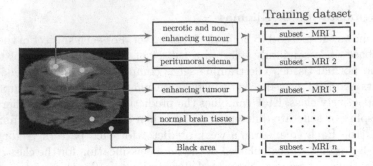

Fig. 7. Method for obtaining the training set.

4   Experiments and Results

We evaluate our method using real patient data obtained from BraTS 2018 challenge. The experiment details are explained as follows. The experimental setup for gliomas segmentation is given as follows:

- Sparse dictionaries are estimated separately over 128 filters from 500000 randomly sampled patches whose size is $3 \times 3 \times 3$. Sparse dictionaries are calculated using the MRI volumes over three levels of a Gaussian pyramid.
- For each MRI volume, its supervoxels are calculated using different sizes of supervoxels. For the first stage, the size of supervoxel is $4 \times 4 \times 4$ voxels in order to delimit the tumour region. For the second stage, the size of supervoxel is $3 \times 3 \times 3$ voxels in order to perform the multilabel classification of the tumour tissues.
- For classification purposes, we use $T = 100$ trees to run the Extremely Random Trees algorithm.

Tables 1 and 2 present the results of the training phase. Tables 3 and 4 presents the results of the validation phase. Tables 5 presents the results of the testing phase. The Mean, StdDev, Median, 25q, and 50q indicate the mean, median, standard deviation, 25th percentile and 75th percentile respectively, for the evaluated metrics. The evaluated metrics are Dice, sensitivity, specificity and 95th percentile Hausdorff for enhancing tumour (ET), whole tumour (WT) and tumour core (TC). In [17] presents the results of this method in the Multimodal Brain Tumour Segmentation Challenge 2018. The method is referred as MexExe Team.

Table 1. Results using the training set for DICE and Sensitivity scores.

	Training phase					
	Dice			Sensitivity		
	ET	WT	TC	ET	WT	TC
Mean	0.5948	0.8398	0.7333	0.7837	0.9185	0.8451
StdDev	0.2520	0.0953	0.1499	0.1673	0.0998	0.1671
Median	0.6751	0.8659	0.7730	0.8284	0.9473	0.9003
25q	0.5213	0.8112	0.6740	0.7309	0.8932	0.8057
50q	0.7665	0.898	0.8425	0.8897	0.9748	0.9613

Table 2. Results using the training set for Specificity and Hausdorff95 scores.

	Training phase					
	Specificity			Hausdorff95		
	ET	WT	TC	ET	WT	TC
Mean	0.9946	0.9835	0.9916	11.69	8.9686	14.6677
StdDev	0.0044	0.0132	0.0067	12.7416	9.0691	11.7043
Median	0.9958	0.9866	0.9933	5.6568	5.3851	11.3578
25q	0.9925	0.9781	0.9893	3	3.3166	6.4031
50q	0.9981	0.9925	0.9963	16.9925	11.0453	20.0199

Table 3. Results using the validation dataset for DICE and Sensitivity scores.

	Validation phase					
	Dice			Sensitivity		
	ET	WT	TC	ET	WT	TC
Mean	0.5719	0.7992	0.6285	0.6836	0.8714	0.7339
StdDev	0.2749	0.1557	0.2590	0.2782	0.1711	0.3083
Median	0.6910	0.8458	0.7103	0.7776	0.9412	0.8697
25q	0.4934	0.7862	0.5365	0.5661	0.8435	0.6204
50q	0.7669	0.8826	0.8196	0.8947	0.9695	0.9615

Table 4. Results using the validation dataset for Specificity and Hausdorff95 scores.

	Validation phase					
	Specificity			Hausdorff95		
	ET	WT	TC	ET	WT	TC
Mean	0.9955	0.9863	0.9919	12.9637	12.4414	17.7162
StdDev	0.0038	0.0094	0.0078	15.7325	19.2750	17.5612
Median	0.9960	0.9883	0.9936	6.5556	6.4031	15.106
25q	0.9942	0.9821	0.9903	3.2008	3.6395	9
50q	0.9980	0.9927	0.9963	17.9093	12.4078	20.7175

Table 5. Results using the testing dataset for Dice and Hausdorff95 scores.

	Testing Phase					
	Dice			Hausdorff95		
	ET	WT	TC	ET	WT	TC
Mean	0.5081	0.7278	0.5778	17.7870	18.5813	22.129
StdDev	0.2864	0.2155	0.2774	22.3555	22.3832	21.4912
Median	0.6061	0.8187	0.6666	8.8011	9.4868	15.4836
25q	0.3424	0.6432	0.4243	3.6055	4.4721	9.4604
50q	0.7234	0.8724	0.7909	22.6933	21.038	26.4602

5   Conclusions

We present an approach for multimodal brain tumour segmentation using and efficient feature encoding technique based on sparse dictionaries and supervoxels. BraTS 2018 database is adopted to evaluated the performance of our method. The method combines the advantages of the supervoxels and sparse coding techniques to generate feature vectors which can be employed to assign each tissue to its respectively class using Extremely Randomized Trees and Conditional Random Field algorithms. According to the numerical results, the performance of our approach can be compared to the performance of other state-of-the-art algorithms which have been evaluated using the same dataset.

Acknowledgements. The first author's research was partially supported by a CONACyT postdoctoral grant and CONACyT Research Grant 256126. The first author would also like to thank the University of Exeter for its hospitality. We thank to the organizing committee of MICCAI BraTS 2018 for help us in the evaluation of the performance of our method.

References

1. Soltaninejad, M., et al.: Automated brain tumour detection and segmentation using superpixel-based extremely randomized trees in flair MRI. Int. J. Comput. Assist. Radiol. Surg. **12**(2), 183–203 (2017)
2. Menze, B.H., et al.: The multimodal brain tumor image segmentation benchmark (brats). IEEE Trans. Med. Imaging **34**(10), 1993–2024 (2015)
3. Havaei, M., Larochelle, H., Poulin, P., Jodoin, P.M.: Within-brain classification for brain tumor segmentation. Int. J. Comput. Assist. Radiol. Surg. **11**(5), 777–788 (2016)
4. Zimeras, S., Gortzis, L., Pylarinou, C.: Shape analysis in radiotherapy and tumor surgical planning using segmentation techniques. In: Sklavos, N., Hübner, M., Goehringer, D., Kitsos, P., et al. (eds.) System-Level Design Methodologies for Telecommunication, pp. 159–173. Springer International Publishing, Cham (2014). https://doi.org/10.1007/978-3-319-00663-5_9
5. Hutt, H., Everson, R., Meakin, J.: 3D intervertebral disc segmentation from MRI using supervoxel-based CRFs. In: Vrtovec, T., et al. (eds.) Computational Methods and Clinical Applications for Spine Imaging, pp. 125–129. Springer International Publishing, Cham (2016)
6. Bakas, S., et al.: Advancing the cancer genome atlas glioma mri collections with expert segmentation labels and radiomic features. Nat. Sci. Data **4**, 170117 (2017)
7. Bakas, S., et al.: Segmentation labels and radiomic features for the pre-operative scans of the TCGA-GBM collection. Cancer Imaging Archive (2017)
8. Bakas, S., et al.: Segmentation labels and radiomic features for the pre-operative scans of the TCGA-LGG collection. Cancer Imaging Archive (2017)
9. Lee, H., Battle, A., Raina, R., Ng, A.Y.: Efficient sparse coding algorithms. In: Advances in Neural Information Processing Systems, pp. 801–808 (2007)
10. Adelson, E.H., Anderson, C.H., Bergen, J.R., Burt, P.J., Ogden, J.M.: Pyramid methods in image processing. RCA Eng. **29**(6), 33–41 (1984)
11. Van den Bergh, M., Boix, X., Roig, G., de Capitani, B., Van Gool, L.: SEEDS: superpixels extracted via energy-driven sampling. In: Fitzgibbon, A., Lazebnik, S., Perona, P., Sato, Y., Schmid, C. (eds.) ECCV 2012. LNCS, vol. 7578, pp. 13–26. Springer, Heidelberg (2012). https://doi.org/10.1007/978-3-642-33786-4_2
12. Boykov, Y., Kolmogorov, V.: An experimental comparison of min-cut/max-flow algorithms for energy minimization in vision. IEEE Trans. Pattern Anal. Mach. Intell. **26**(9), 1124–1137 (2004)
13. Stutz, D., Hermans, A., Leibe, B.: Superpixels: an evaluation of the state-of-the-art. Comput. Vis. Image Underst. **166**, 1–27 (2017)
14. Veksler, O., Boykov, Y., Mehrani, P.: Superpixels and supervoxels in an energy optimization framework. In: Daniilidis, K., Maragos, P., Paragios, N. (eds.) ECCV 2010. LNCS, vol. 6315, pp. 211–224. Springer, Heidelberg (2010). https://doi.org/10.1007/978-3-642-15555-0_16
15. Van den Bergh, M., Boix, X., Roig, G., Van Gool, L.: SEEDS: superpixels extracted via energy-driven sampling. Int. J. Comput. Vis. **111**(3), 298–314 (2015)
16. Geurts, P., Ernst, D., Wehenkel, L.: Extremely randomized trees. Mach. Learn. **63**(1), 3–42 (2006)
17. Bakas, S., Reyes, M., et al.: Identifying the Best Machine Learning Algorithms for Brain Tumor Segmentation, Progression Assessment, and Overall Survival Prediction in the BRATS Challenge. arXiv preprint arXiv:1811.02629 (2018)

Multi-scale Masked 3-D U-Net for Brain Tumor Segmentation

Yanwu Xu[1], Mingming Gong[1,3], Huan Fu[2], Dacheng Tao[2], Kun Zhang[3], and Kayhan Batmanghelich[1(✉)]

[1] Department of Biomedical Informatics,
University of Pittsburgh, Pittsburgh, USA
kayhan@pitt.edu
[2] UBTECH Sydney AI Centre, SIT, FEIT,
The University of Sydney, Sydney, Australia
[3] Philosophy Department, Carnegie Mellon University, Pittsburgh, USA

Abstract. The brain tumor segmentation task aims to classify sub-regions into peritumoral edema, necrotic core, enhancing and non-enhancing tumor core using multimodal MRI scans. This task is very challenging due to its intrinsic high heterogeneity of appearance and shape. Recently, with the development of deep models and computing resources, deep convolutional neural networks have shown their effectiveness on brain tumor segmentation from 3D MRI cans, obtaining the top performance in the MICCAI BraTS challenge 2017. In this paper we further boost the performance of brain tumor segmentation by proposing a multi-scale masked 3D U-Net which captures multi-scale information by stacking multi-scale images as inputs and incorporating a 3-D Atrous Spatial Pyramid Pooling (ASPP) layer. To filter noisy results for tumor core (TC) and enhancing tumor (ET), we train the TC and ET segmentation networks from the bounding box for whole tumor (WT) and TC, respectively. On the BraTS 2018 validation set, our method achieved average Dice scores of 0.8094, 0.9034, 0.8319 for ET, WT and TC, respectively. On the BraTS 2018 test set, our method achieved 0.7690, 0.8711, and 0.7792 dice scores for ET, WT and TC, respectively. Especially, our multi-scale masked 3D network achieved very promising results enhancing tumor (ET), which is hardest to segment due to small scales and irregular shapes.

Keywords: Brain tumor segmentation · Multi-scale · ASPP · U-Net

1 Introduction

Multimodal Brain Tumor Segmentation Challenge (BraTS) [16] provides an excellent platform to boom the development of methods for segmenting tumor regions from 3D MRI scans as well as data [3, 4]. As explained in [16], gliomas are the most common primary brain malignancies, with different degrees of aggressiveness, variable prognosis and various heterogeneous histological sub-regions,

© Springer Nature Switzerland AG 2019
A. Crimi et al. (Eds.): BrainLes 2018, LNCS 11384, pp. 222–233, 2019.
https://doi.org/10.1007/978-3-030-11726-9_20

i.e. peritumoral edema, necrotic core, enhancing and non-enhancing tumor core. In this challenge, our goal is to segment whole tumor (WT), tumor core (TC) and enhancing tumor (ET) from the other patterns. The dataset provided by [16] is composed of annotated low grade gliomas (LGG) and high grade glioblastomas (HGG), where LGG tends to be benign tendencies, while HGG denote the tumors which can grow rapidly and spread fast. For both LGG and HGG, four modals of scanning images are given, including Fluid Attenuation Inversion Recovery (FLAIR), T1-weighted (T1), contrast enhanced T1-weighted (T1ce) and T2-weighted (T2) images. Each modality supplies complementary information and they together provide more complete description of the tumor patterns. For example, the contours of whole tumor detected in FLAIR and T2 are more distinctive from the background than those in T1 and T1ce. Similarly, TC and ET can be easily distinguished from background in T1 and T1ce, the bounding information of which can be restricted by Flair and T2 by segmenting WT first. For instance, as mentioned in [16,21], T2 and FLAIR highlight the tumor with peritumoral edema, designated "whole tumor" as per [16]. T1 and T1ce highlight the tumor without peritumoral edema, designated "tumor core" as per [16]. An enhancing region of the tumor core with hyper-intensity can also be observed in T1ce, designated as "enhancing tumor core" [16].

Nowadays, we have a considerable quantity of diagnostic cases using Magnetic Resonance (MR) images. Moreover, we have the capacity to train very deep neural networks with the development of computing resources from these MR images, which makes automated disease diagnosis possible [2,16]. Automatic brain tumor segmentation can be much faster than manual segmentation; however, due to the irregular characteristics of brain tumor, the possibly subtle distinction between tumor and normal tissue, as well as a high variability in shape, location, and extent across patients, the accuracy of the current brain tumour segmentation algorithms needs further improvement so that they can be deployed in real systems.

There have been many methods for segmenting brain tumor which is detailed in [15]. Recently, the deep convolutional neural networks (CNNs) have shown promising performance in medical image segmentation and other related tasks [9–12,14,21]. DeepMedic [12] is one of the deep model-based method which combines patches with multiple resolutions as inputs to capture fine details and global information. They further introduced an enhancing structure which adds residual connection from previous feature layers. The 3-D U-Net [18] uses a compact encoder-decoder structure, which utilizes the features from several encoder layers twice by concatenating them with the decoder layers. Isensee et al. apply a U-Net based network to capture large scale information by large input patch size [10]. Additional works focus on the modification on the choice of convolutional kernal and loss function, such as the mixture of convolutional kernel and downsampling strategy [8,12]. Coping with unbalanced data, specific loss function [6,19] and sampling strategy [6] are introduced to train networks.

In this work, we focus on extracting multi-scale information from a single patch input instead of using multi-resolution inputs. Our contributions are

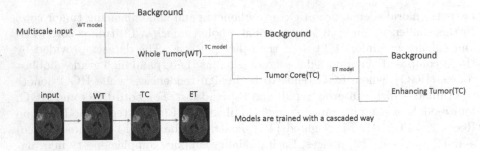

Fig. 1. The diagram depicts the training strategy for three different tumor with a cascaded masked way.

three fold. First, we extend 2D Atrous Spatial Pyramid Pooling (ASPP) [5] to 3D ASPP for extracting multi-scale information from feature maps of the neural network. By making use of the ASPP layer, we are able to enlarge the receptive field and thus capture larger scale information without introducing extra parameters. Second, we adjust basic structure of U-Net for small tumor segmentation by removing subsampling layers in specific layers of U-Net. This could help detect small tumors which are usually ignored in the original U-Net due to too many subsampling layers. Finally, we apply the cascaded masked strategy [21] for tumor segmentation training. Specifically, we segment WT, TC, and ET sequentially and use the bounding box from the former ones to restrict the search space for the following ones. This strategy could help remove false positive detections from the background regions. Our paper is collected in [1].

2   Methods

In this section, we will introduce the details of our method. First, we will describe the data preprocessing and patch extraction methods. Second, we will present the details of our network structure and training strategies.

2.1   Data Preprocess and Patch Extraction

We follow the standard procedure to preprocess the input images. To compensate for the MR inhomogeneity, we apply the bias correction algorithm based on N4ITK library [20] to the T1 and T1ce images. To reduce the effect of the absolute pixel intensities to the model, an intensity normalization step is applied to each volume of all subjects by subtracting the mean and dividing them by the standard deviation so that each MR volume will have a zero mean and unit variance. In practice, as the original uncropped volume is used but the brain only takes the central region, the mean and standard deviation are estimated from the brain area. Because of the GPU memory limitation and insufficient training data, we extract 400 patches per patient with patch size $64 \times 64 \times 64$ and take these patches as network inputs.

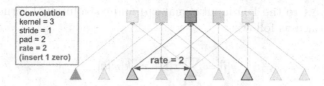

illustration of atrous convolution in 1-D

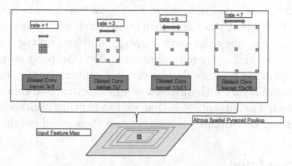

2-D Atrous Spatial Pyramid Pooling (ASPP)

2-D extends to 3-D

2-D of Atrous convolution to 3-D of Atrous convolution

Fig. 2. The proposed extending ASPP layer. By the order of top to bottom, the dilated convolution, 2-D ASPP layer and the extending of 3-D ASPP layer from 2-D ASPP layer are well depicted.

2.2   Cascaded Masked Strategy

To remove false positive detections from background, we apply the cascaded masked strategy as [21]. The training strategy is shown in Fig. 1. By doing so, we can reduce the multiclass segmentation problem as a binary segmentation problem. Specifically, we train the WT network only with WT labeled data. Then we keep the segmented WT tumor as a mask for TC training. Similarly, we set segmented TC as the mask for ET training. Note that we use the groundtruth masks in the training phase, but the predicted masks in the test phase.

2.3   Extended 3-D ASPP Layer

Atrous Spatial Pyramid Pooling (ASPP) is first introduced in [5] for semantic segmentation in 2D natural scene images. ASPP layer consists of multiple scales of Atrous layers, also called dilated convolution layers. Figure 2 shows a one-dimensional Atrous convolutional operation. With the annotation for the output

$y[i]$ with respect to the 1-D input signal $x[i]$ and convolutional kernel $w[k]$, the formula is formed as follows:

$$y[i] = \sum_{k=1}^{K} x[i + r \cdot k] w[k] \qquad (1)$$

Rate r denotes the dilated rate and dilated rate $r = 1$ is the normal convolution. Then the 2-D ASPP is displayed in Fig. 2, we feed feature maps into several Atrous layer with different rates and then combine these feature maps in channel dimension. Finally, we obey the same strategy and extend the 2-D ASPP layer to 3-D ASPP layer which is then applied on U-Net. We propose to apply ASPP layer here for capturing multi-scale objects and context employing multiple 3-D atrous convolutional layers with different sampling rates, and this is implemented with a parallel way. The advantage is that we can capture multi-scale information without introducing additional parameters and thus avoid overfitting.

2.4 Multi-scale Input

Additional, we try to include as much information as possible in the input. We rescale the images by multiple scales and then feed multi-resolution patches as input. Thus we apply a multi-scale input patches rather than only the patches of original size. In this work, the scales chosen are ×0.5, ×1 (original size) and ×2, and these patches are concatenated in channel dimension. By doing so, we can extract global and local information even when the patch size is small.

2.5 Network Structure

Now we can stack the building blocks together to form the final network structure. Our network is based on U-net with 3-D ASPP layer for trade-off between scale information and receptive field as well as memory usage. The designation of network structure is shown in Fig. 3. Compared to the original U-Net, we remove the third downsampling and upsampling layers for WT network and TC network, furthermore, we add the ASPP layer between the encoder and decoder in U-Net. We observe that ET is small with respect to WT and TC and maybe evanescent after downsampling of encoder, which is not able to be recovered by upsampling of decoder. Thus, we only keep the second downsampling and upsampling layers for the ET network.

As for training network, we apply ADAM optimizer [13] and set the parameters of ADAM as $lr = 0.0002$, $\beta_1 = 0.5$ and $\beta_2 = 0.9999$, which is the unified setting. We apply Xavier initialization [7] to initialize the network parameters.

2.6 Loss Function

We adopt the cross entropy loss to train our networks. We classify each voxels to a binary label (1, 0: 1 means tumor and 0 means background), when

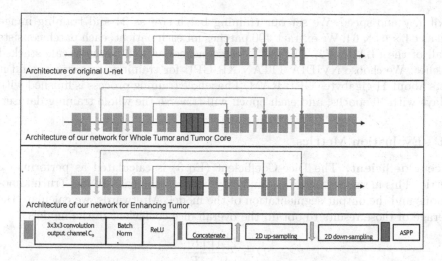

Fig. 3. Our modified U-Net based neural networks. We show the original U-Net on the top and the networks for our contribution is below. The networks trained by whole tumor and tumor core share the same network structure but do not share parameters. As for network designed for enhancing tumor training, we only keep one downsampling layer.

training network for WT, TC and ET separately. The cross entropy loss can be written as

$$loss = \sum (y' \log(y) + (1 - y') \log(1 - y)), \qquad (2)$$

where y' represents ground truth label and y represents predicted label.

3   Experiments

Data. We got all our training data from BraTS web[1] to evaluate our method. The training data consist of 285 patients including segmented masks annotated by human experts. These training data are separated into two categories including HGG and LGG, containing 210 HGG and 75 LGG. There is an unbalance between HGG and LGG, and the data distributions of HGG and LGG are also different, especially for TC and ET. Each patient has four sequences, which are FLAIR, T2, T1, and T1ce. We feed all of the sequences into our network by combining them in channel dimension. Thus, our input data are 5-D, the dimension of which are batch, sequences, width, length, and depth. Regarding the validation data and testing data, they are the same as given training data, however the segmentation labels are not released. We finally receive validation data and testing data which are composed of 66 patients and 191 patients, respectively.

We train our whole network using Pytorch [17], which is a new hybrid front-end seamlessly transitions between eager mode and graph mode to provide both

---

[1] https://www.med.upenn.edu/sbia/brats2018/data.html.

flexibility and speed. We set our training batch size as 24 and training image size as $64 \times 64 \times 64$. We extract 400 patches for each patient, each patch consists of all of the FLAIR, T2, T1, and T1ce sequences as well as multiscale stacked patches. We choose NVIDIA TITAN XP GPU for training our network and it costs about 11 gigabytes GPU RAM. The whole training process is finished with 2 days with 10 epochs, and each epoch will traverse the whole training dataset.

3.1 Evaluation Metrics

Dice Coefficient. The Dice-Coefficient (Eq. 3) is calculated as performance metric. This measure states the similarity between clinical Ground Truth annotations and the output segmentation of the model. Afterwards, we calculate the average of those results to obtain the overall dice coefficient of the models.

$$D = \frac{2|A \bigcap B|}{|A| + |B|} \tag{3}$$

Hausdorff Distance. The Hausdorff Distance (Eq. 4) is mathematically defined as the maximum distance of a set to the nearest point in the other set [15], in other words how close are the segmentation and the expected output.

$$H(A, B) = max(min(d(A, B))) \tag{4}$$

Table 1. Mean values of Dice and Hausdorff measurements of the proposed method on BraTS 2018 validation set. ET, WT, TC denote enhancing tumor core, whole tumor and tumor core, respectively.

	Dice			Hausdorff (mm)		
	ET	WT	TC	ET	WT	TC
Original U-Net	0.739	0.882	0.788	5.329	7.356	10.243
Our network without ASSP layer	0.773	0.899	0.820	4.259	6.374	6.404
Our network with ASSP layer	0.809	0.903	0.832	3.780	6.022	7.091

Segmentation Results. To provide qualitative results of our method, we random choose two segmented images from validation data which are shown in Figs. 4 and 5 as well as in Appendix I. Figure 4 is suspected as one of the HGG data and Fig. 5 is suspected as one of the LGG data. Because the border for Fig. 4 is clear and the red non-enhancing tumor is inside the yellow enhancing tumor core. Furthermore, in Fig. 5, the border for tumor is quite blurred and there is almost no yellow enhancing tumor and it is in accordance with the feature of LGG data from training dataset. As we can observe from Fig. 4, the network with ASPP layer performs better than network without ASPP layer in that network with ASPP layer segments more local information that corresponds to the details shown in original sequences. As shown in Fig. 5, there

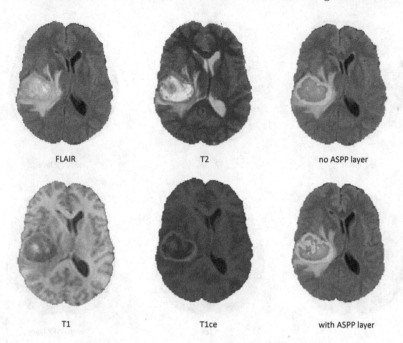

Fig. 4. Segmentation result of the brain tumor (suspected HGG) from a validation image. Green: edema; Red: non-enhancing tumor core; Yellow: enhancing tumor core. On the left, the original images are shown, and on the right, we show the segmented result of network without ASPP layer and network with ASPP layer. (Color figure online)

is a suspected wrong segmented area for red non-enhancing tumor, but the original sequences are blurred as well. In comparison, our network with ASPP layer can perform better on more local details and decrease the wrong classification for each voxel.

We show our quantitative results in Table 1. For comparison with existing methods, we list the result of the original U-Net, our modified U-Net without ASPP layer and our modified U-Net with ASPP layer. As can be seen from Table 1, our baseline of modified U-Net perform much better than the original U-Net in terms of all of the evaluation metrics. If just comparing the effect of ASPP layer, we find that assembling with ASPP layer can help improve TC and ET, especially improving ET by a large margin. However, they almost achieve the same performance on WT. We can also find that our method concentrate on detecting with multi-scale information that can help improve the ability for detecting small tumor area such as ET and TC. In this way, we achieve dice score above 0.8 for ET on testing data (Table 2).

As for testing data, we list the details of the result of mean value, standard deviation, median, 25% ranking and 75% ranking of Dice score and Hausdorff distance. Due to possible overfit on the validation data, we achieve a relatively lower performance on testing data; however our method still obtains rank 9th out of all the submitted methods on testing data.

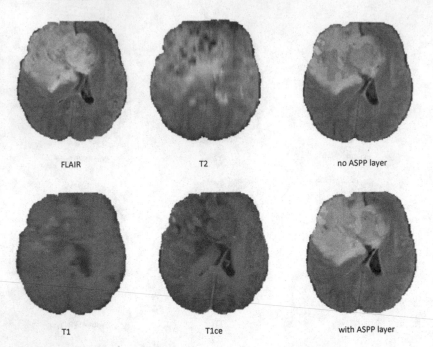

Fig. 5. Segmentation result of the brain tumor (suspected LGG) from a validation image. Green: edema; Red: non-enhancing tumor core; Yellow: enhancing tumor core. On the left, the original images are shown, and on the right, we show the segmented result of network without ASPP layer and network with ASPP layer. (Color figure online)

Table 2. Dice and Hausdorff measurements of the proposed method on BraTS 2017 testing set. EN, WT, TC denote enhancing tumor core, whole tumor and tumor core, respectively.

	Dice			Hausdorff (mm)		
	ET	WT	TC	ET	WT	TC
Mean	0.769	0.871	0.779	4.799	9.523	7.186
Standard deviation	0.240	0.129	0.274	9.293	16.822	10.900
Median	0.842	0.915	0.900	2.000	3.464	3.162
25 quantile	0.747	0.860	0.758	1.414	2.236	2.0000
75 quantile	0.892	0.939	0.936	3.000	6.364	7.280

4   Conclusions

We proposed a multi-scale neural network with a cascaded masked training structure for segmenting glioma subregions from multi-modal brain MR images. Our method receives as input multi-scale 3D patches extracted from the dataset volumes and we train three networks separately based on our cascaded mask strategy. To further incorporate multi-scale information, we also incorporate the 3D ASPP layer which contains filter with various receptive field size without introducing many additional parameters. Our method achieves good results in the BraTS challenge. Future work would be incorporating attention in the network to aggregate multi-scale information.

A   More Example

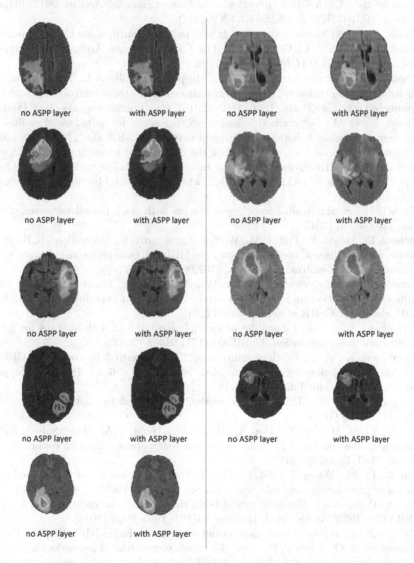

no ASPP layer   with ASPP layer   no ASPP layer   with ASPP layer

no ASPP layer   with ASPP layer   no ASPP layer   with ASPP layer

no ASPP layer   with ASPP layer   no ASPP layer   with ASPP layer

no ASPP layer   with ASPP layer   no ASPP layer   with ASPP layer

no ASPP layer   with ASPP layer

References

1. Bakas, S., et al.: Identifying the best machine learning algorithms for brain tumor segmentation, progression assessment, and overall survival prediction in the BRATS challenge. ArXiv e-prints, November 2018
2. Bakas, S., et al.: Advancing the cancer genome atlas glioma MRI collections with expert segmentation labels and radiomic features. Sci. Data **4**, 170117 (2017)
3. Bakas, S., et al.: Segmentation labels and radiomic features for the pre-operative scans of the TCGA-GBM collection. The Cancer Imaging Archive 2017. https://doi.org/10.7937/K9/TCIA.2017.KLXWJJ1Q
4. Bakas, S., et al.: Segmentation labels and radiomic features for the pre-operative scans of the TCGA-LGG collection. The Cancer Imaging Archive 2017. https://doi.org/10.7937/K9/TCIA.2017.GJQ7R0EF
5. Chen, L., Papandreou, G., Kokkinos, I., Murphy, K., Yuille, A.L.: DeepLab: semantic image segmentation with deep convolutional nets, atrous convolution, and fully connected crfs. CoRR abs/1606.00915 (2016). http://arxiv.org/abs/1606.00915
6. Fidon, L., et al.: Generalised Wasserstein dice score for imbalanced multi-class segmentation using holistic convolutional networks. CoRR abs/1707.00478 (2017)
7. Glorot, X., Bengio, Y.: Understanding the difficulty of training deep feedforward neural networks. In: Proceedings of the International Conference on Artificial Intelligence and Statistics (AISTATS 2010). Society for Artificial Intelligence and Statistics (2010)
8. Havaei, M., et al.: Brain tumor segmentation with deep neural networks. CoRR abs/1505.03540 (2015)
9. Isensee, F., Jaeger, P., Full, P.M., Wolf, I., Engelhardt, S., Maier-Hein, K.H.: Automatic cardiac disease assessment on cine-MRI via time-series segmentation and domain specific features. CoRR abs/1707.00587 (2017)
10. Isensee, F., Kickingereder, P., Wick, W., Bendszus, M., Maier-Hein, K.H.: Brain tumor segmentation and radiomics survival prediction: contribution to the brats 2017 challenge. CoRR abs/1802.10508 (2018)
11. Kamnitsas, K., et al.: Ensembles of multiple models and architectures for robust brain tumour segmentation. CoRR abs/1711.01468 (2017)
12. Kamnitsas, K., et al.: Efficient multi-scale 3D CNN with fully connected CRF for accurate brain lesion segmentation. Med. Image Anal. **36**, 61–78 (2017). https://doi.org/10.1016/j.media.2016.10.004
13. Kingma, D.P., Ba, J.: Adam: a method for stochastic optimization. CoRR abs/1412.6980 (2014)
14. Li, X., Chen, H., Qi, X., Dou, Q., Fu, C.W., Heng, P.A.: H-denseunet: hybrid densely connected UNet for liver and tumor segmentation from ct volumes. IEEE Trans. Med. Imaging (2018)
15. Liu, J., Li, M., Wang, J., Wu, F., Liu, T., Pan, Y.: A survey of MRI-based brain tumor segmentation methods. Tsinghua Sci. Technol. **19**(6), 578–595 (2014)
16. Menze, B.H., et al.: The multimodal brain tumor image segmentation benchmark (BRATS). IEEE Trans. Med. Imaging **34**(10), 1993–2024 (2015)
17. Paszke, A., et al.: Automatic differentiation in pytorch. In: NIPS-W (2017)
18. Ronneberger, O., Fischer, P., Brox, T.: U-net: convolutional networks for biomedical image segmentation. CoRR abs/1505.04597 (2015). http://arxiv.org/abs/1505.04597

19. Sudre, C.H., Li, W., Vercauteren, T., Ourselin, S., Jorge Cardoso, M.: Generalised dice overlap as a deep learning loss function for highly unbalanced segmentations. In: Cardoso, M.J., et al. (eds.) DLMIA/ML-CDS -2017. LNCS, vol. 10553, pp. 240–248. Springer, Cham (2017). https://doi.org/10.1007/978-3-319-67558-9_28

20. Tustison, N.J., et al.: N4ITK: improved N3 bias correction. IEEE Trans. Med. Imaging **29**(6), 1310–1320 (2010). http://dblp.uni-trier.de/db/journals/tmi/tmi29.html#TustisonACZEYG10

21. Wang, G., Li, W., Ourselin, S., Vercauteren, T.: Automatic brain tumor segmentation using cascaded anisotropic convolutional neural networks. CoRR abs/1709.00382 (2017). http://arxiv.org/abs/1709.00382

No New-Net

Fabian Isensee[1]([⊠]), Philipp Kickingereder[2], Wolfgang Wick[3],
Martin Bendszus[2], and Klaus H. Maier-Hein[1]

[1] Division of Medical Image Computing, German Cancer Research Center (DKFZ),
Heidelberg, Germany
f.isensee@dkfz-heidelberg.de
[2] Department of Neuroradiology, University of Heidelberg Medical Center,
Heidelberg, Germany
[3] Neurology Clinic, University of Heidelberg Medical Center, Heidelberg, Germany

Abstract. In this paper we demonstrate the effectiveness of a well trained U-Net in the context of the BraTS 2018 challenge. This endeavour is particularly interesting given that researchers are currently besting each other with architectural modifications that are intended to improve the segmentation performance. We instead focus on the training process arguing that a well trained U-Net is hard to beat. Our baseline U-Net, which has only minor modifications and is trained with a large patch size and a Dice loss function indeed achieved competitive Dice scores on the BraTS2018 validation data. By incorporating additional measures such as region based training, additional training data, a simple postprocessing technique and a combination of loss functions, we obtain Dice scores of 77.88, 87.81 and 80.62, and Hausdorff Distances (95th percentile) of 2.90, 6.03 and 5.08 for the enhancing tumor, whole tumor and tumor core, respectively on the test data. This setup achieved rank two in BraTS2018, with more than 60 teams participating in the challenge.

Keywords: CNN · Brain tumor · Glioblastoma · U-Net · Dice loss

1 Introduction

Quantitative assessment of brain tumors provides valuable information and therefore constitutes an essential part of diagnostic procedures. Automatic segmentation is attractive in this context, as it allows for faster, more objective and potentially more accurate description of relevant tumor parameters, such as the volume of its subregions. Due to the irregular nature of tumors, however, the development of algorithms capable of automatic segmentation remains challenging.

The brain tumor segmentation challenge (BraTS) [1] aims at encouraging the development of state of the art methods for tumor segmentation by providing a large dataset of annotated low grade gliomas (LGG) and high grade glioblastomas (HGG). The BraTS 2018 training dataset, which consists of 210 HGG and 75 LGG cases, was annotated manually by one to four raters and

© Springer Nature Switzerland AG 2019
A. Crimi et al. (Eds.): BrainLes 2018, LNCS 11384, pp. 234–244, 2019.
https://doi.org/10.1007/978-3-030-11726-9_21

all segmentations were approved by expert raters [2–4]. For each patient a T1 weighted, a post-contrast T1-weighted, a T2-weighted and a Fluid-Attenuated Inversion Recovery (FLAIR) MRI was provided. The MRI originate from 19 institutions and were acquired with different protocols, magnetic field strengths and MRI scanners. Each tumor was segmented into edema, necrosis and non-enhancing tumor and active/enhancing tumor. The segmentation performance of participating algorithms is measured based on the DICE coefficient, sensitivity, specificity and 95th percentile of Hausdorff distance.

It is unchallenged by now that convolutional neural networks (CNNs) dictate the state of the art in biomedical image segmentation [5–10]. As a consequence, all winning contributions to recent BraTS challenges were exclusively build around CNNs. One of the first notably successful neural network for brain tumor segmentation was DeepMedic, a 3D CNN introduced by Kamnitsas et al. [5]. It comprises a low and a high resolution pathway that capture semantic information at different scales and recombines them to predict a segmentation based on precise local as well as global image information. Kamnitsas et al. later enhanced their architectures with residual connections for BraTS 2016 [11]. With the success of encoder-decoder architectures for semantic segmentation, such as FCN [12,13] and most notably the U-Net [14], it is unsurprising that these architectures are used in the context of brain tumor segmentation as well. In BraTS 2017, all winning contributions were at least partially based on encoder-decoder networks. Kamnitsas et al. [9], who were the clear winner of the challenge, created an ensemble by combining three different network architectures, namely 3D FCN [12], 3D U-Net [14,15] and DeepMedic [5], trained with different loss functions (Dice loss [16,17] and crossentropy) and different normalization schemes. Wang et al. [10] used a FCN inspired architecture, enhanced with dilated convolutions [13] and residual connections [18]. Instead of directly learning to predict the regions of interest, they trained a cascade of networks that would first segment the whole tumor, then given the whole tumor the tumor core and finally given the tumor core the enhancing tumor. Isensee et al. [6] employed a U-Net inspired architecture that was trained on large input patches to allow the network to capture as much contextual information as possible. This architecture made use of residual connections [18] in the encoder only, while keeping the decoder part of the network as simple as possible. The network was trained with a multiclass Dice loss and deep supervision to improve the gradient flow.

Recently, a growing number of architectural modifications to encoder-decoder networks have been proposed that are designed to improve the performance of the networks for their specific tasks [6,7,10,17,19–22]. Due to the sheer number of such variants, it becomes increasingly difficult for researchers to keep track of which modifications extend their usefulness over the few datasets they are typically demonstrated on. We have implemented a number of these variants and found that they provide no additional benefit if integrated into a well trained U-Net. In this context, our contribution to the BraTS 2018 challenge is intended to demonstrate that such a U-Net, without using significant architectural alterations, is capable of generating competitive state of the art segmentations.

2   Methods

In the following we present the network architecture and training schemes used for our submission. As hinted in the previous paragraph, we will use a 3D U-Net architecture that is very close to its original publication [15] and optimize the training procedure to maximize its performance on the BraTS 2018 training and validation data.

2.1   Preprocessing

With MRI intensity values being non standardized, normalization is critical to allow for data from different institutes, scanners and acquired with varying protocols to be processed by one single algorithm. This is particularly true for neural networks where imaging modalities are typically treated as color channels. Here we need to ensure that the value ranges match not only between patients but between the modalities as well in order to avoid initial biases of the network. We found the following workflow to work well. We normalize each modality of each patient independently by subtracting the mean and dividing by the standard deviation of the brain region. The region outside the brain is set to 0. As opposed to normalizing the entire image including the background, this strategy will yield comparative intensity values within the brain region irrespective of the size of the background region around it.

2.2   Network Architecture

U-Net [14] is a successful encoder-decoder network that has received a lot of attention in the recent years. Its encoder part works similarly to a traditional classification CNN in that it successively aggregates semantic information at the expense of reduced spatial information. Since in segmentation, both semantic as well as spatial information are crucial for the success of a network, the missing spatial information must somehow be recovered. U-Net does this through the decoder, which receives semantic information from the bottom of the 'U' (see Fig. 1) and recombines it with higher resolution feature maps obtained directly from the encoder through skip connections. Unlike other segmentation networks, such as FCN [12] and previous iterations of DeepLab [13] this allows U-Net to segment fine structures particularly well.

Our network architecture is an instantiation of the 3D U-Net [15] with minor modifications. Following our successful participation in 2017 [6], we stick with our design choice to process patches of size $128 \times 128 \times 128$ with a batch size of two. Due to the high memory consumption of 3D convolutions with large patch sizes, we implemented our network carefully to still allow for an adequate number of feature maps. By reducing the number of filters right before upsampling and by using inplace operations whenever possible, this results in a network with 30 feature channels at the highest resolution, which is nearly double the number we could train with in our previous model (using the same 12 GB NVIDIA Titan X GPU). Due to our choice of loss function, traditional ReLU activation functions

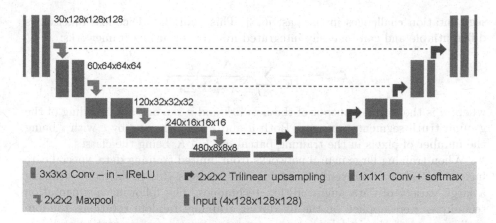

Fig. 1. We use a 3D U-Net architecture with minor modifications. It uses instance normalization [23] and leaky ReLU nonlinearities and reduces the number of feature maps before upsampling. Feature map dimensionality is noted next to the convolutional blocks, with the first number being the number of feature channels.

did not reliably produce the desired results, which is why we replaced them with leaky ReLUs (leakiness 10^{-2}) throughout the entire network. With a small batch size of 2, the exponential moving averages of mean and variance within a batch learned by batch normalization [24] are unstable and do not reflect the feature map activations at test time very well. We found instance normalization [23] to provide more consistent results and therefore used it to normalize all feature map activations (between convolution and nonlinearity). For an overview over our segmentation architecture, please refer to Fig. 1.

2.3 Training Procedure

Our network architecture is trained with randomly sampled patches of size $128 \times 128 \times 128$ voxels and batch size 2. We refer to an epoch as an iteration over 250 batches and train for a maximum of 500 epochs. The training is terminated early if the exponential moving average of the validation loss ($\alpha = 0.95$) has not improved within the last 60 epochs. Training is done using the ADAM optimizer with an initial learning rate $lr_{init} = 1 \cdot 10^{-4}$, which is reduced by factor 5 whenever the above mentioned moving average of the validation loss has not improved in the last 30 epochs. We regularize with a l2 weight decay of 10^{-5}.

One of the main challenges with brain tumor segmentation is the class imbalance in the dataset. While networks will train with crossentropy loss function, the resulting segmentations may not be ideal in the sense of the Dice score they obtain. Since the Dice scores is one of the most important metrics based upon which contributions are ranked, it is imperative to optimize this metric. We achieve that by using a soft Dice loss for the training of our network. While several formulations of the Dice loss exist in the literature [16,17,25], we prefer to use a multi-class adaptation of [16] which has given us good results in segmentation challenges in the past [6,8]. This multiclass Dice loss function is differentiable and can be easily integrated into deep learning frameworks:

$$\mathcal{L}_{dc} = -\frac{2}{|K|} \sum_{k \in K} \frac{\sum_i u_i^k v_i^k}{\sum_i u_i^k + \sum_i v_i^k} \tag{1}$$

where u is the softmax output of the network and v is a one hot encoding of the ground truth segmentation map. Both u and v have shape i by c with i being the number of pixels in the training patch and $k \in K$ being the classes.

When training large neural networks from limited training data, special care has to be taken to prevent overfitting. We address this problem by utilizing a large variety of data augmentation techniques. The following augmentation techniques were applied on the fly during training: random rotations, random scaling, random elastic deformations, gamma correction augmentation and mirroring. Data augmentation was done with our own in-house framework which is publically available at https://github.com/MIC-DKFZ/batchgenerators.

The fully convolutional nature of our network allows to process arbitrarily sized inputs. At test time we therefore segment an entire patient at once, alleviating problems that may arise when computing the segmentation in tiles with a network that has padded convolutions. We furthermore use test time data augmentation by mirroring the images and averaging the softmax outputs.

2.4 Region Based Prediction

Wang et al. [10] use a cascade of CNNs to segment first the whole tumor, then the tumor core and finally the enhancing tumor. We believe the cascade and their rather complicated network architecture to be of lesser importance, but the fact that they did not learn the labels (enhancing tumor, edema, necrosis) but instead optimized the regions that are finally evaluated in the challenge directly to be key to their good performance in last years challenge. For this reason we will also train a version of our model where we replace the final softmax with a sigmoid and optimize the three (overlapping) regions (whole tumor, tumor core and enhancong tumor) directly with the Dice loss.

2.5 Cotraining

285 training cases is a lot for medical image segmentation, but may still not be enough to prevent overfitting entirely. We therefore also experiment with cotraining on additional public and institutional data. For public data, we chose to use the BraTS data made available in the context of the Medical Segmentation Decathlon (http://medicaldecathlon.com). This dataset comprises 484 cases with ground truth segmentations collected from older BraTS challenges.

Cotraining is done for only two datasets at a time. Given that the label definitions between BraTS 2018 and the other datasets may differ, we use separate segmentation layers ($1 \times 1 \times 1$ convolution) at the end, which act as a supervised version of m heads [26]. During training, each segmentation layer only receives gradients from examples of its corresponding dataset. The losses of both layers are averaged to obtain the total loss of a minibatch. The rest of the network weights are shared.

2.6 Postprocessing

One of the most challenging parts in the BraTS challenge data is distinguishing small blood vessels in the tumor core region (that must be labeled either as edema of as necrosis) from enhancing tumor. This is particularly detrimental for LGG patients that may have no enhancing tumor at all. The BraTS challenge awards a Dice score of 1 if a label is absent in both the ground truth and the prediction. Conversely, only a single false positive voxel in a patient where no enhancing tumor is present in the ground truth will result in a Dice score of 0. Therefore we replace all enhancing tumor voxels with necrosis if the total number of predicted enhancing tumor is less than some threshold. This threshold is chosen for each experiment independently by optimizing the mean Dice (using the above mentioned convention) on the BraTS 2018 training cases.

2.7 Dice and Cross-entropy

While being widely popular and providing state of the art results on many medical segmentation challenges, the Dice loss has some downsides, such as badly calibrated softmax probabilities (basically binary 0-1 predictions) and occasional convergence issues (if the true positive term is too small for rare classes) compared to the negative log-likelihood loss (also referred to as cross-entroy loss function). We therefore also experiment with using these losses in conjunction by using both a Dice as well as a negative log-likelihood term and simply adding them together to form the total loss (unweighted sum).

3 Experiments and Results

We designed our training scheme by running a five fold cross-validation on the 285 training cases of BraTS 2018. If additional data is used, the additional training cases are split into five folds as well and used for co-training. Training set results are summarized in Table 3, validation set results can be found in Table 2. Unless noted otherwise, validation set results were obtained by using the five networks from the training cross-validation as an ensemble. For consistency with other publications, all reported values were computed by the online evaluation platform (https://ipp.cbica.upenn.edu/).

Due to the relatively small size of the validation set (66 cases vs 285 training cases) we base our main analysis on the cross-validation results. We are confident that conclusions drawn from the training set are more robust and will generalize well to the test set.

Results on the BraTS2018 training data are summarized in Table 3. We refer to our basic U-Net that was trained on BraTS2018 training data with

Table 1. Results on BraTS 2018 training data (285 cases). All results were obtained by running a five fold cross-validation. Metrics were computed by the online evaluation platform.

	Dice			HD95		
	enh.	whole	core	enh.	whole	core
Isensee et al. (2017) [6]	70.69	89.51	82.76	6.24	6.04	6.95
baseline	73.43	89.76	82.17	4.88	5.86	7.11
baseline + reg	73.81	90.02	82.87	5.01	6.26	6.48
baseline + reg + cotr (dec)	75.94	91.33	85.28	4.29	4.82	5.05
baseline + reg + cotr (dec) + post	**78.68**	91.33	85.28	3.49	**4.82**	**5.05**
baseline + reg + cotr (dec) + post + DC&CE	78.62	**91.75**	**85.69**	**2.84**	4.88	5.11
baseline + reg + cotr (inst) + post + DC&CE	76.32	90.35	84.36	3.74	5.64	5.98
baseline + reg + post + DC&CE	76.78	90.30	83.55	3.66	5.36	6.03

Table 2. Results on BraTS2018 validation data (66 cases). Results were obtained by using the five models from the training set cross-validation as an ensemble. Metrics were computed by the online evaluation platform.

	Dice			HD95		
	enh.	whole	core	enh.	whole	core
baseline	79.59	90.80	84.32	3.12	4.79	8.16
baseline + reg + cotr (dec) + post + DC&CE (*)	80.46	91.21	85.77	2.52	4.38	6.73
baseline + reg + cotr (inst) + post + DC&CE (**)	80.95	91.15	86.6	2.44	5.02	6.73
baseline + reg + post + DC&CE	80.66	90.92	85.22	2.74	5.83	7.20
ensemble of (*) and (**)	80.87	91.26	86.34	2.41	4.27	6.52

large input patches and a Dice loss function as *baseline*. With Dice scores of 73.43/89.76/82.17 (enh/whole/core) on the training set this baseline model is by itself already very strong, especially when compared to the model of Isensee et al. [6] that achieved the third place in BraTS2017 (the training data for both challenges is identical, allowing a direct comparison of the models). Adding region based training (*reg*) improved the Dice scores of both the enhancing tumor as well as the tumor core. When training with decathlon data (*cotr (dec)*), we gain two Dice points in enhancing tumor and minor improvements for the tumor core. Our postprocessing, which is targeted at correcting false positive enhancing tumor predictions in LGG patients has a substantial impact on enhancing tumor Dice. On the training set it increases the mean enhancing tumor Dice by almost three points. Using the sum Dice and cross-entropy as a loss function yields yet another small improvement. Interestingly, using our institutional data for cotraining yields much worse results on the training set. In order to isolate the impact of additional training data we added the model *baseline + reg + post + DC&CE* to the table.

While the model that uses institutional data performed worse on the training set, it was slightly better on the validation set (see Table 2). We explain this discrepancy by the possibility that the Dice and Hausdorff distance scores obtained from the training set cross-validation may be overestimated when cotraining with decathlon data. Since any potential case correspondences between decathlon data and BraTS2018 is unknown due to the naming scheme of the decathlon cases, we cannot exclude the possibility that cases that are currently in the validation split for BraTS 2018 appear in the training split of the decathlon data (albeit with different ground truth segmentations). This uncertainty, along with the strong performance of the model cotrained with institutional data on the validation set led us to the decision to submit an ensemble of these two models. The ensemble achieves Dice scores of 80.87/91.26/86.34 (enh/whole/core) and Hausdorff distances of 2.41/4.27/6.52 on the validation set. For comparison, we also included the validation set result achieved with no additional training data.

Figure 2 shows a qualitative example segmentation. The patient shown is taken from the validation set (CBICA_AZA_1). As can be seen in the middle (t1ce), there are several blood vessels close to the enhancing tumor. Segmentation CNNs typically struggle to correctly differentiate between such vessels and actual enhancing tumor. This is most likely due to a) a difficulty in detecting tube-like structures b) few training cases where these vessels are an issue c) the use of Dice loss functions that does not sufficiently penalize false segmentations of vessels due to their relatively small size. In the case shown here, our model correctly segmented the vessels as background.

Test set results (as communicated by the organizers of the challenge) are presented in Table 3. We used an ensemble of the two models that were trained with institutional and decathlon data for our final submission. Each of these models is in turn an ensemble of five models resulting from the corresponding cross-validation, resulting in a total of 10 predictions for each test case. Our algorithm achieved the second place out of 64 participating teams. We compare

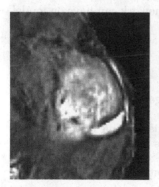 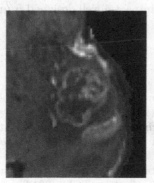 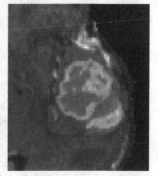

Fig. 2. Qualitative results. The case shown here is patient CBICA_AZA_1 from the validation set. Left: flair, middle: t1ce, right: our segmentation. Enhancing tumor is shown in yellow, necrosis in turquoise and edema in violet. (Color figure online)

Table 3. Test set results of NVDLMED, the winner of BraTS2018, and our method, which achieved the second place.

		Dice			Hausd. dist.		
		enh.	whole	core	enh.	whole	core
NVDLMED	Mean	76.64	88.39	81.54	3.77	5.90	4.81
	StdDev	25.57	11.83	24.99	8.61	10.01	7.52
	Median	84.41	92.06	91.67	1.73	3.16	2.45
MIC-DKFZ	Mean	77.88	87.81	80.62	2.90	6.03	5.08
	StdDev	23.93	12.89	25.02	3.85	9.98	8.09
	Median	84.94	91.79	90.72	1.73	3.16	2.83

our results to the winning contribution by Myronenko et al. (team NVDLMED). While our model had strong results for enhancing tumor, NVDLMED outperformed our approach in both tumor core and whole tumor. Please refer to [27] for a detailed summary of the challenge results.

4 Discussion

In this paper we demonstrated that a generic U-Net architecture that has only minor modifications can obtain very competitive segmentation, if trained correctly. While our base model is already quite strong, enhancing its training procedure by using region-based training, cotraining with additional training data, postprocessing to target false positive enhancing tumor detection as well as a combination of Dice and cross-entropy loss, increases its performance substantially. For our final submission we used an ensemble of a model cotrained with public and another cotrained with institutional data. Despite using only a generic U-Net architecture, our approach achieved the second place in the BraTS2018 challenge, underligning the impact a well designed framework can have on model training.

References

1. Menze, B.H., et al.: The multimodal brain tumor image segmentation benchmark (BRATS). IEEE TMI **34**(10), 1993–2024 (2015)
2. Bakas, S., et al.: Advancing The Cancer Genome Atlas glioma MRI collections with expert segmentation labels and radiomic features. Nat. Sci. Data **4**, 170117 (2017)
3. Bakas, S., et al.: Segmentation labels and radiomic features for the pre-operative scans of the TCGA-GBM collection. TCIA (2017)
4. Bakas, S., et al.: Segmentation labels and radiomic features for the pre-operative scans of the TCGA-LGG collection. TCIA (2017)
5. Kamnitsas, K., et al.: Efficient multi-scale 3D CNN with fully connected CRF for accurate brain lesion segmentation. MIA **36**, 61–78 (2017)

6. Isensee, F., Kickingereder, P., Wick, W., Bendszus, M., Maier-Hein, K.H.: Brain tumor segmentation and radiomics survival prediction: contribution to the BRATS 2017 challenge. In: Crimi, A., Bakas, S., Kuijf, H., Menze, B., Reyes, M. (eds.) BrainLes 2017. LNCS, vol. 10670, pp. 287–297. Springer, Cham (2018). https://doi.org/10.1007/978-3-319-75238-9_25
7. Li, X., Chen, H., Qi, X., Dou, Q., Fu, C.-W., Heng, P.A.: H-DenseUNet: Hybrid densely connected UNet for liver and liver tumor segmentation from CT volumes. arXiv preprint arXiv:1709.07330 (2017)
8. Isensee, F., Jaeger, P.F., Full, P.M., Wolf, I., Engelhardt, S., Maier-Hein, K.H.: Automatic cardiac disease assessment on cine-MRI via time-series segmentation and domain specific features. In: Pop, M., et al. (eds.) STACOM 2017. LNCS, vol. 10663, pp. 120–129. Springer, Cham (2018). https://doi.org/10.1007/978-3-319-75541-0_13
9. Kamnitsas, K., et al.: Ensembles of multiple models and architectures for robust brain tumour segmentation. In: Crimi, A., Bakas, S., Kuijf, H., Menze, B., Reyes, M. (eds.) BrainLes 2017. LNCS, vol. 10670, pp. 450–462. Springer, Cham (2018). https://doi.org/10.1007/978-3-319-75238-9_38
10. Wang, G., Li, W., Ourselin, S., Vercauteren, T.: Automatic brain tumor segmentation using cascaded anisotropic convolutional neural networks. In: Crimi, A., Bakas, S., Kuijf, H., Menze, B., Reyes, M. (eds.) BrainLes 2017. LNCS, vol. 10670, pp. 178–190. Springer, Cham (2018). https://doi.org/10.1007/978-3-319-75238-9_16
11. Kamnitsas, K., et al.: DeepMedic for brain tumor segmentation. In: Crimi, A., Menze, B., Maier, O., Reyes, M., Winzeck, S., Handels, H. (eds.) BrainLes 2016. LNCS, vol. 10154, pp. 138–149. Springer, Cham (2016). https://doi.org/10.1007/978-3-319-55524-9_14
12. Long, J., Shelhamer, E., Darrell, T.: Fully convolutional networks for semantic segmentation. In: Proceedings of the IEEE Conference on Computer Vision and Pattern Recognition, pp. 3431–3440 (2015)
13. Chen, L.-C., Papandreou, G., Kokkinos, I., Murphy, K., Yuille, A.L.: DeepLab: semantic image segmentation with deep convolutional nets, atrous convolution, and fully connected CRFs. IEEE Trans. Pattern Anal. Mach. Intell. 40(4), 834–848 (2018)
14. Ronneberger, O., Fischer, P., Brox, T.: U-Net: convolutional networks for biomedical image segmentation. In: Navab, N., Hornegger, J., Wells, W.M., Frangi, A.F. (eds.) MICCAI 2015. LNCS, vol. 9351, pp. 234–241. Springer, Cham (2015). https://doi.org/10.1007/978-3-319-24574-4_28
15. Çiçek, Ö., Abdulkadir, A., Lienkamp, S.S., Brox, T., Ronneberger, O.: 3D U-Net: learning dense volumetric segmentation from sparse annotation. In: Ourselin, S., Joskowicz, L., Sabuncu, M.R., Unal, G., Wells, W. (eds.) MICCAI 2016. LNCS, vol. 9901, pp. 424–432. Springer, Cham (2016). https://doi.org/10.1007/978-3-319-46723-8_49
16. Drozdzal, M., Vorontsov, E., Chartrand, G., Kadoury, S., Pal, C.: The importance of skip connections in biomedical image segmentation. In: Carneiro, G., et al. (eds.) LABELS/DLMIA -2016. LNCS, vol. 10008, pp. 179–187. Springer, Cham (2016). https://doi.org/10.1007/978-3-319-46976-8_19
17. Milletari, F., Navab, N., Ahmadi, S.-A.: V-Net: Fully convolutional neural networks for volumetric medical image segmentation. In: International Conference on 3D Vision, pp. 565–571. IEEE (2016)
18. He, K., Zhang, X., Ren, S., Sun, J.: Identity mappings in deep residual networks. In: Leibe, B., Matas, J., Sebe, N., Welling, M. (eds.) ECCV 2016. LNCS, vol. 9908, pp. 630–645. Springer, Cham (2016). https://doi.org/10.1007/978-3-319-46493-0_38

19. Jégou, S., Drozdzal, M., Vazquez, D., Romero, A., Bengio, Y.: The one hundred layers tiramisu: fully convolutional densenets for semantic segmentation. In: 2017 IEEE Conference on Computer Vision and Pattern Recognition Workshops (CVPRW), pp. 1175–1183. IEEE (2017)
20. Oktay, O., et al.: Attention U-Net: learning where to look for the pancreas. arXiv preprint arXiv:1804.03999 (2018)
21. Roy, A.G., Navab, N., Wachinger, C.: Concurrent spatial and channel squeeze & excitation in fully convolutional networks. arXiv preprint arXiv:1803.02579 (2018)
22. Kayalibay, B., Jensen, G., van der Smagt, P.: CNN-based segmentation of medical imaging data. arXiv preprint arXiv:1701.03056 (2017)
23. Ulyanov, D., Vedaldi, A., Lempitsky, V.: Instance normalization: The missing ingredient for fast stylization. arXiv preprint arXiv:1607.08022 (2016)
24. Ioffe, S., Szegedy, C.: Batch normalization: Accelerating deep network training by reducing internal covariate shift. arXiv preprint arXiv:1502.03167 (2015)
25. Sudre, C.H., Li, W., Vercauteren, T., Ourselin, S., Jorge Cardoso, M.: Generalised dice overlap as a deep learning loss function for highly unbalanced segmentations. In: Cardoso, M.J., et al. (eds.) DLMIA/ML-CDS.-2017. LNCS, vol. 10553, pp. 240–248. Springer, Cham (2017). https://doi.org/10.1007/978-3-319-67558-9_28
26. Lee, S., Purushwalkam, S., Cogswell, M., Crandall, D., Batra, D.: Why M heads are better than one: Training a diverse ensemble of deep networks. arXiv preprint arXiv:1511.06314 (2015)
27. Bakas, S., et al.: Identifying the best machine learning algorithms for brain tumor segmentation, progression assessment, and overall survival prediction in the brats challenge. arXiv preprint arXiv:1811.02629 (2018)

3D-ESPNet with Pyramidal Refinement for Volumetric Brain Tumor Image Segmentation

Nicholas Nuechterlein$^{(\boxtimes)}$ and Sachin Mehta

University of Washington, Seattle, WA 98195, USA
{nknuecht,sacmehta}@cs.washington.edu

Abstract. Automatic quantitative analysis of structural magnetic resonance (MR) images of brain tumors is critical to the clinical care of glioma patients, and for the future of advanced MR imaging research. In particular, automatic brain tumor segmentation can provide volumes of interest (VOIs) to scale the analysis of advanced MR imaging modalities such as perfusion-weighted imaging (PWI), diffusion-weighted imaging (DTI), and MR spectroscopy (MRS), which is currently hindered by the prohibitive cost and time of manual segmentations. However, automatic brain tumor segmentation is complicated by the high heterogeneity and dimensionality of MR data, and the relatively small size of available datasets. This paper extends ESPNet, a fast and efficient network designed for vanilla 2D semantic segmentation, to challenging 3D data in the medical imaging domain [11]. Even without substantive pre- and post-processing, our model achieves respectable brain tumor segmentation results, while learning only 3.8 million parameters. 3D-ESPNet achieves dice scores of 0.850, 0.665, and 0.782 on whole tumor, enhancing tumor, and tumor core classes on the test set of the 2018 BraTS challenge [1–4,12]. Our source code is open-source and available at https://github.com/sacmehta/3D-ESPNet.

Keywords: Glimoa · BraTS · ESPNet · CNN · Semantic segmentation

1 Introduction

Glioma is the most common primary brain tumor. Due to glioma's highly heterogeneous appearance, extent, and shape, segmentation of brain tumors in MR volumes is one of the most challenging tasks in neuroradiology [7]. This is compounded by the sparsity of data and the heterogeneity incurred by differing scanner models and manufacturers, imaging sites, variation in clinical standards and protocols, and the noise introduced by the movement of patients' heads during scans. At every clinical visit, glioma patients generally receive standard

---

N. Nuechterlein and S. Mehta—Equally contributed.

© Springer Nature Switzerland AG 2019
A. Crimi et al. (Eds.): BrainLes 2018, LNCS 11384, pp. 245–253, 2019.
https://doi.org/10.1007/978-3-030-11726-9_22

of care FLAIR, post-contrast T1-weighted (T1ce), T2, and T1 MR sequences, each of which is described by a distinct volume. These sequences give distinct and complementary information about the tumor extent and composition.

Automated brain tumor segmentation also ranks among the most difficult problems in medical image analysis. The notion that massive amounts of data are required to train deep networks is widely held. Not only are MR scans scarce, they are high dimensional (e.g. $240 \times 240 \times 155 \times 4$) and contain high class imbalances (e.g. $\geq 95\%$ background class). Thus, naive models are predisposed to exhibit extreme background bias.

In similar biomedical domains, patchwise approaches have helped address problems of data shortages and dimensionality. Ciresan et al. proposed a sliding-window method to segment electron microscopic images of the brain, which both localized the problem and exaggerated the dataset [6,14]. Ronneberger et al.'s 2D encoder-decoder network, U-Net, outperformed Ciresan's method [14]. U-Net is a fully convolutional network (FCN) where the traditional pooling operations in the contracting (encoding) path are mirrored by upsampling operations in the symmetric expanding (decoding) path. Skip connections are passed from encoding blocks on the contracting path to same-level decoding blocks in the expanding path.

While some success has been reached using 2D FCNs, like U-Net, these models ignore crucial 3D spatial context, which is undesirable given that most clinical imaging data are volumetric. However, even among 3D FCNs such as DeepMedic, a previous winner of the BraTS competition, fine spatial information is discarded in pooling [9]. This motivates our interest in U-Net's skip connections and, in particular, the architecture of Milletari et al.'s 3D extension of U-Net, V-Net. V-Net benchmarked well on the "PROMISE2012" challenge, where it gave impressive segmentations of MR prostate scans after training on only 50 examples [13].

ESPNet is a faster, more efficient take on U-Net's encoder-decoder architecture [11]. In this paper, we seek to extend and benchmark ESPNet on 3D medical imaging data.

We outline our paper as follows. Section 2 describes our network architecture. We report our methods in Sect. 3. Experimental results are given in Sect. 4. Finally, we close with a discussion of limitations and future directions for our work in Sect. 5.

2   Network Architecture

Our network is an end-to-end system consisting of 3D-ESPNet followed by pyramidal refinement, as shown in Fig. 1. We describe the main building block of our architecture, the ESP module, and, later, 3D-ESPNet's segmentation architecture and pyramidal refinement.

2.1   ESP Module

The Efficient Spatial Pyramid (ESP) module, shown in Fig. 2, is an efficient convolutional module proposed in [11]. The module is based on the RSTM

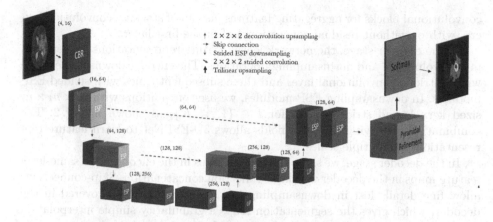

Fig. 1. 3D-ESPNet with pyramidal refinement. 3D-ESPNet's encoder is shown on the left; the decoder is shown on the right with pyramidal refinement. Parentheses give the channel dimensions of incoming and outgoing feature maps. The CBR block consists of a convolutional block followed by batch normalization and ReLU. Light-blue feature maps in the decoder indicate concatenation by long-range, skip connections. Light-blue feature maps in the encoder indicate strided ESP models for downsampling. Arrows are defined in the legend. (Color figure online)

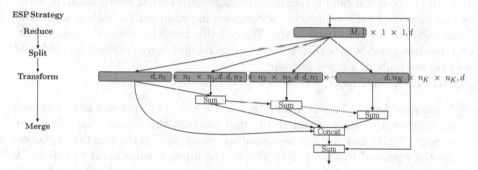

Fig. 2. The Efficient Spatial Pyramid (ESP) module. The blocks in blue represent 3D convolutional layers and are denoted as (# input channels, effective receptive field, # output channels). The ESP module takes an input feature map with M channels and produces an output feature map with N channels, where $d = \frac{N}{K}$ and K represents the number of parallel branches. (Color figure online)

(Reduce-Split-Transform-Merge) strategy and allows the aggregation of the information from a large effective receptive field while learning fewer parameters. We extend the ESP block by replacing its spatial 2D convolutions with volumetric 3D convolutions.

2.2 3D-ESPNet Structure

3D-ESPNet is an encoder-decoder network that extends U-Net [14]. The primary distinction between 3D-ESPNet and U-Net is that 3D-ESPNet employs efficient convolutional blocks for aggregating features instead of stacking convolution layers (with or without residual connections) after the first layer.

In the encoder stage, the network learns feature representations by performing convolutional and downsampling operations. The encoder downsamples once with a strided convolutional layer and three subsequent times with strided ESP modules. In downsampling ESP modules, we use convolutions with $n_i \times n_i \times n_i$ sized kernels and stride of two, for $i \in \{1, \ldots, K\}$, as shown in Fig. 2. The combination of varying receptive fields allows 3D-ESPNet to learn feature representations at multiple scales.

In the decoder stage, we share the feature maps in the encoder with same-level feature maps in the decoder via skip-connection concatenation. Skip-connections allow fine details lost in downsampling in the encoder to be recovered in the decoder, which gives the segmentation maps a granularity simple interpolation cannot achieve. The decoder uses $3 \times 3 \times 3$ deconvolution kernels to upsample the encoder output once, followed by trilinear upsampling layer to return to the resolution at the networks second level. The feature maps of the final ESP module in the decoder are passed into the pyramidal refinement module. The block diagram of 3D-ESPNet is shown in Fig. 1.

Pyramidal Refinement: Pyramid-based approaches sub-sample either the feature maps or the convolutional kernel to learn global contextual information. Inspired by the success of such approaches for segmenting complex 2D scenes, we extend these modules for volumetric data. We call this module pyramidal refinement. Our module combines both feature map-based and convolutional kernel-based pooling methods in a novel fashion.

Pyramidal refinement, shown in Fig. 4, consists of three layers:

- *Projection Layer:* This is a standard $3 \times 3 \times 3$ convolutional layer followed by batch normalization and ReLU that projects the feature maps from the previous ESP block to C-dimensional space, where C is the number of classes.
- *Spatial Pyramid Pooling (SPP) Block*: The input feature maps to this block are low dimensional ($C = 4$). We sub-sample them using convolutional kernels of different sizes and merge their output using sum operations. This is similar to the ASPP block except that we do not use dilated convolutions [5].
- *PSP Block:* A PSP block, sketched in Fig. 3b, is based on the principle of *split-pool-transform-upsample* [15]. *Split:* A PSP block distributes the input feature maps across four parallel branches. *Pool:* Each branch downsamples the feature maps using a different pooling rate. *Transform:* The downsampled feature maps are transformed using point-wise convolutions. *Upsample:* The transformed feature maps are upsampled to the same resolution as the input feature maps using bilinear interpolation. *Merge:* The upsampled feature maps are concatenated with the input feature maps to produce the output feature maps.

Pyramidal refinement is followed by a classification layer. This final layer pools the feature maps using another SPP block and then upsamples by a factor of two using trilinear interpolation. Two convolutional layers are stacked on top of the upsampled feature maps before a softmax.

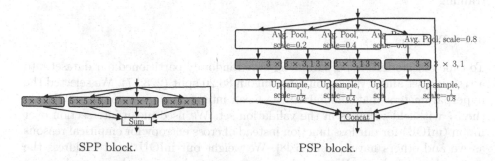

SPP block.   PSP block.

Fig. 3. Pooling modules used in a pyramidal refinement block. Here, a convolutional layer is represented as (kernel size, dilation rate).

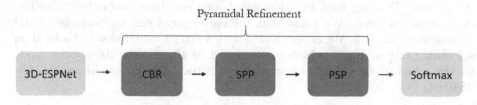

Fig. 4. Pyramidal-refinement. After the second upsampling operation in the 3D-ESPNet decoder, the feature maps are passed through a CBR block, a spatial pyramid pooling block (SPP), and a pyramid pooling module (PSP) at 1/4 resolution. We then upsample to input resolution using trilinear interpolation and compress and pass the feature maps through a softmax to obtain a prediction.

3   Methods

3.1   Data

We train on the Multimodal Brain Tumor Segmentation Challenge (BraTS) 2018 training set, which provides 285 multi-institutional pre-operative multimodal MR tumor scans, each consisting of T1, post-contrast T1-weighted (T1ce), T2, and FLAIR volumes [1–4, 12]. Each case is annotated with the following voxel labels: enhancing tumor, peritumoral edema, background, and necrotic core and non-enhancing tumor. Necrotic core and non-enhancing tumor share a single label. These data are co-registered to the standard MNI anatomical template, interpolated to the same resolution, and skull-stripped. Ground-truth segmentations are manually drawn and approved by neuroradiologists.

3.2   Preprocessing

We used minimal preprocessing. We performed min-max normalization. We also cropped each volume to remove any padding around the brain common in every modality; this allowed us to double our batch size to four, which stabilized training.

3.3   Training

To tune our model's hyperparameters, we randomly partitioned our dataset into a training set and a validation set using an 80:20 split (228:57). We selected the hyperparameters that maximized the mean intersection over union (mIOU) on the 57 withheld volumes in the validation set. We used mean intersection over union (mIOU) for our loss function instead of cross entropy for empirical reasons as we and others have observed [8]. We weight our mIOU loss to address the severe class imbalance. We used data augmentation heavily including scaling and random flips.

We implemented our model in PyTorch. We trained at full resolution on all modalities on an NVIDIA Titan X using a batch size of four. We trained for 300 epochs. Training took less than five hours; test time evaluation takes less than twenty seconds. We found that the optimizer Adam outperformed SGD with momentum [10]. We experimented with learning rate decay and settled on a learning rate of $10e^{-4}$, which we decreased to $10e^{-5}$ after 200 epochs. Code for this adaptation of ESPNet is available at https://github.com/sacmehta/3D-ESPNet.

4   Results

Results on the BraTS 2018 online test and validation sets are shown in Table 1. Visual inspection reveals out model's flexible performance on difficult cases such as gliomas that cross the corpus callosum–so-called butterfly gliomas–shown in Figs. 5 and 6. However, our method lacks some of the granularity present in

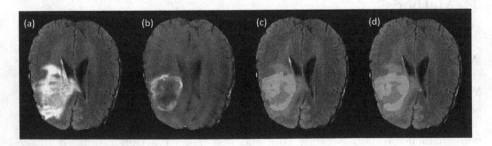

Fig. 5. A butterfly high-grade glioma. (a) FLAIR sequence; (b) T1ce sequence; (c) network prediction; (d) ground truth segmentation.

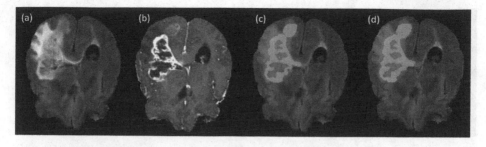

Fig. 6. A second butterfly high-grade glioma. (a) FLAIR sequence; (b) T1ce sequence; (c) network prediction; (d) ground truth segmentation.

the ground truth segmentation. It is clear in the examples provided that our network's predictions are too smooth, especially in Fig. 5, where the predicted non-enhancing and necrotic class is the correct size and in the correct position, but the segmentation does not follow the sharp contours of the gyri outlined in the ground truth. In Fig. 6, we notice that our network tends not to predict necrotic or non-enhancing tumor outside of the tumor-enhancing ring. However, our model is able to handle gaping holes inside tumors filled with cerebrospinal fluid (CSF) just as a resection cavity would appear. This is shown in Fig. 8. These cavities differ from a typical necrotic core on the T2 sequences of a tumor as CSF shows extreme hyperintensity. This robustness is crucial for segmenting post-operative scans which can contain large resection cavities (Figs. 7 and 9).

Table 1. Results obtained on BraTS 2018 online test set are shown in bold. Results obtained on BraTS 2018 online validation set are shown in parenthesis. Sensitivity and specificity results were not given for the online test set.

3D-ESPNet	Dice Score		Sensitivity		Specificity		Hausdorff95	
Whole tumor	**0.850**	(0.883)	-	(0.934)	-	(0.990)	**9.598**	(5.461)
Enhancing tumor	**0.665**	(0.737)	-	(0.831)	-	(0.997)	**5.497**	(5.295)
Tumor core	**0.782**	(0.814)	-	(0.821)	-	(0.997)	**8.668**	(7.850)

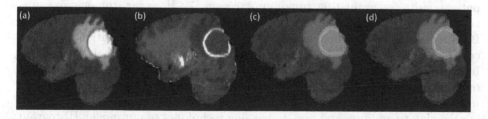

Fig. 7. A sagittal view of a high-grade glioma. (a) FLAIR sequence; (b) T1ce sequence; (c) network prediction; (d) ground truth segmentation.

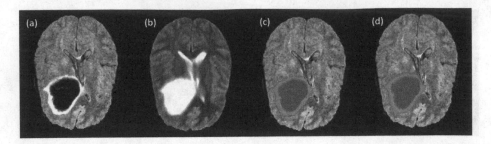

Fig. 8. Low-grade glioma showing bright CSF fluid in ventricles and tumor cavity on the T2 sequence. (a) FLAIR sequences; (b) T2 sequence; (c) network prediction; (d) ground truth segmentation.

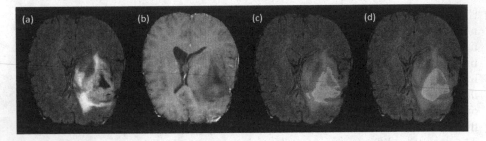

Fig. 9. Low-grade glioma. (a) FLAIR sequence; (b) T1 sequence; (c) network prediction; (d) ground truth segmentation.

5   Discussion

We propose a fast and efficient network for semantic brain tumor segmentation. 3D-ESPNet with pyramidal refinement achieves a respectable 0.850 dice score for whole tumor segmentation on the 2018 BraTS online test set without substantial pre- or post-processing, while learning only 3.8 million parameters.

Brain tumor segmentation has its place in clinic, though neuroradiologist and neuro-oncologists usually limit its use to quantifying volumetric changes in tissue types (edema, enhancing tissue, non-enhancing or necrotic tissue) between patient visits for evaluating tumor progression [7]. However, tumor segmentation is essential to the analysis of advanced MR imaging (DWI, DTI, MRSI). Because such segmentation is usually done manually, segmentation time and cost prevent advanced MR imaging studies from being done at scale. Automatic brain tumor segmentation will allow such advanced imaging studies to be done on massive datasets and, therefore, avail themselves of strong ML analysis and more definitive conclusions.

We plan to add pre- and post-processing techniques to our model. Histogram equalization and N4BiasFieldCorrection might better prepare the training data, and adding a conditional random field after the classifier may help eliminate spurious tumor predictions. We achieved a dice score of 0.850 on the whole tumor class, but work remains to be done on the individual classes. Better hyperparameter tuning and non-linear data augmentation may also improve our performance.

References

1. Bakas, S., et al.: Advancing the cancer genome atlas glioma MRI collections with expert segmentation labels and radiomic features. Sci. Data **4**, 170117 (2017)
2. Bakas, S., et al.: Segmentation labels and radiomic features for the pre-operative scans of the TCGA-GBM collection. The Cancer Imaging Archive (2017). https:// doi.org/10.7937/K9/TCIA.2017.KLXWJJ1Q
3. Bakas, S., et al.: Segmentation labels and radiomic features for the pre-operative scans of the TCGA-LGG collection. The Cancer Imaging Archive (2017). https:// doi.org/10.7937/K9/TCIA.2017.GJQ7R0EF
4. Bakas, S., et al.: Identifying the best machine learning algorithms for brain tumor segmentation, progression assessment, and overall survival prediction in the brats challenge. arXiv preprint arXiv:1811.02629 (2018)
5. Chen, L.C., Papandreou, G., Kokkinos, I., Murphy, K., Yuille, A.L.: Deeplab: semantic image segmentation with deep convolutional nets, atrous convolution, and fully connected CRFs. IEEE Trans. Pattern Anal. Mach. Intell. **40**(4), 834–848 (2018)
6. Ciresan, D.C., Gambardella, L.M., Giusti, A., Schmidhuber, J.: Deep neural networks segment neuronal membranes in electron microscopy images. In: NIPS, pp. 2852–2860 (2012)
7. Fink, J.R., Muzi, M., Peck, M., Krohn, K.A.: Multimodality brain tumor imaging: MR imaging, PET, and PET/MR imaging. J. Nucl. Med. **56**(10), 1554–1561 (2015)
8. Kamnitsas, K., et al.: Ensembles of multiple models and architectures for robust brain tumour segmentation. In: Crimi, A., Bakas, S., Kuijf, H., Menze, B., Reyes, M. (eds.) BrainLes 2017. LNCS, vol. 10670, pp. 450–462. Springer, Cham (2018). https://doi.org/10.1007/978-3-319-75238-9_38
9. Kamnitsas, K., et al.: Deepmedic for brain tumor segmentation. In: BrainLes@MICCAI (2016)
10. Kingma, D.P., Ba, J.: Adam: a method for stochastic optimization. arXiv preprint arXiv:1412.6980 (2014)
11. Mehta, S., Rastegari, M., Caspi, A., Shapiro, L., Hajishirzi, H.: Espnet: efficient spatial pyramid of dilated convolutions for semantic segmentation. arXiv preprint arXiv:1803.06815 (2018)
12. Menze, B.H., et al.: The multimodal brain tumor image segmentation benchmark (brats). IEEE Trans. Med. Imaging **34**(10), 1993 (2015)
13. Milletari, F., Navab, N., Ahmadi, S.A.: V-net: fully convolutional neural networks for volumetric medical image segmentation. In: 2016 Fourth International Conference on 3D Vision (3DV), pp. 565–571 (2016)
14. Ronneberger, O., Fischer, P., Brox, T.: U-Net: convolutional networks for biomedical image segmentation. In: Navab, N., Hornegger, J., Wells, W.M., Frangi, A.F. (eds.) MICCAI 2015. LNCS, vol. 9351, pp. 234–241. Springer, Cham (2015). https://doi.org/10.1007/978-3-319-24574-4_28
15. Zhao, H., Shi, J., Qi, X., Wang, X., Jia, J.: Pyramid scene parsing network. In: IEEE Conference on Computer Vision and Pattern Recognition (CVPR), pp. 2881–2890 (2017)

3D U-Net for Brain Tumour Segmentation

Raghav Mehta$^{(\boxtimes)}$ and Tal Arbel

McGill University, Montreal QC, Canada
{raghav,arbel}@cim.mcgill.ca

Abstract. In this work, we present a 3D Convolutional Neural Network (CNN) for brain tumour segmentation from Multimodal brain MR volumes. The network is a modified version of the popular 3D U-net [13] architecture, which takes as input multi-modal brain MR volumes, processes them at multiple scales, and generates a full resolution multi-class tumour segmentation as output. The network is modified such that there is a better gradient flow in the network, which in turn should allow the network to learn better segmentation. The network is trained end-to-end on BraTS [1–5] 2018 Training dataset using a weighted Categorical Cross Entropy (CCE) loss function. A curriculum on class weights is employed to address the class imbalance issue. We achieve competitive segmentation results on BraTS [1–5] 2018 Testing dataset with Dice scores of 0.706, 0.871, and 0.771 for enhancing tumour, whole tumour, and tumour core, respectively (Docker container of the proposed method is available here: https://hub.docker.com/r/pvgcim/pvg-brats-2018/).

Keywords: Tumour segmentation · Deep learning · Brain MRI

1 Introduction

Automatic quantitative analysis of brain tumours assists in better and faster diagnosis procedure and surgical planning. Development of accurate and reliable tumour segmentation from multi-modal MRI remains a challenging task due to many sources of variability, including: tumour types, shapes and sizes, intensity and contrast difference in MR images, etc. Classical approaches include Multi Atlas segmentation, probabilistic graphical models like Markov Random Field (MRF) [6] and Conditional Random Field (CRF), Random Forest (RF) [7]. These have been successfully used for the task of tumour segmentation. Methods based on generative models have also been explored [8] for tumour segmentation.

Inspired by the success of deep learning in many tasks related to natural images like semantic segmentation [10], object detection [11], and classification [12], many deep learning based approaches have been proposed for various tasks in medical images like segmentation [13], synthesis [14], and classification [15]. Various CNN architectures have been explored for brain tumour segmentation

© Springer Nature Switzerland AG 2019
A. Crimi et al. (Eds.): BrainLes 2018, LNCS 11384, pp. 254–266, 2019.
https://doi.org/10.1007/978-3-030-11726-9_23

which either explicitly [9,16] or implicitly [17,18] model global and local image context. These architectures either take MR images at multiple resolutions as input [9,16] or process single resolution MR images at multiple scales [17,18]. One of the advantages of deep learning based approaches over classical segmentation methods like MRF, RF etc. is that they don't require any hand-crafted features because the networks are trained in end-to-end manner with appropriate loss functions. In recent BraTS challenges [1], deep learning based approaches have outperformed classical methods.

In this work, we develop a modified version of the popular 3D U-net [13] architecture for brain tumour segmentation task on BraTS 2018 datasets. The U-net architecture has been successfully applied to many medical imaging segmentation tasks, such as liver and lesion segmentation [19], retinal layer segmentation [20], organ segmentation [21] etc. In this paper, the 3D U-net is trained using Categorical Cross Entropy (CCE) loss function on BraTS 2018 training dataset and a curriculum on class weights is employed to address class imbalance [26]. We achieved competitive results on BraTS 2018 [5] validation and testing datasets with Dice scores of 0.788, 0.909, and 0.825 on validation dataset, and 0.706, 0.871, and 0.771 on testing dataset for enhancing tumour, whole tumour, and tumour core, respectively.

2 Method

A flowchart of the 3D U-net architecture can be seen in Fig. 1. The network takes as input full 3D volumes of all available sequences of a patient and generates multi-class segmentation of tumours into sub-types, at the same resolution. The 3D U-net is similar to the one proposed in [13], with some modifications. The U-net consists of 4 resolution steps for both encoder and decoder paths. At the start, we use 2 consecutive 3D convolutions of size $3 \times 3 \times 3$ with k filters, where k denotes the user-defined initial number of convolution filters (10). Each step in

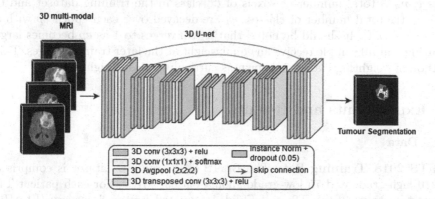

Fig. 1. 3D U-net CNN architecture takes as input four full 3D MR image sequences, and generates the multi-class segmentation of the tumour into sub-types.

the encoder path consists of 2 3D convolutions of size $3 \times 3 \times 3$ with $k * 2^n$ filters, where n denotes the U-net resolution step. This is followed by average pooling of size $2 \times 2 \times 2$. We chose average pooling instead of max pooling as it allows better gradient flow between consecutive layers. At the end of each encoder step, instance normalization [22] is applied, followed by dropout [23] with 0.05 probability. Instance normalization was preferred over batch normalization due to memory constraints, as we were able to fit only one volume at a time in the available GPU memory. In the decoder path at each step, 3D transposed convolution of size $3 \times 3 \times 3$ is applied, with $2 \times 2 \times 2$ stride and $k * 2^n$ filters for the upsampling task. The output of the transposed convolution is concatenated with the corresponding output of the encoder path. We chose transposed convolution as it allows the network to learn an optimal interpolation function instead of a pre-defined interpolation function in the case of standard upsampling. This is, once again, followed by instance normalization and Dropout with 0.05 probability. Finally, 2 3D convolution of size $3 \times 3 \times 3$ with $k * 2^n$ filters are applied. Rectified linear unit is chosen as a non-linearity function for every convolution layer. The last layer has C filters, where C denotes the total number of classes. This is followed by SoftMax non-linearity.

2.1 Loss Function

We optimize weighted Categorical Cross Entropy (CCE) loss function during training. The equation for the same is given below.

$$CCE^i = -\sum_n w_n^i \sum_l t_{n,l}^i \log p_{n,l}^i \tag{1}$$

$$w_n^i = w_l * y_n^i \qquad \text{where,} \ w_l = \left(\frac{\sum_{k=0}^{k=C} m_k}{m_l}\right) * r^{ep} + 1, \tag{2}$$

where, w_n^i and w_l denote the weight for voxel n of volume i and the weight of class l. m_l is total number of voxels of lth class in the training dataset and C denotes the total number of classes. w_l are decayed over each epoch ep with a rate of $r \in [0,1]$. It should be noted that w_l converges to 1 as ep becomes large ensuring that all sample receive an equal weight at the later training stages. This method of weighting classes is known as curriculum class weighting [26].

3   Experiments and Results

3.1   Data

BraTS 2018 Training Set: The BraTS 2018 training dataset is comprised of 210 high-grade and 75 low-grade glioma patient MRIs. For each patient T1, T1 post contrast (T1c), T2, and Fluid Attenuated Inverse Recovery (FLAIR) MR volumes, along with expert tumour segmentation are provided. Each brain tumour is manually delineated into 3 classes: edema, necrotic/non-enhancing core, and enhancing tumour core [1–5].

BraTS 2018 Validation Set: The BraTS 2018 validation dataset is comprised of 66 patient MRIs. For each patient T1, T1c, T2, and FLAIR MR volumes are provided. No expert tumour segmentation masks are provided and the grade of each glioma is not specified [1–5].

BraTS 2018 Testing Set: The BraTS 2018 testing dataset is comprised of 191 patient MRIs. Similar to validation dataset, here for each patient T1, T1c, T2, and FLAIR MR volumes are provided but expert tumour segmentation masks are not provided. The grade of each glioma is also not specified [1–5].

3.2 Pre-processing

The BraTS challenge provides isotropic, skull-stripped, and co-registered MR volumes. We follow this up with a few pre-processing steps. The intensity of volumes were re-scaled using mean subtraction, divided by the standard deviation, and re-scaled from 0 to 1 and were cropped to $184 \times 200 \times 152$.

3.3 5-Fold Cross Validation

We performed 5-fold cross validation on the training dataset. The BraTS 2018 training dataset is randomly split into five folds with 57 patient dataset each such that each fold contains 42 high-grade patients and 15 low-grade patients. We train our network 5 times such that 4 folds are used to train the network and the remaining fold is used to validate the network.

Please note that we use total five networks, obtained by the corresponding cross-validation, as an ensemble to predict segmentation for BraTS 2018 validation and testing datasets. We view this ensemble as bagging [25], which has been shown to improve performance over a single model.

Parameters. In our network, we used initial number of filters $k = 20$ and number of filters in the last layer $C = 4$. We optimize the loss function in Eq. (1) using Adam [24] with a learning rate of 0.001 and batch size of 1. The network is trained for total 240 epochs. Learning rate is decayed by the factor of 0.75 after every 50 epochs. The decay rate r in Eq. (2) is set to 0.95. We regularize the model using data augmentation, where at each training iteration a random affine transformation is applied to the MR volumes and the corresponding segmentation mask. Random translation, rotation, scaling and shear transformations are applied, where the range of transformations is sampled from a uniform distribution of $[-5, 5]$, $[-3°, 3°]$, $[-0.1, 0.1]$, and $[-0.1, 0.1]$, respectively. Volumes are also randomly flipped left to right.

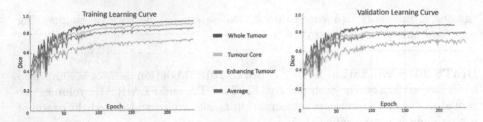

Fig. 2. Training (Left) and Validation (Right) Dice Scores as a function of number of epochs for one of the five cross-validation folds.

Learning Curves. Figure 2 shows an example of evolution of various Dice scores (Tumour, Enhance, Core, and Average) as a function of number of epochs for one of the 5 cross-validation fold.

4   Discussion

4.1   Quantitative Results

Our method performed well, resulting in Dice scores of 0.788, 0.909, and 0.825 (BraTS 2018 validation dataset), and 0.706, 0.871, and 0.771 (BraTS 2018 testing dataset) for the enhancing tumours, whole tumours, and tumour cores, respectively. Tables 1, 2, and 3 show the results of our method based on different evaluation metric statistics, provided by the challenge organizers. The results are based on a number of experiments on the following BraTS 2018 datasets: 5-fold cross validation on the training dataset, and tests on the validation dataset and the testing dataset. The results indicate that the proposed method performs very well on the whole tumours and tumour cores, with relatively lower performance for enhancing tumours. This was expected as enhancing tumours rely heavily on the T1c images, and present similarly to other enhancements on those images. For other tumour sub-types, other modalities assist in the segmentation.

Table 1. Evaluation metric statistics for 5-fold cross validation on BraTS 2018 training dataset for enhancing tumour (ET), whole tumour (WT), and tumour core (TC).

	Dice			Sensitivity			Specificity			Hausdorff-95		
	ET	WT	TC	ET	WT	TC	ET	WT	TC	ET	WT	TC
Mean	0.690	0.888	0.793	0.774	0.880	0.802	0.998	0.995	0.996	7.251	6.600	7.941
StdDev	0.294	0.094	0.206	0.245	0.118	0.210	0.004	0.006	0.007	13.318	11.215	11.805
Median	0.817	0.918	0.876	0.861	0.913	0.879	0.999	0.996	0.999	2.237	3.606	4.062
25quantile	0.641	0.878	0.748	0.709	0.850	0.723	0.997	0.994	0.996	1.414	2.236	2.236
75quantile	0.878	0.941	0.926	0.935	0.958	0.942	0.999	0.998	0.999	5.385	6.557	9.327

Table 2. Evaluation metric statistics for BraTS 2018 validation dataset for enhancing tumour (ET), whole tumour (WT), and tumour core (TC).

	Dice			Sensitivity			Specificity			Hausdorff-95		
	ET	WT	TC	ET	WT	TC	ET	WT	TC	ET	WT	TC
Mean	0.788	0.909	0.825	0.824	0.911	0.811	0.998	0.995	0.998	3.520	4.923	8.316
StdDev	0.233	0.059	0.179	0.222	0.082	0.212	0.004	0.004	0.002	4.992	8.154	13.521
Median	0.869	0.921	0.902	0.893	0.933	0.901	0.999	0.996	0.999	1.732	2.914	3.240
25quantile	0.809	0.894	0.773	0.824	0.880	0.711	0.998	0.994	0.998	1.414	2.000	2.000
75quantile	0.911	0.951	0.945	0.942	0.964	0.958	0.999	0.998	0.999	3.000	4.970	8.658

Table 3. Evaluation metric statistics for BraTS 2018 testing dataset for enhancing tumour (ET), whole tumour (WT), and tumour core (TC).

	Dice			Hausdorff-95		
	ET	WT	TC	ET	WT	TC
Mean	0.706	0.871	0.771	4.145	6.547	8.316
StdDev	0.286	0.139	0.269	5.321	11.806	8.119
Median	0.820	0.914	0.894	2.236	3.162	3.081
25quantile	0.683	0.868	0.750	1.414	2.236	2.059
75quantile	0.878	0.943	0.931	3.162	5.385	7.106

4.2   Qualitative Results

Figures 3 and 4 show examples of slices with the resulting segmentation labels for high-grade and low-grade glioma patients from one fold of the experiments on the BraTS 2018 training dataset. We can observe that the network performs much better on high-grade glioma cases. This can be attributed to the fact that we have more training examples of high-grade glioma cases as compared to low-grade glioma cases. Example slices with the predicted segmentation labels on the BraTS 2018 validation and testing datasets can be seen in Figs. 5, 6, and 7.

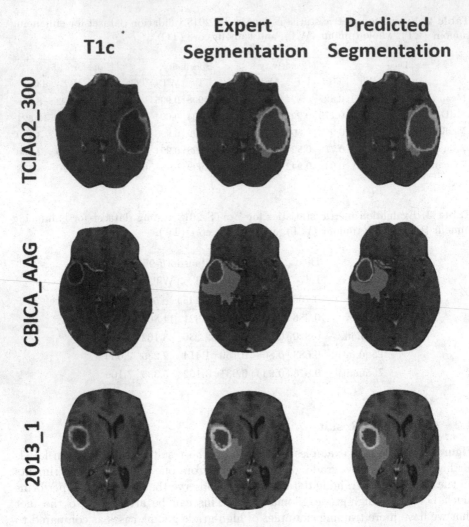

Fig. 3. Examples of high-grade glioma segmentation results for BraTS 2018 training dataset. On T1c MR volume (Column 1), Expert Segmentation (Column 2) and Predicted Segmentation (Column 3) are overlaid. The green label is edema, the red label is non-enhancing or necrotic tumour code, and the yellow label is enhancing tumour core. (Color figure online)

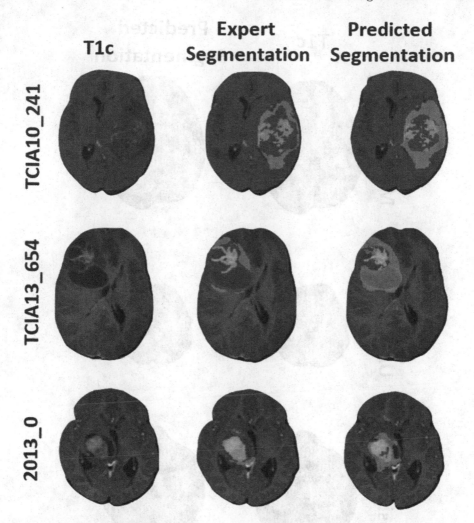

Fig. 4. Examples of low-grade glioma segmentation results for BraTS 2018 training dataset. On T1c MR volume (Column 1), Expert Segmentation (Column 2) and Predicted Segmentation (Column 3) are overlaid. The green label is edema, the red label is non-enhancing or necrotic tumour code, and the yellow label is enhancing tumour core. (Color figure online)

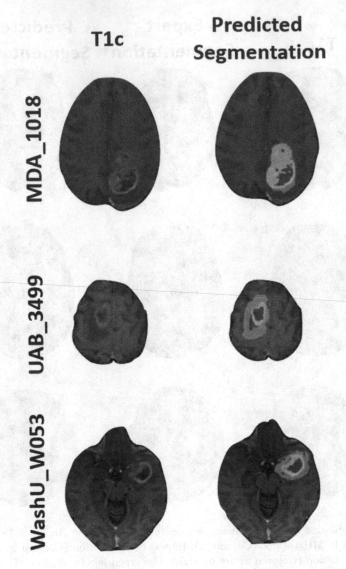

Fig. 5. Examples of segmentation results for BraTS 2018 validation dataset. On T1c MR volume (Column 1) predicted segmentation (Column 2) is overlaid. The green label is edema, the red label is non-enhancing or necrotic tumour code, and the yellow label is enhancing tumour core. (Color figure online)

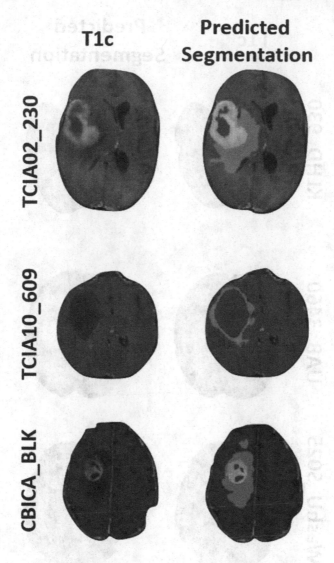

Fig. 6. Examples of segmentation results for BraTS 2018 validation dataset. On T1c MR volume (Column 1) predicted segmentation (Column 2) is overlaid. The green label is edema, the red label is non-enhancing or necrotic tumour code, and the yellow label is enhancing tumour core. (Color figure online)

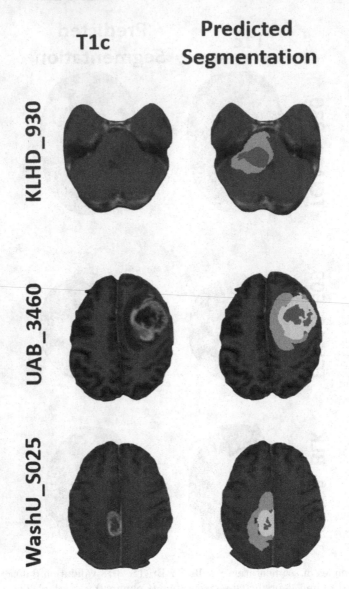

Fig. 7. Examples of segmentation results for BraTS 2018 testing dataset. On T1c MR volume (Column 1) predicted segmentation (Column 2) is overlaid. The green label is edema, the red label is non-enhancing or necrotic tumour code, and the yellow label is enhancing tumour core. (Color figure online)

5   Conclusion

In this work, we demonstrated how a simple CNN network like 3D U-net [13] can be successfully applied for the task of tumour segmentation. U-net process the input multi-modal MR images at multiple scales, which allows it to learn local and global context necessary for tumour segmentation. The network was trained using a curriculum on class weights to address class imbalance, showing competitive results for brain tumour segmentation on BraTS 2018 [5] testing dataset. Our method performed well and we got following Dice scores for enhancing tumour, whole tumour, and tumour core on BraTS 2018 [5] validation and testing datasets: 0.788, 0.909, and 0.825 (validation dataset), and 0.706, 0.871, and 0.771 (testing dataset). But our method showed degradation in performance on the testing dataset in the categories of Enhancing Tumours (ET) and Tumour Core (TC).

Acknowledgment. This work was supported by a Canadian Natural Science and Engineering Research Council (NSERC) Collaborative Research and Development Grant (CRDPJ 505357 - 16) and Synaptive Medical. We gratefully acknowledge the support of NVIDIA Corporation for the donation of the Titan X Pascal GPU used for this research.

References

1. Menze, B.H., et al.: The multimodal brain tumour image segmentation benchmark (BRATS). IEEE TMI **34**(10), 1993 (2015)
2. Bakas, S., et al.: Advancing the cancer genome atlas glioma MRI collections with expert segmentation labels and radiomic features. Sci. Data **4**, 170117 (2017)
3. Bakas, S., et al.: Segmentation labels and radiomic features for the pre-operative scans of the TCGA-GBM collection. Cancer Imaging Arch. **286** (2017)
4. Bakas, S., et al.: Segmentation labels and radiomic features for the pre-operative scans of the TCGA-LGG collection. Cancer Imaging Arch. (2017)
5. Bakas, S., Reyes, M., et al.: Identifying the best machine learning algorithms for brain tumor segmentation, progression assessment, and overall survival prediction in the BRATS challenge. arXiv preprint arXiv:1811.02629 (2018)
6. Subbanna, N., et al.: Iterative multilevel MRF leveraging context and voxel information for brain tumour segmentation in MRI. In: Proceedings of the IEEE CVPR, pp. 400–405 (2014)
7. Zikic, D., et al.: Context-sensitive classification forests for segmentation of brain tumour tissues. In: Proc MICCAI-BraTS, pp. 1–9 (2012)
8. Menze, B.H., van Leemput, K., Lashkari, D., Weber, M.-A., Ayache, N., Golland, P.: A generative model for brain tumor segmentation in multi-modal images. In: Jiang, T., Navab, N., Pluim, J.P.W., Viergever, M.A. (eds.) MICCAI 2010. LNCS, vol. 6362, pp. 151–159. Springer, Heidelberg (2010). https://doi.org/10.1007/978-3-642-15745-5_19
9. Kamnitsas, K., et al.: Efficient multi-scale 3D CNN with fully connected CRF for accurate brain lesion segmentation. Med. Image Anal. **36**, 61–78 (2017)
10. Long, J., et al.: Fully convolutional networks for semantic segmentation. In: Proceedings of the IEEE CVPR, pp. 3431–3440 (2015)
11. Ren, S., et al.: Faster R-CNN: towards real-time object detection with region proposal networks. IEEE Trans. Pattern Anal. Mach. Intell. **6**, 1137–1149 (2017)
12. Krizhevsky, A., et al.: ImageNet classification with deep convolutional neural networks. In: NIPS, pp. 1097–1105 (2012)

13. Çiçek, Ö., Abdulkadir, A., Lienkamp, S.S., Brox, T., Ronneberger, O.: 3D U-Net: learning dense volumetric segmentation from sparse annotation. In: Ourselin, S., Joskowicz, L., Sabuncu, M.R., Unal, G., Wells, W. (eds.) MICCAI 2016. LNCS, vol. 9901, pp. 424–432. Springer, Cham (2016). https://doi.org/10.1007/978-3-319-46723-8_49

14. Chartsias, A., et al.: Multimodal MR synthesis via modality-invariant latent representation. IEEE TMI **37**(3), 803–814 (2018)

15. Mazurowski, M.A., et al.: Deep learning in radiology: an overview of the concepts and a survey of the state of the art. arXiv preprint arXiv:1802.08717 (2018)

16. Havaei, M., et al.: Brain tumour segmentation with deep neural networks. Med. Image Anal. **35**, 18–31 (2017)

17. Havaei, M., Guizard, N., Chapados, N., Bengio, Y.: HeMIS: hetero-modal image segmentation. In: Ourselin, S., Joskowicz, L., Sabuncu, M.R., Unal, G., Wells, W. (eds.) MICCAI 2016. LNCS, vol. 9901, pp. 469–477. Springer, Cham (2016). https://doi.org/10.1007/978-3-319-46723-8_54

18. Shen, H., Wang, R., Zhang, J., McKenna, S.J.: Boundary-aware fully convolutional network for brain tumor segmentation. In: Descoteaux, M., Maier-Hein, L., Franz, A., Jannin, P., Collins, D.L., Duchesne, S. (eds.) MICCAI 2017. LNCS, vol. 10434, pp. 433–441. Springer, Cham (2017). https://doi.org/10.1007/978-3-319-66185-8_49

19. Christ, P.F., et al.: Automatic liver and lesion segmentation in CT using cascaded fully convolutional neural networks and 3D conditional random fields. In: Ourselin, S., Joskowicz, L., Sabuncu, M.R., Unal, G., Wells, W. (eds.) MICCAI 2016. LNCS, vol. 9901, pp. 415–423. Springer, Cham (2016). https://doi.org/10.1007/978-3-319-46723-8_48

20. Roy, A.G., et al.: ReLayNet: retinal layer and fluid segmentation of macular optical coherence tomography using fully convolutional networks. Biomed. Opt. Express **8**(8), 3627–3642 (2017)

21. Roth, H.R., et al.: Hierarchical 3D fully convolutional networks for multi-organ segmentation. arXiv preprint arXiv:1704.06382 (2017)

22. Ulyanov, D., et al.: Instance normalization: the missing ingredient for fast stylization. arXiv preprint arXiv:1607.08022

23. Srivastava, S., et al.: Dropout: a simple way to prevent neural networks from overfitting. JMLR **15**(1), 1929–1958 (2014)

24. Kingma, D.P., Ba, J.: Adam: a method for stochastic optimization. arXiv preprint arXiv:1412.6980

25. Breiman, L.: Bagging predictors. Mach. Learn. **24**(2), 123–140 (1996)

26. Jesson, A., Arbel, T.: Brain tumor segmentation using a 3D FCN with multi-scale loss. In: Crimi, A., Bakas, S., Kuijf, H., Menze, B., Reyes, M. (eds.) BrainLes 2017. LNCS, vol. 10670, pp. 392–402. Springer, Cham (2018). https://doi.org/10.1007/978-3-319-75238-9_34

Automatic Brain Tumor Segmentation with Contour Aware Residual Network and Adversarial Training

Hao-Yu Yang[1,2]($\boxtimes$) and Junlin Yang[1]

[1] Yale University, New Haven, CT 06511, USA
hao-yu.yang@yale.edu
[2] Cura Cloud Cooperation, Seattle, WA 98104, USA

Abstract. Localizing brain tumor and identifying different subtypes of tissues play a crucial role in treatment assessment and management of gliomas. In this paper, we present a contour-aware 3D convolution neural network (CNN) with adversarial training for segmenting gliomas and an ensemble of models for overall survival prediction. For the segmentation task, contour loss and adversarial loss are added as an auxiliary information in addition to the pixel-wise classification loss to ensure the segmentation results mimic the contours of the ground truth annotation. We employed both random-forest-based and neural-network-based regression scores for predicting overall survival time. Hand-crafted imaging feature incorporated with the non-imaging feature is employed. The proposed method was evaluated on the BraTS 2018 dataset and achieved competitive results for both segmentation and survival prediction tasks. We demonstrate that raw segmentation results can be improved by incorporating extra constraints in contours and adversarial training.

Keywords: Neural networks · Adversarial training · Contour aware

1 Introduction

Gliomas are one of the most common types of brain tumors. The segmentation of gliomas and intra-tumor substructures are essential to progression monitor and treatment assessment of the disease. [16] It also serves as an essential step for radiomics applications such as survival prediction.

High-Grade Gliomas (HGG) is an aggressive tumor subtype of glial cells. Patients with HGG exhibit highly differing prognosis. Life expectancy after initial diagnosis is immensely important to patients and their caregivers [6]. The accuracy of OS prediction is highly dependent on the gilomas segmentation results.

Previous studies have found successes in automated brain tumor segmentation. Among existing methods, Convolution Neural Networks (CNN) based methods have achieved state-of-the-art performance, especially with U-Net type

© Springer Nature Switzerland AG 2019
A. Crimi et al. (Eds.): BrainLes 2018, LNCS 11384, pp. 267–278, 2019.
https://doi.org/10.1007/978-3-030-11726-9_24

architecture [10]. The rise of residual CNNs [8] made the training of neural networks with extremely deep architectures possible, which in turns led to higher segmentation accuracy. Combination of these methods granted neural networks more representational power to capture the variations in complex data [14]. The results of the segmentation can then be utilized in overall survival (OS) prediction. OS prediction in previous works involved the combination of radiomic features and machine learning models.

There are several difficulties that lay in these tasks. First, the gilomas vary significantly in shape, size, and location. Second, the normal brain anatomy structure varies from patients to patients as well, making anomaly detection more difficult. Furthermore, the proportion of brain tumors to normal brain tissue is quite low, resulting in extreme class imbalance for tumor segmentation. To address these challenges, we propose a contour-aware neural network with adversarial training.

Active contour is a well-established theoretical approach for robust image segmentation [17]. Incorporating contour information in deep neural networks demonstrated the model capability for segmentation can be improved [5]. Generative adversarial network (GAN) [7] were originally introduced as a generative model that is capable of generating new samples from intractable data distribution such as images. Recent researches [11] demonstrated the potential of using adversarial networks on image segmentation task.

In this work, we've tackled two tasks in 2018 Multimodal Brain Tumor Segmentation Challenge (BraTS). For the tumor segmentation task, we employed an end-to-end trainable adversarial network that consists of a segmentor and a discriminator. Our segmentor is a 3D Residual U-Net designed to be contour-aware by adding contour constraint in the training procedure. A discriminator network is trained alongside the segmentation network to provide auxiliary supervision. By incorporating a discriminator network that attempts to distinguish true annotations from predicted segmentation, it enforces CNNs to learn long range spatial label contiguity. The classification loss from the discrimination network provides gradients for the generating realistic segmentation results. We achieved an average dice score of 0.79 across three subtypes of gliomas. For survival prediction task, we proposed an ensemble of models including convolutional neural network and random forest. Hand-crafted radiomics features and non-imaging clinical features are both incorporated in the random forest model. The details of our methods and results evaluated on the BraTS 2018 datasets are introduced in the following sections.

2   Method

2.1   Data

The proposed model is trained and evaluated on the 2018 BraTS challenge dataset [1–4,12] dataset. The training set contained a total of 285 cases. Each case includes 4 MRI modalities: T1-weighted (T1), T1-weighted with gadolinium-enhancing contrast (T1ce), T2-weighted (T2) and Fluid Attenuation Inversion Recovery(FLAIR). Example of the training data can be found in Fig. 1. Among the 285 cases, 210 cases are listed as High-Grade Gliomas(HGG) and 75 cases Low-Grade Gliomas (LGG).

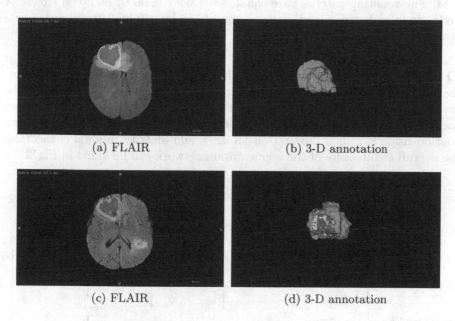

(a) FLAIR   (b) 3-D annotation

(c) FLAIR   (d) 3-D annotation

Fig. 1. Example of training data and manual annotation of the 2018 BraTS challenge

2.2 Gliomas Segmentation

For the Gliomas segmentation task, participants are asked to design an automatic system that produces pixel-wise labels for different sub-regions of brain lesion. Pixel-wise annotation labels that include Gd-enhancing tumor (ET), peritumoral edema(ED) and necrotic/non-enhancing tumor(NCR/NET) were provided alongside the image data. Figure 1 shows an example MRI from BraTS 2018 dataset with tumor sub-regions labeled in different colors. The segmentation results are evaluated by several matrices including Dice score, Hausdorff distance, sensitivity and specificity of the respective tumor tissue.

2.3 Preprocess

Appropriate image preprocessing is essential for the following computing. The official dataset provided by the organizers have been skull-striped and co-registered. Bias correction in [15] is performed on raw data before segmentation. Intensity normalization is also applied for better performance. More details on preprocessing can be found in [13].

Due to the 3D and multi-modality nature of MRI and memory limitations of the current GPU, it is unfeasible to directly feed entire volumes to the proposed network. During the training phase, a single volume is broken down to smaller patches using a moving window. We set the stride size of the sliding window at 24. The resulting patches have equal dimensions in 96 by 96 by 96 voxels. A single modality image with the dimension of 155 by 240 by 240 will generate a total of 72 3-D patches.

3D Residual Unet. The backbone of the segmentation network is a 3D U-net. The U-net consisted of both down sampling and up sampling pathways. The downsampling pathway is made up of multiple residual blocks. Each residual block contained three $3 \times 3 \times 3$ convolution layer, batch normalization, activation function using Rectified Linear Unit (ReLU), and identity residual connection. The overall architecture of the segmentation network can be found in Fig. 2.

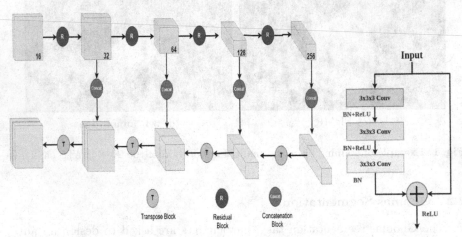

(a) Schematic of the segmentation network. Each circle rep-   (b) Residual block
resents a different set of operations.

Fig. 2. Architecture of the 3-D residual U-net

Contour Aware. In order to obtain a robust segmentation that captures the highly varying multi-class tumor contours, we added an auxiliary constraining factor derived from the ground truth segmentation. During training, the model derives an elastic snake (also known as active contour) [9] for each ground truth mask (Fig. 3).

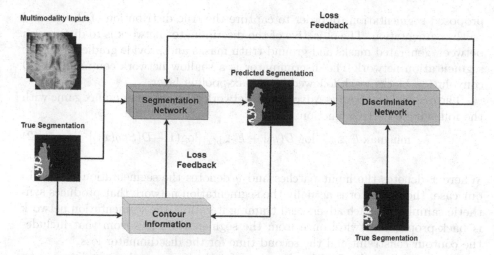

Fig. 3. Overview of the training pipeline for the segmentation task

The total energy of the deformable spine can be written as:

$$E_{snake} = \int_0^1 E_{snake}(v(s))ds$$
$$= \int_0^1 (E_{internal}(v(s)) + E_{image}(v(s)) + E_{contour}(v(s))ds \quad (1)$$

where the set of n points v_i denotes the snake itself. The total energy of the snake is a combination of internal elastic energy and the inherent energy of the image.

In traditional computer vision framework, boundary finding tasks involving active contours usually have objectives that minimize the total energy of the snake. Segmentation masks can be directly obtained by using the snake alone. However, these methods need higher-level supervision for the actual contours. In our case, the snake acts as an supervision and not the main segmentation method. Therefore the target of the snake is the ground truth label and the neural-network predictions. In other words, active contour was performed on the segmentation results and not the image of interest. We used the Hausdorff distance between ground truth contours and prediction contours as a measure of dissimilarity. The objective is to force the neural network to generate predictions that have similar contours to human annotated images. The Hausdorff distance is then added as a loss term to the final objective function.

Adversarial Training. The adversarial training pipeline includes two separate neural networks: the aforementioned residual U-net and an auxiliary discriminator network. In each iteration, the segmentation network will generate a proposed segmentation. In order to capture the true distribution of the ground truth segmentation. The objective of the discriminator network is to distinguish between generated masks and ground truth masks and provide gradients for the segmentation network The discriminator is a shallow network containing 3 3D convolution blocks, each followed by a max-pooling layer.

The original Generative Adversarial Network [7] plays an minimax game with the following objective function:

$$\min_{Seg} \max_{D} E_{y \sim p_{(y)}}[\log D(y)] + E_{x \sim p(x)}[\log(1 - D(Seg(x)))] \tag{2}$$

Where x denotes the input patches and y denotes the segmentation masks. In our case, the generator is actually the segmentation network that produces synthetic samples. At each adversarial training iteration, the segmentation network is back-propagated twice: once from the segmentation loss from that includes the contour constraint and the second time for the discriminator loss.

To summarize, the final loss function for the segmentation pipeline can be written as:

$$\mathcal{L} = \lambda_a \cdot \mathcal{L}_{adver} + \lambda_c \cdot \mathcal{L}_{contour} + \lambda_D \cdot \mathcal{L}_{Dice} \tag{3}$$

where $\mathcal{L}_{adver}$ is the adversarial loss, $\mathcal{L}_{contour}$ is the Hausdorff distance and $\mathcal{L}_{Dice}$ is the negative dice score. λ are the weighting coefficient for each loss. The weighting coefficients were initialized at 1 for all terms. Online normalization was performed at each iteration. Online normalization takes the standard deviation of each individual loss term up to the current iteration and the weighting coefficients were adjusted accordingly. This ensures that the same magnitude of across three loss terms and that the model doesn't skew towards any particular loss.

2.4   Overall Survival Prediction

The second part of the BraTS challenge concerned the prediction of OS of HGG patients using pre-operative scans. For this task, we proposed an ensemble of models including Convolution Network-based regression network and hand-crafted-feature-based random forest model. The overall pipeline of the model can be found in Fig. 6. During the validation and test phase, we assign different weight to combine the above models according to training loss to get the final prediction.

Regression Network. The segmentation results are essential for predicting survival time as it contains extracted information about the tumor tissue. We've designed the regression network to take the four modalities MRI volumes and the tumor segmentation prediction from the first task as input. The regression network contained 5 residual blocks similar to ones in the segmentation network. The last layer of the regression network, however, contained a single node with no activation function. This allowed the network to directly output the survival days (Fig. 4).

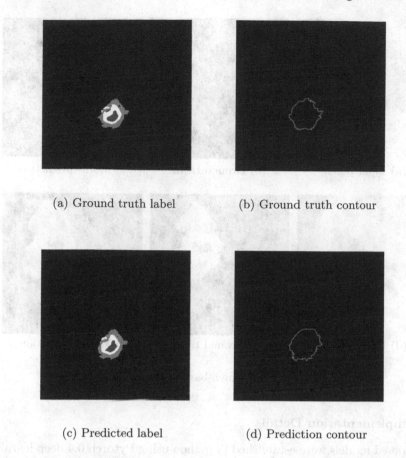

(a) Ground truth label   (b) Ground truth contour

(c) Predicted label   (d) Prediction contour

Fig. 4. Visualized comparison of ground truth contours and prediction contours

Random Forest. Random Forest is a powerful model that can be used for both classification and regression. Here, we utilized hand-crafted image features extracted from the tumor segmentation mask, including the relative size and number of different subtype tumors as inputs for the random forest model. All image-derived features are calculated in 2D and scaled before being fed to the model. The age of the patients is also incorporated as a non-imaging feature. We set the number of estimators in the random forest as 300 (Fig. 5).

Ensemble. The variability of a single model can be quite high. In order to reduce the prediction variance, we ensemble results from regression network and random forests at inference phase. The final prediction is the mean survival time of both models.

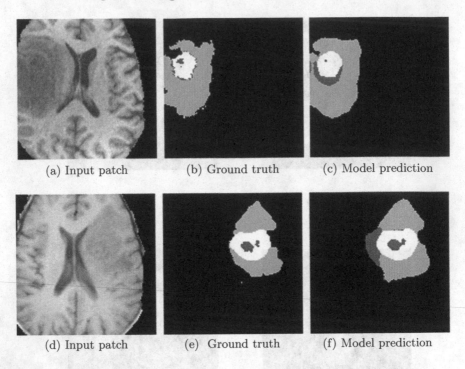

(a) Input patch   (b) Ground truth   (c) Model prediction

(d) Input patch   (e) Ground truth   (f) Model prediction

Fig. 5. Visualization of patch-based segmentation results

2.5 Implementation Details

All proposed models were established in python using Pytorch 0.4 deep learning framework. The experiments with GPU training was done on NVIDIA Tesla V100. Different learning rates were set for the segmentation model and the discriminator network to avoid diminished gradients caused by discriminator network over-powering its counterpart in early epochs. At the beginning of training, the learning rate of 0.001 was assigned to the segmentation network and 0.0005 for the discriminator network. Learning rate decay will be activated if there were no improvements in 10 consecutive epochs. A single decay will reduce the learning rate to 0.8 of the previous iteration. The batch size was set at 8 patches per batch. Learning rate of 0.0001 was set for the survival regression model. The designated number of epochs for both models were 500. Training will be terminated early if there have been no improvements for 30 consecutive epochs. The training for the segmentation pipeline including both the segmentation network and the discriminator network took roughly 36 hours on 2 Tesla V100. As for the survival prediction, 6 hours were needed to train both the regression network and random forest.

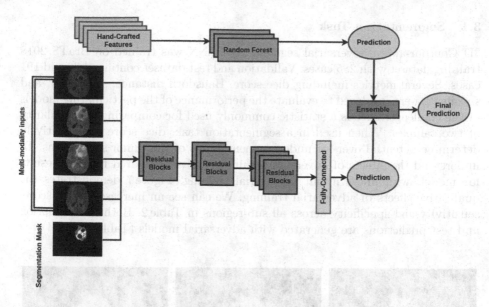

Fig. 6. Schematic overview of the OS prediction task

3   Results

We have trained and evaluated both the segmentation and survival prediction model on the BraTS 2018 dataset. The following Table 1 show the results of the training, validation and test dataset according to the BraTS 2018 on-line evaluation system. We also discuss the effects of the employing adversarial training in this section.

Table 1. Hand-crafted imaging features for random forest model

No.	Hand-Crafted Feature	No.	Hand-Crafted Feature
1	sum of ET area	6	mean intensity of ET
2	sum of ED area	7	mean intensity of ED
3	sum of NCR area	8	mean intensity of NCR
4	sum of all lesion area	9	eccentricity of ET
5	Patient Age	10	eccentricity of ED

No.	Hand-Crafted Feature
11	eccentricity of NCR
12	lesion normal ratio

3.1   Segmentation Task

3D Contour-aware adversarial segmentation CNN was trained on BraTS 2018 training dataset with 285 cases. Validation and test dataset contained 66 and 191 cases. Several metrics including dice score, Hausdorff distance, sensitivity, and specificity were employed to evaluate the performance of the participating models in the challenge. Dice is a statistic commonly used for comparing the similarity of two samples. When used in a segmentation task, dice score can effectively determine potential over- or under-segmentation's of the tumor sub-regions. To understand the effects of adversarial training, we've compared training results for models with and without adversarial training. Figure 7 demonstrates the qualitative effects of adversarial training. We can see an increase of dice score, sensitivity and specificity across all sub-regions in Table 2. Both the validation and test predictions are generated with adversarial models (Tables 3 and 4).

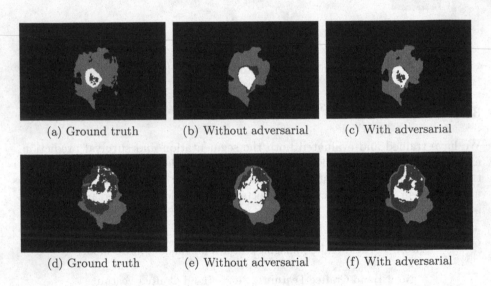

(a) Ground truth   (b) Without adversarial   (c) With adversarial

(d) Ground truth   (e) Without adversarial   (f) With adversarial

Fig. 7. Qualitative comparison of adversarial training

Table 2. Effects of adversarial training on the training dataset

	DICE			Sensitivity			Specificity			Hausdorff95		
	ET	WT	TC	ET	WT	TC	ET	WT	TC	ET	WT	TC
Without adversarial	0.812	0.910	0.877	0.845	0.961	0.898	0.996	0.994	0.997	3.239	4.718	6.022
With adversarial	0.835	0.941	0.914	0.886	0.871	0.956	0.997	0.994	0.999	3.117	4.276	5.810

Table 3. Segmentation task evaluation on the BraTS 2018 validation dataset

DICE			Sensitivity			Specificity			Hausdorff95		
ET	WT	TC	ET	WT	TC	ET	WT	TC	ET	WT	TC
0.792	0.903	0.854	0.816	0.930	0.853	0.998	0.993	0.998	3.367	5.460	6.588

Table 4. Segmentation task evaluation on the BraTS 2018 test dataset

	DICE			Hausdorff95		
	ET	WT	TC	ET	WT	TC
Mean	0.722	0.869	0.789	5.202	8.051	7.474
StdDev	0.273	0.127	0.24981	9.666	13.920	11.990
Median	0.823	0.909	0.89392	2.236	3.605	3.316
25 quantile	0.704	0.861	0.782	1.414	2.236	2.000
75 quantile	0.880	0.937	0.933	3.316	5.873	7.035

3.2 Survival Prediction Task

The dataset for the survival prediction task is a subset of the segmentation dataset. The survival dataset consisted of 163 and 53 for training and validation respectively. The number of cases expected on the online evaluation for training, validation, and test was 59, 28 and 77 cases.

The results for overall survival prediction are evaluated in two different evaluation schemes, classification, and regression. For classification, each patient is classified as one of the three survival based groups including long, short and middle and accuracy is used to evaluate the predictions. For regression principles, the Mean Square Error (MSE) is computed for each patient. Table 5 shows the results evaluated by the above two principles on training, validation and test dataset.

Table 5. Survival prediction results on the BraTS training, validation and test set

	Accuracy	MSE	medianSE	stdSE	SpearmanR
Training	0.475	44724.102	12996	79142.403	0.666
Validation	0.261	155914.087	43681	275558.255	−0.382
Test	0.195	329619.775	57973.203	863455.278	−0.015

4 Conclusion

In this paper, we've proposed a novel method for accurate pixel-wise classification and survival prediction. For the segmentation task, we've developed an adversarial and contour-aware training paradigm for 3D Residual U-Net.

We show that by introducing extra constraints via contours and adversarial training to the model, the neural network was able to produce predictions that highly resembles the ground truth and fine-tune predictions according to subtle inconsistencies. For Survival prediction, we proposed an ensemble of models including direct regression using a neural network and hand-crafted features-based random forest model to predict survival days.

References

1. Bakas, S., et al.: Advancing the cancer genome atlas glioma MRI collections with expert segmentation labels and radiomic features. Nature Sci. Data **4**, 170117 (2017)
2. Bakas, S., et al.: Segmentation labels and radiomic features for the pre-operative scans of the TCGA-GBM collection (2017)
3. Bakas, S., et al.: Segmentation labels and radiomic features for the pre-operative scans of the TCGA-LGG collection. In: The Cancer Imaging Archive (2017)
4. Bakas, S., et al.: Identifying the Best Machine Learning Algorithms for Brain Tumor Segmentation, Progression Assessment, and Overall Survival Prediction in the BRATS Challenge. ArXiv e-prints, November 2018
5. Chen, H., Qi, X., Yu, L., Heng, P.-A.: DCAN: deep contour-aware networks for accurate gland segmentation. In Proceedings of the IEEE Conference on Computer Vision and Pattern Recognition, pp. 2487–2496 (2016)
6. Cheng, J.-X., Zhang, X., Liu, B.-L.: Health-related quality of life in patients with high-grade glioma. Neuro-oncology **11**(1), 41–50 (2009)
7. Goodfellow, I., et al.: Generative adversarial nets. In: Ghahramani, Z., Welling, M., Cortes, C., Lawrence, N.D., Weinberger, K.Q. (eds.) Advances in Neural Information Processing Systems 27, pp. 2672–2680. Curran Associates Inc (2014)
8. He, K., Zhang, X., Ren, S., Sun, J.: Deep residual learning for image recognition. CoRR abs/1512.03385 (2015)
9. Kass, M., Witkin, A., Terzopoulos, D.: Snakes: active contour models. Int. J. Comput. Vis. **1**(4), 321–331 (1988)
10. Li, X., Chen, H., Qi, X., Dou, Q., Fu, C., Heng, P.: H-DenseUNet: Hybrid densely connected UNet for liver and liver tumor segmentation from CT volumes. CoRR abs/1709.07330 (2017)
11. Luc, P., Couprie, C., Chintala, S., Verbeek, J.: Semantic segmentation using adversarial networks. CoRR abs/1611.08408 (2016)
12. Menze, B.H., et al.: The multimodal brain tumor image segmentation benchmark (brats). IEEE Trans. Med. Imaging **34**(10), 1993–2024 (2015)
13. Nyúl, L.G., Udupa, J.K., Zhang, X.: New variants of a method of MRI scale standardization. IEEE Trans. Med. Imaging **19**(2), 143–150 (2000)
14. Quan, T.M., Hildebrand, D.G.C., Jeong, W.: Fusionnet: A deep fully residual convolutional neural network for image segmentation in connectomics. CoRR abs/1612.05360 (2016)
15. Tustison, N.J., et al.: N4ITK: Improved N3 bias correction. IEEE Trans. Med. Imaging **29**(6), 1310–1320 (2010)
16. Wen, P.Y., Kesari, S.: Malignant gliomas in adults. New Engl, J. Med. **359**(5), 492–507 (2008)
17. Yushkevich, P.A., et al.: User-guided 3D active contour segmentation of anatomical structures: significantly improved efficiency and reliability. Neuroimage **31**(3), 1116–1128 (2006)

Brain Tumor Segmentation Using an Ensemble of 3D U-Nets and Overall Survival Prediction Using Radiomic Features

Xue Feng[1]([envelope]) [iD], Nicholas Tustison[2], and Craig Meyer[1,2]

[1] Biomedical Engineering, University of Virginia,
Charlottesville, VA 22903, USA
xf4j@virginia.edu
[2] Radiology and Medical Imaging, University of Virginia,
Charlottesville, VA 22903, USA

Abstract. Accurate segmentation of different sub-regions of gliomas including peritumoral edema, necrotic core, enhancing and non-enhancing tumor core from multimodal MRI scans has important clinical relevance in diagnosis, prognosis and treatment of brain tumors. However, due to the highly heterogeneous appearance and shape, segmentation of the sub-regions is very challenging. Recent development using deep learning models has proved its effectiveness in the past several brain segmentation challenges as well as other semantic and medical image segmentation problems. Most models in brain tumor segmentation use a 2D/3D patch to predict the class label for the center voxel and variant patch sizes and scales are used to improve the model performance. However, it has low computation efficiency and also has limited receptive field. U-Net is a widely used network structure for end-to-end segmentation and can be used on the entire image or extracted patches to provide classification labels over the entire input voxels so that it is more efficient and expect to yield better performance with larger input size. Furthermore, instead of picking the best network structure, an ensemble of multiple models, trained on different dataset or different hyper-parameters, can generally improve the segmentation performance. In this study we propose to use an ensemble of 3D U-Nets with different hyper-parameters for brain tumor segmentation. Preliminary results showed effectiveness of this model. In addition, we developed a linear model for survival prediction using extracted imaging and non-imaging features, which, despite the simplicity, can effectively reduce overfitting and regression errors.

Keywords: Brain tumor segmentation · Ensemble · 3D U-Net · Deep learning · Survival prediction · Linear regression

1 Introduction

Gliomas are the most common primary brain malignancies, with different degrees of aggressiveness, variable prognosis and various heterogeneous histological sub-regions, i.e. peritumoral edema, necrotic core, enhancing and non-enhancing tumor core. This intrinsic heterogeneity of gliomas is also portrayed in their radiographic phenotypes, as

© Springer Nature Switzerland AG 2019
A. Crimi et al. (Eds.): BrainLes 2018, LNCS 11384, pp. 279–288, 2019.
https://doi.org/10.1007/978-3-030-11726-9_25

their sub-regions are depicted by different intensity profiles disseminated across multimodal MRI (mMRI) scans, reflecting differences in tumor biology. Quantitative analysis of imaging features such as volumetric measures after manual/semi-automatic segmentation of the tumor region has shown advantages in image-based tumor phenotyping over traditionally used clinical measures such as largest anterior-posterior, transverse, and inferior-superior tumor dimensions on a subjectively-chosen slice [1, 2]. Such phenotyping may enable assessment of reflected biological processes and assist in surgical and treatment planning. To compare and evaluate different automatic segmentation algorithms, the Multimodal Brain Tumor Segmentation Challenge (BraTS) 2018 was organized using multi-institutional pre-operative MRI scans for the segmentation of intrinsically heterogeneous brain tumor sub-regions [3, 4]. More specifically, the dataset used in this challenge includes multiple-institutional clinically-acquired pre-operative multimodal MRI scans of glioblastoma (GBM/HGG) and low-grade glioma (LGG) containing (a) native (T1) and (b) post-contrast T1-weighted (T1Gd), (c) T2-weighted (T2), and (d) Fluid Attenuated Inversion Recovery (FLAIR) volumes [5, 6]. 285 training volumes with annotated GD-enhancing tumor, peritumoral edema and necrotic and non-enhancing tumor. Furthermore, to pinpoint the clinical relevance of this segmentation task, BraTS'18 also included the task to predict patient overall survival from images together with the patient age and resection status. To tackle these two tasks, this study is performed with two goals: (1) provide pixel-by-pixel label maps for the three sub-regions and background; (2) estimate the survival days.

Convolutional neural network (CNN) based models have proven their effectiveness and superiority over traditional medical image segmentation algorithms and are quickly becoming the mainstream in BraTS challenges. Due to the highly heterogeneous appearance and shape of brain tumors, small patches are usually extracted to predict the class for the center voxel. To improve model performance, multi-scale patches with different receptive field sizes are often used in the model [7]. In contrast, U-Net is a widely used convolutional network structure that consists of a contracting path to capture context and a symmetric expanding path that enables precise localization with 3D extension [8, 9]. It can be used on the entire image or extracted patches to provide class labels for all input voxels when padding is used. Furthermore, instead of picking the best network structure, an ensemble of multiple models, trained on different dataset or different hyper-parameters, can generally improve the segmentation performance over a single model due to the averaging effect. In this study we propose to use an ensemble of 3D U-Nets with different hyper-parameters trained on non-uniformly extracted patches for brain tumor segmentation. During testing, a sliding window approach is used to predict class labels with adjustable overlap to improve accuracy. With the segmentation labels, we will develop a linear model for survival prediction using extracted imaging features and additional non-imaging features since the linear models can effectively reduce overfitting and thus regression errors.

2   Methods

For the brain tumor segmentation task, the steps in our proposed method include pre-processing of the images, patch extraction, training multiple models using a generic 3D U-Net structure with different hyper-parameters, deployment of each model for full volume prediction and final ensemble modeling. For the survival task, the steps include feature extraction, model fitting, and deployment. Details are described as follows.

2.1   Image Pre-processing

To compensate for the MR inhomogeneity, the bias correction algorithm based on N4ITK was applied to the T1, T1Gd images, T2 and flair images [10]. A smooth inhomogeneity field due to variations in coil sensitivity was estimated and compensated from the images. A non-local means denoising method was then used to reduce noise after bias correction [11]. The implementations on ITK [12] were used with a Python wrapper from Nipype [13]. Python-based parallel execution with multiple threads was used to accelerate the two steps. The processed images were stored for future usage. Figure 1 shows the original T1 image (left), image with only bias correction (center) and image with bias correction and denoising (right). The signal-to-noise ratio (SNR) of the image is increased with the denoising method, which could potentially help improving the segmentation accuracy and robustness against noise.

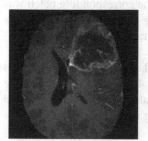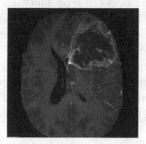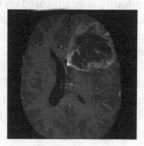

Fig. 1. Original T1 image (left), image with only bias correction (center) and image with bias correction and denoising (right). The right image has improved SNR.

As MR images do not have standard pixel intensity values, to reduce the effects from different contrasts and different subjects, each 3D image was normalized to 0 to 1 separately by subtracting the min values and divided by the pixel intensity range. After normalization, for each subject, images of all contrast were fused to form the last dimension so that the whole input image size becomes $155 \times 240 \times 240 \times 4$.

2.2   Non-uniform Patch Extraction

For simplicity, we will use foreground to denote all tumor pixels and background to denote the rest. There are several challenges in directly using the whole images as the input to a 3D U-Net: (1) the memory of a moderate GPU is often 12 Gb so that in order to fit the model into the GPU, the network needs to greatly reduce the number of features and/or the layers, which often leads to a significant drop in performance as the expressiveness of the network is much reduced; (2) the training time will be greatly prolonged since more voxels contribute to calculation of the gradients at each step and the number of steps cannot be proportionally reduced during optimization; (3) as the background voxels dominate the whole image, the class imbalance will cause the model to focus on background if trained with uniform loss, or prone to false positives if trained with weighted loss that favors the foreground voxels. Therefore, to more effectively utilize the training data, smaller patches were extracted from each subject. As the foreground labels contain much more variability and are the main targets to segment, more patches from the foreground voxels should be extracted.

In implementation, during each epoch, a random patch was extracted from each subject using non-uniform probabilities. The valid patch centers were first calculated by removing edges to make sure each extracted patch was completely within the whole image. The probability of each valid patch center $p_{i,j,k}$ was calculated using the following equation:

$$p_{i,j,k} = \frac{s_{i,j,k}}{\sum_{i,j,k} s_{i,j,k}} \tag{1}$$

in which $s_{i,j,k} = 1$ for all voxels with maximal intensity lower than the 1st percentile, $s_{i,j,k} = 6$ for all foreground voxels and $s_{i,j,k} = 3$ for the rest. The patch center was then randomly selected based on the calculated probability and the corresponding patch was extracted. Since normal brain images are symmetric along the left-right direction, a random flip along this direction was made after patch extraction. No other augmentation was applied.

Before training, the per-input-channel mean and standard deviation of extracted patches were calculated by running the extraction process 400 times, with each time using a randomly selected training subject. The extracted patches were then subtracted with the mean and divided by the standard deviation along each input channel.

2.3   Network Structure and Training

A 3D U-Net based network was used as the general structure, as shown in Fig. 2. Zero padding was used to make sure the spatial dimension of the output is the same with the input. For each encoding block, a VGG like network with two consecutive 3D convolutional layers with kernel size 3 followed by the activation function and batch norm layers were used. The parametric rectilinear function (PReLU), given as:

$$f(x) = \max(0, x) - \alpha \max(0, -x) \tag{1}$$

was used with trainable parameter α as the activation function. The number of features was doubled while the spatial dimension was halved with every encoding block, as in conventional U-Net structure. To improve the expressiveness of the network, a large number of features were used in the first encoding block. Dropout with ratio 0.5 was added after the last encoding block. Symmetric decoding blocks were used with

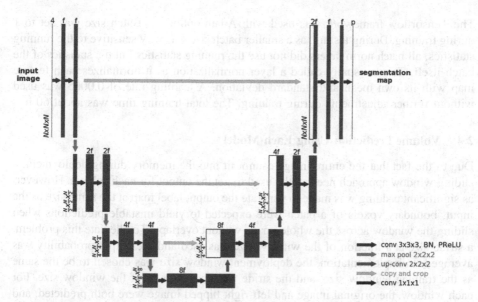

Fig. 2. 3D U-Net structure with 3 encoding and 3 decoding blocks.

skip-connections from corresponding encoding blocks. Features were concatenated to the de-convolution outputs. The extracted segmentation map of the input patch was expanded to the multi-class the ground truth labels (3 foreground classes and the background). Weighted/non-weighted cross entropy was used as the loss function.

The number of encoding/decoding blocks, the weights in the loss function and the patch size were chosen as the tunable hyper-parameters when constructing multiple models. Due to memory limitations, for a larger patch size, the number of features needs to be reduced. In current implementation, due to constraint in computational resources, six models were trained, with detailed parameters shown in Table 1. N denotes the input size, M denotes the number of encoding/decoding blocks and f denotes the input features at the first layer. For weighted loss, 1.0 was used for background and 2.0 was used for each class of foreground voxels.

Training was performed on a Nvidia Titan Xp GPU with 12 Gb memory. 640 epochs were used. As mentioned earlier, during each epoch, only one patch was extracted every subject. Subject orders were randomly permuted every epoch.

Table 1. Detailed parameters for all 6 3D U-Net models.

Model #	M	N	f	Loss type
1	3	64	96	Uniform
2	3	64	96	Weighted
3	4	64	96	Uniform
4	4	96	96	Weighted
5	3	80	64	Uniform
6	3	80	64	Weighted

The Tensorflow framework was used with Adam optimizer. Batch size was set to 1 during training. During testing, as a smaller batch size was very sensitive to the running statistics, all batch norm layers did not use the running statistics but the statistics of the batch itself. This is usually called a layer normalization as it normalizes each feature map with its own mean and standard deviation. A learning rate of 0.0005 was used without further adjustments during training. The total training time was about 60 h.

2.4   Volume Prediction Using Each Model

Due to the fact that the entire image cannot fit into the memory during deployment, a sliding window approach needs to be used to get the output for each subject. However, as significant padding was made to generate the output label map at the same size as the input, boundary voxels of a patch were expected to yield unstable predictions when sliding the window across the whole image without overlaps. To alleviate this problem, a stride size at a fraction of the window size was used and the output probability was averaged. In implementation, the deployment window size was chosen to be the same as the training window size, and the stride was chosen as ½ of the window size. For each window, the original image and left-right flipped image were both predicted, and the average probability after flipping back the output of the flipped input was used as the output. Therefore, each voxel, except for a few on the edge, will be predicted 16 times when sliding across all directions. Although smaller stride sizes can be used to further improve the accuracy with more averages, the deployment time will be increased 8 times for every ½ reduction of the window size and thus will quickly become unmanageable. Using the parameters as mentioned on the same GPU, it took about 1 min to generate the output for the entire volume per subject. Instead of performing a thresholding on the probability output to get the final labels, the direct probability output was saved for each model to the disk.

2.5   Ensemble Modeling

The ensemble modeling process was rather straightforward. The probability output of all classes from each model was read from the disk and the final probability was calculated via simple averaging. The class with the highest probability was selected as the final segmentation label of each voxel.

2.6   Survival Prediction

To predict the post-surgery survival time measured in days, extracted images features and non-image features were used to construct a linear regression model. 6 image features were calculated from the ground truth label maps during training and the predicted label maps during validation. For each foreground class, the volume (V) by summing up the voxels and the surface area (S) by summing up the magnitude of the gradients along three directions were obtained, as described in the following equations

$$V_{ROI} = \sum\nolimits_{i,j,k} s_{i,j,k} \qquad (2)$$

$$S_{ROI} = \sum\nolimits_{i,j,k} s_{i,j,k} \sqrt{(\frac{\partial s}{\partial i})^2 + (\frac{\partial s}{\partial j})^2 + (\frac{\partial s}{\partial k})^2} \qquad (3)$$

in which ROI denotes a specific foreground class and $s_{i,j,k} = 1$ for voxels that are classified to belong to this ROI and $s_{i,j,k} = 0$ otherwise.

Age and resection status were used as non-imaging clinical features. As there were two classes of resection status and many missing values of this status, a two-dimensional feature vector was used to represent the status, given as GTR: (1, 0), STR: (0, 1) and NA: (0, 0). A linear regression model after normalizing the input features to zero mean and unit standard deviation was fit with the training data. As the input feature size is 9, the risk for overfitting is greatly reduced.

3 Results

3.1 Brain Tumor Segmentation

All 285 training subjects were used in the training process. 66 subjects were provided as validation. The dice indexes, sensitivities and specificities, 95 Hausdorff distances of the enhanced tumor (ET), whole tumor (WT) and tumor core (TC) were automatically calculated after submitting to the CBICA's Image Processing Portal. With multiple submissions, we were able to compare the performances of each individual model and the final ensemble.

Table 2 shows the mean dice scores and 95 Hausdorff distances of ET, WT and TC for the 6 individual models and the ensemble of them. Sensitivity and specificity are highly correlated with the dice indexes so that they are not included. The best performance of each evaluation metric is highlighted. All 3D U-Net models perform similar but the ensemble of them has the overall best performances as compared with each individual model. It is also noticed that weighted cross-entropy loss has high sensitivity but lower specificity compared with the uniform counterpart, which is likely due to the fact that by assigning more weights to the foreground, the network tends to be more aggressive in assigning foreground labels.

Table 2. Performances of each individual model and the ensemble

Model #	Dice_ET	Dice_WT	Dice_TC	Dist_ET	Dist_WT	Dist_TC
1	0.7688	0.9015	0.8237	4.1270	4.5437	**5.5226**
2	0.7677	0.9066	0.8248	4.2218	6.4637	8.8593
3	0.7695	0.9040	0.8306	7.1372	8.9214	11.4460
4	0.7707	0.8990	0.8104	**3.1454**	6.0081	6.9814
5	0.7863	0.9078	0.8217	4.1894	4.5704	6.2030
6	0.7616	0.8917	0.8149	4.2222	4.1053	6.9598
Ensemble	**0.7917**	**0.9094**	**0.8362**	4.0186	**3.8009**	5.6451

3.2 Survival Prediction

All 163 training subjects with survival data were used in the training process. The training coefficient of determination was 0.259. 28 cases were evaluated after submitting to the CBICA's Image Processing Portal. The accuracy was 0.321, MSE was 99115.86, median SE was 77757.86, std SE was 104291.596 and Spearman Coefficient was 0.264. The performance on the validation dataset is not as accurate as other top teams in this task, however, our method won the 1st place in the testing dataset, which is likely due to significant overfitting of other teams in validation. The final result is encouraging and shows that a linear model is robust against overfitting.

4 Discussion and Conclusions

In this paper we developed a brain tumor segmentation method using an ensemble of 3D U-Nets. Bias correction and denoising were used as pre-processing. 6 networks were trained with different number of encoding/decoding blocks, input patch sizes and different weights for loss. The preliminary results showed an improvement with ensemble modeling. For survival prediction, we used a simple linear regression by combining radiomics features from images such as volumes and surface areas of each sub-region and non-imaging clinical features.

For segmentation, it is noted that the median metrics are significantly higher than the mean metrics. For example, the median dice indexes were 0.867, 0.923 and 0.904 for ET, WT and TC in the final ensembled model. It makes sense in that the theoretical maximum dice index is 1 and minimum dice index is 0. However, we noted that in several cases, the dice indexes are as low as 0 for ET and TC and 0.6 for WT. It is mostly due to the low sensitivity meaning that the model is not able to recognize the corresponding tumor regions. The possible reason for these failed regions is that their characteristics deviate a lot from the training dataset. This is also encouraging in that for majority of the cases, the segmentation quality is very high.

In the 3D U-Net model, we found that the batch norm layer was helpful in improving the model stability and performance. However, different with the canonical application of the batch norm layer, in which the batch statistics is used in training and the global statistics is used in deployment, it performed much better with batch statistics in deployment than global statistics. Since the batch size is 1, a per-channel normalization is actually performed by subtracting its own mean. One possible explanation could be that by doing such normalization, the model focuses on the differences of neighboring pixels in one channel and ignores the absolute values, which may help the segmentation process. However, further investigation is needed to figure out the exact reason.

Compared with the patch-based model that only predicts the center pixel, when predicting the segmentation label maps for the full patch, different pixels are very likely to have different effective receptive field sizes due to the zero padding in the edge. We argue that a pixel should still be able to be predicted even based on partial receptive field, which, for the very edge pixel, corresponds to only half of the maximal receptive field.

Furthermore, the significant overlap in the sliding windows during deployment can improve the accuracy with more averages.

In the current implementation, 6 networks were trained due to limitations in computation time. It is expected with more networks, the results can be further improved, although the marginal improvement is expected to decrease.

For the survival prediction task, since it is very likely to overfit with such a small dataset and we argue that as many other features may play more important roles in overall survival such as histological and genetic features but unfortunately, they are not available in this challenge, a linear regression model was the safest option to minimize the test errors, although at the cost of its expressiveness. Further exploration of those additional features through clinical collaboration is expected to improve the accuracy of survival prediction.

In conclusion, we developed an ensemble of 3D U-Nets for brain tumor segmentation. The network hyper-parameters are varied to obtain multiple trained models. A linear regression model was also developed for the survival prediction task. Our survival prediction model won the 1st place in the final stage of the competition. The code is available at https://github.com/xf4j/brats18. The paper that summarizes the challenge is available at [14].

References

1. Kumar, V., et al.: Radiomics: the process and the challenges. Magn. Reson. Imaging **30**, 1234–1248 (2012)
2. Gillies, R.J., Kinahan, P.E., Hricak, H.: Radiomics: images are more than pictures, they are data. Radiology **278**, 563–577 (2015)
3. Menze, B.H., et al.: The multimodal brain tumor image segmentation benchmark (BRATS). IEEE Trans. Med. Imaging **34**(10), 1993–2024 (2015)
4. Bakas, S., et al.: Advancing The Cancer Genome Atlas glioma MRI collections with expert segmentation labels and radiomic features. Nat. Sci. Data **4**, Article no. 170117 (2017)
5. Bakas, S., et al.: Segmentation labels and radiomic features for the pre-operative scans of the TCGA-GBM collection, The Cancer Imaging Archive (2017). https://doi.org/10.7937/k9/tcia.2017.klxwjj1q
6. Bakas, S., et al.: Segmentation labels and radiomic features for the pre-operative scans of the TCGA-LGG collection, The Cancer Imaging Archive (2017). https://doi.org/10.7937/k9/tcia.2017.gjq7r0ef
7. Kamnitsas, K., et al.: Efficient multi-scale 3D CNN with fully connected CRF for accurate brain lesion segmentation. Med. Image Anal. **36**, 61–78 (2017)
8. Ronneberger, O., Fischer, P., Brox, T.: U-Net: convolutional networks for biomedical image segmentation, arXiv preprint, arXiv:1505.04597 (2015)
9. Cicek, O., Abdulkadir, A., Lienkamp, S.S., Brox, T., Ronneberger, O.: 3D U-Net: learning dense volumetric segmentation from sparse annotation, arXiv preprint, arXiv:1606.06650 (2016)
10. Tustison, N.J., et al.: N4ITK: improved N3 bias correction. IEEE Trans. Med. Imaging **29**(6), 1310–1320 (2010)
11. Manjon, J.V., Coupe, P., Marti-Bonmati, L., Collins, L., Robles, M.: Adaptive non-local means denoising of MR images with spatially varying noise levels. J. Magn. Reson. Imaging **31**(1), 192–203 (2010)

12. ITK Homepage. https://itk.org/
13. Nipype Homepage. https://nipype.readthedocs.io/en/latest/
14. Bakas, S., et al.: Identifying the best machine learning algorithms for brain tumor segmentation, progression assessment, and overall survival prediction in the BRATS challenge, arXiv preprint, arXiv:1811.02629 (2018)

A Novel Domain Adaptation Framework for Medical Image Segmentation

Amir Gholami[1]([✉]), Shashank Subramanian[2], Varun Shenoy[1],
Naveen Himthani[2], Xiangyu Yue[1], Sicheng Zhao[1], Peter Jin[1],
George Biros[2], and Kurt Keutzer[1]

[1] University of California Berkeley, Berkeley, USA
amirgh@berkeley.edu
[2] The University of Texas at Austin, Austin, USA

Abstract. We propose a segmentation framework that uses deep neural networks and introduce two innovations. First, we describe a biophysics-based domain adaptation method. Second, we propose an automatic method to segment white matter, gray matter, glial matter and cerebrospinal fluid, in addition to tumorous tissue. Regarding our first innovation, we use a domain adaptation framework that combines a novel multispecies biophysical tumor growth model with a generative adversarial model to create realistic looking synthetic multimodal MR images with known segmentation. These images are used for the purpose of training time data augmentation. Regarding our second innovation, we propose an automatic approach to enrich available segmentation data by computing the segmentation for healthy tissues. This segmentation, which is done using diffeomorphic image registration between the BraTS training data and a set of pre-labeled atlases, provides more information for training and reduces the class imbalance problem. Our overall approach is not specific to any particular neural network and can be used in conjunction with existing solutions. We demonstrate the performance improvement using a 2D U-Net for the BraTS'18 segmentation challenge. Our biophysics based domain adaptation achieves better results, as compared to the existing state-of-the-art GAN model used to create synthetic data for training.

Keywords: Segmentation · Neural network · Machine learning ·
Glioblastoma multiforme · Tumor growth models · Image registration

1 Introduction

Automatic segmentation methods have the potential to provide accurate and reproducible labels leading to improved tumor prognosis and treatment planning, especially for cases where access to expert radiologists is limited.

A. Gholami and S. Subramanian—Equally contributed.

© Springer Nature Switzerland AG 2019
A. Crimi et al. (Eds.): BrainLes 2018, LNCS 11384, pp. 289–298, 2019.
https://doi.org/10.1007/978-3-030-11726-9_26

In the BraTS competition [4], we seek to segment multimodal MR images of glioma patients. Common brain MRI modalities include post-Gadolinium T1 (used to enhance contrast and visualization of the blood-brain barrier), T2 and FLAIR (to highlight different tissue fluid intensities), and T1. We use the data for these four modalities to generate the segmentations using a methodology that we outline below.

In most image classification tasks deep neural networks (DNNs) have been a very powerful technique that tends to outperform other approaches and BraTS is no different. From past BraTS competitions two main DNN architectures have emerged: DeepMedic [15] and U-Net [20]. How can we further improve this approach? Most research efforts have been on further improving these architectures, as well as coupling them with post-processing and ensembling techniques. In our work here, we propose a framework to work around the relatively small training datasets used in the BraTS competition. Indeed, in comparison to other popular classification challenges like ImageNet [5] (which consists of one million images for training), the BraTS training set contains only 285 instances (multimodal 3D MR images), a number that is several orders of magnitude smaller than the typical number of instances required for DNNs to work well. These observations have motivated this work, whose contributions we summarize below.

Related work: Recently, deep learning approaches using convolutional neural networks (CNNs) have demonstrated excellent performance in semantic segmentation tasks in medical imaging. Seminal works for segmentation stem from fully-convolutional networks (FCNs) [14]. U-Net [20] is another popular architecture for medical segmentation, which merges feature maps from the contracting path of an FCN to its expanding path to preserve local contextual information. Multi-scale information is often incorporated by using parallel convolutional pathways of various resolutions [12] or by using dilated convolutions and cascading network architectures [23]. Post-processing and ensemble methods are also usually used after training with these models. The most commonly used post processing step is Conditional Random Fields (CRF) [12], which has been found to significantly reduce false positives and sharpen the segmentation. Ensembling is also very important to reduce overfitting with deep neural networks. The winning algorithm of the Multimodal Brain Tumor Image Segmentation Benchmark (BraTS) challenge in 2017 was based on Ensembles of Multiple Models and Architectures (EMMA) [11], which bagged a heterogeneous collection of networks (including DeepMedic (winner of ISLES 2015 [15]), U-Nets and FCNs) to build a robust and generic segmentation model.

There are established techniques to address training with small datasets, such as regularization, or ensembling, which was the approach taken by the winning team of BraTS'17. However, in this paper we propose an orthogonal method to address this problem.

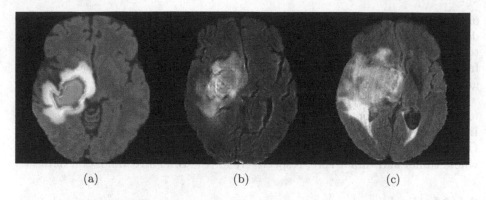

(a)   (b)   (c)

Fig. 1. Domain adaptation results: (a) represents a synthetic FLAIR brain image, (b) represents the domain adapted synthetic FLAIR image, (c) represents the real BraTS FLAIR image. As we can see from the intensity distributions, the values in the adapted images are qualitatively closer to the real images.

Contributions: Our main contributions are as follows:

1. **Data augmentation:** We propose a biophysics-based domain adaptation strategy to add synthetic tumor-bearing MR images to the training examples. There have been many notable works to simulate tumor growth (see [6–8,10,13,19]). We use an in-house PDE based multispecies tumor growth model [22] to simulate synthetic tumors. Since simulated data does not contain the correct intensity distribution of a real MR image, we train an auxiliary neural network to transform the simulated images to match real MRIs. This network gets a multimodal input and transforms this data to match the distribution of BraTS images by imposing certain cycle consistency constraints. As we will show, this is a very promising approach.

2. **Extended segmentation:** We extend the segmentation to the healthy parenchyma. This is done in two steps. First, we segment the training dataset using an atlas-based ensemble registration (using an in-house diffeomorphic registration code). Second, we train our DNN network to segment both tumor and healthy tissue (Four classes: cerebrospinal fluid, gray and white matter, glial matter). Our approach adds important information about healthy tissue delineation, which is actually used by radiologists. It also reduces the inherent class imbalance problem.

Our data augmentation strategy is different from the recent work of [21], which uses GANs to automatically generate data. To the best of our knowledge, our work here is the first to use a biophysics-based domain adaptation framework for automatic data generation, and our approach achieves five percentage points higher dice score as compared to [21], even though we use a 2D neural network architecture (which has a sub-optimal performance as compared to the 3D network used in [21]).

Limitations: Currently, our framework only supports 2D domain transformations. Hence, we are limited to transforming 3D brains slice-by-slice and using only 2D neural network architectures. This is sub-optimal as 3D CNNs can demonstrably utilize volumetric medical imaging data more efficiently, leading to better and more robust performance (see [9,11,12]). Hence, extending our framework to 3D is the focus of our future work and can potentially lead to greater improvements in performance.

The outline of the paper is as follows. In Sect. 2, we discuss the methodology for domain adaptation (Sect. 2.1), and the whole brain segmentation (Sect. 2.2). In Sect. 2.2 we present preliminary results for the BraTS'18 challenge [1–3,15]. Our method achieves a Dice score of [79.15, 90.81, 81.91] for enhancing tumor, whole tumor and tumor core, respectively for the BraTS'18 validation dataset.

2   Methods

2.1   Domain Adaptation

As mentioned above, one of the main challenges in medical imaging is the scarcity of training data. To address this issue, we use a novel domain adaptation strategy and generate synthetic tumor-bearing MR images to enrich the training dataset. This is performed by first solving an in-house PDE based multispecies tumor model using an atlas brain [22]. This model captures the time evolution of enhancing and necrotic tumor concentrations along with tumor-induced brain edema. The governing equations for the model are reaction-diffusion-advection equations for the tumor species along with a diffusion equation for oxygen and other nutrients. We couple this model with linear elasticity equations with variable elasticity material properties to simulate the deformation of surrounding brain tissue due to tumor growth, also known as "mass effect". However, this data cannot be used directly due to the difference in intensity distributions between a BraTS MRI scan and a synthetic MRI scan. Directly using synthetic data during the training process will adversely guide the neural network to learn features which do not exist in a real MR image, resulting in poor performance.

To address this issue, we use CycleGAN [24] to perform domain adaptation from the generated synthetic data to the real BraTS images. This is done by learning a mapping $G : X \to Y$ such that the distribution of images from $G(X)$ is indistinguishable from the distribution Y using an adversarial loss. Here, X is the simulated tumor data, and Y is the corresponding data which matches the BraTS distribution. Because this mapping is highly under-constrained, it is coupled with an inverse mapping $F : Y \to X$ and a cycle consistency loss is introduced to enforce $F(G(X)) \approx X$ (and vice versa).

For training the domain adaptation network, we first computationally simulate synthetic tumors in a healthy brain atlas, located approximately at the whole tumor center taken from each BraTS image. Hence, every synthetic tumorous brain is paired with the corresponding data from a real BraTS image. Then, we perform a pre-processing step to transform our synthetic results to intensities. We produce a segmentation map for every tissue (healthy and tumorous) class and sample intensities for each class from a real MRI scan. We assign these sampled intensities to every voxel in our synthetic segmentation map to finally obtain

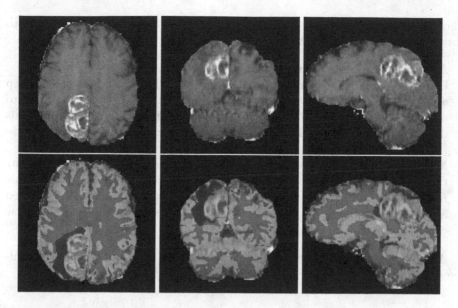

Fig. 2. (*Top row*): The original T1ce image for Brats18_TCIA02_135_1 training data is shown for different views (axial, coronal and sagittal). (*Bottom row*): The corresponding extended segmentation for healthy cells computed by solving a 3D registration problem with a segmented atlas. We overlay the BraTS tumor segmentation with the registered segmentation to get the final results.

our synthetic MRI scans. Then, we train with these pre-processed synthetic MRI scans and their corresponding BraTS images. Samples of our adaptation results are shown Fig. 1, which demonstrate an almost indistinguishable adaptation of the simulated data with the real images.

2.2 Whole Brain Segmentation with Healthy Tissues

An orthogonal approach that we propose for data augmentation, is an extended segmentation BraTS training data. That is, we segment the healthy parenchyma into gray/white matter, cerebrospinal fluid, and glial cells. The delineation of these healthy tissues contain important information which is actually used by radiologists. For example, the delineation of the tissues could be compressed due to tumor growth in the confined space of the brain. Providing this information to the classifier can help in better segmenting tumorous regions. However, such data is not readily available in the BraTS training dataset, since labelling the tumorous regions itself is laborious, let alone annotating full healthy tissues which is orders of magnitude more time consuming. We propose a novel automated approach to compute this information through image registration. In our method, we only need one (or preferably a few) fully segmented brains.

Then given an input 3D brain, we perform the following automatic steps to obtain the extended segmentation:

1. Affine registration of each atlas image to the brats image.
2. Diffeomorphic registration of each atlas image to the BraTS image: This step aims to find a deformation map that would "translate" a healthy atlas to match the structure of a given BraTS training example. We compute this deformation by solving a PDE-constrained optimization problem. We refer to [16–18] for details on solving this optimization problem.
3. Majority voting to fuse labels of all deformed atlases to get the final healthy tissue segmentation: The votes are weighted with the quality of diffeomorphic registration measured by the L_2 norm of the residual between each deformed atlas and brats image. This ensures the highest weight for the deformed atlas closest to the BraTS image.

We show an exemplary segmentation for an MRI scan from the BraTS training data in Fig. 2.

3   Setup

Here, we describe our setup and then report our segmentation results in the subsequent section on BraTS'18 dataset.

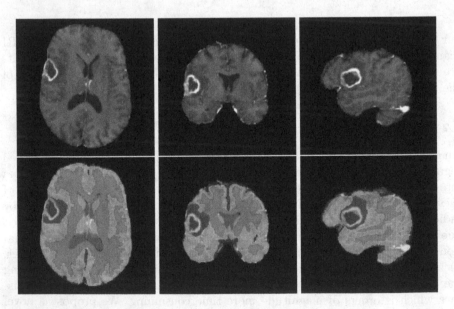

Fig. 3. (*Top row*): The original T1ce validation MR-image for Brats18_CBICA_ABT_1 data is shown for different views (axial, coronal and sagittal). (*Bottom row*): The corresponding segmentation result for healthy cells computed by the neural network. (Color figure online)

Baseline Network for Healthy and Tumor Segmentation. We first obtain the healthy tissue segmentation for all the BraTS training data using the image registration method (with 22 healthy atlases) discussed above, and use the fine grained data to train a neural network. Given that our current domain adaptation framework only supports 2D transformations, we follow a two stage segmentation routine by first localizing the tumor location(s), and then creating 2D slices/crops around the tumor and passing it to a 2D U-Net[1]. We use fixed sizes for our crops (specifically 48×48, 96×96 and 144×144). This is to ensure no loss of information due to strided operations when we go deeper in the neural network. We train our network using a five-fold validation split of the training data with ADAM optimizer and ensemble the splits to obtain the baseline results. We show the healthy segmentation for a validation MRI scan in Fig. 3.

Data Augmentation Through Domain Adaptation. To augment our data with domain adaptation results, we simulate a synthetic tumor in our atlas corresponding to the whole tumor center of mass of every BraTS training image. We transfer the synthetic brain to the BraTS domain for every axial slice. Hence, our augmented dataset consists of approximately twice the amount of training brains.

Our final neural network is a 2D multi-view U-Net (with the tumor localization strategy described above) with data augmentation using domain adaptation. We train three U-Nets corresponding to the axial, sagittal or coronal view of the MRI scan and ensemble them (similar to the multi-view fusion method outlined in [23]). This is done in order to avoid noisy segmentations and reduce the class imbalance inherent in the BraTS dataset. As before, we train five-fold cross-validation splits and ensemble them to avoid overfitting to the training data.

Table 1. We report the BraTS'18 results for our method for both the baseline model and the final 2D network. Our final submission to the validation portal is highlighted. The last row shows the dice scores for BRaTS'18 testing dataset. Even though we use a sub-optimal 2D network, but we can still achieve significant improvement with the proposed framework.

	Dice Score		
	EN	WT	TC
Baseline (Validation)	73.86	89.49	79.94
Proposed (Validation)	79.15	90.81	81.91
Proposed (Testing)	70.96	87.11	76.87

---

[1] For the localization task, we use a simple 3D U-Net [9], with ten layers and multiclass dice loss.

4   Results

We trained the framework using the BraTS'18 data. The fine-grained segmentation result from the first stage 3D U-Net is shown in Fig. 3. As one can see, this involves both the tumor segmentation, shown in red/yellow/green, as well as

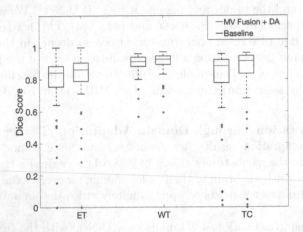

Fig. 4. Box plot for the final model's dice score on the BraTS'18 validation data is shown. This model achieves a mean dice score of (79.15, 90.81, 81.91) percent for (WT, TC, EN), respectively.

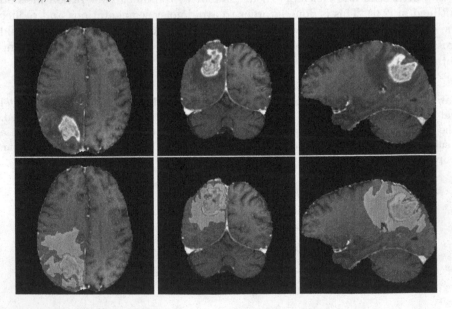

Fig. 5. (*Top row*): The original T1ce validation MR-image for Brats18_CBICA_AAM_1 data is shown for different views (axial, coronal and sagittal). (*Bottom row*): The corresponding tumor segmentation result from the final 2D network.

healthy structure of the brain shown in purple/cyan/gray/dark blue. This data is used for localizing the tumor boundaries. We then use this data and create multi-view slices around the tumor bearing region. Then, this data is passed through the second stage 2D U-Net which was trained along with the domain adaptation data, and fused together to obtain the final segmentation as shown in Fig. 5.

We show quantitative values for the Dice score in Table 1, with the corresponding box plots shown in Fig. 4. The baseline network has a dice score of [73.86, 89.49, 79.94] for Enhancing Tumor (ET), Whole Tumor (WT), and Tumor Core (TC). Using our proposed data augmentation framework leads to a dice score of [79.15, 90.81, 81.91]. These could be further improved by using a 3D network instead of a 2D one, by developing a 3D domain adaptation framework.

5   Conclusion

We presented a new framework for biophysics-based medical image segmentation. Our contributions include an automatic healthy tissue segmentation of the BraTS dataset, and a novel Generative Adversarial Network to enrich the training dataset using a model to generate synthetic phenomenological structures of a glioma. We demonstrated that our approach yields promising results on the BraTS'18 validation dataset. Our framework is not specific to a particular model, and could be used with other proposed neural networks for the BraTS challenge. Extending our domain adaptation framework to 3D can potentially lead to better performance and is the focus of our future work.

References

1. Bakas, S., et al.: Advancing the cancer genome atlas glioma MRI collections with expert segmentation labels and radiomic features. Nat. Sci. Data **4**, 170117 (2017)
2. Bakas, S., et al.: Segmentation Labels for the Pre-operative Scans of the TCGA-GBM Collection (2017). http://doi.org/10.7937/k9/tcia.2017.klxwjj1q
3. Bakas, S., et al.: Segmentation Labels for the Pre-operative Scans of the TCGA-LGG Collection (2017). http://doi.org/10.7937/k9/tcia.2017.gjq7r0ef
4. Bakas, S., et al.: Identifying the best machine learning algorithms for brain tumor segmentation, progression assessment, and overall survival prediction in the brats challenge. arXiv preprint arXiv:1811.02629 (2018)
5. Deng, J., Dong, W., Socher, R., Li, L.J., Li, K., Fei-Fei, L.: ImageNet: a large-scale hierarchical image database. In: IEEE Conference on Computer Vision and Pattern Recognition, CVPR 2009, pp. 248–255. IEEE (2009)
6. Gholami, A.: Fast algorithms for biophysically-constrained inverse problems in medical imaging. Ph.D. thesis, The University of Texas at Austin (2017)
7. Hawkins-Daarud, A., Rockne, R.C., Anderson, A.R.A., Swanson, K.R.: Modeling tumor-associated edema in gliomas during anti-angiogenic therapy and its imapct on imageable tumor. Front. Oncol. **3**, 66 (2013)
8. Hawkins-Daarud, A., van der Zee, K.G., Tinsley Oden, J.: Numerical simulation of a thermodynamically consistent four-species tumor growth model. Int. J. Numer. Methods Biomed. Eng. **28**(1), 3–24 (2012)

9. Isensee, F., Kickingereder, P., Wick, W., Bendszus, M., Maier-Hein, K.H.: Brain tumor segmentation and radiomics survival prediction: contribution to the BRATS 2017 challenge. CoRR abs/1802.10508 (2018). http://arxiv.org/abs/1802.10508

10. Ivkovic, S., et al.: Direct inhibition of myosin II effectively blocks glioma invasion in the presence of multiple motogens. Mol. Biol. Cell **23**(4), 533–542 (2012)

11. Kamnitsas, K., et al.: Ensembles of multiple models and architectures for robust brain tumour segmentation. CoRR abs/1711.01468 (2017). http://arxiv.org/abs/1711.01468

12. Kamnitsas, K., et al.: Efficient multi-scale 3d CNN with fully connected CRF for accurate brain lesion segmentation. Med. Image Anal. **36**, 61–78 (2017). http://www.sciencedirect.com/science/article/pii/S1361841516301839

13. Lima, E., Oden, J., Hormuth, D., Yankeelov, T., Almeida, R.: Selection, calibration, and validation of models of tumor growth. Math. Models Methods Appl. Sci. **26**(12), 2341–2368 (2016)

14. Long, J., Shelhamer, E., Darrell, T.: Fully convolutional networks for semantic segmentation. CoRR abs/1411.4038 (2014). http://arxiv.org/abs/1411.4038

15. Maier, O., et al.: ISLES 2015 - a public evaluation benchmark for ischemic stroke lesion segmentation from multispectral MRI. Med. Image Anal. **35**, 250–269 (2017)

16. Mang, A., Biros, G.: A semi-Lagrangian two-level preconditioned Newton-Krylov solver for constrained diffeomorphic image registration. SIAM J. Sci. Comput. **39**(6), B1064–B1101 (2017)

17. Mang, A., Gholami, A., Biros, G.: Distributed-memory large deformation diffeomorphic 3d image registration. In: SC16: International Conference for High Performance Computing, Networking, Storage and Analysis (2016)

18. Mang, A., Gholami, A., Davatzikos, C., Biros, G.: CLAIRE: a distributed-memory solver for constrained large deformation diffeomorphic image registration. arXiv preprint arXiv:1808.04487 (2018)

19. Oden, J.T., et al.: Toward predictive multiscale modeling of vascular tumor growth. Arch. Comput. Methods Eng. **23**(4), 735–779 (2016)

20. Ronneberger, O., Fischer, P., Brox, T.: U-Net: convolutional networks for biomedical image segmentation. CoRR abs/1505.04597 (2015). http://arxiv.org/abs/1505.04597

21. Shin, H.-C., et al.: Medical image synthesis for data augmentation and anonymization using generative adversarial networks. In: Gooya, A., Goksel, O., Oguz, I., Burgos, N. (eds.) SASHIMI 2018. LNCS, vol. 11037, pp. 1–11. Springer, Cham (2018). https://doi.org/10.1007/978-3-030-00536-8_1

22. Subramanian, S., Gholami, A., Biros, G.: Simulation of glioblastoma growth using a 3d multispecies tumor model with mass effect. arXiv preprint arXiv:1810.05370

23. Wang, G., Li, W., Ourselin, S., Vercauteren, T.: Automatic brain tumor segmentation using cascaded anisotropic convolutional neural networks. CoRR abs/1709.00382 (2017). http://arxiv.org/abs/1709.00382

24. Zhu, J.Y., Park, T., Isola, P., Efros, A.A.: Unpaired image-to-image translation using cycle-consistent adversarial networks. In: 2017 IEEE International Conference on Computer Vision (ICCV) (2017)

Context Aware 3D CNNs for Brain Tumor Segmentation

Siddhartha Chandra[1]([✉]), Maria Vakalopoulou[1,2], Lucas Fidon[1,3],
Enzo Battistella[1,2], Théo Estienne[1,2], Roger Sun[1,2], Charlotte Robert[2],
Eric Deutsch[2], and Nikos Paragios[1,3]

[1] CVN, CentraleSupélec, Université Paris-Saclay, Paris, France
robinchandra19@gmail.com
[2] Gustave Roussy Institute, Paris, France
[3] TheraPanacea, Paris, France

Abstract. In this work we propose a novel deep learning based pipeline for the task of brain tumor segmentation. Our pipeline consists of three primary components: (i) a preprocessing stage that exploits histogram standardization to mitigate inaccuracies in measured brain modalities, (ii) a first prediction stage that uses the V-Net deep learning architecture to output dense, per voxel class probabilities, and (iii) a prediction refinement stage that uses a Conditional Random Field (CRF) with a bilateral filtering objective for better context awareness. Additionally, we compare the V-Net architecture with a custom 3D Residual Network architecture, trained on a multi-view strategy, and our ablation experiments indicate that V-Net outperforms the 3D ResNet-18 with all bells and whistles, while fully connected CRFs as post processing, boost the performance of both networks. We report competitive results on the BraTS 2018 validation and test set.

Keywords: Brain tumor segmentation ·
3-D fully convolutional CNNs · Fully-connected CRFs

1 Introduction

Cancer is currently the second leading cause of death worldwide with overall 14.1 million new cases and 8.2 million deaths in 2012 [12]. Brain tumors, with gliomas being one of the most frequent malignant types, are among the most aggressive and dangerous types of cancer [5]. According to recent classifications malignant gliomas are classified into four WHO grades. From these low grade gliomas (LGG), including grade I and II are considered as relatively slow-glowing while high grade gliomas (HGG), including grade III and grade IV glioblastoma are more aggressive with the average survival time of approximately 1 year for patients with glioblastoma (GBM) [13,21]. Besides being very aggressive, gliomas are very costly to treat, so accurately diagnosing of them at early stages is very important.

© Springer Nature Switzerland AG 2019
A. Crimi et al. (Eds.): BrainLes 2018, LNCS 11384, pp. 299–310, 2019.
https://doi.org/10.1007/978-3-030-11726-9_27

Multimodality magnetic resonance imaging (MRI) is the primary method of screening and diagnosis for gliomas. However, due to inconsistency and diversity of MRI acquisition parameters and sequences, there are large differences in appearance, shape and intensity ranges, adding variability to the one that gliomas can have between different patients. Currently, tumor regions are segmented manually by radiologists, but this process is very time consuming while the inter-observer agreement between them is considerably low. In order to address all these challenges, the multimodal brain tumor segmentation challenge (BraTS) [1–3,22] is organized annually, in order to highlight efficient approaches and the way forward for the accurate segmentation of brain tumors.

Currently, the emergence of deep learning as disruptive innovation method in the field of computer vision has impacted significantly the medical imaging community, with numerous architectures being proposed addressing task-specific problems. Fully Convolutional Networks (FCN) [20] and their extension to 3D [14,23] are among the most commonly used architectures, boosting considerably the accuracies of the semantic segmentation. Inspired by these recent advances of deep learning, in this paper we exploit 3D CNNs coupled with fully-connected Conditional Random Fields for segmentation of brain tumor. More specifically, we compare two popular network architectures: V-Net [23] and 3D Residual-Nets [16] (ResNet), trained using a multi-view strategy, and provide preliminary results which indicate that V-Net architecture is better suited for dense-per-voxel brain tumor segmentation.

In the next sections, we discuss our contributions in detail, and we report our performance on the Training, Validation and Test Dataset of BraTS 2018.

2   Context-Aware 3D Networks

In this section, we give an overview of the different methods and strategies (Figs. 1 and 2) we follow and we discuss in detail the different components of our pipeline.

2.1   Preprocessing Using Histogram Standardization

MRI is the most popular medical imaging tool to capture the images of the brain and other internal organs. It is preferred due to its non-invasive nature and its ability to capture diverse types of tissues and physiological processes. It measures the response of body tissues to high-frequency radio waves when placed in a strong magnetic field, to produce images of the internal organs. MRI scans typically suffer with a bias due to artefacts produced by inhomogeneity in the magnetic field or small movements made by the patient during acquisition. Since MRI intensities are expressed in arbitrary units and may vary across acquisitions, this bias can adversely impact segmentation algorithms. State-of-the-art approaches typically employ bias correction strategies to pre-process the data corresponding to different modalities in order to mitigate this bias.

After careful comparison of existing bias-correction literature, we decide to use the recently proposed histogram standardizing approach [24] for bias correction. The authors in [24] propose a two phase algorithm that exploits the statistics of the different modalities in a dataset to transform the dataset in a manner where similar intensities correspond to similarity in the tissue semantics. We pre-process all our data in this work using this strategy.

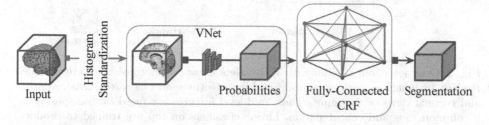

Fig. 1. A schematic overview of our approach. We first perform bias correction in the input brain volume using histogram standardization. A V-Net architecture is then trained on these data to deliver first phase of segmentation prediction. Further, we use a bilateral filtering performing fully-connected CRF to post-process our network predictions.

2.2 V-Net for 3D Semantic Segmentation

The first prediction stage in this work uses the V-Net architecture introduced by Milletari *et al.* in [23]. The V-Net architecture is a 3D fully convolutional neural network which can be trained end-to-end to deliver dense, per voxel class probabilities. The V-Net architecture has been exploited in literature for a variety of 3D segmentation tasks. Further, we use the generalized dice overlap loss as presented in [28] to optimise V-Net, which is a surrogate for the Dice coefficient used for evaluation. Using this loss function for training alleviates the need to compensate for the imbalance between the number of training samples for the different classes. We encourage the readers to refer to the original paper [23] for details on the network architecture.

2.3 Custom 3D ResNets for Semantic Segmentation

In this section, we discuss the 3D ResNets, as an alternative to the V-Net architecture described above. Residual networks were introduced by He *et al.* in [16]. ResNets ease the training of networks by adding 'residual' connections to the network architecture. These residual connections induce a short-cut connection of identity mapping without adding any extra parameters or computational complexity, thereby recasting the original mapping $F(x)$ as $F(x) = F'(x) + x$. We encourage the readers to refer to the original paper [16] for details.

ResNets are the building blocks of the majority of approaches on a variety of computer vision image segmentation benchmarks [11,14,31,32], and thus were

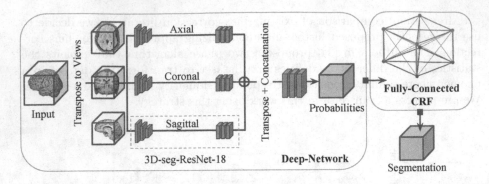

Fig. 2. Overview of our pipeline with 3D ResNets. Our network consists of three parallel ResNet-18 branches, each computing mid-level features on one of the axial, sagittal and coronal views of the input. These mid-level features are fused by transposing to a common view and concatenation. Linear classifiers on top are trained to produce probabilities for each of whole tumor, tumor core and enhancing tumor categories. Further, we use a bilateral filtering performing fully-connected CRF to post-process our network predictions.

a natural starting point in this work. However, the vanilla ResNets lack certain desirable characteristics which make their application to the task of brain tumor segmentation challenging. For the BraTS 2018 benchmark, we addressed these challenges by extending the 3D ResNet architecture from [14]. We briefly discuss these challenges one by one and describe our strategies to overcome them.

Network Stride. Approaches to semantic segmentation use 'fully-convolutional networks' (FCNs) [19,20] which are networks composed entirely of stacks of convolution operations, thereby producing per-patch outputs which spatially correspond to patches in the input image. A major challenge that presents itself in the use of FCNs is the *network stride*, also referred to as the *downsampling factor*. The output activations delivered by FCNs are smaller in spatial size than the input image due to the repeated max-pooling and convolutional strides. Thus, obtaining a labeling that is the same size as the input image requires upsampling of the output scores via interpolation, resulting in quantization and approximation errors and over-smooth predictions which do not capture the finer details in the input.

The downsampling factor of ResNets, like other popular network architectures such as [18,27] is 32. This means that each output unit corresponds to a 32×32 patch in the input image. For the BraTS 2018 data where the size of the input volume is $240 \times 240 \times 155$, the vanilla ResNet delivers outputs of the size $8 \times 8 \times 5$. A popular approach to reduce the downsampling factor is using a deconvolution filter which is a backwards convolution operation to upsample the output, as proposed by Long and Shelhamer in [20]. However, this results in an increased number of parameters, which will lead to overfitting for smaller datasets like the BraTS 2018 dataset where obtaining pixel-accurate ground truth is tedious.

In this work, we use atrous convolutions proposed by Chen *et al.* [7]. The atrous algorithm introduces holes in the convolution kernel, thereby allowing us to reduce the loss in spatial resolution without any increase in the number of parameters. Authors in [30] use the same operation, rebranding it as 'dilated convolutions'. With a strategic use of atrous convolutions, we reduce the downsampling factor of ResNets to 4. This amounts to an output of size $60 \times 60 \times 39$ for the BraTS 2018 data.

Context Awareness. Standard deep networks do not have a built-in capacity to estimate the scale of the input [8]. This limitation becomes especially crippling for brain tumor segmentation where the scales of the whole tumor, tumor core and enhancing tumor categories depend on a variety of factors, therefore estimating the correct scale of tumors is a challenging task. Approaches typically address this shortcoming by feeding the input to the network at different scales and averaging the network responses across scales [8,9]. A number of recent methods have proposed using feature pyramids [7,10,32] which instead capture features at multiple scales. The feature pyramids are finally fused into a single feature map via element-wise maximization, averaging or concatenation. In this work, we use the atrous spatial pyramid pooling (ASPP) approach proposed in [10]. ASPP uses a stack of convolutional filters with increasing degrees of dilation, thereby simulating filtering at multiple sampling rates and receptive fields. This captures visual context at multiple scales and leads to performance boosts for a variety of segmentation benchmarks [7]. The features at different scales are fused via averaging. This strategy enhances the context-awareness of the network.

Richness of Features (Network Depth) vs Training/Inference Speed. Deeper networks typically learn richer, more meaningful features as indicated by performance boosts over shallower networks [15,16]. However, an increase in depth also increases training/inference time because the network represents a sequential directed acyclic graph and prior activations need to be computed before subsequent ones.

3D FCNs are much slower than their 2D counterparts. Unlike 2D convolutions which have benefitted from both software and hardware level optimizations, 3D convolutions still involve slow computations as the research into their optimization is in its infancy. To allow fast experimentation and validation, the network architecture design needs careful consideration.

The authors in [31] demonstrate that decreasing the depth and increasing the width of ResNets leads to both better accuracy and reduced training/testing time. Inspired by them, rather than using very deep ResNets, we use the smallest residual network ResNet-18 in our experiments. To increase the width of the network, we use a multi-view fusion architecture where our network has three branches, each computing features on one of axial, sagittal and coronal views.

The features from the three views are transposed to a common view and concatenated, and linear classifiers for the three categories whole tumor, tumor core and enhancing tumor are trained on the fused features. This increases the speed at which the network operates as the activations of the three branches of the network can be computed in parallel. Here, we want to emphasize that each of these three branches is using the 3D input, in contrast to the 2.5D methods [26]. Further our preliminary experiments indicate that this multi-view fusion leads to better performance on a validation set, compared to deeper variants: ResNet-34 and ResNet-50. Our approach is described in Fig. 2.

2.4 Fully-Connected Conditional Random Fields

Fully convolutional deep networks such as V-Net and ResNets that produce per-voxel predictions consist of several downsampling phases followed by several upsampling phases. These phases involve quantization and approximations due to which these pipelines typically produce oversmooth predictions which do not capture the finer details in the input data. To address this limitation, we follow up the first pass of prediction using the network with a post-processing refinement pass. The refinement of the network prediction is done using a fully-connected Conditional Random Field (CRF). The fully-connected CRF performs bilateral filtering to refine the predictions made by our network, and uses the objective function proposed in [17]. Precisely, the CRF expresses the energy of a fully-connected CRF model as the sum of unary and pairwise potentials given by

$$E_I(\mathbf{l}) = \sum_i \psi_u(l_i) + \sum_i \sum_{j<i} \psi_p(l_i, l_j),\tag{1}$$

where

$$\psi_p(l_i, l_j) =$$

$$\mu(l_i, l_j) \sum_{m=1}^{K} \underbrace{w_m^1 \exp(-\frac{|s_i - s_j|^2}{2\theta_\alpha^2} - \frac{|p_i - p_j|^2}{2\theta_\beta^2})}_{\text{appearance}} + \underbrace{w_m^2 \exp(-\frac{|s_i - s_j|^2}{2\theta_\gamma^2})}_{\text{smoothness}}.\tag{2}$$

Here $\mathbf{l} = \{l_i\}$ denotes the labels for all the pixels indexed by i coming from a set of candidate labels $l_i \in \{1, 2, \dots, L\}$. ψ_u denotes the image dependent unary potentials, and the image dependent pairwise potentials $\psi_p(l_i, l_j)$ are expressed by the product of a label compatibility function μ and a weighted sum over Gaussian kernels. The pixel intensities are expressed using the 4 modalities in the input data $p_i = (flair, t1, t2, t1ce)$ and spatial positions are simply the coordinates in $3D$ space $s_i = (\mathbf{x}, \mathbf{y}, \mathbf{z})$. These are used together to define the appearance kernel, and the spatial positions alone are used to define the smoothness kernel. The appearance kernel tries to assign the same class labels to nearby pixels with similar intensity, and the hyperparameters θ_α and θ_β control the degrees of nearness and similarity. The smoothness kernel aims to remove small isolated regions. The model parameters $(\theta_\alpha, \theta_\beta, \theta_\gamma, w_m^1, w_m^2)$ are set by doing parameter sweeps using a validation set.

Having discussed our method in detail, we now delve into the experimental details and results in the next section.

3   Experiments and Results

3.1   Training Protocol

As described in Sect. 2, we use histogram standardization [24] for data pre-processing.

V-Net. We train the V-Net from scratch on randomly cropped 3D patches, as presented in Fig. 1, of size $128 \times 128 \times 128$ voxels. We do not employ any other form of data augmentation. Our network takes as input all 4 input modalities (*flair, t1, t2, t1ce*) and is trained using the generalized Dice loss [28] to output class probabilities for the 3 classes in the dataset alongside 2 additional classes (void and background/healthy tissue). We use the standard stochastic gradient descent algorithm for training, with a weight-decay of $1 \times e^{-5}$ and momentum of 0.9. We use a polynomially decaying learning rate policy, with a starting learning-rate of $1 \times e^{-4}$ and we train for $10K$ iterations. Our implementation uses the pytorch [25] library.

ResNet-18. As described in Sect. 2 and in Fig. 2, the three branches of our 3D ResNet-18 are initialized from the 3D ResNet-18 network from [14] which was trained for action recognition in videos. We augment the first convolutional layer (conv1) of the network from [14] with an additional input channel since we have 4 modalities (flair, t1, t2, t1ce), as opposed to 3 channels in natural images (r, g, b). We train our networks with randomly sampled input patches of size $97 \times 97 \times 97$, and our network outputs predictions of size $25 \times 25 \times 25$. The input brain volume is preprosessed by subtracting the per-image mean for each modality independently. We use the weighted Softmax Cross-Entropy loss to train our network for three classes: whole tumor, tumor core, and enhancing tumor. The weights for these three classes are 5,10,10 respectively. We use random flipping across the axial axis, and random scaling of the input between scales $0.25 - 2.5$ for data augmentation. We use the standard stochastic gradient descent algorithm for training, with a weight-decay of $1 \times e^{-5}$ and momentum of 0.9. We use a polynomially decaying learning rate policy, with a starting learning-rate of $1 \times e^{-4}$ and we train for $100K$ iterations. Our implementation is based on the Caffe2 library.

Fully Connected CRF. Our CRF parameters (Sect. 2.4) are estimated using the validation set of 85 patients. We use these probabilities as unary terms along hand-crafted pairwise terms (Sect. 2.4) for the CRF post-processing.

3.2 Testing Protocol

For results on the training set, we use a random train-test split of 200 − 85 patients respectively. For results on the validation and test sets, we use all the 285 training patients to train, and evaluate on the 66 validation, and 191 test patients.

V-Net. Our testing is done in a sliding window fashion on 3D patches of size $128 \times 128 \times 128$ voxels and predictions of ovelapping voxels are obtained via averaging.

Table 1. Results on BraTS 2018 Validation dataset.

Method	Dice			Sensitivity			Specificity			Hausdorff95		
	ET	WT	TC	ET	WT	TC	ET	WT	TC	ET	WT	TC
ResNet	0.740	0.868	0.801	0.771	0.811	0.769	0.991	0.992	0.997	5.312	4.971	9.891
ResNet+CRF	0.741	0.872	0.799	0.795	0.829	0.789	0.997	0.994	0.997	5.575	5.038	9.588
V-Net	0.766	0.896	0.810	0.821	0.909	0.815	0.992	0.989	0.952	7.211	6.541	7.821
V-Net+CRF	**0.767**	**0.901**	**0.813**	**0.839**	**0.916**	**0.819**	**0.998**	**0.994**	**0.997**	**7.569**	**6.68**	**7.630**

Table 2. Results on BraTS 2018 Test dataset.

Label	Dice			Hausdorff95		
	ET	WT	TC	ET	WT	TC
Mean	0.61824	0.82991	0.73334	24.93432	20.45375	26.48868
StdDev	0.3083	0.16348	0.27445	33.86977	26.42336	31.0645
Median	0.75368	0.88719	0.85481	4.12311	6.16441	8.66025
25 quantile	0.48567	0.82071	0.65831	2.20361	3.60555	3.0
75 quantile	0.84363	0.92246	0.91996	49.78338	28.60328	47.69619

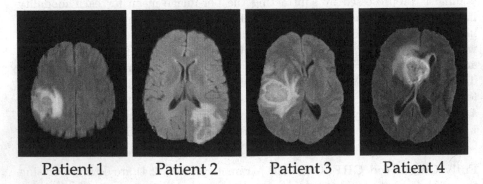

Patient 1   Patient 2   Patient 3   Patient 4

Fig. 3. Example segmentations on the Brats 2018 Validation set delivered by our approach for four patients. Green: edema, Red: non-enhancing tumor core; Yellow: enhancing tumor core. (Color figure online)

ResNet-18. Our testing is done in a sliding window fashion on 3D patches of size $97 \times 97 \times 97$ voxels and predictions of ovelapping voxels are obtained via averaging. We use multi-scale testing alongside flipping along the axial plane and average the probabilities delivered by the network.

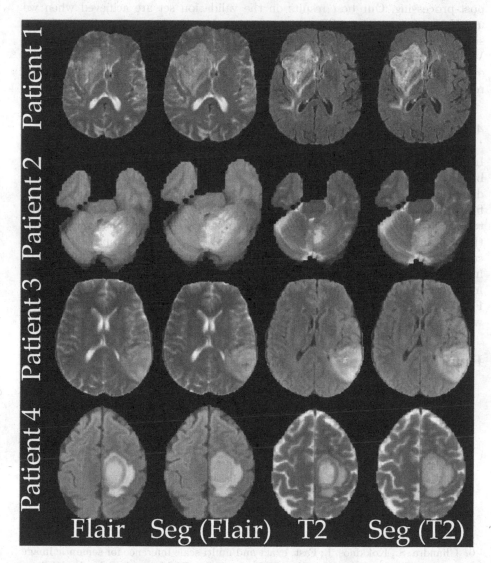

Fig. 4. Example of segmentations corresponding to four different patients, superimposed on Flair and T2 modalities on the Brats 2018 Test set delivered by our approach. Green: edema, Red: non-enhancing tumor core; Yellow: enhancing tumor core. (Color figure online)

3.3   Results

Our results on the BraTS 2018 Validation and Test datasets are tabulated in Tables 1 and 2 respectively. On the Validation set, we compare the two network architectures we considered, ResNet-18 and V-Net, with and without using CRF post-processing. Our best results on the validation set are achieved when we use V-Net followed by CRF post-processing. For this reason our final submission Table 2 for the test dataset of BraTS 2018 have been performed using the V-Net architecture. These results were generated by the evaluation server on the official BraTS 2018 website and have been also summarized in [4]. Qualitative results are shown in Figs. 3 and 4.

4   Conclusions and Future Work

In this work, we have described a novel deep-learning architecture for automatic brain tumor segmentation. More specifically, our pipeline uses histogram standardization for input bias correction and uses the V-Net architecture for the first phase of segmentation prediction. We also describe a fully-connected CRF to refine the network outputs in a post-processing step, while we also investigate the use of the multiview approach to fuse 3D features. Our approach delivers competitive results on the BraTS 2018 dataset. In the future, we would like to incorporate spatial pyramids for richer feature representation, and adapt techniques that perform data augmentation in a natural way as presented in [29]. Finally, we will try to investigate techniques that integrate CRFs into the network training [6].

References

1. Bakas, S., et al.: Advancing the cancer genome atlas glioma MRI collections with expert segmentation labels and radiomic features. Sci. Data **4**, 9 (2017)
2. Bakas, S., et al.: Segmentation labels and radiomic features for the pre-operative scans of the TCGA-GBM collection, July 2017
3. Bakas, S., et al.: Segmentation labels and radiomic features for the pre-operative scans of the TCGA-LGG collection, July 2017
4. Bakas, S., Reyes, M., et al.: Identifying the best machine learning algorithms for brain tumor segmentation, progression assessment, and overall survival prediction in the brats challenge. CoRR, abs/1811.02629 (2018)
5. Holland, E.C.: Progenitor cells and GLIOMA formation. Curr. Opin. Neurol. **14**, 683–688 (2002)
6. Chandra, S., Kokkinos, I.: Fast, exact and multi-scale inference for semantic image segmentation with deep gaussian CRFs. In: Leibe, B., Matas, J., Sebe, N., Welling, M. (eds.) ECCV 2016. LNCS, vol. 9911, pp. 402–418. Springer, Cham (2016). https://doi.org/10.1007/978-3-319-46478-7_25
7. Chen, L., Papandreou, G., Schroff, F., Adam, H.: Rethinking atrous convolution for semantic image segmentation. CoRR, abs/1706.05587 (2017)
8. Chen, L., Yang, Y., Wang, J., Xu, W., Yuille, A.L.: Attention to scale: Scale-aware semantic image segmentation. In: CVPR (2016)

9. Chen, L.-C., Papandreou, G., Kokkinos, I., Murphy, K., Yuille, A.L.: Semantic image segmentation with deep convolutional nets and fully connected CRFs. arXiv preprint arXiv:1412.7062 (2014)
10. Chen, L.-C., Papandreou, G., Kokkinos, I., Murphy, K., Yuille, A.L.: Deeplab: semantic image segmentation with deep convolutional nets, atrous convolution, and fully connected CRFs. arXiv:1606.00915 (2016)
11. Chen, L.-C., Papandreou, G., Murphy, K., Yuille, A.L.: Weakly and semi-supervised learning of a deep convolutional network for semantic image segmentation. In: ICCV (2015)
12. Ferlay, J., et al.: Cancer incidence and mortality worldwide (2013)
13. Hadziahmetovic, M., Shirai, K., Chakravarti, A.: Recent advancements in multi-modality treatment of gliomas. Future Oncol. **7**(10), 1169–1183 (2011)
14. Hara, K., Kataoka, H., Satoh, Y.: Can spatiotemporal 3D CNNs retrace the history of 2D CNNs and imagenet? In: CVPR (2018)
15. He, K., Gkioxari, G., Dollár, P., Girshick, R.: Mask R-CNN. In: ICCV (2017)
16. He, K., Zhang, X., Ren, S., Sun, J.: Deep residual learning for image recognition. In: CVPR (2016)
17. Krähenbühl, P., Koltun, V.: Efficient inference in fully connected CRFs with gaussian edge potentials. In: NIPS (2011)
18. Krizhevsky, A., Sutskever, I., Hinton, G.E.: Imagenet classification with deep convolutional neural networks. In: NIPS (2012)
19. Lecun, Y., Bottou, L., Bengio, Y., Haffner, P.: Gradient-based learning applied to document recognition. In: Proceedings of the IEEE (1998)
20. Long, J., Shelhamer, E., Darrell, T.: Fully convolutional networks for semantic segmentation. In: CVPR, pp. 3431–3440 (2015)
21. Louis, D.N., et al.: The 2016 world health organization classification of tumors of the central nervous system: a summary. Acta Neuropathol. **131**(6), 803–820 (2016)
22. Menze, B.H., et al.: The multimodal brain tumor image segmentation benchmark (brats). IEEE Trans. Med. Imaging **34**(10), 1993 (2015)
23. Milletari, F., Navab, N., Ahmadi, S.-A.: V-net: fully convolutional neural networks for volumetric medical image segmentation. In: 2016 Fourth International Conference on 3D Vision (3DV), pp. 565–571. IEEE (2016)
24. Nyúl, L.G., Udupa, J.K., Zhang, X.: New variants of a method of MRI scale standardization. IEEE Trans. Med. Imaging **19**(2), 143–150 (2000)
25. Paszke, A., et al.: Automatic differentiation in pytorch. In: NIPS-W (2017)
26. Roth, H.R., et al.: A new 2.5D representation for lymph node detection using random sets of deep convolutional neural network observations. In: Golland, P., Hata, N., Barillot, C., Hornegger, J., Howe, R. (eds.) MICCAI 2014. LNCS, vol. 8673, pp. 520–527. Springer, Cham (2014). https://doi.org/10.1007/978-3-319-10404-1_65
27. Simonyan, K., Zisserman, A.: Very deep convolutional networks for large-scale image recognition. In: ICLR (2015)
28. Sudre, C.H., Li, W., Vercauteren, T., Ourselin, S., Jorge Cardoso, M.: Generalised dice overlap as a deep learning loss function for highly unbalanced segmentations. In: Cardoso, M.J., et al. (eds.) DLMIA/ML-CDS -2017. LNCS, vol. 10553, pp. 240–248. Springer, Cham (2017). https://doi.org/10.1007/978-3-319-67558-9_28
29. Vakalopoulou, M., et al.: AtlasNet: multi-atlas non-linear deep networks for medical image segmentation. In: Frangi, A.F., Schnabel, J.A., Davatzikos, C., Alberola-López, C., Fichtinger, G. (eds.) MICCAI 2018. LNCS, vol. 11073, pp. 658–666. Springer, Cham (2018). https://doi.org/10.1007/978-3-030-00937-3_75

30. Yu, F., Koltun, V.: Multi-scale context aggregation by dilated convolutions. In: ICLR (2016)
31. Zagoruyko, S., Komodakis, N.: Wide residual networks. In: BMVC (2016)
32. Zhao, H., Shi, J., Qi, X., Wang, X., Jia, J.: Pyramid scene parsing network. CoRR, abs/1612.01105 (2016)

3D MRI Brain Tumor Segmentation
Using Autoencoder Regularization

Andriy Myronenko[✉]

NVIDIA, Santa Clara, CA, USA
amyronenko@nvidia.com

Abstract. Automated segmentation of brain tumors from 3D magnetic resonance images (MRIs) is necessary for the diagnosis, monitoring, and treatment planning of the disease. Manual delineation practices require anatomical knowledge, are expensive, time consuming and can be inaccurate due to human error. Here, we describe a semantic segmentation network for tumor subregion segmentation from 3D MRIs based on encoder-decoder architecture. Due to a limited training dataset size, a variational auto-encoder branch is added to reconstruct the input image itself in order to regularize the shared decoder and impose additional constraints on its layers. The current approach won 1st place in the BraTS 2018 challenge.

1 Introduction

Brain tumors are categorized into primary and secondary tumor types. Primary brain tumors originate from brain cells, whereas secondary tumors metastasize into the brain from other organs. The most common type of primary brain tumors are gliomas, which arise from brain glial cells. Gliomas can be of low-grade (LGG) and high-grade (HGG) subtypes. High grade gliomas are an aggressive type of malignant brain tumor that grow rapidly, usually require surgery and radiotherapy and have poor survival prognosis. Magnetic Resonance Imaging (MRI) is a key diagnostic tool for brain tumor analysis, monitoring and surgery planning. Usually, several complimentary 3D MRI modalities are acquired - such as T1, T1 with contrast agent (T1c), T2 and Fluid Attenuation Inversion Recover (FLAIR) - to emphasize different tissue properties and areas of tumor spread. For example the contrast agent, usually gadolinium, emphasizes hyperactive tumor subregions in T1c MRI modality.

Automated segmentation of 3D brain tumors can save physicians time and provide an accurate reproducible solution for further tumor analysis and monitoring. Recently, deep learning based segmentation techniques surpassed traditional computer vision methods for dense semantic segmentation. Convolutional neural networks (CNN) are able to learn from examples and demonstrate state-of-the-art segmentation accuracy both in 2D natural images [6] and in 3D medical image modalities [19].

Multimodal Brain Tumor Segmentation Challenge (BraTS) aims to evaluate state-of-the-art methods for the segmentation of brain tumors by providing

© Springer Nature Switzerland AG 2019
A. Crimi et al. (Eds.): BrainLes 2018, LNCS 11384, pp. 311–320, 2019.
https://doi.org/10.1007/978-3-030-11726-9_28

a 3D MRI dataset with ground truth tumor segmentation labels annotated by physicians [2–5,18]. This year, BraTS 2018 training dataset included 285 cases (210 HGG and 75 LGG), each with four 3D MRI modalities (T1, T1c, T2 and FLAIR) rigidly aligned, resampled to $1 \times 1 \times 1$ mm isotropic resolution and skull-stripped. The input image size is $240 \times 240 \times 155$. The data were collected from 19 institutions, using various MRI scanners. Annotations include 3 tumor subregions: the enhancing tumor, the peritumoral edema, and the necrotic and non-enhancing tumor core. The annotations were combined into 3 nested subregions: whole tumor (WT), tumor core (TC) and enhancing tumor (ET), as shown in Fig. 2. Two additional datasets without the ground truth labels were provided for validation and testing. These datasets required participants to upload the segmentation masks to the organizers' server for evaluations. The validation dataset (66 cases) allowed multiple submissions and was designed for intermediate evaluations. The testing dataset (191 cases) allowed only a single submission, and was used to calculate the final challenge ranking.

In this work, we describe our semantic segmentation approach for volumetric 3D brain tumor segmentation from multimodal 3D MRIs, which won the BraTS 2018 challenge. We follow the encoder-decoder structure of CNN, with asymmetrically large encoder to extract deep image features, and the decoder part reconstructs dense segmentation masks. We also add the variational autoencoder (VAE) branch to the network to reconstruct the input images jointly with segmentation in order to regularize the shared encoder. At inference time, only the main segmentation encode-decoder part is used.

2   Related Work

Last year, BraTS 2017, top performing submissions included Kamnitsas et al. [13] who proposed to ensemble several models for robust segmentation (EMMA), and Wang et al. [21] who proposed to segment tumor subregions in cascade using anisotropic convolutions. EMMA takes advantage of an ensemble of several independently trained architectures. In particular, EMMA combined DeepMedic [14], FCN [16] and U-net [20] models and ensembled their segmentation predictions. During training they used a batch size of 8, and a crop of $64 \times 64 \times 64$ 3D patch. EMMA's ensemble of different models demonstrated a good generalization performance winning the BraTS 2017 challenge. Wang et al. [21] second place paper took a different approach, by training 3 networks for each tumor subregion in cascade, with each subsequent network taking the output of the previous network (cropped) as its input. Each network was similar in structure and consists of a large encoder part (with dilated convolutions) and a basic decoder. They also decompose the $3 \times 3 \times 3$ convolution kernel into intra-slice ($3 \times 3 \times 1$) and inter-slice ($1 \times 1 \times 3$) kernel to save on both the GPU memory and the computational time.

This year, BraTS 2018 top performing submission (in addition to the current work) included Isensee et al. [12] in the 2nd place, McKinly et al. [17] and Zhou et al. [23], who shared the 3rd place. Isensee et al. [12] demonstrated that a generic U-net architecture with a few minor modifications is enough to achieve competitive performance. The authors used a batch size of 2 and a crop size of $128 \times 128 \times 128$. Furthermore, the authors used an additional training data from their own institution (which yielded some improvements for the enhancing tumor dice).

McKinly et al. [17] proposed a segmentation CNN in which a DenseNet [11] structure with dilated convolutions was embedded in U-net-like network. The authors also introduce a new loss function, a generalization of binary cross-entropy, to account for label uncertainty. Finally, Zhou et al. [23] proposed to use an ensemble of different networks: taking into account multi-scale context information, segmenting 3 tumor subregions in cascade with a shared backbone weights and adding an attention block.

Compared to the related works, we use the largest crop size of $160 \times 192 \times 128$ but compromise the batch size to be 1 to be able to fit network into the GPU memory limits. We also output all 3 nested tumor subregions directly after the sigmoid (instead of using several networks or the softmax over the number of classes). Finally, we add an additional branch to regularize the shared encoder, used only during training. We did not use any additional training data and used only the provided training set.

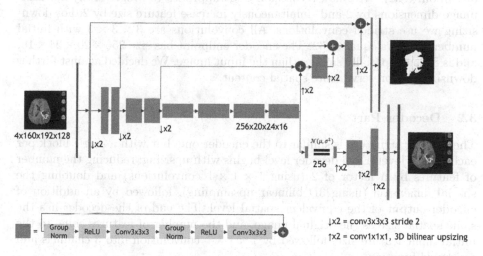

Fig. 1. Schematic visualization of the network architecture. Input is a four channel 3D MRI crop, followed by initial $3 \times 3 \times 3$ 3D convolution with 32 filters. Each green block is a ResNet-like block with the GroupNorm normalization. The output of the segmentation decoder has three channels (with the same spatial size as the input) followed by a sigmoid for segmentation maps of the three tumor subregions (WT, TC, ET). The VAE branch reconstructs the input image into itself, and is used only during training to regularize the shared encoder. (Color figure online)

3   Methods

Our segmentation approach follows encoder-decoder based CNN architecture with an asymmetrically larger encoder to extract image features and a smaller decoder to reconstruct the segmentation mask [6,7,9,19,20]. We add an additional branch to the encoder endpoint to reconstruct the original image, similar to auto-encoder architecture. The motivation for using the auto-encoder branch is to add additional guidance and regularization to the encoder part, since the training dataset size is limited. We follow the variational auto-encoder (VAE) approach to better cluster/group the features of the encoder endpoint. We describe the building parts of our networks in the next subsections (see also Fig. 1).

3.1   Encoder Part

The encoder part uses ResNet [10] blocks, where each block consists of two convolutions with normalization and ReLU, followed by additive identity skip connection. For normalization, we use Group Normalization (GN) [22], which shows better than BatchNorm performance when batch size is small (bath size of 1 in our case). We follow a common CNN approach to progressively downsize image dimensions by 2 and simultaneously increase feature size by 2. For downsizing we use strided convolutions. All convolutions are $3 \times 3 \times 3$ with initial number of filters equal to 32. The encoder endpoint has size $256 \times 20 \times 24 \times 16$, and is 8 times spatially smaller than the input image. We decided against further downsizing to preserve more spatial content.

3.2   Decoder Part

The decoder structure is similar to the encoder one, but with a single block per each spatial level. Each decoder level begins with upsizing: reducing the number of features by a factor of 2 (using $1 \times 1 \times 1$ convolutions) and doubling the spatial dimension (using 3D bilinear upsampling), followed by an addition of encoder output of the equivalent spatial level. The end of the decoder has the same spatial size as the original image, and the number of features equal to the initial input feature size, followed by $1 \times 1 \times 1$ convolution into 3 channels and a sigmoid function.

3.3   VAE Part

Starting from the encoder endpoint output, we first reduce the input to a low dimensional space of 256 (128 to represent mean, and 128 to represent std). Then, a sample is drawn from the Gaussian distribution with the given mean and std, and reconstructed into the input image dimensions following the same architecture as the decoder, except we don't use the inter-level skip connections from the encoder here. The VAE part structure is shown in Table 1.

Table 1. VAE decoder branch structure, where GN stands for group normalization (with group size of 8), Conv - $3 \times 3 \times 3$ convolution, Conv1 - $1 \times 1 \times 1$ convolution, AddId - addition of identity/skip connection, UpLinear - 3D linear spatial upsampling, Dense - fully connected layer

Name	Ops	Repeat	Output size
VD	GN, ReLU, Conv (16) stride 2, Dense (256)	1	256×1
VDraw	Sample $\sim \mathcal{N}(\mu(128), \sigma^2(128))$	1	128×1
VU	Dense, ReLU, Conv1, UpLinear	1	$256 \times 20 \times 24 \times 16$
VUp2	Conv1, UpLinear	1	$128 \times 40 \times 48 \times 32$
VBlock2	GN, ReLU, Conv, GN, ReLU, Conv, AddId	1	$128 \times 40 \times 48 \times 32$
VUp1	Conv1, UpLinear	1	$64 \times 80 \times 96 \times 64$
VBlock1	GN, ReLU, Conv, GN, ReLU, Conv, AddId	1	$64 \times 80 \times 96 \times 64$
VUp0	Conv1, UpLinear	1	$32 \times 160 \times 192 \times 128$
VBlock0	GN, ReLU, Conv, GN, ReLU, Conv, AddId	1	$32 \times 160 \times 192 \times 128$
Vend	Conv1	1	$4 \times 160 \times 192 \times 128$

3.4 Loss

Our loss function consists of 3 terms:

$$\mathbf{L} = \mathbf{L}_{dice} + 0.1 * \mathbf{L}_{L2} + 0.1 * \mathbf{L}_{KL} \tag{1}$$

$\mathbf{L}_{dice}$ is a soft dice loss [19] applied to the decoder output p_{pred} to match the segmentation mask p_{true}:

$$\mathbf{L}_{dice} = \frac{2 * \sum p_{true} * p_{pred}}{\sum p_{true}^2 + \sum p_{pred}^2 + \epsilon} \tag{2}$$

where summation is voxel-wise, and ϵ is a small constant to avoid zero division. Since the output of the segmentation decoder has 3 channels (predictions for each tumor subregion), we simply add the three dice loss functions together.

$\mathbf{L}_{L2}$ is an L2 loss on the VAE branch output I_{pred} to match the input image I_{input}:

$$\mathbf{L}_{L2} = ||I_{input} - I_{pred}||_2^2 \tag{3}$$

$\mathbf{L}_{KL}$ is standard VAE penalty term [8,15], a KL divergence between the estimated normal distribution $\mathcal{N}(\mu, \sigma^2)$ and a prior distribution $\mathcal{N}(0, 1)$, which has a closed form representation:

$$\mathbf{L}_{KL} = \frac{1}{N} \sum \mu^2 + \sigma^2 - \log \sigma^2 - 1 \tag{4}$$

where N is total number of image voxels. We empirically found a hyper-parameter weight of 0.1 to provide a good balance between dice and VAE loss terms in Eq. 1.

3.5 Optimization

We use Adam optimizer with initial learning rate of $\alpha_0 = 1e-4$ and progressively decrease it according to:

$$\alpha = \alpha_0 * \left(1 - \frac{e}{N_e}\right)^{0.9} \tag{5}$$

where e is an epoch counter, and N_e is a total number of epochs (300 in our case). We use batch size of 1, and draw input images in random order (ensuring that each training image is drawn once per epoch).

3.6 Regularization

We use L2 norm regularization on the convolutional kernel parameters with a weight of $1e - 5$. We also use the spatial dropout with a rate of 0.2 after the initial encoder convolution. We have experimented with other placements of the dropout (including placing dropout layer after each convolution), but did not find any additional accuracy improvements.

3.7 Data Preprocessing and Augmentation

We normalize all input images to have zero mean and unit std (based on non-zero voxels only). We apply a random (per channel) intensity shift ($-0.1..0.1$ of image std) and scale ($0.9..1.1$) on input image channels. We also apply a random axis mirror flip (for all 3 axes) with a probability 0.5.

4 Results

We implemented our network in Tensorflow [1] and trained it on NVIDIA Tesla V100 32 GB GPU using BraTS 2018 training dataset (285 cases) without any additional in-house data. During training we used a random crop of size $160 \times 192 \times 128$, which ensures that most image content remains within the crop area. We concatenated 4 available 3D MRI modalities into the 4 channel image as an input. The output of the network is 3 nested tumor subregions (after the sigmoid).

We report the results of our approach on BraTS 2018 validation (66 cases) and the testing sets (191 cases). These datasets were provided with unknown glioma grade and unknown segmentation. We uploaded our segmentation results to the BraTS 2018 server for evaluation of per class dice, sensitivity, specificity and Hausdorff distances.

Aside from evaluating a single model, we also applied test time augmentation (TTA) by mirror flipping the input 3D image axes, and averaged the output of the resulting 8 flipped segmentation probability maps. Finally, we ensembled a set of 10 models (trained from scratch) to further improve the performance.

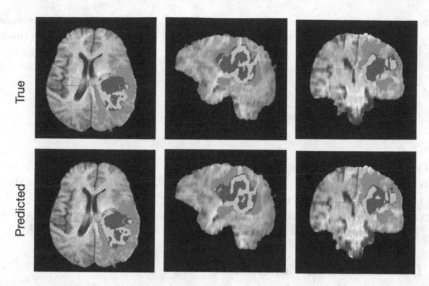

Fig. 2. A typical segmentation example with true and predicted labels overlaid over T1c MRI axial, sagittal and coronal slices. The whole tumor (WT) class includes all visible labels (a union of green, yellow and red labels), the tumor core (TC) class is a union of red and yellow, and the enhancing tumor core (ET) class is shown in yellow (a hyperactive tumor part). The predicted segmentation results match the ground truth well. (Color figure online)

Table 2 shows the results of our model on the BraTS 2018 validation dataset. At the time of initial short paper submission (Jul 13, 2018), our dice accuracy performance was second best (team name NVDLMED[1]) for all of the 3 segmentation labels (ET, WT, TC). The TTA only marginally improved the performance, but the ensemble of 10 models resulted in 1% improvement, which is consistent with the literature results of using ensembles.

For the testing dataset, only a single submission was allowed. Our results are shown in Table 3, which won the 1st place at BraTS 2018 challenge.

Table 2. BraTS 2018 validation dataset results. Mean Dice and Hausdorff measurements of the proposed segmentation method. EN - enhancing tumor core, WT - whole tumor, TC - tumor core.

Validation dataset	Dice			Hausdorff (mm)		
	ET	WT	TC	ET	WT	TC
Single model	0.8145	0.9042	0.8596	3.8048	4.4834	8.2777
Single model + TTA	0.8173	0.9068	0.8602	3.8241	4.4117	6.8413
Ensemble of 10 models	0.8233	0.9100	0.8668	3.9257	4.5160	6.8545

---

[1] https://www.cbica.upenn.edu/BraTS18/lboardValidation.html.

Table 3. BraTS 2018 testing dataset results. Mean Dice and Hausdorff measurements of the proposed segmentation method. EN - enhancing tumor core, WT - whole tumor, TC - tumor core.

Testing dataset	Dice			Hausdorff (mm)		
	ET	WT	TC	ET	WT	TC
Ensemble of 10 models	0.7664	0.8839	0.8154	3.7731	5.9044	4.8091

Time-wise, each training epoch (285 cases) on a single GPU (NVIDIA Tesla V100 32 GB) takes 9 min. Training the model for 300 epochs takes 2 days. We've also trained the model on NVIDIA DGX-1 server (that includes 8 V100 GPUs interconnected with NVLink); this allowed to train the model in 6 h (7.8x speed up). The inference time is 0.4 s for a single model on a single V100 GPU.

5   Discussion and Conclusion

In this work, we described a semantic segmentation network for brain tumor segmentation from multimodal 3D MRIs, which won the BraTS 2018 challenge. While experimenting with network architectures, we have tried several alternative approaches. For instance, we have tried a larger batch size of 8 to be able to use BatchNorm (and take advantage of batch statistics), however due to the GPU memory limits this modification required to use a smaller image crop size, and resulted in worse performance. We have also experimented with more sophisticated data augmentation techniques, including random histogram matching, affine image transforms, and random image filtering, which did not demonstrate any additional improvements. We have tried several data post-processing techniques to fine tune the segmentation predictions with CRF [14], but did not find it beneficial (it helped for some images, but made some other image segmentation results worse). Increasing the network depth further did not improve the performance, but increasing the network width (the number of features/filters) consistently improved the results. Using the NVIDIA Volta V100 32 GB GPU we were able to double the number of features compared to V100 16 GB version. Finally, the additional VAE branch helped to regularize the shared encoder (in presence of limited data), which not only improved the performance, but helped to consistently achieve good training accuracy for any random initialization. Our BraTS 2018 testing dataset results are 0.7664, 0.8839 and 0.8154 average dice for enhanced tumor core, whole tumor and tumor core, respectively.

References

1. Abadi, M., et al.: TensorFlow: large-scale machine learning on heterogeneous systems (2015). https://www.tensorflow.org/, software available from tensorflow.org
2. Bakas, S., et al.: Segmentation labels and radiomic features for the pre-operative scans of the TCGA-GBM collection. Cancer Imaging Arch. (2017). https://doi.org/10.7937/K9/TCIA.2017.KLXWJJ1Q

3. Bakas, S., et al.: Segmentation labels and radiomic features for the pre-operative scans of the TCGA-LGG collection. Cancer Imaging Arch. (2017). https://doi.org/10.7937/K9/TCIA.2017.GJQ7R0EF

4. Bakas, S., et al.: Advancing the cancer genome atlas glioma MRI collections with expert segmentation labels and radiomic features. Sci. Data **4**, 170117 (2017)

5. Bakas, S., Reyes, M., et al.: Identifying the best machine learning algorithms for brain tumor segmentation, progression assessment, and overall survival prediction in the BRATS challenge. In: arXiv:1811.02629 (2018)

6. Chen, L.C., Zhu, Y., Papandreou, G., Schroff, F., Adam, H.: Encoder-decoder with atrous separable convolution for semantic image segmentation. arXiv:1802.02611 (2018)

7. Chollet, F.: Xception: deep learning with depthwise separable convolutions. In: IEEE Conference on Computer Vision and Pattern Recognition, pp. 1800–1807 (2017)

8. Doersch, C.: Tutorial on variational autoencoders. arxiv e-print (2016). http://arxiv.org/abs/1606.05908

9. He, K., Gkioxari, G., Dollar, P., Girshick, R.: Mask R-CNN. In: Proceedings of the International Conference on Computer Vision (ICCV) (2017)

10. He, K., Zhang, X., Ren, S., Sun, J.: Identity mappings in deep residual networks. In: Leibe, B., Matas, J., Sebe, N., Welling, M. (eds.) ECCV 2016. LNCS, vol. 9908, pp. 630–645. Springer, Cham (2016). https://doi.org/10.1007/978-3-319-46493-0_38

11. Huang, G., Liu, Z., van der Maaten, L., Weinberger, K.Q.: Densely connected convolutional networks. In: Proceedings of the IEEE Conference on Computer Vision and Pattern Recognition, pp. 2261–2269 (2017)

12. Isensee, F., Kickingereder, P., Wick, W., Bendszus, M., Maier-Hein, K.H.: No new-net. In: Crimi, A., Bakas, S., Kuijf, H., Keyvan, F., Reyes, M., van Walsum, T. (eds.): BrainLes 2018. LNCS, vol. 11384, pp. 234–244. Springer, Cham (2019)

13. Kamnitsas, K., et al.: Ensembles of multiple models and architectures for robust brain tumour segmentation. In: Crimi, A., Bakas, S., Kuijf, H., Menze, B., Reyes, M. (eds.) BrainLes 2017. LNCS, vol. 10670, pp. 450–462. Springer, Cham (2018). https://doi.org/10.1007/978-3-319-75238-9_38

14. Kamnitsas, K., et al.: Efficient multi-scale 3D CNN with fully connected CRF for accurate brain lesion segmentation. Med. Image Anal. **36**, 61–78 (2016)

15. Kingma, D.P., Welling, M.: Auto-encoding variational Bayes. In: The International Conference on Learning Representations (ICLR) (2014)

16. Long, J., Shelhamer, E., Darrell, T.: Fully convolutional networks for semantic segmentation. In: The IEEE Conference on Computer Vision and Pattern Recognition (CVPR), pp. 3431–3440 (2015)

17. McKinley, R., Meier, R., Wiest, R.: Ensembles of densely-connected CNNs with label-uncertainty for brain tumor segmentation. In: Crimi, A., Bakas, S., Kuijf, H., Keyvan, F., Reyes, M., van Walsum, T. (eds.): BrainLes 2018. LNCS, vol. 11384, pp. 456–465. Springer, Cham (2019)

18. Menze, B.H., et al.: The multimodal brain tumor image segmentation benchmark (BRATS). IEEE Trans. Med. Imaging **34**(10), 1993–2024 (2015)

19. Milletari, F., Navab, N., Ahmadi, S.A.: V-Net: fully convolutional neural networks for volumetric medical image segmentation. In: Fourth International Conference on 3D Vision (3DV) (2016)

20. Ronneberger, O., Fischer, P., Brox, T.: U-Net: convolutional networks for biomedical image segmentation. In: Navab, N., Hornegger, J., Wells, W.M., Frangi, A.F. (eds.) MICCAI 2015. LNCS, vol. 9351, pp. 234–241. Springer, Cham (2015). https://doi.org/10.1007/978-3-319-24574-4_28

21. Wang, G., Li, W., Ourselin, S., Vercauteren, T.: Automatic brain tumor segmentation using cascaded anisotropic convolutional neural networks. In: Crimi, A., Bakas, S., Kuijf, H., Menze, B., Reyes, M. (eds.) BrainLes 2017. LNCS, vol. 10670, pp. 178–190. Springer, Cham (2018). https://doi.org/10.1007/978-3-319-75238-9_16

22. Wu, Y., He, K.: Group normalization. In: Ferrari, V., Hebert, M., Sminchisescu, C., Weiss, Y. (eds.) ECCV 2018. LNCS, vol. 11217, pp. 3–19. Springer, Cham (2018). https://doi.org/10.1007/978-3-030-01261-8_1

23. Zhou, C., Chen, S., Ding, C., Tao, D.: Learning contextual and attentive information for brain tumor segmentation. In: Crimi, A., Bakas, S., Kuijf, H., Keyvan, F., Reyes, M., van Walsum, T. (eds.): BrainLes 2018. LNCS, vol. 11384, pp. 497–507. Springer, Cham (2019)

voxel-GAN: Adversarial Framework for Learning Imbalanced Brain Tumor Segmentation

Mina Rezaei[✉], Haojin Yang, and Christoph Meinel

Hasso Plattner Institute, Berlin, Germany
{mina.rezaei,haojin.yang,christoph.meinel}@hpi.de

Abstract. We propose a new adversarial network, named voxel-GAN, to mitigate imbalanced data problem in brain tumor semantic segmentation where the majority of voxels belong to a healthy region and few belong to tumor or non-health region. We introduce a 3D conditional generative adversarial network (cGAN) comprises two components: a segmentor and a discriminator. The segmentor is trained on 3D brain MR or CT images to learn the segmentation label's in voxel-level, while the discriminator is trained to distinguish a segmentor output, coming from the ground truth or generated artificially. The segmentor and discriminator networks simultaneously train with new weighted adversarial loss to mitigate imbalanced training data issue. We show evidence that the proposed framework is applicable to different types of brain images of varied sizes. In our experiments on BraTS-2018 and ISLES-2018 benchmarks, we find improved results, demonstrating the efficacy of our approach.

Keywords: 3D generative adversarial network ·
Learning imbalanced data

1 Introduction

Brain imaging studies using magnetic resonance imaging (MRI) or computed tomography (CT) provides an important information for disease diagnosis and treatment planning [6]. One of the major challenges in brain tumor segmentation is unbalanced training data which the majority of the voxel healthy and only fewer voxels are non-healthy or a tumor. A model learned from class imbalanced training data is biased towards the majority class. The predicted results of such networks have low sensitivity, showing the ability of not correctly predicting non-healthy classes. In medical applications, the cost of misclassification of the minority class could be more than the cost of misclassification of the majority class. For example, the risk of not detecting tumor could be much higher than referring to a healthy subject to doctors.

The problem of class imbalanced has been recently addressed in diseases classification, tumor recognition, and tumor segmentation. Two types of approaches proposed in the literature: data-level and algorithm-level approaches.

© Springer Nature Switzerland AG 2019
A. Crimi et al. (Eds.): BrainLes 2018, LNCS 11384, pp. 321–333, 2019.
https://doi.org/10.1007/978-3-030-11726-9_29

At data-level, the objective is to balance the class distribution through re-sampling the data space [21], by including SMOTE (Synthetic Minority Over-sampling Technique) of the positive class [10] or by under-sampling of the nega-tive class [19]. However, these approaches often lead to remove some important samples or add redundant samples to the training set.

Algorithm-level based solutions address class imbalanced problem by modify-ing the learning algorithm to alleviate the bias towards majority class. Examples are cascade training [8,33,36], training with cost-sensitive function [40], such as Dice coefficient loss [12,35,38], and asymmetric similarity loss [16] that modify-ing the training data distribution with regards to the misclassification cost.

Here, we study the advantage of mixing adversarial loss with weighted cat-egorical cross-entropy and weighted $\ell 1$ losses in order to mitigate the nega-tive impact of the class imbalanced. Moreover, we train voxel-GAN simulta-neously with semantic segmentation masks and inverse class frequency segmen-tation masks, named complementary segmentation labels. Assume, Y is true segmentation label annotated by expert and $\bar{Y}$ is complementary label where the $P(\bar{Y} = i \mid Y = j), i \neq j \in \{0, 1, ..., c - 1\}$, and c is a number of semantic segmentation class labels. The complementary label $\bar{Y}$ is a reverse label for the background labels. Then, our network train with both true segmentation mask Y and complementary segmentation mask $\bar{Y}$ at the same time.

Automating brain tumor segmentation is challenging task due to the high diversity in the appearance of tissues among different patients, and in many cases, the similarity between healthy and non-healthy tissues. Numerous auto-matic approaches have been developed to speed up medical image segmenta-tion [6,25]. We can roughly divide the current automated algorithms into two categories: those based on generative models and those based on discriminative models.

Generative probabilistic approaches build the model based on prior domain knowledge about the appearance and spatial distribution of the different tissue types. Traditionally, generative probabilistic models have been popular where simple conditionally independent Gaussian models [13] or Bayesian learning [32] are used for tissue appearance. On the contrary, discriminative probabilistic models, directly learn the relationship between the local features of images and segmentation labels without any domain knowledge. Traditional discrim-inative approaches such as SVMs [2,9], random forests [23], and guided random walks [11] have been used in medical image segmentation. Deep neural net-works (DNNs) are one of the most popular discriminative approaches, where the machine learns the hierarchical representation of features without any hand-crafted features [22]. In the field of medical image segmentation, Ronneberger et al. [37] presented a fully convolutional neural network, named UNet, for seg-menting neuronal structures in electron microscopic stacks.

Recently, GANs [15] have gained a lot of momentum in the research frater-nities. Mirza et al. [26] extended the GANs framework to the conditional setting by making both the generator and the discriminator network class conditional. Conditional GANs have the advantage of being able to provide better represen-tations for multi-modal data generation since there is a control on the modes of the data being generated. This makes cGANs suitable for image semantic segmentation task, where we condition on an observed image and generate a corresponding output image.

Unlike previous works on cGANs [18,27,34,36,41], we investigate the 3D MR or CT images into 3D semantic segmentation. Summarizing, the main contributions of this paper are:

- We introduce voxel-GAN, a new adversarial framework that improves semantic segmentation accuracy.
- Our proposed method mitigates imbalanced training data with biased complementary labels in task of semantic segmentation.
- We study the effect of different losses and architectural choices that improve semantic segmentation.

The rest of the paper is organized as follows: in the next section, we explain our proposed method for learning brain tumor segmentation, while the detailed experimental results are presented in Sect. 3. We conclude the paper and give an outlook on future research in Sect. 4.

2   voxel-GAN

In a conventional generative adversarial network, generative model G tries to learn a mapping from random noise vector z to output image y; $G : z \rightarrow y$. Meanwhile, a discriminative model D estimates the probability of a sample coming from the training data x_{real} rather than the generator x_{fake}. The GAN objective function is a two-player mini-max game like Eq. (1).

$$\min_{G} \max_{D} V(D, G) = E_y[logD(y)] + \\ E_z[log(1 - D(G(z)))]$$

(1)

Similar conditional GAN [26]; in our proposed voxel-GAN, segmentor network takes 3D multimodal MR or CT images x and Gaussian vector z, and outputs a 3D semantic segmentation; The discriminator takes the segmentor output $S(x, z)$ and the ground truth annotated by an expert y_{seg} and outputs a confidence value $D(x)$ of whether a 3D object input x is real or synthetic. The training procedure is similar to two-player mini-max game as shown in Eq. (2).

$$\mathcal{L}_{adv} \leftarrow \min_{S} \max_{D} V(D, S) = E_{x,y_{seg}}[logD(x, y_{seg})] + \\ E_{x,z}[log(1 - D(x, S(x, z)))]$$

(2)

In this work, similar to the work of Isola et al. [18], we used Gaussian noise z in the generator alongside the input data x. As discussed by Isola et al. [18], in training procedure of conditional generative model from conditional distribution $P(y|x)$, that would be better to model produces more than one sample y, from each input x. When the generator G, takes plus input image x, random vector z, then $G(x, z)$ can generate as many different values for each x as there are values of z. Specially for medical image segmentation, the diversity of image acquisition methods (e.g., MRI, fMRI, CT, ultrasound), regarding their settings (e.g., echo time, repetition time), geometry (2D vs. 3D), and differences in hardware (e.g., field strength, gradient performance) can result in variations in the appearance of body organs and tumour shapes [17], thus learning random vector z with input image x makes network robust against noise and act better in the output samples. This has been confirmed by our experimental results using datasets having a large range of variation.

To mitigate the problem of unbalanced training samples, the segmentor loss is weighted same as Eq. (3) to reduce effect of class voxel frequencies for the whole training dataset.

$$w_i = \begin{cases} avg\{f_i\}\{0 < i < c\}/f_{max}, & \text{if } i \text{ is max frequency} \\ 1, & \text{otherwise} \end{cases} \tag{3}$$

$$\mathcal{L}_{L1}(S) = E_{x,z} \parallel y_{seg} - S((x * w), z) \parallel \tag{4}$$

The segmentor loss Eq. (4) is mixed with $\ell 1$ term to minimize the absolute difference between the predicted value and the existing largest value. Previous studies [36,41] on cGANs have shown the success of mixing the cGANs objective with $\ell 1$ distance. Hence, the $\ell 1$ objective function takes into account CNNs feature differences between the predicted segmentation and the ground truth segmentation and resulting in fewer noises and smoother boundaries.

The final objective function for semantic segmentation of brain tumors $\mathcal{L}_{seg}$ calculated by adversarial loss and additional segmentor $\ell 1$ loss as follows:

$$\mathcal{L}_{seg}(D, S) = \mathcal{L}_{adv}(D, S) + \mathcal{L}_{L1}(S) \tag{5}$$

2.1   Segmentor Network

As shown in Fig. 1, the segmentor architecture is two, 3D fully convolutional encoder-decoder network that predicts a label for each voxel. The first encoder takes $64 \times 64 \times 64$ of multi-modal MRI or CT images at same time as different channel input. Last decoder outputs 3D images with size $64 \times 64 \times 64$. Similar to UNet [37], we added the skip connections between each layer i and layer $n - i$, where n is the total number of layers in each encoder and decoder part. Each skip connection simply concatenates all channels at layer i with those at layer $n - i$. Moreover, we concatenate the bottleneck features and last convolutional decoder to capture better feature representation.

2.2   Discriminator Network

As depicted in Fig. 1, the discriminator is 3D fully convolutional encoder network which classifies whether a predicted voxel label belongs to right class. Similar to the pix-to-pix [18], we use path-GAN as discriminator with setting of voxel

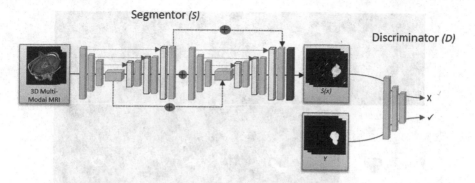

Fig. 1. The architecture of the proposed voxel-GAN consists of a segmentor network S and a discriminative network D. S takes 3D multi modal images as a condition and generates the 3D semantic segmentation as outputs, D determines whether those outputs are real or fake. We use two modified 3D UNet architecture as a segmentor network in order to capture local and global features extracted in bottleneck and last convolutional decoder. Here, D is 3D fully convolutional encoder.

to voxel analysis. More specifically, the discriminator is trained to minimize the average negative cross-entropy between predicted and the true labels.

Then, the segmentor and the discriminator networks are trained through back propagation corresponding to a two-player mini-max game. We use categorical cross entropy [29] as an adversarial loss. As mentioned before, we weighted loss to only attenuate healthy voxel impact in training and testing time.

3   Experiments

We validated the performance of our voxel-GAN on two recent medical imaging challenges: real patient data obtained from the MICCAI 2018, MRI brain tumor segmentation (BraTS) [3–5,25] and CT brain lesion segmentation challenge (ISLES-2018) [20,24].

3.1   Datasets and Pre-processing

The first experiment is carried out on real patient data obtained from BraTS2018 challenge [3–5,25]. The BraTS2018 released data in three subsets train, validation, and test comprising 289, 68, and 191 MR images respectively in four multisite modalities of T1, T2, T1ce, and Flair which the annotated file provided only for the training set. The challenge is semantic segmentation of complex and heterogeneously located of tumor(s) on highly imbalanced data. Pre-processing is an important step to bring all subjects in similar distributions, we applied z-score normalization on four modalities with computing the mean and stdev of the brain intensities. We also applied bias field correction introduced by Nyúl et al. [30]. Figure 2 shows an 2D slice of prepocessed images (our network takes 3D images).

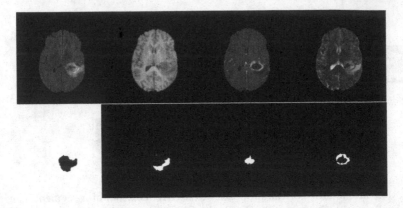

Fig. 2. The brain MR image, from Brats 2018 after pre-processing. We extracted complementary mask from inverse of ground truth file annotated by medical expert, presented in the first column. Other binary masks extracted from ground truth file in columns 2–4 respectively are whole tumor, enhanced tumor, and core of tumor which they are used by the discriminator. The 5–8 columns are a slice of example 3D input of the segmentor.

In second experiment, We applied the ISLES2018 benchmark which contains 94 computer tomography (CT) and MRI training data in six modalities of CT, 4DPWI, CBF, CBV, MTT, Tmax, and annotated ground truth file. The examined patients were suffering from different brain cancers. The challenging part is binary segmentation of unbalance labels. Here, pre-processing is carried out in a slice-wise fashion. We applied Hounsfield unit (HU) values, which were windowed in the range of [30, 100] to get soft tissues and contrast. Furthermore, we applied histogram equalization to increase the contrast for better differentiation of abnormal lesion tissue.

To prevent over fitting, we added data augmentation to each datasets such as randomly cropped, re-sizing, scaling, rotation between −10 and 10 degree, and Gaussian noise applied on training and testing time for both datasets.

3.2   Implementation

Configuration: Our proposed method is implemented based on a Keras library [7] with back-end Tensorflow [1] supporting 3D convolutional network and is publicly available[1]. All training and experiments were conducted on a workstation equipped with a multiple GPUs. The learning rate was initially set to 0.0001. The Adadelta optimizer is used in both the segmentor and the discriminator that continues learning even when many updates have been done. The model is trained for up to 200 epochs on each dataset separately. We used Adadelta as an optimizer for cGAN network.

---

[1] https://github.com/HPI-DeepLearning/VoxelGAN.

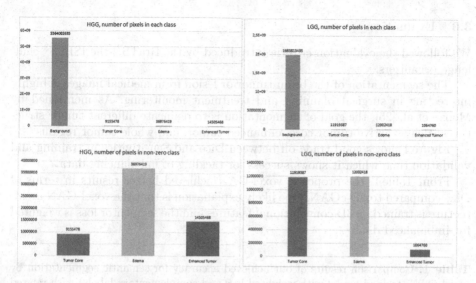

Fig. 3. The number of pixels for each tumor classes represents how imbalanced is training data in detail of two subsets: high and low grade glioma brain tumor on BraTS2018.

Network Architecture: In this work, a segmentor network is a modified UNet architecture that we designed two UNet architecture with sharing circumvent bottlenecks and last fully convolutional layer in decoder part. The UNet architecture allows low-level features to shortcut across the network. Motivated by previous studies on interpreting encoder-decoder networks [31], that show the bottleneck features carried local features and fully convolutional up-sampling encoder represented global features, we concatenate circumvent bottlenecks and last fully convolutional layer to capture more important features.

Our discriminator is fully convolutional Markovian PatchGAN classifier [18] which only penalizes structure at the scale of image patches. Unlike, the Path-GAN discriminator introduced by Isola et al. [18] which classified each N N patch for real or fake, we have achieved better results for task of semantic segmentation in voxel level $1 \times 1 \times d$ we consider $N = 1$ and different $d = 64, 32, 16$, and 8. We used categorical cross entropy [29] as an adversarial loss with combination of $\ell 1$ loss in generator network.

Regarding the highly imbalance datasets as shown in Fig. 3, minority voxels with lesion label are not trained as well as majority voxels with non-lesion label. Therefore, we weighted only non-lesion classes to be in same average of lesion or tumor(s) classes. Tables 1 and 2 describe our achieved results with and without weighting loss on BraTS2018.

3.3   Evaluation

We followed the evaluation criteria introduced by the BraTS[2], the ISLES[3] challenge organizers.

The segmentation of the brain tumor or lesion from medical images is highly interesting in surgical planning and treatment monitoring. As mentioned by Menze et al. [25], the goal of segmentation is to delineate different tumor structures such as active tumor core, enhanced tumor, and whole tumor regions.

Figure 4 shows good trade-off between Dice and Sensitivities in training and validation time which it shows success for tackling of unbalancing data.

From Table 1, the proposed voxel-GAN achieved better results in terms of Dice compared to 2D-cGAN. One likely explanation is that the voxel-GAN architecture is trained on 3D convolutional features and the segmentor loss is weighted for imbalanced data.

Table 1. Comparison results of our achieved accuracy for semantic segmentation by voxel-GAN (trained model with weighted loss and complementary labels) with related work and top ranked team, in terms of Dice, sensitivity, specificity, and Hausdorff distance on five fold cross validation after 80 epochs while the reported results in second and third rows are after 200 epochs. WT, ET, and CT are abbreviation of whole tumor, enhanced tumor, and core of tumor regions respectively.

Methods	Dice			Hdff			Sen			Spec		
	WT	ET	CT	WT	ET	CT	WT	ET	CT	WT	ET	CT
Voxel-GAN	0.84	0.63	0.79	6.41	7.1	10.38	0.86	0.74	0.78	0.99	0.99	0.99
cGAN [34]	0.81	0.61	0.64	7.30	9.22	12.04	0.75	0.61	0.55	0.99	0.99	0.99
Cycle-GAN [14]	0.90	0.78	0.81	2.50	4.5	5.4	0.89	0.89	0.81	0.99	0.99	0.99
Ensemble of 10 3D-Models [28]	0.91	0.82	0.86	3.9	4.5	6.8	-	-	-	-	-	-
3D UNet + TTA [39]	0.87	0.75	0.78	4.5	5.9	8.0	-	-	-	-	-	-

Unlike previous works [14,28,39], we start training from scratch and even after 200 epochs our results are not as good as top ranked team. From Table 1, two top ranked team used ensemble of pre-trained models. Ensemble networks provides good solution for imbalanced data by modifying the training data distribution with regards to the different misclassification costs. In future we will focus on training voxel-GAN with one segmentor from scratch and many different pre-trained discriminators.

---

[2] http://www.med.upenn.edu/sbia/brats2018/evaluation.html.
[3] https://www.smir.ch/ISLES/Start2018.

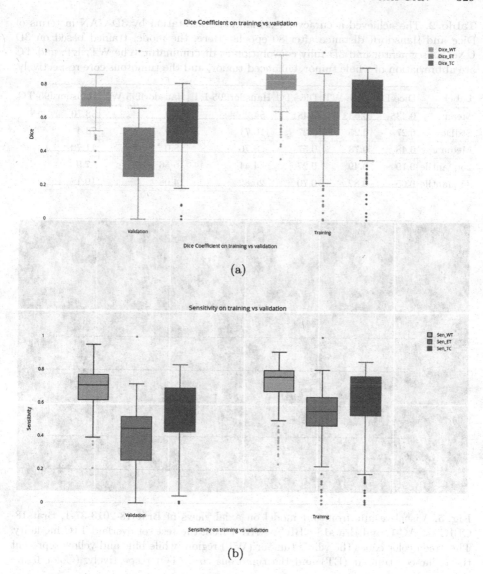

Fig. 4. The achieved accuracy obtained by voxel-GAN in terms of Dice and sensitivity at training and validation time on BraTS-2018.

Table 2. The achieved accuracy for semantic segmentation by 3D-GAN in terms of Dice and Hausdorff distance after 80 epochs. Here, the model trained based on 3D UNet as segmentor and 3D fully convolution as discriminator. The WT, ET, and TC are abbreviation of whole tumor, enhanced tumor, and the tumorous core respectively.

Label	Dice-ET	Dice-WT	Dice-TC	Hausdorff95-ET	Hausdorff95-WT	Hausdorff95-TC
Mean	0.438	0.633	0.481	54.2	12.9	33.70
StdDev	0.27	0.25	0.27	116.71	14.9	78.4
Median	0.48	0.73	0.57	8.76	8.0	11.70
25quantile	0.19	0.49	0.27	4.41	5.56	7.9
75quantile	0.65	0.82	0.70	20.82	14.08	19.1

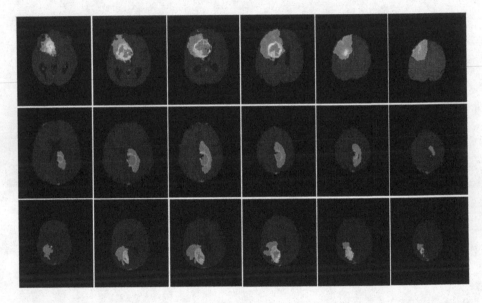

Fig. 5. Visual results from our model on axial views of Brats18-2013-37-1, Brats18-CBICA-AAC-1, and Brats18-CBICA-AAK-1 from the test set overlaid T1C modality. The green color codes the whole tumour (WT) region, while blue and yellow represent the enhanced tumour (ET) and the tumorous core (TC) respectively. (Color figure online)

Table 3. The achieved accuracy for semantic segmentation on ISLES dataset by voxel-GAN in terms of Dice, Hausdorff distance, Precision, and Recall on five fold cross validation after 200 epochs.

	Dice	Hausdorff	Precision	Recall
voxel-GAN	0.83	9.3	0.81	0.78

4 Conclusion

In this paper, we presented a new 3D conditional generative adversarial architecture, named voxel-GAN, that mitigates the issue of unbalanced data for the brain lesion or tumor segmentation. To this end, we proposed a segmentor network and a discriminator network where the first segments the voxel label, and the later classifies whether the segmented output is real or fake. Moreover, we analyzed an effects of different losses and architectural choices that help to improve semantic segmentation results. We validated our framework on CT ISLES2018 and MRI BraTS-2018 images for lesion and tumor semantic segmentation. In the future, we plan to investigate ensemble network based on voxel-GAN but with many pre-trained discriminator networks for semantic segmentation task.

References

1. Abadi, M., et al.: Tensorflow: a system for large-scale machine learning. In: OSDI, vol. 16, pp. 265–283 (2016)
2. Afshin, M., et al.: Regional assessment of cardiac left ventricular myocardial function via MRI statistical features. IEEE Trans. Med. Imaging **33**(2), 481–494 (2014)
3. Bakas, S., Akbari, H.: Segmentation labels and radiomic features for the pre-operative scans of the TCGA-GBM collection. Cancer Imaging Arch. **286** (2017). https://doi.org/10.7937/K9/TCIA.2017.KLXWJJ1Q
4. Bakas, S., et al.: Segmentation labels and radiomic features for the pre-operative scans of the TCGA-LGG collection. Cancer Imaging Arch. **286** (2017). https://doi.org/10.7937/K9/TCIA.2017.GJQ7R0EF
5. Bakas, S., et al.: Advancing The Cancer Genome Atlas glioma MRI collections with expert segmentation labels and radiomic features. Nat. Sci. Data (2017). https://doi.org/10.1038/sdata.2017.117
6. Bakas, S., Reyes, M., Menze, B. et al.: Identifying the best machine learning algorithms for brain tumor segmentation, progression assessment, and overall survival prediction in the brats challenge. arXiv preprint arXiv:1811.02629 (2018)
7. Chollet, F., et al.: Keras (2015)
8. Christ, P.F., et al.: Automatic liver and tumor segmentation of CT and MRI volumes using cascaded fully convolutional neural networks. CoRR abs/1702.05970 (2017). http://arxiv.org/abs/1702.05970
9. Ciecholewski, M.: Support vector machine approach to cardiac SPECT diagnosis. In: Aggarwal, J.K., Barneva, R.P., Brimkov, V.E., Koroutchev, K.N., Korutcheva, E.R. (eds.) IWCIA 2011. LNCS, vol. 6636, pp. 432–443. Springer, Heidelberg (2011). https://doi.org/10.1007/978-3-642-21073-0_38
10. Douzas, G., Bacao, F.: Effective data generation for imbalanced learning using conditional generative adversarial networks. Expert Syst. Appl. **91**, 464–471 (2018)
11. Eslami, A., Karamalis, A., Katouzian, A., Navab, N.: Segmentation by retrieval with guided random walks: application to left ventricle segmentation in MRI. Med. Image Anal. **17**(2), 236–253 (2013)
12. Fidon, L., et al.: Generalised wasserstein dice score for imbalanced multi-class segmentation using holistic convolutional networks. In: Crimi, A., Bakas, S., Kuijf, H., Menze, B., Reyes, M. (eds.) BrainLes 2017. LNCS, vol. 10670, pp. 64–76. Springer, Cham (2018). https://doi.org/10.1007/978-3-319-75238-9_6

13. Fischl, B., et al.: Sequence-independent segmentation of magnetic resonance images. Neuroimage **23**, S69–S84 (2004)
14. Gholami, A., et al.: A novel domain adaptation framework for medical image segmentation. CoRR abs/1810.05732 (2018). http://arxiv.org/abs/org/abs/1810.05732
15. Goodfellow, I.J., et al.: Generative Adversarial Networks. ArXiv e-prints (2014)
16. Hashemi, S.R., Salehi, S.S.M., Erdogmus, D., Prabhu, S.P., Warfield, S.K., Gholipour, A.: Tversky as a loss function for highly unbalanced image segmentation using 3D fully convolutional deep networks. CoRR abs/1803.11078 (2018). http://arxiv.org/abs/1803.11078
17. Inda, M.d.M., Bonavia, R., Seoane, J., et al.: Glioblastoma multiforme: a look inside its heterogeneous nature. Cancers **6**(1), 226–239 (2014)
18. Isola, P., Zhu, J.Y., Zhou, T., Efros, A.A.: Image-to-image translation with conditional adversarial networks. In: The IEEE Conference on Computer Vision and Pattern Recognition (CVPR), July 2017
19. Jang, J., et al.: Medical image matching using variable randomized undersampling probability pattern in data acquisition. In: 2014 International Conference on Electronics, Information and Communications (ICEIC), pp. 1–2, January 2014. https://doi.org/10.1109/ELINFOCOM.2014.6914453
20. Kistler, M., Bonaretti, S., Pfahrer, M., Niklaus, R., Büchler, P.: The virtual skeleton database: an open access repository for biomedical research and collaboration. J. Med. Internet Res. **15**(11) (2013)
21. Kohli, M.D., Summers, R.M., Geis, J.R.: Medical image data and datasets in the era of machine learning—whitepaper from the 2016 C-MIMI meeting dataset session. J. Digit. Imaging **30**(4), 392–399 (2017)
22. LeCun, Y., Bengio, Y., Hinton, G.: Deep learning. Nature **521**(7553), 436–444 (2015)
23. Mahapatra, D.: Automatic cardiac segmentation using semantic information from random forests. J. Digit. Imaging **27**(6), 794–804 (2014)
24. Maier, O., et al.: ISLES 2015-a public evaluation benchmark for ischemic stroke lesion segmentation from multispectral MRI. Med. Image Anal. **35**, 250–269 (2017)
25. Menze, B.H., et al.: The multimodal brain tumor image segmentation benchmark (brats). IEEE transactions on medical imaging **34**(10), 1993–2024 (2015)
26. Mirza, M., Osindero, S.: Conditional generative adversarial nets. CoRR abs/1411.1784 (2014)
27. Moeskops, P., Veta, M., Lafarge, M.W., Eppenhof, K.A.J., Pluim, J.P.W.: Adversarial training and dilated convolutions for brain MRI segmentation. CoRR abs/1707.03195 (2017). http://arxiv.org/abs/1707.03195
28. Myronenko, A.: 3D MRI brain tumor segmentation using autoencoder regularization. arXiv preprint arXiv:1810.11654 (2018)
29. Nasr, G.E., Badr, E., Joun, C.: Cross entropy error function in neural networks: forecasting gasoline demand. In: FLAIRS Conference, pp. 381–384 (2002)
30. Nyúl, L.G., Udupa, J.K., Zhang, X.: New variants of a method of MRI scale standardization. IEEE Trans. Med. Imaging **19**(2), 143–150 (2000)
31. Palade, V., Neagu, D.-C., Patton, R.J.: Interpretation of trained neural networks by rule extraction. In: Reusch, B. (ed.) Fuzzy Days 2001. LNCS, vol. 2206, pp. 152–161. Springer, Heidelberg (2001). https://doi.org/10.1007/3-540-45493-4_20
32. Pohl, K.M., Fisher, J., Grimson, W.E.L., Kikinis, R., Wells, W.M.: A Bayesian model for joint segmentation and registration. NeuroImage **31**(1), 228–239 (2006)

33. Rezaei, M., Yang, H., Meinel, C.: Instance tumor segmentation using multitask convolutional neural network. In: 2018 International Joint Conference on Neural Networks (IJCNN), pp. 1–8, July 2018. https://doi.org/10.1109/IJCNN.2018. 8489105

34. Rezaei, M., et al.: A conditional adversarial network for semantic segmentation of brain tumor. In: Crimi, A., Bakas, S., Kuijf, H., Menze, B., Reyes, M. (eds.) BrainLes 2017. LNCS, vol. 10670, pp. 241–252. Springer, Cham (2018). https:// doi.org/10.1007/978-3-319-75238-9_21

35. Rezaei, M., Yang, H., Meinel, C.: Deep neural network with l2-norm unit for brain lesions detection. In: Liu, D., Xie, S., Li, Y., Zhao, D., El-Alfy, E.S. (eds.) ICONIP 2017. LNCS, vol. 10637, pp. 798–807. Springer, Cham (2017). https://doi.org/10. 1007/978-3-319-70093-9_85

36. Rezaei, M., Yang, H., Meinel, C.: Whole heart and great vessel segmentation with context-aware of generative adversarial networks. In: Maier, A., Deserno, T., Handels, H., Maier-Hein, K., Palm, C., Tolxdorff, T. (eds.) Bildverarbeitung für die Medizin 2018. Informatik aktuell, pp. 353–358. Springer, Heidelberg (2018). https://doi.org/10.1007/978-3-662-56537-7_89

37. Ronneberger, O., Fischer, P., Brox, T.: U-Net: convolutional networks for biomedical image segmentation. In: Navab, N., Hornegger, J., Wells, W.M., Frangi, A.F. (eds.) MICCAI 2015. LNCS, vol. 9351, pp. 234–241. Springer, Cham (2015). https://doi.org/10.1007/978-3-319-24574-4_28

38. Sudre, C.H., Li, W., Vercauteren, T., Ourselin, S., Jorge Cardoso, M.: Generalised dice overlap as a deep learning loss function for highly unbalanced segmentations. In: Cardoso, M., et al. (eds.) DLMIA/ML-CDS -2017. LNCS, vol. 10553, pp. 240–248. Springer, Cham (2017). https://doi.org/10.1007/978-3-319-67558-9_28

39. Wang, G., Li, W., Ourselin, S., Vercauteren, T.: Automatic brain tumor segmentation using convolutional neural networks with test-time augmentation. arXiv preprint arXiv:1810.07884 (2018)

40. Xu, J., Schwing, A.G., Urtasun, R.: Tell me what you see and i will show you where it is. In: Proceedings of the IEEE Conference on Computer Vision and Pattern Recognition, pp. 3190–3197 (2014)

41. Xue, Y., Xu, T., Zhang, H., Long, L.R., Huang, X.: Segan: adversarial network with multi-scalel1 loss for medical image segmentation. CoRR abs/1706.01805 (2017)

Brain Tumor Segmentation and Survival Prediction Using a Cascade of Random Forests

Szidónia Lefkovits[1]([✉]) [iD], László Szilágyi[2] [iD], and László Lefkovits[2] [iD]

[1] Department of Computer Science,
University of Medicine, Pharmacy, Sciences and Technology, Tîrgu Mureş, Romania
szidonia.lefkovits@science.upm.ro
[2] Department of Electrical Engineering,
Sapientia Hungarian University of Transylvania, Tîrgu Mureş, Romania
lefkolaci@ms.sapientia.ro

Abstract. Brain tumor segmentation is a difficult task due to the strongly varying intensity and shape of gliomas. In this paper we propose a multi-stage discriminative framework for brain tumor segmentation based on BraTS 2018 dataset. The framework presented in this paper is a more complex segmentation system than our previous work presented at BraTS 2016. Here we propose a multi-stage discriminative segmentation model, where every stage is a binary classifier based on the random forest algorithm. Our multi-stage system attempts to follow the layered structure of tumor tissues provided in the annotation protocol. In each segmentation stage we dealt with four major difficulties: feature selection, determination of training database used, optimization of classifier performances and image post-processing. The framework was tested on the evaluation images from BraTS 2018. One of the most important results is the determination of the tumor ROI with a sensitivity of approximately 0.99 in stage I by considering only 16% of the brain in the subsequent stages. Based on the segmentation obtained we solved the survival prediction task using a random forest regressor. The results obtained are comparable to the best ones presented in previous BraTS Challenges.

Keywords: Multi-stage classifier · Random forest · Feature selection · Variable importance · MRI brain tumor segmentation

1 Introduction

Image processing is a powerful tool for computer-aided diagnosis especially in the medical field. The most important advantage of medical imaging is the fact that

The work of S. Lefkovits in this article was supported by the Communitas Foundation and the work of L. Lefkovits in this article was supported by Sapientia Foundation-Institute for Scientific Research (KPI).

© Springer Nature Switzerland AG 2019
A. Crimi et al. (Eds.): BrainLes 2018, LNCS 11384, pp. 334–345, 2019.
https://doi.org/10.1007/978-3-030-11726-9_30

examination performed non-intrusively. MR imaging and diagnosis is increasingly being used for medical investigation. This article is restricted to MRI brain imaging and provides a framework for the automated brain tumor segmentation method proposed, delimiting different types of tumors in multi-modal MRI images. Automatic systems based on machine learning overcomes the laborious, lengthy work of segmentation done manually by experts. It is replicable and much faster than the segmentation performed by experts, which might be fairly different. The most important advantage of such a system is that it can lend assistance in determining the correct diagnosis, surgery or treatment plan and monitor the evolution of the disease. A comprehensive study of all the participating methods at BraTS 2018 Challenges is described in paper [6].

The framework presented in this paper is an extended segmentation system of greater complexity based on our model presented at MICCAI-BraTS 2016 [15]. This model was built on a feature extraction algorithm [14] and single-staged random forest (RF) [11] classifiers with optimized parameters. The random forest approach was used in few systems presented at BraTS 2017 [10,17–19]. The segmentation results obtained showed that the tumor region is well detected, but the contours of the whole tumor and the interior tumor tissues are not well delimited. The source of the aforementioned errors could be the choice of training samples used, the unbalanced database provided, and its enormous size. These three factors cannot be counteracted by a single-stage RF classifier. Another deficiency in our previous model is that it considered almost any spatial relationship between the tumor tissues, according to the annotation protocol described in [12,16].

In the current work we propose a multi-stage classifier based on the random forest algorithm. In our current experiments we attempt to circumvent the deficiencies of our old framework and improve the segmentation results.

The proposed framework is built around the model given in Fig. 1. The stage I classifier detects the tumorous zone from the entire 3D MRI image. This phase is tuned to have extremely good sensitivity. It considers the tumor zone to be the goal of detection, and therefore this is the positive segmentation zone. Thus, it is able to delimit the image ROI containing the tumor with a sensitivity of approximately 0.99, which is only around 16% of the entire brain. This means that our ROI considered in the subsequent stages is a highly reduced region. In stage II we developed a classifier that is able to separate the images in the two types given, i.e., LGG and HGG. Consequently, at this point our method is split into two structurally similar branches, because the classifiers are trained differently only on LGG or HGG images. In stage III the WT (whole tumor) classifier delimits the whole tumor from healthy brain tissue. In stage IV the role of the classifier is the determination of the ET (enhancing tumor) region. In the case of HGG, this region is a considerable part of the tumor and includes necrotic tissue regions too. The necrotic tissue inside ET is labelled with the same class number as the non-enhancing tumor. These two tissue types had different annotations at BraTS 2016. In the case of LGG, the segmentation of ET is more difficult because of its small size, and it can also be confused with

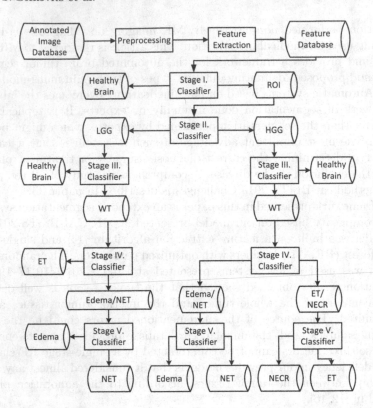

Fig. 1. Discriminative model proposed for segmentation (Each stage-classifier is binary as described in Subsect. 2.1.)

other tissues such as vessels. Stage V tries to delimit the edema from the non-enhancing tumor. Because of the similar visual aspects of the two tissues, this segmentation step is error-sensitive.

The use of binary classifiers for all these classification decisions follows from the annotation protocol. It states that "the various tissue elements (edema, non-enhancing, enhancing, necrosis) usually follow an outside-inside sequence therefore one should start from the outside and delineate regions within the previous layer. Due to this « Mozart kugel » appearance it is enough to always delimitate what is outside and internal border should not be delimitated [12]".

The rest of the paper is organized as follows: in Sect. 2 the proposed cascaded binary classification model is described, followed by Sect. 3 presenting the validation and test results both for the segmentation and survival tasks. Finally, conclusions are drawn and discussion and further improvements are proposed.

2 Method

2.1 Segmentation

The delimitation of the brain tumor from the healthy tissues can be achieved by a voxel-wise segmentation. To solve this task we propose a multi-stage discriminative model based mainly on the random forest algorithm and its facilities. Voxel-wise segmentation starts with the construction of the feature database obtained from the annotated image database. The feature database generation process is identical both for the segmentation (classification) and the training phases, as well. It consists of the following steps: preprocessing, local feature definition and extraction (Figs. 1 and 2).

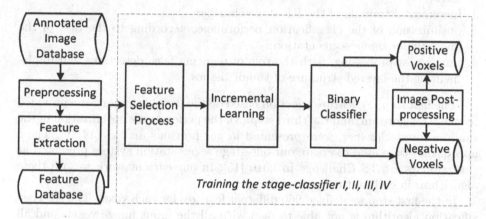

Fig. 2. Discriminative model proposed for training each stage

The database used in our segmentation made up of was the training and validation databases created for the BraTS 2018 Challenge [4]. The training set consists of 75 low-grade and 210 high-grade MRI brain images. The image data consists of 4 modalities T1, T1c, T2 and FLAIR, acquired from 19 different MRI scanners using different protocols [4,7]. All the images had been segmented manually by several experts, and the average annotation is in fact the ground truth given in the database. The modalities are co-registered, interpolated to the same resolution and skull-striped. The annotated regions [8,9] are labeled in 4 different classes: 0 for background and healthy tissue, 1 for NCR/NET (necrotic and/or non-enhancing tumor), label 2 for ED (the edema) and label 4 for ET (the enhancing tumor).

During preprocessing we handled three important artifacts: inhomogeneity correction, noise filtering and intensity standardization. For inhomogeneity reduction in MR images, we applied the N4 filter implemented in the ITK package [1]. The anisotropic filtering from the same package was used for noise reduction. Intensity normalization was done by histogram linear transformation in such a way that the first and third quartiles had predefined values.

In voxel-wise segmentation it is necessary to define a set of intensity- and local neighboring features. The following features were extracted: first order operators (mean, standard deviation, max, min, median, Sobel, gradient); higher order operators (laplacian, difference of gaussian, entropy, curvatures, kurtosis, skewness); texture features (Gabor filter); spatial context features (symmetry, projection, neighborhoods), – the same as in our previous work.

The segmentation workflow given in Fig. 1 requires nine binary classifiers. Each classifier is trained and evaluated on its own feature database during its training process (Fig. 2). The global training consists of five training stages and each stage is composed of the following four steps:

1. feature selection based on variable importance [13] provided by the random forest;
2. incremental training of the RF stage-classifier;
3. optimization of the classification performance according to the task of the given tumor tissue segmentation;
4. image post-processing, with the role of reducing false detections and implementing the layered structure of tumor tissues.

The first step (1), feature selection based on the variable importance provided by the RF algorithm, and the third step (3), the performance optimization of the random forest classifier, were presented in our previous articles [14,15]. These approaches were used to create our one-stage segmentation system presented at the previous BraTS Challenge in 2016 [15]. In our current work we use these algorithms in each of the five stages.

In the first step we defined 960 different features for each voxel. The RF classification algorithm is not able to deal with all the input image voxels and all 960 features previously defined, due to hardware and software limits. Therefore, this large amount of data was handled by taking advantage of the random forest variable importance evaluation. Our idea was to implement an iterative feature selection algorithm presented in [14]. The main idea of the algorithm is to evaluate the variable importance several times on a randomly chosen part of the feature database (Fig. 3). If the OOB error of the forest ensemble was below a certain threshold then the variable importance was taken into consideration and cumulated. Averaging the variable importances in the iterations the algorithm was able to eliminate the most unimportant 20–40% of variables in each run.

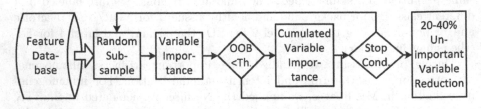

Fig. 3. Feature selection algorithm

In random forest approaches the training set is usually created out of the existing annotated images by random subsampling. In the case of BraTS 2018 the annotated image set contains 285 MR images and each image is made up of about 1,500,000 voxels, which means about 450 million samples. This huge database is, in addition, extremely unbalanced. In consequence we must obtain a well-defined database for training our random forest classifier. The solution to this is the incremental learning procedure that consists of enlarging the current training set by incrementally adding incorrectly classified random subsamples. In the second step (2), this incremental learning is repeated several times until the classification performances are adequate or the upper limit of hardware and/or software is reached. The flowchart of the incremental learning is given in Fig. 4.

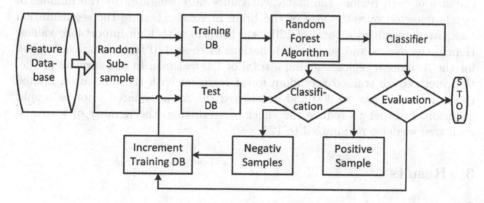

Fig. 4. Incremental learning

The classifier performance optimization (step 3) is in strong correlation with the segmentation task. This assumes the correct choice of training parameters. The random forest classification performance can be tuned via three important parameters: m_{tires}– the number of randomly chosen features used as a splitting criterion in each node of the trees; the n_{trees}– the number of trees in the forest; n_{nodes}– the maximum number of nodes in each tree. These parameters determine the size of the random forest ensemble. the segmentation performances, training time and system complexity, as well. In our experiments [15] these durations can be drastically reduced without any loss in segmentation accuracy.

The last step (4), after the training of each stage-classifier (Fig. 2), is an image post-processing step to do with the segmentation goal of the current stage. Here we managed to incorporate some knowledge about the tumor, such as the number of distinct tumors in a brain, one tumor is a connected zone within the healthy brain tissue, the tumor core is inside the edema, the enhancing tumor is a connected zone inside the whole tumor, etc. By applying this post-processing step we succeeded to eliminate the most of the false detections and improve the quality of segmentation.

2.2 Survival Prediction

Survival prediction has a considerable role, especially from a medical perspective, as well as that of the life expectancy of the patient. It helps in monitoring the effects of the medication and treatment applied. This prediction has to be correlated with the disease state and physical well-being of the patient. In this task the only information available were the MRI scans and the age of a small number of patients (59). In the case of this reduced dataset, prediction becomes difficult and leads to a high margin of error. In our survival prediction approach, we evaluated first-order statistics of feature values used in the segmentation phase. The mean and standard deviation of features were computed in the three segmented regions: edema, enhancing tumor and non-enhancing tumor. In order to include the size of each region, the statistical values were weighted by the number of voxels detected over the size of the brain in voxels. During the segmentation task, we determined a total of 120 local features with high importance values. Hence, the means and standard deviations of these 120 features are computed for the 3 tumoral regions, giving a total of 720 features. In our survival prediction method, we trained a random forest regressor with these features, limited the number of trees to 300 and considered the mean squared error as a split criterion. In order to reduce the effect of overfitting, the number of leaves on each tree was also maximized to 128.

3 Results

3.1 Segmentation

The proposed discriminative model is quite laborious and the proposed classifiers have to be tuned separately (Fig. 1). For training we used well-chosen samples, provided by incremental learning, from the entire BraTS 2018 training dataset. In training, beginning with Stage II, we created different classifiers for HGG and LGG as explained above. The results obtained were evaluated on both sets, namely, the complete training and validation sets, and on the test set within the challenge.

The stage I classifier determines the ROI (region of interest) that contains the tumor region with a high probability. This binary classifier was trained on the whole brain in order to delimit the healthy region from the tumoral region. In this step we used an incrementally trained classifier and applied a post-processing step consisting of a region dilation of 3 voxels. In addition, the two most important connected zones were taken into consideration. The results of the last three incremental steps are given in Table 1 and the improvement brought by the post-processing in Table 2. In this way the ROI obtained is about twice as large as the whole tumor, but the sensitivity reached on the complete training set is 0.989. The correct determination of this ROI has a crucial role in the subsequent stages. Table 1 and Fig. 5a show the average of the sensitivity after the binary segmentation and post-processing in stage I. The ROI obtained reduces the image region

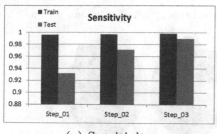

(a) Sensitivity

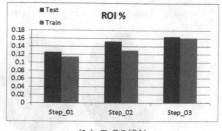

(b) ROI(%)

Fig. 5. Stage I segmentation

Table 1. Incremental training of Stage I segmentation

Stage I	Training			Test		
	Sens.	PPV	ROI%	Sens.	PPV	ROI%
Step_01	0.9959	0.5184	0.1143	0.9319	0.508	0.1257
Step_02	0.9971	0.4526	0.1301	0.9712	0.444	0.1523
Step_03	0.9984	0.4299	0.1615	0.9898	0.4257	0.1642

and, implicitly, the feature database, by about 8 times (Fig. 5b). This allows us to create a more precise classifier in the next stages.

In stage II the images are classified into two types with regard to segmentation. In the HGG images, the enhancing tumor is a considerable part of the whole tumor and may include some necrotic tissue. In LGG images, the greater part of the tumor consists of edema and non-enhancing tumor. By applying this kind of LGG-HGG separation, we could reduce the effects of unbalanced data (especially the ET in LGG). The image classes obtained correspond in a proportion of more than 95% to the medical LGG-HGG classification given for the training set. The subsequent stages (III, IV and V) are trained differently for LGG and HGG images.

The stage III classifier is applied only on the ROI. Its segmentation task is to delimit the remainder of healthy tissue from the WT. In this stage the segmentation with post-processing creates two disjunct regions, considering the tumor zone a connected region inside the healthy tissue. The Dice scores obtained for the whole tumor (WT) are 0.911 on the training and 0.885 on the validation sets (Table 3).

The stage IV classifier is applied only inside the WT region, and its task is to delimit the enhancing tumor. In the case of HGG images, the ET forms a significant connected region including some necrotic tissue. In the case of LGG images, the region is only a small piece in the WT and may easily be confused with vessels. If the segmentation obtains an ET region of less than 100 voxels, it will be neglected and considered to be a vessel. If the ET is near vessels, false detections are often obtained.

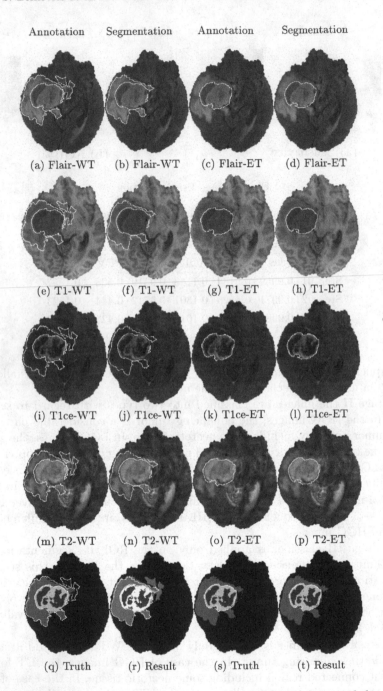

Annotation   Segmentation   Annotation   Segmentation

(a) Flair-WT   (b) Flair-WT   (c) Flair-ET   (d) Flair-ET

(e) T1-WT   (f) T1-WT   (g) T1-ET   (h) T1-ET

(i) T1ce-WT   (j) T1ce-WT   (k) T1ce-ET   (l) T1ce-ET

(m) T2-WT   (n) T2-WT   (o) T2-ET   (p) T2-ET

(q) Truth   (r) Result   (s) Truth   (t) Result

Fig. 6. Segmentation example (Rows represent the 4 MRI modalities and the segmentation obtained. Column 1 is the annotated contour of edema; column 2 is the segmentation result of it. Column 3 is the annotation contour of the ET⊃NECR; column 4 is the segmentation results of it.)

Table 2. Post-processing: Improvements in sensitivity

	Sensitivity	PPV	ROI%	Dice
Classif.	0.963	0.581	0.112	0.707
Step_03	0.990	0.426	0.164	0.574

Table 3. Segmentation results of Stage IV & V on the validation database

Label		Dice			Hausdorff95		
		ET	WT	TC	ET	WT	TC
Training	Mean	0.880	0.913	0.911	3.433	4.721	5.434
HGG	StdDev	0.067	0.064	0.061	6.973	7.580	7.244
Training	Mean	0.808	0.916	0.908	3.057	2.725	2.907
LGG	StdDev	0.320	0.045	0.120	5.974	2.050	3.399
Validation	Mean	0.801	0.883	0.786	5.811	7.410	11.511
HGG	StdDev	0.161	0.082	0.180	9.418	12.644	14.141
Validation	Mean	0.489	0.891	0.415	10.265	4.741	15.858
LGG	StdDev	0.420	0.081	0.285	8.864	2.951	9.875
Final	Mean	0.730	0.885	0.702	6.349	6.803	12.499
Validation	StdDev	0.275	0.081	0.259	9.392	11.230	13.349
Final	Mean	0.684	0.830	0.657	6.186	9.180	11.649
Test	StdDev	0.302	0.193	0.308	9.394	13.062	12.670

The stage V classifier has the most difficult task working on the WT, excluding the ET obtained in stage IV. It has to delimit the edema from the non-enhancing tumor tissues. In the case of HGG images, this stage contains another classifier that finds the necrotic tissues inside the ET. The results obtained on the training validation and test images are given in Table 3. A visual segmentation sample is depicted in Fig. 6.

3.2   Survival Prediction

The results obtained in survival prediction are in strong correlation with the segmentation performances. Concerning the MSE parameter (Mean Squared Error), meaning the squared difference in number of days, we managed to take the first place on the validation database, as shown on the leaderboard [5] and in Table 4. The individual results of the test have no basis for comparison to the other teams owing to the lack of their results. The comparative study will be published by the organizers of BraTS 2018 [6].

Table 4. Survival prediction of the validation database

Team	Cases Eval.	Acc.	MSE	median SE	std SE	Spearman R
lefko	28	0.429	76081.29	24352	111586.8	0.342
Ranking		13	1	4	2	5

4   Conclusion and Discussion

In this paper we developed a five-stage discriminative model for brain tumor segmentation based on multi-modal MRI data. Our five-stage model implements the layered tissue structure by adequate training of binary classifiers and image post-processing in each segmentation stage. In each stage we attempted to solve the four important issues concerning discriminative models. Our results show that binary classifiers are very efficient for the layered segmentation task. One of the most important results is the determination of a ROI that has to enclose the whole tumor with a very high probability. In stage I, the sensitivity attained is 0.989, with a PPV of 0.426. This step reduces the size of the feature database by about 8 times and provides a reliable ROI for the next segmentation stages. Furthermore, the LGG-HGG separation increased the Dice score by 2%. The WT segmentation reached a Dice score of about 0.885 both on the training and validation sets [5]. This result is comparable to the most well-performing methods. In the test set, the reported Dice decreased by 5%, to 0.83. Analysing the test set, we came to the conclusion that the test set contained many HGG images with different visual aspects compared to the training or validation images. The finals results of the survival task in correlation with segmentation performances will be published soon by the BraTS organizers. In our opinion, the MSE score is much more relevant than the accuracy that considers three disjunct time periods as a crisp set (less than 10 months, between 10–15 months, more than 15 months). The system developed is a complex implementation using a large variety of software packages and modules such as ITK in C++ [1], Java, ImageJ and Fiji with Trainable Weka Segmentation [2], the random forest package from R [3], Matlab for performance evaluation and image conversion. Our system is quite complex and still we are working on its dockerized version.

References

1. ITK. http://www.itk.org/
2. Trainable segmentation. http://imagej.net/Scripting_the_Trainable_Segmentation
3. The R project for statistical computing. https://www.r-project.org/
4. Multimodal Brain Tumor Segmentation Challenge 2018. https://www.med.upenn.edu/sbia/brats2018/
5. Multimodal Brain Tumor Segmentation Challenge Evaluation Website. https://www.cbica.upenn.edu/BraTS18/
6. Bakas, S., Reyes, M., et al.: Identifying the best machine learning algorithms for brain tumor segmentation, progression assessment, and overall survival prediction in the BRATS challenge. arXiv preprint arXiv:1811.02629 (2018)

7. Bakas, S., et al.: Advancing the cancer genome atlas glioma MRI collections with expert segmentation labels and radiomic features. Sci. Data **4**, 170117 (2017)

8. Bakas, S., et al.: Segmentation labels and radiomic features for the pre-operative scans of the TCGA-GBM collection. Cancer Imaging Arch. (2017). https://doi.org/10.7937/K9/TCIA.2017.KLXWJJ1Q

9. Bakas, S., et al.: Segmentation labels and radiomic features for the pre-operative scans of the TCGA-LGG collection. Cancer Imaging Arch. (2017). https://doi.org/10.7937/K9/TCIA.2017.GJQ7R0EF

10. Bharath, H.N., Colleman, S., Sima D.M., Van Huffel, S.: Tumor segmentation from multi-parametric MRI using random forest with superpixel and tensor based feature extraction. In Proceedings of MICCAI-BRATS Challenge (2017)

11. Breiman, L.: Random forests. Mach. Learn. **45**(1), 5–32 (2001)

12. Jakab, A.: Segmenting brain tumors with the slicer 3D software manual for providing expert segmentations for the BRATS tumor segmentation challenge. University of Debrecen/ETH Zürich (2012)

13. Goldstein, B.A., Polley, E.C., Briggs, F.: Random forests for genetic association studies. Stat. Appl. Genet. Mol. Biol. **10**(1), 1544–6115 (2011)

14. Lefkovits, L., Lefkovits, S., Emerich, S., Vaida M.F.: Random forest feature selection approach for image segmentation. In: Proceedings of the 9th International Conference on Machine Vision (ICMV), vol. 10341, p. 1034117 (2016)

15. Lefkovits, L., Lefkovits, S., Szilágyi, L.: Brain tumor segmentation with optimized random forest. In: Crimi, A., Menze, B., Maier, O., Reyes, M., Winzeck, S., Handels, H. (eds.) BrainLes 2016. LNCS, vol. 10154, pp. 80–99. Springer, Cham (2016). https://doi.org/10.1007/978-3-319-55524-9_9

16. Menze, B.H., Jakab, A., Bauer, S., et al.: The multimodal brain tumor image segmentation benchmark (BRATS). IEEE Trans. Med. Imaging **34**(10), 1993–2024 (2015)

17. Phophalia, A., Maji, P.: Multimodal brain tumor segmentation using ensemble of forest method. In: Crimi, A., Bakas, S., Kuijf, H., Menze, B., Reyes, M. (eds.) BrainLes 2017. LNCS, vol. 10670, pp. 159–168. Springer, Cham (2018). https://doi.org/10.1007/978-3-319-75238-9_14

18. Revanuru, K., Shah N.: Fully automatic brain tumour segmentation using random forests and patient survival prediction using XGBoost. In: Proceedings of MICCAI-BRATS (2017)

19. Soltaninejad, M., Zhang, L., Lambrou, T., Yang, G., Allinson, N., Ye, X.: MRI brain tumor segmentation using random forests and fully convolutional networks. In: Procedings of MICCAI-BRATS (2017)

Automatic Segmentation of Brain Tumor Using 3D SE-Inception Networks with Residual Connections

Hongdou Yao, Xiaobing Zhou[✉], and Xuejie Zhang

School of Information Science and Engineering, Yunnan University,
Kunming 650091, People's Republic of China
zhouxb@ynu.edu.cn

Abstract. Nowadays, there are various kinds of methods in medical image segmentation tasks, in which Cascaded FCN is an effective one. The idea of this method is to convert multiple classification tasks into a sequence of two categorization tasks, according to a series of sub-hierarchy regions of multi-modal Magnetic Resonance Images. We propose a model based on this idea, by combining the mainstream deep learning models for two dimensional images and modifying the 2D model to adapt to 3D medical image data set. Our model uses the Inception model, 3D Squeeze and Excitation structures, and dilated convolution filters, which are well known in 2D image segmentation tasks. When segmenting the whole tumor, we set the bounding box of the result, which is used to segment tumor core, and the bounding box of tumor core segmentation result will be used to segment enhancing tumor. We not only use the final output of the model, but also combine the results of intermediate output. In MICCAI BraTs 2018 gliomas segmentation task, we achieve a competitive performance without data augmentation.

Keywords: 3D-SE-Inception-ResNet · Cascaded FCN · Anisotropic · Medical image segmentation

1 Introduction

Image segmentation has always been a challenging task in the field of computer vision. Especially in medical image field, multi-modal Magnetic Resonance Images can be used to segment human body pathological tissue. Many medical committees such as MICCAI, have always been focusing on the evaluation of state-of-the-art methods for the segmentation of brain tumors in multi-modal Magnetic Resonance Imaging (MRI) scans. In 2D image processing fields, many effective models were proposed. AlexNet, presented by Krizhevsky et al. [12], won the image classification task of ImageNet 2012. Since then the method of deep learning has aroused researchers' attention. Later, Deep learning models have been kept explosive growth. VGGNet [17], used a series of small convolution filters to substitute for large convolution filters. GoogleNet [19] proposed

© Springer Nature Switzerland AG 2019
A. Crimi et al. (Eds.): BrainLes 2018, LNCS 11384, pp. 346–357, 2019.
https://doi.org/10.1007/978-3-030-11726-9_31

a multi scale concept, by using different size filters to extract information, and its improved version Inception [18], creatively used $1 * 1$ convolution filters to reduce the number of model parameters, while ensuring the model depth without increasing the parameters of the model. Squeeze and Excitation Networks [11], a kind of attention mechanism, introduced the attention mechanism into the spatial dimension, further improving the performance of the model. However, using multi-modal Magnetic Resonance Images to segment human tissue has been very challenging. Because medical image data is more complex than ordinary image data, both plane information and spatial information should be considered. So some researchers try to solve the problem of medical image segmentation by using deep learning method. In the first attemp, the modified variants of 2D CNN was adopted, by using aggregated adjacent slices [6] or orthogonal planes [15,16], but this method did not take into account space information, it couldn't segment object accurately. Recently, a variety of 3D models had been developed to segment objects from volumetric data and gained competitive performance. For examples, 3D U-Net [8] allows end-to-end training and testing for volumetric image segmentation. VoxResNet [5], a deep voxelwise residual network, improves the volumetric segmentation performance by seamlessly integrating the low-level image appearance features, implicit shape information and high-level context together.

The contribution of this paper are four-fold. First, we combine the mainstream segmentation models of 2D CNNs [13] and modified Inception structure to deal with 3D images. In the process of designing the model, we also consider the computation performance, and design two kinds of Inception layer, which are named as Lower Inception and Higher Inception. Second, we apply the 3D Squeeze and Excitation structure to our model. Third, we use multi-scale filters to downsample the 3D feature maps, the loss of valid information can be better reduced when resizing the 3D feature maps. Fourth, our model uses the residual connection to make sure the information can be transferred better and the training process of the model can be accelerated.

2   Methods

2.1   Cascaded Framework

The cascaded framework [7,22] is designed to simplify segmentation problems. We use triple cascaded networks to segment substructures of brain tumor, each network can be seen as a binary segmentation network. While the first network segments the whole tumor task according to the MRI, a bounding box of the whole tumor is obtained. The region of the input images is cropped based on the bounding box, and the cropped result is used as the input of the second network to segment tumor core. After segmenting tumor core, another smaller bounding box is obtained. The image region is resized according to the smaller bounding box of the tumor core. Then the resized image region is used as the input of the third network to segment the enhancing tumor core. During the training phase, the bounding boxes are decided by the ground truth. In the testing stage, the bounding boxes are generated based on the segmentation results.

2.2   Neural Networks Architecture

The overall architecture of the model we proposed is shown in Fig. 1. It includes inception layers, SE structures, reduction layers and residual connection. High-Level Inception uses dilated convolution. The model contains a great deal of 2D image mainstream model structures. Considering the huge advantages of their own structures in 2D images, modifying them to adapt the 3D medical image data can have better effects.

Low-Level Inception. The Low-level Inception structure is shown in Fig. 2. We use $1 * 3 * 3$, $1 * 5 * 5$ and $1 * 7 * 7$ convolution filters to better capture the information of feature maps early in the networks. Why do we design model like this? There are several model design principles [20]. The first principle is to avoid representational bottlenecks, especially early in the network. Any feed-forward networks can be seen as an acyclic graph from input to output. Once the model is defined, the flow direction of information will be decided. When the information passes the model, information is fading. We use the multi large receptive fields early in the network to avoid bottlenecks with extreme compression. In AlexNet [12], Krizhevsky et al. used the $11 * 11$ receptive fields. However, large convolution filters have a serious shortcoming, i.e., large convolution filters have a huge number of training parameters. The parameters of $7 * 7$ receptive fields are 5 times as much as those of $3 * 3$ receptive fields. But large receptive fields can better capture the space information. We should consider the trade-off between computation performance and model complexity, so we apply the large convolution filters only to the first four layers. We use different multi-scale size filters to better capture the space information, while avoiding representation bottleneck.

3D Squeeze and Excitation Structure. Squeeze and Excitation structure was proposed by Hu et al. [11] in 2017, they used the SENet to get a top performance in the ImageNet 2017. The innovation of this model is to explicitly model the interdependence between feature channels. Specifically, it is important to acquire each characteristic channel automatically through the way of learning, improve the useful features and restrain the small features of the current task in accordance with its importance. Based on this idea, we redefine the squeeze and excitation operation in our model. For any given transformation $F_{tr} : X \rightarrow U, X \in \mathbb{R}^{D' \times W' \times H'}, U \in \mathbb{R}^{D' \times W' \times H'}$. We take F_{tr} as a standard 3D convolution operator. $V = [V_1, V_2, ..., V_C]$ denotes the learned set of filter kernels, where V_C refers to the parameters of the c-th filter. We denote $U = [u_1, u_2, ..., u_c]$ as the output of F_{tr}, where

$$u_c = v_c * X = \sum_{s=1}^{C'} v_c^s * x^s \qquad (1)$$

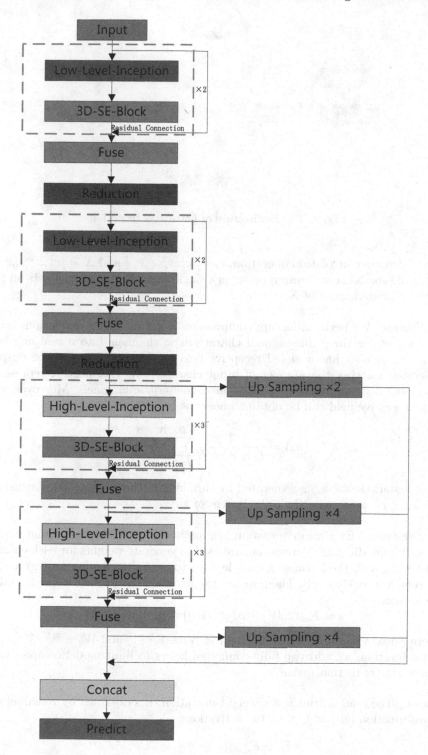

Fig. 1. The architecture of the model we proposed.

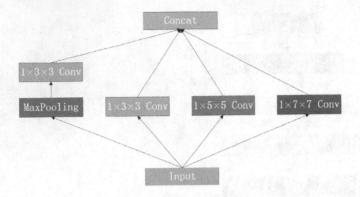

Fig. 2. The architecture of Low-Level Inception

Here $*$ denotes convolution operation, $v_c = [v_c^1, v_c^2, ... v_c^{c'}]$ and $X = [x^1, x^2, ... x^{c'}]$, v_c^s is a 3D spatial kernel, and represents a single channel of v_c, which acts on the corresponding channel of X.

3D Squeeze: We perform feature compression along the space dimension, turning each of the three dimensional characteristic channels into a real number. This real number has a global receptive field to some extent, and the output dimension matches the number of input characteristic channels. It represents the global distribution of responses on characteristic channels. Moreover, the whole receptive field can be obtained near the input layer.

$$z_c = F_{sq}(u_c) = \frac{1}{D \times W \times H} \sum_{i=1}^{D} \sum_{j=1}^{W} \sum_{k=1}^{H} u_c(i, j, k) \tag{2}$$

Here, a statistic $z \in \mathbb{R}^c$ is generated by shrinking U through spatial dimensions $D \times W \times H$, z_c denotes the c-th element of z.

3D Excitation: Excitation operation is a mechanism similar to recurrent neural network's middle gate. Parameters are used to generate weights for each characteristic channel, the parameters are learned to explicitly model the correlation between feature channels. Then, we use the sigmoid activation as a simple gating mechanism:

$$s = F_{ex}(z, W) = \sigma(g(z, W)) = \sigma(W_2 \delta(W_1 z)) \tag{3}$$

where δ refers to the *ReLu* function [13], $W_1 \in \mathbb{R}^{\frac{C}{r} \times C}$ and $W_2 \in \mathbb{R}^{C \times \frac{C}{r}}$. After *ReLu* function, we add two fully-connected layers to limit model complexity. r denotes the reduction ration.

Output: The final output is a reweight operation. It's obtained by rescaling the transformation output U with the activations:

$$\widetilde{x}_c = F_{scale}(u_c, s_c) = s_c \cdot u_c \tag{4}$$

where $\widetilde{X} = [\widetilde{x_1}, \widetilde{x_2}, ..., \widetilde{x_c}]$, and $F_{scale}(u_c, s_c)$ refers to channel-wise multiplication between the feature map $u_c \in \mathbb{R}^{D \times W \times H}$ and the scalar s_c. 3D SE structure is a kind of attention mechanism that can pay attention to 3D channels relationship. c denotes the channels, r denotes ration (In our model, the ration $r = 4$, 8 and 16, we tested the model separately). The SE structure is shown in Fig. 3.

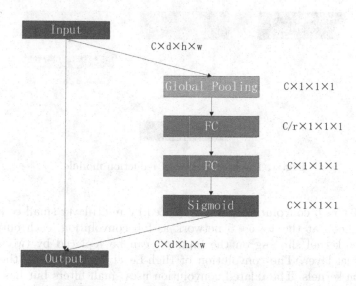

Fig. 3. The architecture of 3D squeeze and excitation

Reduction Structure. Reduction structure is used for reduction feature maps. As mentioned before, using multi-scale can capture more spatial information. Different variants of this blocks (with various number of filters) can be set by users, here we set the number of m, n, o, k and I as 8. As shown in Fig. 4, reduction structure can use multi-scale convolution to capture the information from input feature maps. m, n, o, k and i can be set arbitrarily. We consider the simplified model, so set all the variables to the same number 8 and use $1*1*1$ convolution. In the design principles we mentioned earlier [20], the second principle is intent to let the spatial aggregation be done over lower dimensional embeddings without affecting representational power. Considering that these signals are easy to be compressed, dimensionality reduction will speed up the learning process. We redesign the reduction structure according to this idea.

High-Level Inception. The High-level Inception structure is shown in Fig. 5. The third principle is to factorize a large convolution kernel into smaller ones. Convolutions with large filters have a huge computation complexity. For example, in the case of the same number of convolution kernel, $1*5*5$ convolution is $25/9 = 2.78$ times more computationally complex than that of $1*3*3$. But simply reducing the size of the convolution core will cause information loss.

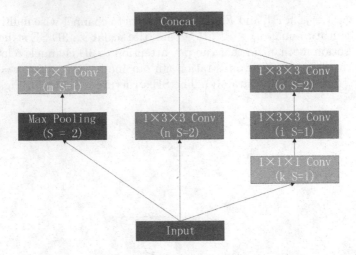

Fig. 4. The architecture of reduction module

However $1*5*5$ convolution can be replaced by multi-layer small convolution networks. Look at the $1*5*5$ network as full convolution, each output is a convolution kernel slipping on the input, it can be replaced by two $1*3*3$ convolutional layer. The convolution of High-Level Inception uses the dilated convolution kernels. The dilated convolution uses small filters but has a larger receptive fields, without increasing the parameters. We set the dilation rate 1, 2, 3 and 3, 2, 1 corresponding to each High-Level Inception layers in order. The High-Level Inception architecture we designed can be seen in Fig. 5.

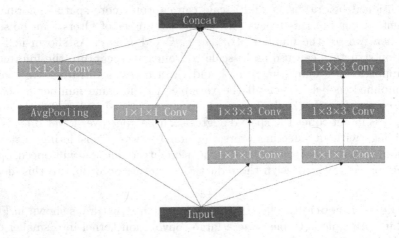

Fig. 5. The architecture of High-Level Inception

Residual Connection. ResNet was put forward in 2015 by He et al. [10], it won the first place in the classification competition of ImageNet. With the increasing of network depth, the problem of the disappearance of the gradient is becoming more and more obvious. The training of the network has become quite difficult. The basic idea of ResNet is to introduce "shortcut connection" that can skip one or more layers. ResBlock can be defined as:

$$y = F(x, w_i) + x \tag{5}$$

Here x and y are the input and output vectors of the layers considered. The function $F(x, w_i)$ represents the residual mapping to be learned. If the dimensions of x and F don't equal, we can perform a linear projection W_s by the shortcut connections to match the dimensions:

$$y = F(x, w_i) + W_s x \tag{6}$$

W_s is used only when matching dimensions.

Prediction and Fusion. In the prediction phase, we not only use the final result but also use the intermediate output results, and concatenate them as the final prediction result. In the training phase, each neural network is trained in axial, sagittal and coronal views. During the test phase, predictions are fused to get the final segmentation. We average the softmax outputs in these cascade networks. Fusion structure is a simple $3 * 1 * 1$ convolution, as one can see the green block in the Fig. 1. The overall model decomposes $3 * 3 * 3$ convolution kernels to $1 * 3 * 3$ convolution and $3 * 1 * 1$ convolution, $1 * 3 * 3$ convolutions are used to extract the datasets features and $3 * 1 * 1$ convolutions are used to fusion the datasets spatial features.

3   Experiments and Results

Brain tumor segmentation is a challenging task, which has attracted a lot of attentions in the past few years. We use the BRATS 2018 dataset [1,2], which is composed of multiple segmentation subproblems. The whole tumor region is identified in a set of multi-modal images, tumor core areas and active tumor regions [4,14].

Medical Image Data. Brats 2018 dataset contains real volumes of 210 high-grade and 75 low-grade glioma subjects. For each patient, T1Gd, T1, T2, FLAIR and Ground Truth MR volumes are available. These 285 subjects are used in training set, and there are 66 other subjects as the validation dataset. Considering the unbalance distribution of the training data, we expand the LGG dataset 3 times based on the original one, during the training data loading process, each LGG dataset copies and reloads 3 times. When training the network, we randomly choose 5 subjects as the input. All of these volume average size is

$155 * 240 * 240$, we resize the volume and extract the voxel of specified shape in the middle volume as the final training input. The biggest black box outside represents the source MRI data set, and the middle gray bounding box represents the valid volumes (include human brain tissue), red point is the core of the target size train patch. In the valid volumes bounding box, dotted line box random crops with the center of the red point, it's used to train our neural networks, we train three cascaded anisotropic networks, use the different patch size to train the different network. We extract the (26, 120, 120) patch size for training the whole tumor segmentation network, (26, 72, 72) patch size for training the tumor core segmentation network and (26, 48, 48) patch size for training the enhancing tumor segmentation network. The details are shown in Fig. 6.

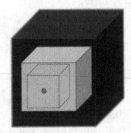

Fig. 6. Data preprocessing details sketch map (Color figure online)

Training Details. Our network is implemented in Tensorflow and NiftyNet, no external data was used during the training. We use Adam optimizer to train, and use PReLu [9] as the activation, set the batch-size = 5, weight decay = 10^{-7}, learning rate = 10^{-3}, max-iteration = $20k$. We train on the GTX 1080Ti GPU. For the data pre-processing, the images are normalized by the mean and standard deviation. And we use the Dice coefficient as the model loss function.

Segmemtaion Results. In order to test the influence of parameter r on the model, we perform three groups of experiments. However, the experiments show that too large or too small parameter r can not get the best result, a moderate parameter r can achieve a better result. More details are shown in Tables 1, 2 and 3. The result of Table 2 is the best among all of them. From the perspective of SE structure, parameter r relates to the number of the first fully connected layer ($fc = c/r \times 1 \times 1 \times 1$), when we give the parameter r a small number, the number of the first FC layer will be quite large, it will increase computation complexity, makes the model hard to train, as shown in Table 1. But if we set the parameter a large number, the number of the first FC layer will be small, it will make the model difficult to learn the channel characteristics better, as can be seen in Table 3. At present, there is no authoritative idea on how to set the parameter r. You can only adjust the parameter r according to the result of experiments. It is regrettable that we failed to submit our best results before the deadline. Table 4 shows the our official scores computed by the organizer of the challenge. Besides, we also test our model on the Brats 2015 dataset with good results. The detail results of our model are shown as Table 5.

Table 1. Table shows the result of our model predict (ration = 4).

Data Set	Dice			Sensitivity			Specificity			Hausdorff95		
	WT	TC	ET	WT	TC	ET	WT	TC	ET	WT	TC	ET
Training	0.729	0.885	0.834	0.805	0.924	0.879	0.998	0.992	0.996	5.657	16.999	7.082
Validation	0.784	0.878	0.807	0.814	0.935	0.844	0.998	0.991	0.997	4.380	19.034	9.408

Table 2. Table shows the result of our model predict (ration = 8).

Data Set	Dice			Sensitivity			Specificity			Hausdorff95		
	WT	TC	ET	WT	TC	ET	WT	TC	ET	WT	TC	ET
Training	0.773	0.910	0.872	0.832	0.929	0.886	0.998	0.994	0.997	3.738	6.938	4.644
Validation	0.798	0.901	0.813	0.818	0.933	0.831	0.998	0.993	0.997	4.158	6.371	8.840

Table 3. Table shows the result of our model predict (ration = 16).

Data Set	Dice			Sensitivity			Specificity			Hausdorff95		
	WT	TC	ET	WT	TC	ET	WT	TC	ET	WT	TC	ET
Training	0.768	0.910	0.868	0.822	0.918	0.879	0.998	0.994	0.997	3.974	6.878	4.841
Validation	0.796	0.903	0.818	0.810	0.928	0.820	0.998	0.993	0.998	3.971	6.255	8.371

Table 4. Performance of proposed method on Test Dataset (model ration = 4).

Label	Dice-ET	Dice-WT	Dice-TC	Hausdorff95-ET	Hausdorff95-WT	Hausdorff95-TC
Mean	0.724	0.864	0.772	5.353	9.131	8.115
StdDev	0.277	0.138	0.263	10.431	14.717	12.041
Median	0.828	0.909	0.889	2.236	3.317	3.742
25quantile	0.710	0.857	0.727	1.414	2.236	2
75quantile	0.879	0.938	0.928	3.317	6.164	8.108

Table 5. Brats 2015 test set results: we rank 8th on the Brats 2015 leaderboard.

Data Set	Dice			Positive			Sensitivity			Rank		
	WT	TC	ET	WT	TC	ET	WT	TC	ET	WT	TC	ET
Test	0.86	0.73	0.63	0.85	0.82	0.61	0.89	0.71	0.70	14.25	15.25	32.50

4 Conclusion

The results of all participants can be seen in [3], compared with other participants, our results can achieve more competitive performance than many of them. As is shown above, setting the model parameter r = 8 can achieve better results than others. We don't perform enough parameter adjustment experiments and don't use other optimization algorithms. When processing data, we only use single volume size. In the future, we plan to integrate convolution CRFs [21] and self-attention of 2D image segmentation into our model.

Acknowledgements. We'd like to thank the team of Wang [22] for their source code, with which we can focus more on the innovation of the model and do our work more easily. And we also thank the NiftyNet team, the deep learning tools they developed enables us construct our model more efficiently. This work was supported by the Natural Science Foundations of China under Grants No. 61463050, No. 61702443, No. 61762091, the NSF of Yunnan Province under Grant No. 2015FB113, the Project of Innovative Research Team of Yunnan Province under Grant No. 2018HC019.

References

1. Bakas, S., et al.: Segmentation labels and radiomic features for the pre-operative scans of the TCGA-GBM collection. Cancer Imaging Arch. (2017). https://doi.org/10.7937/K9/TCIA.2017.KLXWJJ1Q
2. Bakas, S., et al.: Segmentation labels and radiomic features for the pre-operative scans of the TCGA-LGG collection. Cancer Imaging Arch. (2017). https://doi.org/10.7937/K9/TCIA.2017.GJQ7R0EF
3. Bakas, S., Reyes, M., et al.: Identifying the best machine learning algorithms for brain tumor segmentation, progression assessment, and overall survival prediction in the brats challenge. arXiv preprint arXiv:1811.02629 (2018)
4. Bakas, S., et al.: Advancing the Cancer Genome Atlas Glioma MRI collections with expert segmentation labels and radiomic features. Sci. Data **4**, 170117 (2017)
5. Chen, H., Dou, Q., Yu, L., Heng, P.A.: Voxresnet: deep voxelwise residual networks for volumetric brain segmentation. arXiv preprint arXiv:1608.05895 (2016)
6. Chen, H., Yu, L., Dou, Q., Shi, L., Mok, V.C., Heng, P.A.: Automatic detection of cerebral microbleeds via deep learning based 3D feature representation. In: 2015 IEEE 12th International Symposium on Biomedical Imaging (ISBI), pp. 764–767. IEEE (2015)
7. Christ, P.F., et al.: Automatic liver and lesion segmentation in CT using cascaded fully convolutional neural networks and 3D conditional random fields. In: Ourselin, S., Joskowicz, L., Sabuncu, M.R., Unal, G., Wells, W. (eds.) MICCAI 2016. LNCS, vol. 9901, pp. 415–423. Springer, Cham (2016). https://doi.org/10.1007/978-3-319-46723-8_48
8. Çiçek, Ö., Abdulkadir, A., Lienkamp, S.S., Brox, T., Ronneberger, O.: 3D U-net: learning dense volumetric segmentation from sparse annotation. In: Ourselin, S., Joskowicz, L., Sabuncu, M.R., Unal, G., Wells, W. (eds.) MICCAI 2016. LNCS, vol. 9901, pp. 424–432. Springer, Cham (2016). https://doi.org/10.1007/978-3-319-46723-8_49

9. He, K., Zhang, X., Ren, S., Sun, J.: Delving deep into rectifiers: surpassing human-level performance on imagenet classification. In: Proceedings of the IEEE International Conference on Computer Vision, pp. 1026–1034 (2015)

10. He, K., Zhang, X., Ren, S., Sun, J.: Deep residual learning for image recognition. In: Proceedings of the IEEE Conference on Computer Vision and Pattern Recognition, pp. 770–778 (2016)

11. Hu, J., Shen, L., Sun, G.: Squeeze-and-excitation networks. arXiv preprint arXiv:1709.01507 7 (2017)

12. Krizhevsky, A., Sutskever, I., Hinton, G.E.: Imagenet classification with deep convolutional neural networks. In: Advances in Neural Information Processing Systems, pp. 1097–1105 (2012)

13. Long, J., Shelhamer, E., Darrell, T.: Fully convolutional networks for semantic segmentation. In: Proceedings of the IEEE conference on Computer Vision and Pattern Recognition, pp. 3431–3440 (2015)

14. Menze, B.H., et al.: The multimodal brain tumor image segmentation benchmark (BRATS). IEEE Trans. Med. Imaging 34(10), 1993 (2015)

15. Prasoon, A., Petersen, K., Igel, C., Lauze, F., Dam, E., Nielsen, M.: Deep feature learning for knee cartilage segmentation using a triplanar convolutional neural network. In: Mori, K., Sakuma, I., Sato, Y., Barillot, C., Navab, N. (eds.) MICCAI 2013. LNCS, vol. 8150, pp. 246–253. Springer, Heidelberg (2013). https://doi.org/10.1007/978-3-642-40763-5_31

16. Roth, H.R., et al.: A new 2.5D representation for lymph node detection using random sets of deep convolutional neural network observations. In: Golland, P., Hata, N., Barillot, C., Hornegger, J., Howe, R. (eds.) MICCAI 2014. LNCS, vol. 8673, pp. 520–527. Springer, Cham (2014). https://doi.org/10.1007/978-3-319-10404-1_65

17. Simonyan, K., Zisserman, A.: Very deep convolutional networks for large-scale image recognition. arXiv preprint arXiv:1409.1556 (2014)

18. Szegedy, C., Ioffe, S., Vanhoucke, V., Alemi, A.A.: Inception-v4, inception-resnet and the impact of residual connections on learning. In: AAAI, vol. 4, p. 12 (2017)

19. Szegedy, C., et al.: Going deeper with convolutions. In: Proceedings of the IEEE Conference on Computer Vision and Pattern Recognition, pp. 1–9 (2015)

20. Szegedy, C., Vanhoucke, V., Ioffe, S., Shlens, J., Wojna, Z.: Rethinking the inception architecture for computer vision. In: Proceedings of the IEEE Conference on Computer Vision and Pattern Recognition, pp. 2818–2826 (2016)

21. Teichmann, M.T., Cipolla, R.: Convolutional CRFs for semantic segmentation. arXiv preprint arXiv:1805.04777 (2018)

22. Wang, G., Li, W., Ourselin, S., Vercauteren, T.: Automatic brain tumor segmentation using cascaded anisotropic convolutional neural networks. In: Crimi, A., Bakas, S., Kuijf, H., Menze, B., Reyes, M. (eds.) BrainLes 2017. LNCS, vol. 10670, pp. 178–190. Springer, Cham (2018). https://doi.org/10.1007/978-3-319-75238-9_16

S3D-UNet: Separable 3D U-Net for Brain Tumor Segmentation

Wei Chen, Boqiang Liu[✉], Suting Peng, Jiawei Sun, and Xu Qiao

Department of Biomedical Engineering,
School of Control Science and Engineering, Shandong University, Jinan, China
chenypic@mail.sdu.edu.cn, bqliu@sdu.edu.cn

Abstract. Brain tumor is one of the leading causes of cancer death. Accurate segmentation and quantitative analysis of brain tumor are critical for diagnosis and treatment planning. Since manual segmentation is time-consuming, tedious and error-prone, a fully automatic method for brain tumor segmentation is needed. Recently, state-of-the-art approaches for brain tumor segmentation are built on fully convolutional neural networks (FCNs) using either 2D or 3D convolutions. However, 2D convolutions cannot make full use of the spatial information of volumetric medical image data, while 3D convolutions suffer from high expensive computational cost and memory demand. To address these problems, we propose a novel Separable 3D U-Net architecture using separable 3D convolutions. Preliminary results on BraTS 2018 validation set show that our proposed method achieved a mean enhancing tumor, whole tumor, and tumor core Dice scores of 0.74932, 0.89353 and 0.83093 respectively. Finally, during the testing stage we achieved competitive results with Dice scores of 0.68946, 0.83893, and 0.78347 for enhancing tumor, whole tumor, and tumor core, respectively.

Keywords: Separable · Segmentation · BraTS ·
Convolutional neural networks

1 Introduction

Image segmentation, especially semantic segmentation, is a fundamental and classic problem in computer vision. It refers to partitioning an image into several disjoint semantically meaningful parts and classifying each part into a pre-determined class. In the application of brain tumor segmentation, the task includes the division of several sub-regions, such as GD-enhancing tumor, peritumoral edema, and the necrotic and non-enhancing tumor core [1]. Accurate segmentation and quantitative analysis of brain tumor are critical for diagnosis and treatment planning. Generally, manual segmentation of brain tumor is known to be time-consuming, tedious and error-prone. Therefore, there is a strong need for a fully automatic method for brain tumor segmentation. However, brain tumor segmentation is a challenging task because MR images are typically acquired using various protocols and magnet strengths, which results in the non-standard range of MR images. In addition, brain tumors can appear anywhere in the brain, and their shape and size vary greatly. Furthermore, the intensity

© Springer Nature Switzerland AG 2019
A. Crimi et al. (Eds.): BrainLes 2018, LNCS 11384, pp. 358–368, 2019.
https://doi.org/10.1007/978-3-030-11726-9_32

profiles of tumor regions are largely overlapped with healthy parts. Due to the challenge of brain tumor segmentation and the broad medical prospect, many researchers have proposed various methods to solve the problem of brain tumor segmentation.

Brain tumor segmentation methods can be divided into different categories according to different principles [2]. Broadly, these methods can be divided into two major categories: generative methods and discriminative methods. Generative methods typically rely on the prior information about the appearance of both healthy tissues and tumors. The proposed models often regard the task of segmentation as a problem of a posteriori distribution estimation. On the contrary, discriminative methods use very little prior information and typically rely on a large number of low-level image features to learn the distribution from the annotated training images.

More recently, due to the success of convolutional neural networks (CNNs), great progress has been made in the field of computer vision. At the same time, many deep learning based brain tumor segmentation methods have been proposed and achieved great success. Havaei et al. [3] proposed a two-pathway architecture with a local pathway and a global pathway, which can simultaneously exploit both local features and more global contextual features. Kamnitsas et al. [4] proposed an efficient fully connected multi-scale CNN architecture named deepmedic that uses 3D convolution kernels and reassembles a high resolution and a low resolution pathway to obtain the segmentation results. Furthermore, they used a 3D fully connected conditional random field to effectively remove false positives. Isensee et al. [5] proposed 3D U-Net, which carefully modified the popular U-Net architecture and used a dice loss function to cope with class imbalance. They achieved competitive results on the BraTS 2017 testing data. Kamnitsas et al. [6] introduced EMMA, an ensemble of multiple models and architectures including deepmedic, FCNs and U-Net. Due to the heterogeneous collection of networks, the model is insensitive to independent failures of each component and has good generalization performance. They won first place in the final testing stage of the BraTS 2017 challenge among more than 50 teams.

Although so many achievements have been made, the progress of medical image analysis is slower than that of static images, and a key reason is the 3D properties of medical images. This problem also occurs in the tasks of video understanding. To solve this problem, Xie et al. [7] proposed S3D model by replacing 3D convolutions with spatiotemporal-separable 3D convolutions. This model significantly improved on the previous state-of-the-art 3D CNN model in terms of efficiency.

Inspired by S3D architecture for video classification and the state-of-the-art U-Net architecture for medical image segmentation, we propose a novel framework named S3D-UNet for brain tumor segmentation. To make full use of 3D volumes, we design a new separable 3D convolution by dividing each 3D convolution into three branches in a parallel fashion, each with a different orthogonal view, namely axial, sagittal and coronal. We also propose a separable 3D block that takes advantage of the state-of-the-art residual inception architecture. During the testing stage we achieved competitive results with Dice scores of 0.68946, 0.83893, and 0.78347 for enhancing tumor, whole tumor, and tumor core, respectively [8].

2   Methods

2.1   Dataset

The brain tumor MRI dataset used in this study are provided by BraTS'2018 Challenge [1, 9–11]. The training dataset includes multimodal brain MRI scans of 285 subjects, of which 210 are GBM/HGG and 75 are LGG. Each subject contains four scans: native T1-weighted (T1), post-contrast T1-weighted (T1c), T2-weighted (T2), and T2 Fluid Attenuated Inversion Recovery (FLAIR). All the subjects in the training dataset are provided with ground truth labels, which are segmented manually by one to four raters. Annotations consist of the GD-enhancing tumor (ET - label 4), the peritumoral edema (ED - label 2), and the necrotic and non-enhancing tumor core (NCR/NET - label 1). The validation and testing datasets include multimodal brain MRI scans of 66 subjects and 191 subjects which are similar to the training dataset but have no expert segmentation annotations and the grading information.

2.2   Data Pre-processing

To remove the bias field caused by the inhomogeneity of the magnetic field and the small motions during scanning, the N4ITK bias correction algorithm [12] is first applied to the T1, T1c and T2 scans. The multimodal scans in BraTS 2018 were acquired with different clinical protocols and various scanners from multiple institutions [1], resulting in non-standardized intensity distribution. Therefore, normalization is a necessary stage of processing multi-mode scanning by a single algorithm. We use the histogram matching algorithm [13] to transform each scan to a specified histogram to ensure that all the scans have a similar intensity distribution. We also resize the original image of $240 \times 240 \times 155$ voxels to $128 \times 128 \times 128$ voxels by removing as many zero background as possible. This processing not only can effectively improve the calculation efficiency, but also retain the original image information as much as possible. In the end, we normalize the data to have a zero mean and unit variance.

2.3   Network Architecture

S-3D Convolution Block. Traditional 2D CNNs for computer vision mainly involve spatial convolutions. However, for video applications such as human action, both spatial and temporal information need to be modeled jointly. By using 3D convolution in the convolutional layers of CNNs, discriminative features along both the spatial and the temporal dimensions can be captured. 3D CNNs have been widely used for human action recognition in videos. However, the training of 3D CNN requires expensive computational cost and memory demand, which hinders the construction of a very deep 3D CNN. To mitigate this problem, Xie et al. [7] proposed S3D model by replacing 3D convolutions with spatiotemporal-separable 3D convolutions. Each 3D convolution can be replaced by two consecutive convolutional layers: one 2D convolution to learn spatial features and one 1D convolution to learn temporal features, as shown in Fig. 1 (a). By using separable temporal convolution, they build a new block using inception architecture called "temporal inception block", as shown in Fig. 1(b).

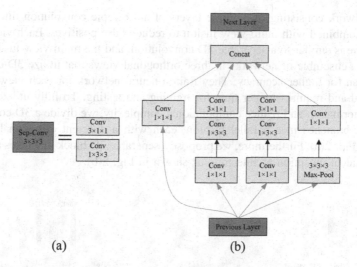

Fig. 1. (a) An illustration of separable 3D convolution. A 3D convolution can be replaced by two consecutive convolutional layers. (b) Temporal separable inception block.

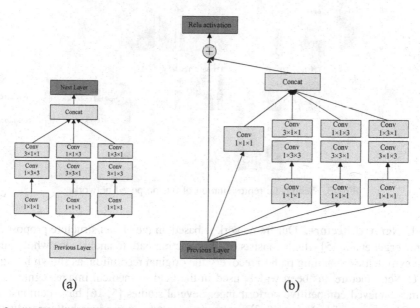

Fig. 2. (a) We divide a 3D convolution into three branches in a parallel fashion. (b) Our proposed S3D block, which takes advantage of the residual inception architecture.

Unlike video data, volumetric medical data have three orthogonal views, namely axial, sagittal and coronal, and each view has important anatomical features. To implement the separable 3D convolution directly, we need to specify which view as the temporal direction. Wang et al. [14] propose a cascaded anisotropic convolutional neural network consisting of multiple layers of anisotropic convolution filters, which are then combined with multi-view fusion to reduce false positives. Each view of this architecture is similar to a separable 3D convolution, and the multi-view fusion can be view as an ensemble of networks in three orthogonal views that utilize 3D contextual information for higher accuracy. They train a neural network for each view, it is not end-to-end and requires longer time for training and testing. To fully utilize 3D contextual information and reduce computational complexity, we divide a 3D convolution into three branches in a parallel fashion, each with a different orthogonal view, as shown in Fig. 2(a). Furthermore, we propose a separable 3D block that takes advantage of the residual inception architecture, as shown in Fig. 2(b).

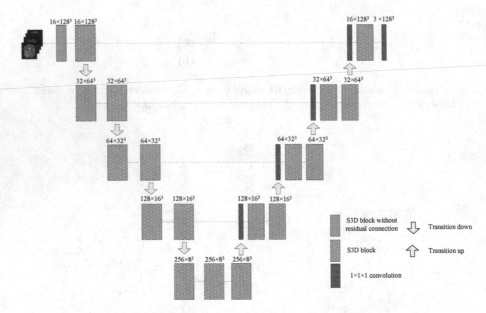

Fig. 3. Schematic representation of our proposed network.

S3D U-Net Architecture. Our framework is based on the U-Net structure proposed by Ronneberger et al. [15] which consists of a contracting path to analyze the whole image and a symmetric expanding path to recovery the original resolution, as shown in Fig. 3. The U-Net structure has been widely used in the field of medical image segmentation and has achieved competitive performance. Several studies [5, 16] have demonstrated that a 3D version of U-Net using 3D volumes as input can produce better results than an entirely 2D architecture.

Just like the U-Net and its extensions, our network has an autoencoder-like architecture with a contracting path and an expanding path, as shown in Fig. 3. The contracting path encodes the increasingly abstract representation of the input, and the expanding path restores the original resolution. Similar to [5], we refer to the depth of the network as level. Higher levels have lower spatial resolution but higher dimensional feature representations and vice versa.

The input to the contracting path is a $128 \times 128 \times 128$ voxel block with 4 channels. The contracting path has 5 levels. Except for the first level, each level consists of two S3D blocks. It is worth noting that each convolution in S3D block is followed by instance normalization [17] and LeakyReLU. Different levels are connected by transition down block to reduce the resolution of the feature maps and double the number of feature channels. Transition down module consists of a $3 \times 3 \times 3$ convolution with stride 2 followed by instance normalization and LeakyReLU. After the contracting path, the size of the feature maps is decreased to $8 \times 8 \times 8$.

In order to recover the input resolution at expanding path, we first adopt a transition up module to upsample the previous feature maps and halve the number of feature channels. Transition up module consists of a transposed $3 \times 3 \times 3$ convolution with stride 2 followed by instance normalization and LeakyReLU. Then the feature maps from contracting path are concatenated with feature maps from expanding path via long skip connections. At each level of expanding path, we use a $1 \times 1 \times 1$ convolution with stride 1 to halve the number of feature channels, followed by two S3D blocks that are the same as in the contracting path. The final segmentation is done by a $1 \times 1 \times 1$ convolutional layer followed by a softmax operation among the objective classes.

Table 1. The distribution of the classes in the training data of BraTS 2018.

	Background	NCR/NET	ED	ET
Percentage	98.88	0.28	0.64	0.20

2.4 Loss Function

The performance of neural network depends not only on the choice of network structure but also on the choice of the loss function [18]. Especially for severe class imbalance, the choice of loss functions becomes more important. Due to the physiological characteristics of brain tumors, the segmentation task has an inherent class imbalance problem. Table 1 illustrates the distribution of the classes in the training data of BraTS 2018. Background (label 0) is overwhelmingly dominant. According to [5], we apply a multiclass Dice loss function to approach this issue. Let R be the one hot coding ground truth segmentation with voxel values r_n^k, where $k \in K$ being the class at voxel $n \in N$. Let P be the output the network with voxel values p_n^k, where $k \in K$ being the class at voxel $n \in N$. The multiclass Dice loss function can be expressed as

$$DL = 1 - \frac{2}{K} \sum_{k \in K} \frac{\sum_n p_n^k r_n^k}{\sum_n p_n^k + \sum_n r_n^k} \tag{1}$$

2.5 Evaluation Metrics

Multiple criteria are computed as performance metrics to quantify the segmentation result. Dice coefficient (Eq. 2) is the most frequently used metric for evaluating medical image segmentation. P_1 is the area that is predicted to be tumor and T_1 is true tumor area. It measures the overlap between the segmentations and ground truth with a value between 0 and 1. The higher the Dice score, the better the segmentation performance.

$$Dice(P,T) = \frac{|P_1 \wedge T_1|}{(|P_1| + |T_1|)/2} \tag{2}$$

Sensitivity and specificity are also commonly used statistical measures. The sensitivity (Eq. 3), also called true positive rate, defined as the proportion of positives that are correctly predicted. It measures the portion of tumor regions in the ground truth that are also predicted as tumor regions by the segmentation method. The specificity (Eq. 4), also called true negative rate, defined as the proportion of negatives that are correctly predicted. It measures the portion of normal tissue regions (T_0) in the ground truth that are also predicted as normal tissue regions (P_0) by the segmentation method.

$$Sens(P,T) = \frac{|P_1 \wedge T_1|}{|T_1|} \tag{3}$$

$$Spec(P,T) = \frac{|P_0 \wedge T_0|}{|T_0|} \tag{4}$$

The Hausdorff Distance (Eq. 5) is used to evaluates the distance between the segmentation boundary and the ground truth boundary. Mathematically, it is defined as the maximum distance of all points p on the surface ∂P_1 of a given volume P_1 to the nearest points t on the surface ∂T_1 of the other given volume T_1.

$$Haus(P,T) = \max\{ \sup_{p \in \partial P_1} \inf_{t \in \partial T_1} d(p,t), \sup_{t \in \partial T_1} \inf_{p \in \partial P_1} d(t,p)\} \tag{5}$$

3   Experiments and Results

The network is trained on a GeForce GTX 1080Ti GPU with a batch size of 1 using PyTorch toolbox. Adam [19] is used as the optimizer with an initial learning rate 0.001 and a 12 weight decay of 1e−8. We evaluate all the cases for training data and validation data using online CBICA portal for BraTS 2018 challenge. The sub-regions considered for evaluation are "enhancing tumor" (ET), "tumor core" (TC), and "whole tumor" (WT).

Table 2 presents the quantitative evaluations with the BraTS 2018 training set via five cross-validation. It shows that the proposed method achieves average Dice scores of 0.73953, 0.88809 and 0.84419 for enhancing tumor, whole tumor and tumor core, respectively. A 3D U-Net without the proposed S3D block is also trained, and the quantitative evaluations with the BraTS 2018 training set are shown in Table 3. It can be seen that the Dice score of enhancing tumor has been significantly improved using S3D block. The corresponding values for BraTS 2018 validation set are 0.74932, 0.89353 and 0.83093, respectively, as shown in Table 4. Examples of the segmentations obtained from the training set using our method are shown in Fig. 4.

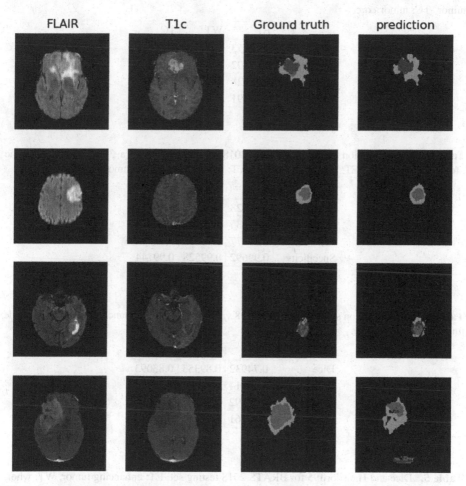

Fig. 4. Examples of segmentation from the of BraTS 2018 training data. red: NCR/NET, green: ED, blue: ET. (the first two rows) Satisfying segmentation. (the last two rows) Unsatisfactory segmentation. In the future, we will adopt some post-processing methods to improve the segmentation performance. (Color figure online)

Table 5 shows the challenge testing set results. Our proposed method achieves average Dice scores of 0.68946, 0.83893 and 0.78347 for enhancing tumor, whole tumor and tumor core, respectively. Compared with the performance of the training and validation sets, the scores are significantly reduced. However, the high median values show that the testing set may contains some difficult cases, resulting in the lower average scores.

Table 2. The evaluation scores for BraTS 2018 training set. ET: enhancing tumor, WT: whole tumor, TC: tumor core.

	ET	WT	TC
Dice	0.73953	0.88809	0.84419
Hausdorff95	4.63102	5.88769	5.66071
Sensitivity	0.78628	0.88069	0.83281
Specificity	0.99791	0.99481	0.9972

Table 3. The evaluation scores for BraTS 2018 training set using a 3D U-Net without the proposed S3D block. ET: enhancing tumor, WT: whole tumor, TC: tumor core.

	ET	WT	TC
Dice	0.68428	0.89912	0.86772
Hausdorff95	5.32635	5.55958	5.10478
Sensitivity	0.81677	0.88683	0.85932
Specificity	0.99692	0.99528	0.99744

Table 4. The evaluation scores for BraTS 2018 validation set. ET: enhancing tumor, WT: whole tumor, TC: tumor core.

	ET	WT	TC
Dice	0.74932	0.89353	0.83093
Hausdorff95	4.43214	4.71646	7.74775
Sensitivity	0.78492	0.92903	0.81606
Specificity	0.99761	0.99274	0.99814

Table 5. Dice and Hausdorff95 for BRATS 2018 testing set. ET: enhancing tumor, WT: whole tumor, TC: tumor core.

	Dice			Hausdorff95		
	ET	WT	TC	ET	WT	TC
Mean	0.68946	0.83893	0.78347	4.51842	9.20202	7.71181
StdDev	0.27809	0.17584	0.2549	8.04775	16.55337	15.64779
Median	0.78848	0.89967	0.89183	2.23607	3.60555	3
25quantile	0.68368	0.83469	0.75508	1.41421	2.23607	2
75quantile	0.84938	0.93011	0.92732	3.31662	6.89116	6.7082

4 Discussion and Conclusion

We propose a S3D-UNet architecture for automatic brain tumor segmentation. In order to make full use of 3D volume information while reducing the amount of calculation, we adopt separable 3D convolutions. For the characteristics of the isotropic resolution of brain tumor MR images, we design a new separable 3D convolution architecture by dividing each 3D convolution into three branches in a parallel fashion, each with a different orthogonal view, namely axial, sagittal and coronal. We also propose a separable 3D block that takes advantage of the state-of-the-art residual inception architecture. Finally, based on separable 3D convolutions, we propose the S3D-UNet architecture using the prevalent U-Net structure.

This network has been evaluated on the BraTS 2018 Challenge testing dataset and achieved an average Dice scores of 0. 68946, 0. 83893 and 0. 78347 for the segmentation of enhancing tumor, whole tumor and tumor core, respectively. Compared with the performance of the training and validation sets, the scores of testing set are lower. This may be due to the difficult cases in testing set because the median values are high. In the future, we will work to enhance the robustness of the network.

For volumetric medical image segmentation, 3D contextual information is an important factor to obtain high-performance results. The straightforward way to capture such 3D context is to use 3D convolutions. However, the use of a large number of 3D convolutions will significantly increase the number of parameters, thus complicating the training process. In the video understanding tasks, the separable 3D convolutions with higher computational efficiency have been adopted. In this paper, we demonstrate that the U-Net with separable 3D convolutions can achieve promising results in the field of medical image segmentation.

In the future work, we will continue to improve the structure of the network and use some post-processing methods such as fully connected conditional random field to further improve the segmentation performance.

Acknowledgment. This work was supported by the Department of Science and Technology of Shandong Province (Grant No. 2015ZDXX0801A01, ZR2014HQ054, 2017CXGC1502), National Natural Science Foundation of China (grant no. 61603218).

References

1. Menze, B.H., et al.: The multimodal brain tumor image segmentation benchmark (BRATS). IEEE Trans. Med. Imaging **34**, 1993–2024 (2015)
2. Gordillo, N., Montseny, E., Sobrevilla, P.: State of the art survey on MRI brain tumor segmentation. Magn. Reson. Imaging **31**, 1426–1438 (2013)
3. Havaei, M., et al.: Brain tumor segmentation with Deep Neural Networks. Med. Image Anal. **35**, 18–31 (2017)
4. Kamnitsas, K., et al.: Efficient multi-scale 3D CNN with fully connected CRF for accurate brain lesion segmentation. Med. Image Anal. **36**, 61–78 (2017)

5. Isensee, F., Kickingereder, P., Wick, W., Bendszus, M., Maier-Hein, K.H.: Brain tumor segmentation and radiomics survival prediction: contribution to the BRATS 2017 challenge. In: Crimi, A., Bakas, S., Kuijf, H., Menze, B., Reyes, M. (eds.) BrainLes 2017. LNCS, vol. 10670, pp. 287–297. Springer, Cham (2018). https://doi.org/10.1007/978-3-319-75238-9_25

6. Kamnitsas, K., et al.: Ensembles of multiple models and architectures for robust brain tumour segmentation. In: Crimi, A., Bakas, S., Kuijf, H., Menze, B., Reyes, M. (eds.) BrainLes 2017. LNCS, vol. 10670, pp. 450–462. Springer, Cham (2018). https://doi.org/10.1007/978-3-319-75238-9_38

7. Xie, S., Sun, C., Huang, J., Tu, Z., Murphy, K.: Rethinking spatiotemporal feature learning for video understanding. arXiv:1712.04851 [cs] (2017)

8. Bakas, S., Reyes, M., et al.: Identifying the best machine learning algorithms for brain tumor segmentation, progression assessment, and overall survival prediction in the BRATS challenge, arXiv preprint arXiv:1811.02629 (2018)

9. Bakas, S., et al.: Advancing The Cancer Genome Atlas glioma MRI collections with expert segmentation labels and radiomic features. Sci. Data. **4**, 170117 (2017)

10. Bakas, S., et al.: Segmentation labels and radiomic features for the pre-operative scans of the TCGA-GBM collection. The Cancer Imaging Archive (2017)

11. Bakas, S., et al.: Segmentation labels and radiomic features for the pre-operative scans of the TCGA-LGG collection. The Cancer Imaging Archive (2017)

12. Tustison, N.J., et al.: N4ITK: improved N3 bias correction. IEEE Trans. Med. Imaging **29**, 1310–1320 (2010)

13. Nyul, L.G., Udupa, J.K., Zhang, X.: New variants of a method of MRI scale standardization. IEEE Trans. Med. Imaging **19**, 143–150 (2000)

14. Wang, G., Li, W., Ourselin, S., Vercauteren, T.: Automatic brain tumor segmentation using cascaded anisotropic convolutional neural networks. In: Crimi, A., Bakas, S., Kuijf, H., Menze, B., Reyes, M. (eds.) BrainLes 2017. LNCS, vol. 10670, pp. 178–190. Springer, Cham (2018). https://doi.org/10.1007/978-3-319-75238-9_16

15. Ronneberger, O., Fischer, P., Brox, T.: U-Net: convolutional networks for biomedical image segmentation. In: Navab, N., Hornegger, J., Wells, W.M., Frangi, A.F. (eds.) MICCAI 2015. LNCS, vol. 9351, pp. 234–241. Springer, Cham (2015). https://doi.org/10.1007/978-3-319-24574-4_28

16. Çiçek, Ö., Abdulkadir, A., Lienkamp, S.S., Brox, T., Ronneberger, O.: 3D U-Net: learning dense volumetric segmentation from sparse annotation. arXiv:1606.06650 [cs] (2016)

17. Ulyanov, D., Vedaldi, A., Lempitsky, V.: Instance normalization: the missing ingredient for fast stylization. arXiv:1607.08022 [cs] (2016)

18. Sudre, C.H., Li, W., Vercauteren, T., Ourselin, S., Cardoso, M.J.: Generalised Dice overlap as a deep learning loss function for highly unbalanced segmentations. arXiv:1707.03237 [cs], vol. 10553, pp. 240–248 (2017)

19. Kingma, D.P., Ba, J.: Adam: a method for stochastic optimization. arXiv preprint arXiv: 1412.6980 (2014)

Deep Learning Radiomics Algorithm for Gliomas (DRAG) Model: A Novel Approach Using 3D UNET Based Deep Convolutional Neural Network for Predicting Survival in Gliomas

Ujjwal Baid[1], Sanjay Talbar[1], Swapnil Rane[2], Sudeep Gupta[2],
Meenakshi H. Thakur[2], Aliasgar Moiyadi[2], Siddhesh Thakur[1],
and Abhishek Mahajan[2(✉)]

[1] Shri Guru Gobind Singhji Institute of Engineering and Technology,
Nanded, India
ujjwalbaid0408@gmail.com
[2] Tata Memorial Hospital, Tata Memorial Centre, Mumbai, India
drabhishek.mahajan@yahoo.in

Abstract. Automated segmentation of brain tumors in multi-channel Magnetic Resonance Image (MRI) is a challenging task. Heterogeneous appearance of brain tumors in MRI poses critical challenges in diagnosis, prognosis and survival prediction. In this paper, we present a novel approach for glioma tumor segmentation and survival prediction with Deep Learning Radiomics Algorithm for Gliomas (DRAG) Model using 3D patch based U-Net model in Brain Tumor Segmentation (BraTS) challenge 2018. Radiomics feature extraction and classification was done on segmented tumor for overall survival (OS) prediction task. Preliminary results of DRAG model on BraTS 2018 validation dataset demonstrated that the proposed method achieved a good performance with Dice scores as 0.88, 0.83 and 0.75 for whole tumor, tumor core and enhancing tumor, respectively. For survival prediction, 57.1% accuracy was achieved on the validation dataset. The proposed DRAG model was one of the top performing models and accomplished third place for OS prediction task in BraTS 2018 challenge.

Keywords: Brain Tumor Segmentation · Gliomas ·
Convolutional Neural Networks · Radiomics · MRI · Radiogenomics ·
Survival prediction

1 Introduction

Glioma is the most frequent primary brain tumor. It originates from glial cells and can be classified in High Grade and Low Grade depending upon the aggressiveness. Gliomas may have different degrees of aggressiveness, variable prognosis and several heterogeneous histological sub-regions. These are described by varying intensity profiles across different Magnetic Resonance Imaging (MRI) modalities, which reflect

© Springer Nature Switzerland AG 2019
A. Crimi et al. (Eds.): BrainLes 2018, LNCS 11384, pp. 369–379, 2019.
https://doi.org/10.1007/978-3-030-11726-9_33

diverse tumor biological properties [9]. De-spite of recent advances in automated algorithms for brain tumor segmentation in multimodal MRI scans, the problem is still a challenging task in medical imaging analysis [6, 7, 12].

Prior to BRATS challenge, researchers tested their proposed algorithms on local datasets and there was no gold standard available for fair evaluation of methods globally. BraTS challenge provided global platform for researchers to evaluate their proposed algorithms on publically available dataset with leaderboard. This year BraTS challenge was divided in two parts 1. Segmentation of brain tumor with intra-tumor parts 2. Prediction of overall survival of the patients in number of days based on imaging features.

Many computational methods based on texture analysis, probabilistic models, active contours, random forests are proposed for tumor segmentation over decades [10]. Several advances were made in active contours where either an initial seed point was mentioned which would grow till the boundaries of the tumor or a bounding box was drawn across the abnormal region which would further confine to tumor boundaries. Figure 1 shows FLAIR, T1, T1ce, and T2 images with intra-tumor parts- Green for Edema, Blue for Enhancing tumor and Red for Tumor Core. Researchers had proposed methods based on Non-Negative Matrix Factorization where a data matrix was generated from MR data which acted as a feature representation. This data matrix was further clustered with Fuzzy C-means clustering algorithm for brain tumor segmentation [14].

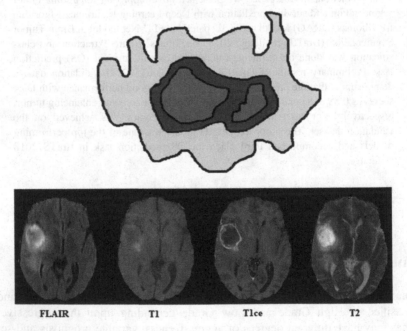

Fig. 1. MRI modalities with intra-tumor parts. Edema in yellow, enhancing tumor in blue and necrotic tumor is shown in red color (Color figure online)

Deep Learning algorithms have been outperforming over all other the state of the art methods for segmentation, classification and detection applications. Researchers have proposed application specific models for which Convolutional Neural Networks (CNN) is the basic building block of the architecture. The advantage of CNN is that it is computationally cheaper compared to Fully Convolutional Networks (FCN). The implementation of CNN is successful because of the advancement in the computational power of the machines. This enables to de-sign neural network models with deep architecture to extract features in an image. Researchers had proposed several pixel classification based approaches for segmentation task where a window was considered around a pixel and the class of the window was the class of center pixel. U-Net based models have outperformed over traditional machine learning methods in bio-medical image segmentation [11]. Recently, there has been an increase in popularity of 3D CNNs which are effective in segmentation task at the expense of additional computational complexity compared to other state of the art algorithms [5].

2   Method

We developed patch based 3D U-Net model for tumor segmentation and evaluated efficiency of radiomic features for OS prediction named as 'Deep Learning Radiomics Algorithm for Gliomas (DRAG) Model'. There was high class imbalance between tumor pixel and rest of the normal brain pixels in the BraTS datasets. This led to biased training of the model as the loss function was affected by normal brain pixels as compared to the tumor pixels. The problem became more challenging during intra-tumor segmentation. To overcome this issue, we adopted a patch-based training approach. Fixed sized 3D patches were extracted from the BraTS dataset which were used for training the network. Details of our approach are given in the section below.

2.1   Dataset

This proposed method was trained and validated on BraTS 2018 training dataset and validation dataset [1–3]. The training dataset included 210 High Grade Glioma (HGG) cases and 75 cases with Low Grade Glioma (LGG) while validation set consisted of 66 cases. For each case, there were four MRI sequences viz. the T1-weighted (T1), T1 with gadolinium enhancing contrast (T1ce), T2-weighted (T2) and FLAIR. All cases had been segmented manually, by four raters and marked annotations were approved by experienced neuro-radiologists into intra-tumor parts like tumor core, enhancing tumor and edema. The MRI data was collected from various institutions and acquired with different protocols, magnetic field strengths and MRI scanners. Furthermore, to pinpoint the clinical relevance of this segmentation task, BraTS 2018 also focused on the prediction of patient overall survival via analysis of radiomic features. For this purpose, the survival data (in days) of 163 cases was provided in training set and 54 cases in validation set. Reference segmentation and OS for validation set was hidden and evaluation was carried out via online evaluation portal.

2.2   Pre-processing

The MRI data in BraTS challenge dataset was already pre-processed which included skull stripping and the data was co-register and re-sampled to 1 mm × 1 mm × 1 mm resolution. The dimensions of each volume were 240 × 240 × 155. The intensity inhomogeneity was addressed with N4ITK tool [13]. All four MRI channel data was normalized to zero mean and unit variance.

2.3   Patch Extraction and Training

The proposed model is modified version of 3D U-Net with 3 down-sampling and 3 up-sampling branches with two back to back convolution layers with kernel size 3. 3D voxels with size 64 × 64 × 64 were extracted randomly from the training data and given as an input to the first layer of the model. Four patches extracted from FLAIR, T1, T2, T1c were concatenated together to form 64 × 64 × 64 × 4 and fed for training to the input layer along with corresponding Ground Truth. Patch extraction was challenging because of the high class imbalance between tumor area and normal brain tissues. During patch extraction, care was taken to include significant tumor area to avoid bias to background and non-tumor pixels. This was done for all the four modalities and ground truth as well. Each layer was followed by ReLU activation and Batch Normalization. No data augmentation was performed during the training of model.

At output 4 probability maps were generated for Necrosis, Edema, Enhancing Tumor and Background (including non-tumor brain pixels). The label was assigned to the map with highest probability amongst all. It was observed that there were some False Positives present in the segmentation output.

3D Connected Component Analysis was done to identify all the segmented components pre-sent in the segmented volume. Threshold value in terms of number of pixels was identified and insignificant small components which false positives were assigned to background label. This reduced false positives significantly. Similarly, to reduce over-segmentation in certain cases a binary brain mask was generated from brain volume and logical AND operation was performed on segmentation output. This improved the accuracy of the segmentation significantly.

2.4   Radiomic Feature Extraction and Training

After segmentation of intra-tumor parts, the next task in BraTS 2018 was to predict the over-all survival of the patients in number of days. For this task, organizers had provided only age details and OS in days which made the task challenging. From the last few years, researchers are working actively on Radiomic Feature extraction for tumor analysis and survival prediction task [4]. In our approach, we computed Radiomic features on FLAIR and T1c volume with different combination of intra-tumor parts as (Figs. 1 and 2):

- Edema, Enhancing tumor and tumor core i.e. Whole tumor (WT)
- Tumor Core and Enhancing tumor (TC+ET)
- Enhancing tumor (ET)

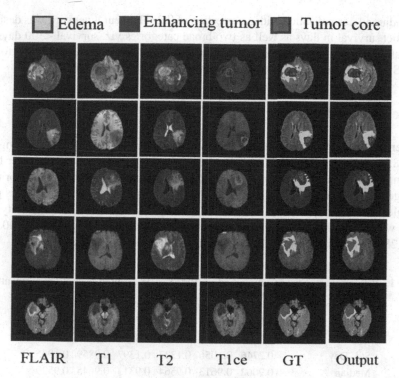

Fig. 2. Sample segmentation results. Each row represents one case. Columns from left to right: FLAIR, T1, T2, T1c, GT and Output. Segmentation labels: Yellow for edema, Blue for enhancing tumor and Red for tumor core. (Color figure online)

We computed first order statistics, shape features, Gray Level Co-occurrence Matrix and Gray Level Run Length Matrix features. We computed 468 features for edema, tumor core and enhancing tumor. These features were used to train the regression model for survival prediction task. We started with 679 variables [678 radiomic variables (113 from each of the different tumor parts for both FLAIR and T1c images) and Age]. The radiomic variables with near perfect correlation (Spearman's correlation coefficient 0.95 or higher, p 0.05 or lower) with each other were excluded with only one of the variables in each set being retained (N = 117). Age and all radiomics variables with no significant autocorrelations (N = 117) were assessed for relationship with survival.

Multi-Layer perceptron was used to train the neural network. Variables which had a statistically significant correlation (N = 56, including age) with survival were included for training the neural network. Results were replicated by setting a random seed. To assess the efficacy of the neural network and to correct over training, if any, we divided the BraTS 2018 training dataset (N = 163) into Training (51.5%), Validation (14.7%) and Testing (33.7%) subsets, randomly using Bernoulli variates. The neural network had two hidden layers with the number of units per layer set to auto, sigmoid activation functions for both the hidden as well as output layers. The variables were re-scaled using adjusted normalization with a correction of 0.2. The neural network was designed to predict survival in days as well as two broad categories viz. survival <300 days and survival >=300 days. All statistical procedures for survival prediction were performed using SPSS for Windows v24 on a standard computer running Windows 10.

3   Result and Discussion

The performance of the proposed method was evaluated on BraTS 2018 training data with 285 cases and validated on 66 cases for segmentation. The validation leader-board gave interesting information about the performance of the different teams' algorithms. Average performance of proposed method on training data and validation data is given in Tables 1 and 2 respectively in terms of Dice Similarity Index and Sensitivity. The model was trained for 50 epochs and needed 48 h for training on NVIDIA P100 GPU with 128 GB system RAM. The framework was developed in Tensorflow [8].

Table 1. Performance of proposed method on BraTS 2018 training dataset for segmentation.

Evaluation metrics	Dice			Sensitivity		
	ET	WT	TC	ET	WT	TC
Mean	0.8002	0.9324	0.9197	0.8951	0.9508	0.9359
Std. Dev.	0.2746	0.1056	0.1327	0.1397	0.0859	0.1005
Median	0.9062	0.9613	0.9564	0.9313	0.9645	0.9529
25 quantile	0.8421	0.9405	0.9303	0.8901	0.9451	0.9328
75 quantile	0.9422	0.9728	0.9687	0.9622	0.9777	0.9697

Table 2. Performance of proposed method on BraTS 2018 validation dataset for segmentation.

Evaluation metrics	Dice			Sensitivity		
	ET	WT	TC	ET	WT	TC
Mean	0.7480	0.8780	0.8266	0.8266	0.9058	0.8186
Std. Dev.	0.2659	0.1345	0.1828	0.2306	0.1413	0.2136
Median	0.8527	0.9179	0.8985	0.9035	0.9436	0.9167
25 quantile	0.7325	0.8665	0.7771	0.8238	0.8953	0.7637
75 quantile	0.8853	0.9419	0.9444	0.9514	0.9720	0.9571

Overall, our approach reached a superior result in the whole tumor segmentation task with an average dice coefficient of 93% over training dataset and 87% over validation dataset. Sample segmentation results for intra-tumor parts are given in Fig. 2 The performance of the proposed approach is given in Table 3 in terms of Dice Coefficient and Hausdorff95 distance.

Table 3. Performance of proposed method on BraTS 2018 test dataset for segmentation.

Evaluation metrics	Dice			Hausdorff95		
	ET	WT	TC	ET	WT	TC
Mean	0.6677	0.8474	0.7687	9.0554	17.2184	14.5728
Std. Dev.	0.3120	0.1699	0.2786	19.8975	28.9190	26.1504
Median	0.8013	0.9049	0.8946	2.2360	3.4641	3.3166
25 quantile	0.6556	0.8336	0.7519	1.4142	2.2360	2.0000
75 quantile	0.8656	0.9404	0.9328	3.6055	9.4604	8.4844

For the prediction of survival categories, the neural network demonstrated an accuracy of 70.2% in the training subset and 62.5% and 63.6% in the validation and testing subsets, respectively. The accuracy was 69.5% for the entire training dataset. The Area Under Curve (AUC) was 0.799 (Figs. 3 and 4). For prediction of survival in days, the proposed model performed better for values in the middle, with lower performance for the values at both extremes. The relative error was 0.842 for the training subset, 0.774 for the validation subset and 0.910 for the testing dataset.

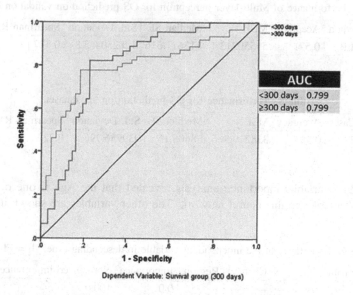

Fig. 3. ROC curve depicting the accuracy of the model for categorizing into <300 and >=300 days survival.

The performance of proposed OS prediction approach is given in Table 5 for 77 cases. The proposed approach stood third for overall Survival Prediction Task in BraTS 2018 Challenge (Table 4).

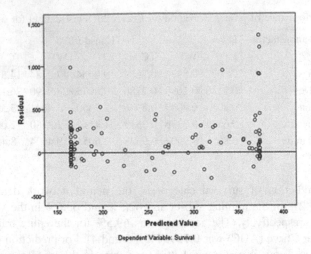

Fig. 4. Residual to predicted scatter plot showing the good fit of the model for survival values in the middle between 200–350 days.

Table 4. Performance of Multi-layer perceptron for OS prediction on validation dataset.

Method	Accuracy	MSE	Median SE	Std. Deviation	Spearman R
MLP	0.571	59550213.1	113611.616	128250465.8	0.427

Table 5. Performance for OS prediction on test dataset.

Cases	Accuracy	MSE	Median SE	Std. Deviation	Spearman R
77	0.558	338219.366	38408.16	939986.796	0.222

Individual variable importance analysis revealed that the Age is one of the most significant variables in this neural network. The other variables are shown in Table 6.

Table 6. Importance of the independent variable in descending order (F = FLAIR)

Variable name	Channel	Region	Importance	Normalized importance (%)
Age	–	–	0.07	100.0
Entropy	F	TC+ET	0.049	70.3
Variance	F	TC+ET	0.04	57.4
Enhance count	F	TC+ET	0.038	54.4
Core count	T1ce	WT	0.034	49.10
Cluster shade	T1ce	TC+ET	0.031	44.7
Edema count	F	TC+ET	0.03	42.8

(*continued*)

Table 6. (*continued*)

Variable name	Channel	Region	Importance	Normalized importance (%)
Dissimilarity	T1ce	TC+ET	0.029	42.3
Difference in entropy	F	ET	0.029	41.2
Variance	T1ce	TC+ET	0.027	39.0
Maximum probability	T1ce	ET	0.026	36.9
Sum of variance	T1ce	ET	0.025	36.3
Homogeneity	T1ce	TC+ET	0.024	34.7
Minimum	T1ce	WT	0.023	32.4
Correlation	T1ce	TC+ET	0.022	31.0
Inverse difference	T1ce	TC+ET	0.021	30.0
Contrast	F	TC+ET	0.021	29.9
Cluster shade	T1ce	ET	0.019	26.7
Correlation	T1ce	TC+ET	0.016	23.0
Variance	T1ce	ET	0.015	21.5
Maximum probability	F	ET	0.013	19.0
Cluster prominence	T1ce	ET	0.013	18.5
Dissimilarity	F	ET	0.013	18.5
Auto-correlation	T1ce	TC+ET	0.013	18.4
Inverse difference	T1ce	ET	0.013	18.4
Sum of squares variance	T1ce	ET	0.012	17.3
Difference in entropy	F	TC+ET	0.012	17.0
Average	F	TC+ET	0.012	16.6
Maximum probability	T1ce	TC+ET	0.011	15.4
Homogeneity	F	TC+ET	0.01	14.7
Difference in entropy	T1ce	TC+ET	0.009	13.1
Mean	F	TC+ET	0.009	12.4
Cluster prominence	T1ce	TC+ET	0.007	10.3
Sum average	T1ce	TC+ET	0.007	10.0
Inverse difference	F	TC+ET	0.006	9.2
Minimum	F	WT	0.005	7.8
Contrast	T1ce	TC+ET	0.005	6.7
Sum of intensities	F	ET	0.004	6.3
Contrast	F	ET	0.003	4.4
Homogeneity	F	ET	0.002	2.4
Contrast	T1ce	ET	0.001	1.9
Dissimilarity	T1ce	ET	0.001	1.7

4   Conclusion

In this study, we proposed a Deep Learning Radiomics Algorithm for Gliomas (DRAG) Model based on 3D U-Net network for brain tumor segmentation. 3D patches were extracted from multi-channel MRI data to train the proposed model. Radiomic features were extracted from FLAIR and T1ce channels for OS prediction task. MLP is trained with these radiomic features to predict the OS in days. The proposed approach achieved third place for OS prediction task in BraTS 2018 challenge [15].

The difference between mean and median in Table 2 indicates that for some cases, our pro-posed approach achieved poor accuracy, which is very close to zero and more analysis is required on this. Prediction of survival without more clinical data and treatment information is challenging and the same is reflected through accuracy of the participants in the leader-board. As the number of cases for OS prediction are less there is a need to develop an efficient feature selection algorithm which will select potential features for accurate OS prediction. Our future goal is to design radiomic features extraction pipeline with deep neural networks.

Acknowledgement. This work was supported by Ministry of Electronics and Information Technology, Govt. of India under Visvesvaraya PhD scheme with implementation reference number: PhD-MLA/4(67/2015-16). Authors are thankful to Center of Excellence in Signal and Image Processing, Shri Guru Gobind Singhji Institute of Engineering and Technology, Nanded and Tata Memorial Centre, Mumbai.

References

1. Bakas, S., et al.: Segmentation labels and radiomic features for the pre-operative scans of the TCGA-GBM collection (2017)
2. Bakas, S. et al.: Segmentation labels and radiomic features for the pre-operative scans of the TCGA-LGG collection (2017)
3. Bakas, S., et al.: Advancing The Cancer Genome Atlas glioma MRI collections with expert segmentation labels and radiomic features. Sci. Data **4**, 1–13 (2017)
4. Kickingereder, P., et al.: Radiomic profiling of glioblastoma: identifying an imaging predictor of patient survival with improved performance over established clinical and radiologic risk models. Radiol. **280**(3), 880–889 (2016)
5. Konstantinos, K., et al.: Efficient multi-scale 3D CNN with fully connected CRF for accurate brain lesion segmentation. Med. Image Anal. **36**, 61–78 (2017)
6. Mahajan, A., et al.: Bench to bedside molecular functional imaging in translational cancer medicine: to image or to imagine? Clin. Radiol. **70**(10), 1060–1082 (2015)
7. Mahajan, A., Moiyadi, A.V., Jalali, R., Sridhar, E.: Radiogenomics of glioblastoma: a window into its imaging and molecular variability. Cancer Imaging **15**(Suppl. 1), 5–7 (2015)
8. Martin, A., et al.: TensorFlow: large-scale machine learning on heterogeneous distributed systems. CoRR (2016)
9. Menze, B.H., et al.: The multimodal brain tumor image segmentation benchmark (BraTS). IEEE Trans. Med. Imaging **34**(10), 1993–2024 (2015)
10. Nelly, G., Eduard, M., Pilar, S.: State of the art survey on MRI brain tumor segmentation. Magn. Reson. Imaging **31**(8), 1426–1438 (2013)

11. Ronneberger, O., Fischer, P., Brox, T.: U-Net: convolutional networks for biomedical image segmentation. In: Navab, N., Hornegger, J., Wells, W.M., Frangi, A.F. (eds.) MICCAI 2015. LNCS, vol. 9351, pp. 234–241. Springer, Cham (2015). https://doi.org/10.1007/978-3-319-24574-4_28

12. Seow, P., Wong, J.H.D., Ahmad-Annuar, A., Mahajan, A., Abdullah, N.A., Ramli, N.: Quantitative magnetic resonance imaging and radiogenomic biomarkers for glioma characterisation: a systematic review. Br. J. Radiol. **91**, 20170930 (2017)

13. Tustison, N.J., et al.: N4ITK: improved N3 bias correction. IEEE Trans. Med. Imaging **29** (6), 1310–1320 (2010)

14. Baid, U., Talbar, S., Talbar, S.: Brain tumor segmentation based on non negative matrix factorization and fuzzy clustering. In: Fifth International Conference on Bio-Imaging (2017)

15. Bakas, S., Reyes, M., et al.: Identifying the best machine learning algorithms for brain tumor segmentation, progression assessment, and overall survival prediction in the BRATS challenge. arXiv preprint arXiv:1811.02629 (2018)

Automatic Brain Tumor Segmentation with Domain Adaptation

Lutao Dai[1], Tengfei Li[2], Hai Shu[3], Liming Zhong[3,4],
Haipeng Shen[1], and Hongtu Zhu[2(✉)]

[1] Faculty of Business and Economics, The University of Hong Kong,
Pok Fu Lam, Hong Kong
[2] The Biomedical Research Imaging Center, Department of Radiology
and Department of Biostatistics, University of North Carolina at Chapel Hill,
Chapel Hill, NC, USA
htzhu@email.unc.edu
[3] Department of Biostatistics, The University of Texas
MD Anderson Cancer Center, Houston, TX, USA
[4] Guangdong Provincial Key Laboratory of Medical Image Processing,
School of Biomedical Engineering, Southern Medical University, Guangzhou, China

Abstract. Deep convolution neural networks, in particular, the encoder-decoder networks, have been extensively used in image segmentation. We develop a deep learning approach for tumor segmentation by combining a modified U-Net and its domain-adapted version (DAU-Net). We divide training samples into two domains according to preliminary segmentation results, and then equip the modified U-Net with domain adaptation structure to obtain a domain invariant feature representation. Our proposed segmentation approach is applied to the BraTS 2018 challenge for brain tumor segmentation, and achieves the mean dice score of 0.91044, 0.85057 and 0.80536 for whole tumor, tumor core and enhancing tumor, respectively, on the challenge's validation data set.

Keywords: Confusion loss · Domain adaptation ·
Encoder-decoder network · Brain tumor · Segmentation

1 Introduction

Image segmentation plays an important role in the accurate diagnosis and efficient treatment of brain tumors. However, segmenting brain tumors, such as glioblastomas and gliomas, is difficult, because of poor tissue contrast, irregular shapes and various appearing locations. Moreover, manual segmentation can be very time-consuming and may have large intra/inter-expert variability. This creates a great need to develop reliable automatic approaches for brain tumor segmentation.

The Brain Tumor Segmentation (BraTS) challenge [2–4,14] is an event to evaluate state-of-the-art methods in automating tumor segmentation on a large

© Springer Nature Switzerland AG 2019
A. Crimi et al. (Eds.): BrainLes 2018, LNCS 11384, pp. 380–392, 2019.
https://doi.org/10.1007/978-3-030-11726-9_34

data set of annotated, high-grade glioblastomas and lower grade gliomas. To foster accurate segmentation, the BraTS 2018 challenge provides multimodal MRI scans of each patient, including native T1-weighted, post-contrast T1-weighted, T2-weighted, and T2 Fluid Attenuated Inversion Recovery (FLAIR) volumes.

Modern deep convolutional networks have exhibited exceptional competitiveness in image segmentation, becoming industrial benchmarks [10,11,13,17]. One widely-used is the encoder-decoder networks with U-shaped architectures, such as SegNet [1], U-Net [7] and DeconvNet [15]. These networks are composed of a convolutional encoder to extract salient features, and a deconvolutional decoder to recover image details. Such architecture has advantages, including flexible input image sizes, consideration of spatial information, and an end-to-end prediction, leading to lower computational cost and higher representation power.

Despite the excellent performance in the 2017 challenge, the state-of-the-art encoder-decoder network of [10] in our model exploration still loses significant segmentation accuracy for part of the BraTS 2018 training set. This is probably because the network primarily captures the key features of well-segmented samples, but misses those of the others. From the transfer learning perspective, as in [19], if treating the well-segmented samples as samples in the "source" domain and the poor-segmented samples in the "target" domain, then the network fails to learn a domain invariant feature representation. This can hence be viewed as the so-called domain adaptation problem, which aims to match the marginal feature distributions of source and target. Inspired by the domain adaptation technique of [19], we add a domain classifier to the modified U-net of [10], together with a confusion loss to learn a domain invariant feature representation for the brain tumor segmentation task. Our proposed network with domain adaptation significantly enhances the segmentation accuracy on the validation set, with mean Dice scores 0.91044, 0.85057 and 0.80536 for whole tumor, tumor core and enhancing tumor, respectively. The scores on the final test set are 0.871, 0.788 and 0.738, respectively, where detailed comparison with all the other participants in this challenge can be found in [5].

2   Data Description

The BraTS 2018 challenge data are collected from three different resources that are denoted as "2013","CBICA", and "TCIA", respectively. The training data set includes 20 high-grade glioma subjects (HGGs) from the group 2013, 88 HGGs from CBIC, and 102 HGGs from TCIA, and also includes 10 from 2013 and 65 from TCIA subjects with low-grade gliomas (LGG) that are less aggressive and infiltrative. Each subject has four modalities of MRI scans, including native T1-weighted (T1), post-contrast T1-weighted (T1Gd), T2-weighted (T2), and T2 FLAIR volumes. All MRI images are registered to a common template with the volume size of $240 \times 240 \times 155$ voxels resampled to 1 mm isotropic resolution. The tumor regions are annotated into three classes: the GD-enhancing tumor (ET, labeled 4), the peritumoral edema (ED, labeled 2), and the necrotic and non-enhancing tumor (NCR/NET, labeled 1).

A validation data set of 66 subjects is also provided for each participating team, but with no HGG/LGG status or tumor labels. The final evaluation of the segmentation approach is conducted on an independent test data set of 191 subjects.

3 Segmentation

In this section, we introduce the details of our framework for brain tumor segmentation. Our model is an ensemble of two base models: a modified U-Net and a U-Net with domain adaptation (DAU-Net). In either model, modalities are treated as channels. Domain adaptation is applied to regulate the feature representation learning process so that the extracted features are more invariant to differences between domains. We also discuss our data preprocessing and post-processing procedures that smooth and optimize the segmentation results. Figure 1 contains the workflow illustration.

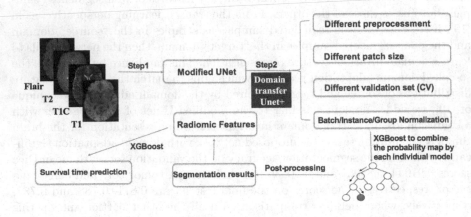

Fig. 1. Segmentation pipeline

3.1 Data Preprocessing

The main purpose of data preprocessing is to bring data to a similar distribution to avoid any initial bias, which is important for data-driven approaches. The provided data has already been skull stripped, co-registered, and resized to uniform resolution. On top of that, we remove the top and bottom 1 percentile of intensity in the brain areas for each image, and normalized the brain intensities by subtracting the mean and dividing the standard deviation. The preprocessing is conducted on brain regions only and independently across modalities and individuals.

3.2 Modified U-Net

Model Description. Our modified U-Net is inspired by [10]. In our model, each level of the encoding pathway consists of a residual block with the same structure.

The first convolution layer of each residual block halves the spatial dimension with a stride of 2 (except for the first residual block), and increases the number of channels to 8×2^n, with n being the level counting from 1. As a result, the stack of 5 residual blocks progressively reduces the spatial dimension of the input tensor by a factor of 16 and learns increasingly abstract feature representations. To increase the prediction resolution, the decoding pathway progressively doubles the spatial dimension on each level by an upsampling layer of scale 2, and eventually recovers the spatial dimension of input data. The feature maps generated by the first four residual blocks are concatenated to decoding pathway of the same level to encourage the gradient flow. We apply group normalization [21] to all normalization layers, because it is more stable given a small batch and yields a better result compared to instance normalization [20]. The group number is 16 for level 1 and 32 for the remaining levels. Moreover, we adopt the idea of deep supervision [12], where output maps of different levels are combined sequentially through element-wise addition to constitute the network's final prediction via the softmax function. We integrate the multiclass dice loss function into our framework, since it can effectively mitigate the problem of class imbalance:

$$L_{\text{dice}} = -\frac{2}{|K|} \sum_{k \in K} \frac{\sum_i u_i^k v_i^k}{\sum_i u_i^k + \sum_i v_i^k}$$

where K is the set of prediction classes, u is the probability maps output by the backbone structure, v is the one-hot encoding of the ground truth, and i is the voxel index.

3.3   DAU-Net

Model Description The backbone structure is the same as the model in Subsection 3.2, except all normalization layers are instance normalization [20], because we observe more boost by domain adaptation with the presence of instance normalization. The domain classifier is appended to the end of the encoding pathway, where the feature representation is the most complex. A $1 \times 1 \times 1$ convolution layer is first applied to significantly reduce the number of channels from 256 to 32, followed by alternating three fully-connected layers of lengths 256, 32, and 1, and two leaky ReLU rectifiers.

Domain Division. We perform two sets of five-fold cross-validation on training data with the backbone structure only. One set follows the data preprocessing procedure described in Subsect. 3.1, while the other set has an additional N4 bias correction processed by ANTs [18]. We compare the differences of Dice coefficient between the segmented tumors of the two sets. Although N4 bias correction has a minimal effect on most samples, it does yield significantly different results for some cases. We pick 75 subjects with the most variations and classified them to a different domain from the rest. The full list of those 75 subjects can be found in the Table 3. Based on the two domains of subjects, the network is shown in Fig. 2.

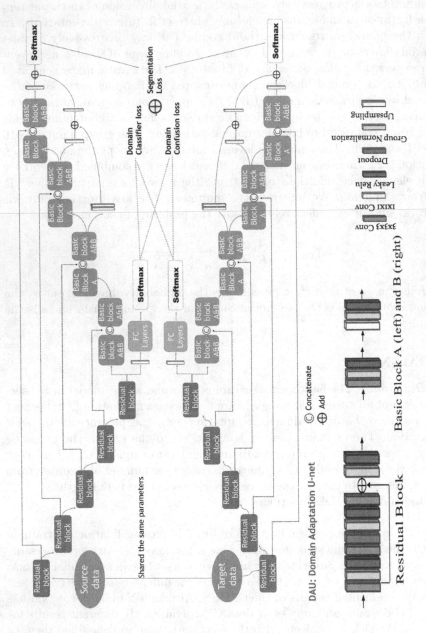

Fig. 2. Domain adaptation network architecture

Training Procedure. The backbone structure and the domain classifier are trained alternatively. In the $2n$-th epoch, the objective function adds an additional confusion loss onto the dice loss function L_{dice}, which is the cross entropy between the predicted domain label and a uniform distribution:

$$L_{2n} = -\frac{2}{|K|} \sum_{k \in K} \frac{\sum_i u_i^k v_i^k}{\sum_i u_i^k + \sum_i v_i^k} - \lambda \sum_d \frac{1}{|D|} \log q_d,$$

where D is the set of domain categories, and q_d is the estimated probability for the d-th domain from the domain classifier. q_d is modeled by a softmax function of the classifier activations $q_d = \text{softmax}(\theta_1^T f(\theta_2))$, where θ_1 includes activation parameters in the fully connected layer and θ_2 includes representation parameters in the modified U-net (Fig. 2). In this step, θ_1 is kept unchanged and parameter of the backbone structure θ_2 is updated. The hyperparameter λ controls the degree of the domain confusion relative to the backbone structure.

In the $(2n + 1)$-th epoch, θ_2 is frozen so that only θ_1 is updated. The domain classifier aims to discriminate samples according to the feature representation output by the encoding pathway. The cross-entropy loss is computed with domain labels as follows:

$$L_{2n+1} = -\sum_d \mathbb{I}[y_D = d] \log q_d.$$

In summary, the two steps update different parts of parameters. By training the model iteratively, both the backbone structure and the domain classifier are optimized. The best domain classifier learned by minimizing L_{2n+1} is expected to still perform poorly on the final domain prediction, due to the confusion loss in L_{2n}. With such a domain classifier, the encoding pathway has incentives to capture the domain-invariant features. This helps to improve the generalizability of the model, since differences in MRI data representation are usually significant. The training was carried out on 4 NVIDIA Titan Xp GPU cards for about 2 days.

3.4 Experiment Configuration

The input tensors of size $128 \times 128 \times 128$ are randomly sampled from brain areas and augmented by random flipping and transpose during each epoch. The training is implemented by PyTorch using the Adam optimizer with the learning rate initially set to be $8 \times v10^{-4}$ and exponentially decaying at a rate of 0.98 every epoch for the modified U-Net, and every two epochs for the DAU-Net. All networks are trained for about 600 epochs, and the ones with the lowest Dice loss of whole tumor are selected as candidates for ensemble.

At the test time, the domain classifier is dropped. The whole brain regions whose dimensions are padded to the nearest multiple of 16 are served as inputs to the modified U-Net, and the returned segmentation maps are subsequently padded with zeros to reach the original dimension.

3.5 Model Ensemble and XGBoost

XGBoost [6], short for extreme gradient boosting, is an implementation of gradient-boosted decision trees designed for speed and performance. The term gradient boosting was proposed by Friedman [8]. It is a tree-based machine learning algorithm that usually dominates structured or tabular datasets on classification and regression predictive modeling problems, and boosting is an ensemble technique where new models are added to correct the errors made by existing models.

We aim to ensemble multiple models trained from deep learning for brain tumor segmentation using XGBoost. Most of the previous literature focused on majority voting and averaging; [9] compared the three ensembling approaches including majority voting, averaging and expectation-maximization; [22] demonstrated that the XGBoost approach can outperform the majority voting label fusion. We here train different models, which differ in preprocessing methods (with or without bias correction), different patch sizes, different splits of training dataset into training and validation parts, and different normalization (instance/group) of the network (modified U-net with group normalization and the DAU-net with instance normalization). We choose the 9 models with top Dice coefficients on the validation set; for each subject in the training dataset, we make predictions using the 9 models; next we calculate the set S of voxels where there exists disagreement for the 9 models; then, for each tumor class and each subject, we randomly choose 1000 voxels without replacement from the set S, and we use the predicted probability of each model for the four tumor classes at each voxel as covariates to predict the true label using XGBoost; to determine the hyperparameters, we split the training dataset into five parts and used a 5-fold cross-validation to optimize the maximal depth, minimum child weight, penalization parameters and learning rate, etc., to minimize the softmax loss function.

3.6 Post-processing

We employ the following post-processing techniques [22] to fill in the holes and delete the small, isolated clusters:

1. Segment the tumor mask into all connected components/clusters. Voxels in clusters whose volume is less than 0.2 times the largest connected cluster volume will be reclassified as non-tumor.
2. Segment the enhancing core mask into all connected components/clusters. Voxels in clusters whose volume is less than 0.01 times the largest connected cluster volume will be reclassified as the necrosis.
3. Fill in the holes within the tumor mask and assign voxels within the holes to necrosis area.

We find that the performance can be improved by applying post-processing on the existing results.

3.7 Survival Prediction

Based on the previous segmentation results, we crop out the tumor core region and extract various radiomic features to predict the survival time for the three modalities Flair, T1 post-contrast, and T2. For each modality, features include: 10 intensity statistics features (such as maximum, minimum, median, and quantiles, skewness, kurtosis, and entropy), 51 shape features (24 Zernike moment based shape descriptors, 21 Hu Moment based shape descriptors, and 6 statistics of local binary patterns), 112 texture features (13 gray-level co-occurrence matrix features, 27 threshold adjacency statistics, and 72 wavelet transform features). We also add the ratio of all tumor class volumes to the whole brain volume as additional features. Similar features are used in [22].

We use XGBoost to predict survival time based on the above radiomic features, together with age. The difference between the prediction here and that in Subsect. 3.5 is the dimension adopted in the survival prediction task is much higher, which will bring about overfitting and high computational complexity. We use a 5-fold cross-validation to select important features and optimize the maximal depth, minimum child weight, penalization parameters and learning rate, and other tuning parameters.

4 Results

4.1 Selection of λ

To investigate the optimum value of λ introduced in L_{2n}, we conduct multiple trials with λ as the only varying parameter (Fig. 3). Within a certain range of λ, there is a clear enhancement of the average dice coefficient for whole tumor and tumor core, whose optimum values are achieved at $\lambda = 0.1$ and $\lambda = 0.075$, respectively. Passing over the optimum point, we can see a clear decline in the average dice coefficient for both. The average dice coefficient of enhancing tumor

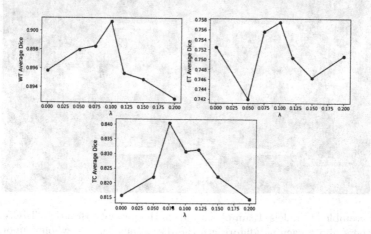

Fig. 3. Dice coefficients with varying λ.

Table 1. Segmentation results on validation data

Method	Dice_ET	Dice_WT	Dice_TC
Phase1	0.75245	0.89571	0.81561
Phase2 Model a	0.75983	0.90397	0.82489
Phase2 Model b	0.76091	0.90616	0.83622
Phase2 Model c	0.74669	0.90349	0.8278
Phase2 Model d	0.74187	0.90435	0.83211
Phase2 Model e	0.75779	0.90733	0.83824
Phase2 Model f	0.76091	0.90420	0.83713
Phase2 Model g	0.76814	0.90574	0.84704
Phase2 Model h	0.75440	0.90594	0.83826
Phase2 Model i	0.78582	0.90491	0.83689
XGBoost+Postprocessing	0.80536	0.91044	0.85057

fluctuates with λ, but its highest peak is at $\lambda = 0.1$, which is very close to the optimal λ's of whole tumor and tumor core. We hence choose $\lambda = 0.1$ for our proposed network. Detailed segmentation results for the validation set, with and without domain adaptation, are shown in Table 1, Phase 1 and Phase 2a, respectively.

4.2   Results of Mean Dice Score and Survival Prediction

The prediction results of the validation set after XGBoost and postprocessing are the best, as shown in the final row (i.e., XGBoost+Postprocessing of Table 1). According to the Wilcoxon signed-rank test, its Dice scores are significantly

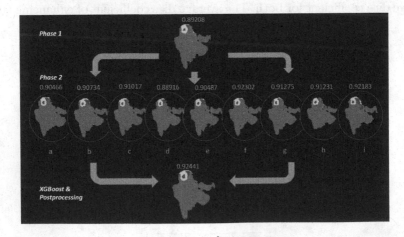

Fig. 4. Ensemble of models. Examples are from the patient "Brats18_CBICA_ALV_1". The numbers above each sub-figure are the dice coefficients of whole tumor for that patient.

larger than those in phases 1 (in the 1st row), whose p-values are less than 0.048, 4×10^{-5}, and 3×10^{-4}, for ET, WT and TC, respectively. We use an example from the validation dataset to illustrate the improvement of ensemble in Fig. 4. The prediction results of the patients' survival time for the validation set are shown in Table 2.

Table 2. The prediction result on validation data

Accuracy	MSE	MedianSE	stdSE	SpearmanR
0.5	99409.107	33754.5	210658.725	0.332

5   Conclusion

We presented our contribution to the BraTS 2018 challenge in this paper. We developed a deep learning approach for the tumor segmentation by combining a modified U-Net and the DAU-Net. Both models were trained with extensive data augmentation. We applied the XGBoost procedure to ensemble our image segmentation predictions. The ensemble of the 9 top-performing models outperformed each individual model on validation data. Due to time constraints, we did not explore the effect of domain adaptation on feature learning, which can explain the improvement on performance. How domain adaptation regulates feature learning will be a promising research topic that sheds lights on a better design of data augmentation as well as a preprocessing pipeline. Moreover, we tried a few ways of defining domains, including dividing the data by gliomas' grades and data source, but have not yet developed a common standard for automatic domain split with good interpretability. Keeping in mind that it is hard to train an effective domain classifier by dividing a small dataset into groups, we also expect the model to have a better result with the introduction of external data. In this case, the data naturally come from different domains. Besides, we tried different hyperparameters in our postprocessing to make improvement of Dice performance on validation dataset, which may cause overfitting.

For survival prediction, we extracted radiomic features based on the segmentation results, but did not include deep-learning features together, due to time constraints. Neural networks for predictions can usually outperform the traditional prediction methods [16], which is worth further exploration.

Acknowlegement. This research was partially supported by Ministry of Science and Technology Major Project of China 2017YFC1310903, University of Hong Kong Stanley Ho Alumni Challenge Fund, University Research Committee Seed Funding Award 104004215, US National Science Foundation grants DMS-1407655, and NIH grants MH086633 and MH116527.

Appendix

Table 3. Selected 75 cases to be another domain

	Name	Type		Name	Type
1	Brats18_2013_11_1	HGG	46	Brats18_2013_0_1	LGG
2	Brats18_2013_19_1	HGG	47	Brats18_2013_15_1	LGG
3	Brats18_2013_22_1	HGG	48	Brats18_2013_16_1	LGG
4	Brats18_2013_25_1	HGG	49	Brats18_2013_1_1	LGG
5	Brats18_2013_4_1	HGG	50	Brats18_2013_9_1	LGG
6	Brats18_CBICA_ABB_1	HGG	51	Brats18_TCIA09_141_1	LGG
7	Brats18_CBICA_ABO_1	HGG	52	Brats18_TCIA09_177_1	LGG
8	Brats18_CBICA_ANG_1	HGG	53	Brats18_TCIA09_255_1	LGG
9	Brats18_CBICA_ANP_1	HGG	54	Brats18_TCIA09_402_1	LGG
10	Brats18_CBICA_AOD_1	HGG	55	Brats18_TCIA09_462_1	LGG
11	Brats18_CBICA_AOH_1	HGG	56	Brats18_TCIA09_493_1	LGG
12	Brats18_CBICA_AOZ_1	HGG	57	Brats18_TCIA09_620_1	LGG
13	Brats18_CBICA_AQA_1	HGG	58	Brats18_TCIA10_130_1	LGG
14	Brats18_CBICA_AQQ_1	HGG	59	Brats18_TCIA10_261_1	LGG
15	Brats18_CBICA_AQR_1	HGG	60	Brats18_TCIA10_266_1	LGG
16	Brats18_CBICA_AQU_1	HGG	61	Brats18_TCIA10_276_1	LGG
17	Brats18_CBICA_ARW_1	HGG	62	Brats18_TCIA10_282_1	LGG
18	Brats18_CBICA_ARZ_1	HGG	63	Brats18_TCIA10_413_1	LGG
19	Brats18_CBICA_ASE_1	HGG	64	Brats18_TCIA10_420_1	LGG
20	Brats18_CBICA_ASH_1	HGG	65	Brats18_TCIA10_442_1	LGG
21	Brats18_CBICA_ATF_1	HGG	66	Brats18_TCIA10_490_1	LGG
22	Brats18_CBICA_AUQ_1	HGG	67	Brats18_TCIA10_628_1	LGG
23	Brats18_CBICA_AWI_1	HGG	68	Brats18_TCIA10_629_1	LGG
24	Brats18_CBICA_AXN_1	HGG	69	Brats18_TCIA10_637_1	LGG
25	Brats18_CBICA_AXQ_1	HGG	70	Brats18_TCIA10_644_1	LGG
26	Brats18_CBICA_AYI_1	HGG	71	Brats18_TCIA13_618_1	LGG
27	Brats18_CBICA_BFP_1	HGG	72	Brats18_TCIA13_621_1	LGG
28	Brats18_CBICA_BHB_1	HGG	73	Brats18_TCIA13_633_1	LGG
29	Brats18_CBICA_BHK_1	HGG	74	Brats18_TCIA13_645_1	LGG
30	Brats18_TCIA01_180_1	HGG	75	Brats18_TCIA13_650_1	LGG
31	Brats18_TCIA01_190_1	HGG			
32	Brats18_TCIA01_411_1	HGG			
33	Brats18_TCIA01_425_1	HGG			
34	Brats18_TCIA02_168_1	HGG			
35	Brats18_TCIA02_226_1	HGG			
36	Brats18_TCIA03_257_1	HGG			
37	Brats18_TCIA04_328_1	HGG			
38	Brats18_TCIA04_343_1	HGG			
39	Brats18_TCIA04_437_1	HGG			
40	Brats18_TCIA05_277_1	HGG			
41	Brats18_TCIA06_165_1	HGG			
42	Brats18_TCIA06_211_1	HGG			
43	Brats18_TCIA06_409_1	HGG			
44	Brats18_TCIA08_278_1	HGG			
45	Brats18_TCIA08_406_1	HGG			

References

1. Badrinarayanan, V., Kendall, A., Cipolla, R.: Segnet: A deep convolutional encoder-decoder architecture for image segmentation. arXiv preprint arXiv:1511.00561 (2015)
2. Bakas, S., et al.: Segmentation labels and radiomic features for the pre-operative scans of the TCGA-GBM collection. In: The Cancer Imaging Archive (2017)
3. Bakas, S., et al.: Segmentation labels and radiomic features for the pre-operative scans of the TCGA-LGG collection. In: The Cancer Imaging Archive (2017)
4. Bakas, S., et al.: Advancing the cancer genome atlas glioma MRI collections with expert segmentation labels and radiomic features. Nature Sci. Data **4**, 170117 (2017)
5. Bakas, S., Reyes, M., Menze, B., et al.: Identifying the best machine learning algorithms for brain tumor segmentation, progression assessment, and overall survival prediction in the BRATS challenge. arXiv preprint arXiv:1811.02629 (2018)
6. Chen, T., Guestrin, C.: Xgboost: a scalable tree boosting system. In: Proceedings of the 22nd ACM SIGKDD International Conference on Knowledge Discovery and Data Mining, pp. 785–794. ACM (2016)
7. Çiçek, Ö., Abdulkadir, A., Lienkamp, S.S., Brox, T., Ronneberger, O.: 3D U-Net: learning dense volumetric segmentation from sparse annotation. In: Ourselin, S., Joskowicz, L., Sabuncu, M.R., Unal, G., Wells, W. (eds.) MICCAI 2016. LNCS, vol. 9901, pp. 424–432. Springer, Cham (2016). https://doi.org/10.1007/978-3-319-46723-8_49
8. Friedman, J.H.: Greedy function approximation: a gradient boosting machine. Ann. Stat. 1189–1232 (2001)
9. Huo, J., Okada, K., Pope, W., Brown, M.: Sampling-based ensemble segmentation against inter-operator variability. In: Medical Imaging 2011: Computer-Aided Diagnosis, vol. 7963, p. 796315. International Society for Optics and Photonics (2011)
10. Isensee, F., Kickingereder, P., Wick, W., Bendszus, M., Maier-Hein, K.H.: Brain tumor segmentation and radiomics survival prediction: contribution to the BRATS 2017 challenge. In: Crimi, A., Bakas, S., Kuijf, H., Menze, B., Reyes, M. (eds.) BrainLes 2017. LNCS, vol. 10670, pp. 287–297. Springer, Cham (2018). https://doi.org/10.1007/978-3-319-75238-9_25
11. Kamnitsas, K., et al.: Ensembles of multiple models and architectures for robust brain tumour segmentation. In: Crimi, A., Bakas, S., Kuijf, H., Menze, B., Reyes, M. (eds.) BrainLes 2017. LNCS, vol. 10670, pp. 450–462. Springer, Cham (2018). https://doi.org/10.1007/978-3-319-75238-9_38
12. Kayalibay, B., Jensen, G., van der Smagt, P.: CNN-based segmentation of medical imaging data. arXiv preprint arXiv:1701.03056 (2017)
13. Li, T., Fan, Z., Ziliang, Z., Hai, S., Hongtu, Z.: A label-fusion-aided convolutional neural network for isointense infant brain tissue segmentation. In: 2018 IEEE 15th International Symposium on Biomedical Imaging, pp. 692–695 (2018)
14. Menze, B.H., et al.: The multimodal brain tumor image segmentation benchmark (brats). IEEE Trans. Med. Imaging **34**(10), 1993 (2015)
15. Noh, H., Hong, S., Han, B.: Learning deconvolution network for semantic segmentation. In: Proceedings of the IEEE International Conference on Computer Vision, pp. 1520–1528 (2015)
16. Notley, S., Magdon-Ismail, M.: Examining the use of neural networks for feature extraction: A comparative analysis using deep learning, support vector machines, and k-nearest neighbor classifiers. arXiv preprint arXiv:1805.02294 (2018)

17. Ronneberger, O., Fischer, P., Brox, T.: U-Net: convolutional networks for biomedical image segmentation. In: Navab, N., Hornegger, J., Wells, W.M., Frangi, A.F. (eds.) MICCAI 2015. LNCS, vol. 9351, pp. 234–241. Springer, Cham (2015). https://doi.org/10.1007/978-3-319-24574-4_28

18. Tustison, N.J., et al.: Large-scale evaluation of ants and freesurfer cortical thickness measurements. Neuroimage **99**, 166–179 (2014)

19. Tzeng, E., Hoffman, J., Darrell, T., Saenko, K.: Simultaneous deep transfer across domains and tasks. In: Proceedings of the IEEE International Conference on Computer Vision, pp. 4068–4076 (2015)

20. Ulyanov, D., Vedaldi, A., Lempitsky, V.: Instance Normalization: The Missing Ingredient for Fast Stylization. arXiv.org, November 2017

21. Wu, Y., He, K.: Group Normalization. arXiv.org, June 2018

22. Zhou, F., Li, T., Li, H., Zhu, H.: TPCNN: two-phase patch-based convolutional neural network for automatic brain tumor segmentation and survival prediction. In: Crimi, A., Bakas, S., Kuijf, H., Menze, B., Reyes, M. (eds.) BrainLes 2017. LNCS, vol. 10670, pp. 274–286. Springer, Cham (2018). https://doi.org/10.1007/978-3-319-75238-9_24

Global Planar Convolutions for Improved Context Aggregation in Brain Tumor Segmentation

Santi Puch[1(✉)], Irina Sánchez[1], Aura Hernández[2],
Gemma Piella[3], and Vesna Prčkovska[1]

[1] QMENTA, Boston, MA, USA
{santi,irina,vesna}@qmenta.com
[2] Computer Vision Center, Universitat Autònoma de Barcelona, Barcelona, Spain
aura@cvc.uab.es
[3] SIMBIOsys, Universitat Pompeu Fabra, Barcelona, Spain
gemma.piella@upf.edu

Abstract. In this work, we introduce the Global Planar Convolution module as a building-block for fully-convolutional networks that aggregates global information and, therefore, enhances the context perception capabilities of segmentation networks in the context of brain tumor segmentation. We implement two baseline architectures (3D UNet and a residual version of 3D UNet, ResUNet) and present a novel architecture based on these two architectures, ContextNet, that includes the proposed Global Planar Convolution module. We show that the addition of such module eliminates the need of building networks with several representation levels, which tend to be over-parametrized and to showcase slow rates of convergence. Furthermore, we provide a visual demonstration of the behavior of GPC modules via visualization of intermediate representations. We finally participate in the 2018 edition of the BraTS challenge with our best performing models, that are based on ContextNet, and report the evaluation scores on the validation and the test sets of the challenge.

Keywords: Brain tumors · 3D fully-convolutional CNN ·
Magnetic resonance imaging · Global planar convolution

1 Introduction

It is estimated that, as of today, 700.000 people in the United States are living with a primary brain tumor, from which 80% are benign and 20% are malignant tumors [14]. Of all malignant brain tumors, 81% are gliomas, which are tumors that originate in glial cells [15]. Glioblastomas are the most common type of glioma, representing 45% of all gliomas; they are one of the most aggressive types of brain tumors, having an estimated 5-year relative survival rate

© Springer Nature Switzerland AG 2019
A. Crimi et al. (Eds.): BrainLes 2018, LNCS 11384, pp. 393–405, 2019.
https://doi.org/10.1007/978-3-030-11726-9_35

of approximately 5%, which means that only 5% of people diagnosed with a glioblastoma will still be alive 5 years after being diagnosed [15].

It is clear that such dismal prognosis requires proper treatment planning and follow-up, which can be greatly improved if proper in-vivo, non-invasive delineation and identification of glioma structures is in place. However, this poses a significant burden on the radiologist: multiple imaging modalities have to be assessed in parallel, as each highlights different regions of the tumor, and the process of delineation in a 3D acquisition is tedious and prone to errors. As a consequence, inter-observer variability has been reported to be a major—if not the largest—factor of inaccuracy in radiation therapy, constituting the weakest link in the radiotherapy chain that goes from diagnosis and consultation, going through 3D imaging and target volume delineation, to treatment delivery [19].

Therefore, automating the delineation and identification process on MR images would accelerate treatment planning and improve treatment follow-up. However, the problem of tumor segmentation poses several challenges, such as blurry or smoothed boundaries, variability of shape, location and extension or heterogeneity of appearance of brain tumors on MR images.

The research community has concentrated efforts in order to address the brain tumor segmentation task, and to this end initiatives like the Brain Tumor Segmentation (BraTS) challenge [12] have made the problem accessible to a larger audience. As a result, a large variety of computational methods have been proposed to automate the delineation of brain tumors. These methods can be broadly categorized in two groups: generative models, which rely on prior knowledge about tissue appearance and distribution; and discriminative models, which directly learn the relationship between the image features and the segmentation labels. Deep Learning approaches, especially Convolutional Neural Networks (CNNs), cover a large portion of the recently proposed discriminative methods. The majority of these works have based their methods in well-known semantic segmentation networks, either using a 2D variant on one or more planes of the brain, or implementing a 3D architecture that takes spatial information in all directions into account. In all these works, several training strategies are leveraged, such as dense training with patches combined with patch sampling schemes, and a distinction between local, refined features and global, coarse features is accomplished via multiresolution approaches or skip connections.

In this work, we introduce the Global Planar Convolution (GPC) module, a fully-convolutional module that enhances the context perception capabilities of segmentation networks in the context of brain tumor segmentation. We first explore different 3D fully-convolutional architectures for brain tumor segmentation, starting with a 3D variation of UNet [17]. We then introduce a variation of such network that incorporates residual elements from [7], that we call ResUNet. We finally refine this architecture by adding GPC modules; we refer to this network as ContextNet. We train these architectures on the 2018 BraTS Challenge dataset, that consists of 210 High Grade Glioma (HGG) cases and 75 Low Grade Glioma (LGG) cases with four MR image modalities and manual annotations of the distinct intra-tumoral structures of interest [1–3,12]. We compare the behavior of the three proposed networks when trained with all the image modalities and when trained only with a subset of them. Then, we show that the addition of Global Planar Convolution modules eliminates the need of building networks with several representation levels (i.e. the set of operations executed at the same spatial resolution level), which are prone to over-parametrization and slow convergence rates. We include a visually guided interpretation of the behavior of GPC modules via visualization of intermediate representations of the network. We finally report the performance of our best performing model—based on ContextNet—and a model ensemble on the BraTS validation set. This last model ensemble is submitted to participate in the 2018 edition of the BraTS challenge.

2 Methods

2.1 Data

The data used in this project originates in the 2018 version of the yearly Multimodal Brain Tumor Segmentation Challenge dataset. This dataset consists of 285 multi-institutional clinically-acquired pre-operative scans. Each multimodal scan is formed by T1, T1-Gd, T2 and T2-FLAIR volumes acquired with various scanners from 19 institutions. All the scans have been segmented manually by one to four raters, and approved by experienced neuroradiologists. The ground-truth labels comprise the enhancing tumor, the peritumoral edema and the necrotic and non-enhancing tumor when present. Each multi-modal scan in the BraTS challenge dataset is co-registered to the same anatomical template, skull-stripped and resampled to $1\,mm^3$ isotropic resolution. Therefore, no further preprocessing is needed.

2.2 CNN Architectures

All the Convolutional Architectures implemented and trained are 3D and fully-convolutional by design, meaning that they can be trained using 3D patches of data, and then they can be used for inference on whole brain volumes.

UNet. This reference architecture proposed initially in [17] is based on [11], with a similar contracting path, but in order to improve the localization capabilities of the network a supplementary expanding path is introduced, in which high-resolution features from the contracting path are combined with upsampled feature maps. This architecture is extended to 3D by replacing all the convolutions, transposed convolutions and max pooling operations by their 3D alternatives, similarly to [5].

In our experiments we use Rectified Linear Unit (ReLU) as activation function and Batch Normalization [8] after convolutions and before activations.

ResUNet. The work of [7] introduced the concept of deep residual learning: instead of stacking a series of layers and letting them learn the desired underlying mapping, these layers can be set to explicitly fit a residual mapping. This residual learning framework not only improved the performance on the image classification task by a wide margin, but also alleviated the degradation problem found when an excessive amount of layers was used.

The motivation of introducing residual layers originates from the fact that the contracting path has a twofold purpose: on one hand it learns increasingly abstract features that encode contextual information necessary to decide *what*, and on the other hand it connects feature maps from lower-level representations with the expanding path to decide *where*. Residual elements allow for an increased number of layers, which has been shown empirically to increase the representational power of the network [7,18], thus helping with the first task (*identification*). They also facilitate learning of identity mappings, which enables the possibility of passing low-level representations throughout the network, thus easing the second task (*delineation*).

The architecture of ResUNet is essentially an extension of UNet with residual elements. The convolutional layers in UNet are replaced by residual layers; concretely, 2 residual layers are used at each resolution level.

Again, as in UNet, we use Rectified Linear Unit (ReLU) as activation function and Batch Normalization after convolutions and before activations.

ContextNet. ContextNet is a novel architecture introduced in this work that aims to enhance the context-awareness capabilities of 3D imaging segmentation architectures. It is build upon the aforementioned ResUNet architecture and includes the proposed Global Planar Convolution modules, inspired by [16]. An overview of the architecture is shown in Fig. 1.

The localization aspect of semantic segmentation networks is addressed by skip-connections and residual elements, as these components let low level representations pass through the network and inform the latest layers about fine-grained spatial details. However, the classification aspect of semantic segmentation networks, that deals with proper identification of the delineated structures, is hindered by the fact that these networks are focused on proper boundary alignment. State-of-the-art classification architectures rely on layers that are globally connected, which in the most extreme case (all the nodes are connected with each other) corresponds to a fully-connected layer. It is clear that such type of operation is not feasible in a fully-convolutional architecture, however we can approximate global connectivity by increasing kernel size in convolutions: in the limit, the kernel is as big as the input feature map, which can be interpreted again as a fully-connected layer. The problem with increasing the kernel size is the computational and memory requirements associated with it, and it is not feasible in the case of 3D CNNs with current accelerated computing hardware.

However, global connectivity can be approximated by constraining the kernel parameters' subspace. Specifically, we can constrain the convolutional kernels to have one dimension less than they would normally have, which in practice

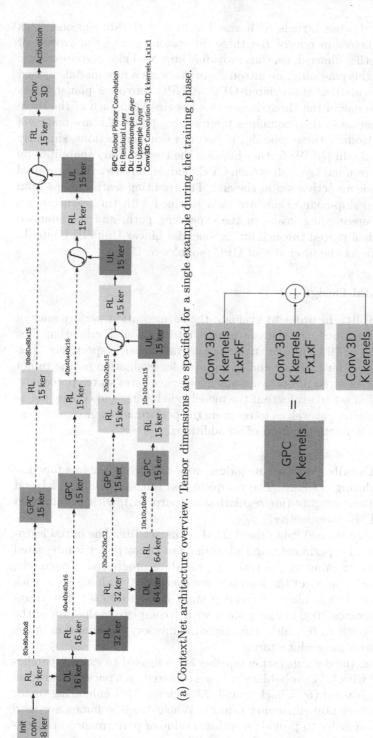

(a) ContextNet architecture overview. Tensor dimensions are specified for a single example during the training phase.

(b) Global Planar Convolution (GPC) module

Fig. 1. ContextNet architecture diagrams

is implemented by having kernels with size 1 in one of the dimensions. This reduction of parameters in one of the three dimensions allows the growth of kernel sizes in the other dimensions, thus providing improved global connectivity.

On the basis of this reasoning, we introduce in this work a new module named *Global Planar Convolution*, abbreviated GPC. A GPC convolves planar filters (i.e. filters in which one of the three dimensions has size 1) in each of the three orthogonal directions, and then combines the resulting planar feature maps via summation. We introduce these modules in between skip-connections, similarly to bottleneck modules in [7]. We further improve the localization capabilities by including an extra residual layer after each GPC module, however these residual layers do not include an activation in the end. The resulting feature maps from each of this altered skip-connections are then summed with the feature maps outputted by the upsampling layers in the expanding path, and the summed feature maps are then passed through an Exponential Linear Unit (ELU) [6]. In our experiments we set the filter size of GPC modules to 15.

2.3   Experimental Design

Local Dataset Split. In order to evaluate the segmentation performance of the proposed architectures, the dataset is split into train and validation sets. We use a split ratio of 70%–30% for training and validation, respectively. This results in 199 subjects for training and 86 subjects for evaluation on the tumor segmentation task. As we do not perform any hyper-parameter search procedure, we use the validation set to evaluate if the model is behaving and converging as expected during training, as well as to compare the performance among different models. Thus, we eliminate the need of an additional test set.

CNN Training Details. We use categorical cross-entropy as the loss function to be minimized during training. The complete loss function includes L1 and L2 penalization of the weights (for regularization purposes), with penalization ratios of 1E-6 and 1E-4, respectively.

All the models are trained using the ADAM optimizer [10]. The initial learning is set to 1E-3 in all experiments, and a learning rate decay policy is integrated in order to stabilize training as the training procedure progressed. Concretely, we use an exponential decay of the learning rate every 1000 training steps with a decay rate of 0.9. The number of training steps is set to 35000. The training procedure alternates 1000 training steps with 1 complete evaluation of the model. Batch size is set to 6, which maximizes the memory consumption in the most memory demanding architectures.

During training, the data ingestion pipeline is configured to extract patches of size $80 \times 80 \times 80$ with 50% probability of being centered on a background voxel and 50% on a tumor voxel (50% background, 20% edema, 15% enhancing tumor and 15% necrosis and non-enhancing tumor). Whole brain volumes are used during evaluation in order to provide a realistic value of performance in a real-world scenario. The CNNs are trained on two different hardware configurations, depending on the availability of computing resources: (1) AWS p2.xlarge instance with a single NVIDIA K80 with 12 GiB of GPU memory; (2) on-premises server with two NVIDIA GeFore GTX 1080 Ti with 11GiB of GPU memory.

Restriction of Availability of Input Modalities. We perform data ablation experiments by restricting the available input modalities at training time, but always maintaining the minimum required modalities to properly identify all structures, namely T1-Gd and FLAIR. The motivation for such experiments is twofold. First, we want to assess the relative contribution of each modality to the overall segmentation, and inspect if some modalities are redundant or indeed provide useful information. Second, it is convenient and even necessary to have models that can work with a restricted number of modalities (as in some clinical cases not all MR sequences are included in the protocol) even if such models with restricted input information do not perform as well as models trained without data restrictions.

Restriction of Number of Representation Levels. We hypothesize that the inclusion of GPC modules enables the network to perceive greater context without the need of having several representation levels. By representation levels we mean the set of operations and feature maps that operate at the same spatial resolution. In order to validate this hypothesis, we train two variations of the ContextNet architecture with as little as 2 or 3 representation levels. The first model, with just 2 representation levels, has 32 and 64 kernels at each convolutional layer before the GPC modules, while the GPC modules still have 15 kernels with the same kernel size as in Sect. 2.2. The second model has 3 representation levels with 16, 32 and 64 kernels at each convolutional layer before the GPC modules, maintaining again the same number of kernels and kernel sizes at the GPC modules and subsequent layers.

Visualization of GPC Feature Maps. We extract intermediate representations from the residual layers around the GPC modules and from the GPC modules themselves on all the representation-level variations of ContextNet. The intent of such experiment is to provide insight about the behavior of the GPC modules, and to link such behavior to the performance of these models, despite being aware of the limitations of this method for network interpretability.

2.4   Evaluation

The primary evaluation score for the segmentation task is the Sørensen-Dice coefficient, usually abbreviated as DICE. In this context, the DICE coefficient compares the similarity between the set of true examples and the set of positive examples:

$$DICE = \frac{2TP}{2TP + FP + FN} \tag{1}$$

The Hausdorff distance is used to evaluate the distance between segmentation boundaries. Results are reported using the 95% quantile of the maximal surface distance between the ground truth P_1 and the predicted segmentation T_1 [13]:

$$Haus(P, T) = max(\sup_{p \in \partial P_1} \inf_{t \in \partial T_1} d(p, t), \sup_{t \in \partial T_1} \inf_{p \in \partial P_1} d(t, p)) \qquad (2)$$

The targets of these evaluation scores are the following tumoral structures:

- **Whole Tumor**: comprises all tumoral structures, i.e. edema, enhancing tumor, non-enhancing tumor and necrosis.
- **Enhancing Tumor**: comprises only the enhancing tumor class.
- **Tumor Core**: encompasses the enhancing tumor, necrosis and non-enhancing tumor, thus excluding edema.

3   Results

Table 1 shows the DICE coefficients on the local validation set for all target structures of all baseline architectures, trained with different data configurations. The best model for whole tumor segmentation is ContextNet trained with all modalities (0.897 DICE score), while the best model for enhancing tumor and tumor core segmentation is ResUNet trained only with T1-Gd, FLAIR and T1, achieving 0.752 and 0.799 DICE scores, respectively. It is particularly remarkable that excluding the T2 from training enables the ResUNet model to better segment the tumor core structures; such behavior can be explained if we consider that the network is encouraged to focus more on structures more noticeable on T1-related modalities (enhancing tumor and tumor core) thanks to the exclusion of redundant information about edema provided by the T2 image (which is clearly visible in FLAIR images).

Table 1. DICE coefficients (avg ± std) of baseline architectures trained with different data configurations. Scores are computed on the local validation set.

	Enhancing tumor	Whole tumor	Tumor core
UNet - all modalities	0.698 ± 0.229	0.847 ± 0.095	0.694 ± 0.235
ResUNet - all modalities	0.739 ± 0.207	0.892 ± 0.064	0.785 ± 0.200
ResUNet - T1-Gd, FLAIR, T1	**0.752 ± 0.193**	0.882 ± 0.080	**0.799 ± 0.171**
ResUNet - T1-Gd, FLAIR	0.723 ± 0.218	0.884 ± 0.070	0.790 ± 0.184
ContextNet - all	**0.752 ± 0.207**	**0.897 ± 0.059**	0.797 ± 0.195
ContextNet - T1-Gd, FLAIR, T1	0.743 ± 0.216	0.881 ± 0.071	0.770 ± 0.211
ContextNet - T1-Gd, FLAIR	0.734 ± 0.231	0.878 ± 0.080	0.770 ± 0.216

We show in Table 2 a comparison of the performance (according to the DICE score) of the different representation level variations of ContextNet, as detailed in Sect. 2.3. We include the scores of ResUNet trained with all modalities, as it is the most valid non-GPC model to be used for comparison. Both ContextNet models with reduced number of representation levels match or even surpass the performance of the ResUNet model at whole tumor and enhancing tumor segmentation. We hypothesize that GPC modules enable the aggregation of contextual information without the need of obtaining a deep representation via several pooling operations, which permits the proper segmentation of large structures such as the whole tumor. At the same time, by reducing the number of coarse features associated with increased representation levels the network can focus on fine details, such as the enhancing tumor. Furthermore, a reduction of representation levels entails a reduction of the number of trainable parameters, as can be seen in Table 2. The downside is that the depth of the network is severely reduced, and much more complex structures such as the necrotic and non-enhancing tissue are harder to segment for such networks.

Table 2. DICE coefficients (avg ± std) of baseline ResUNet and variations of ContextNet with different number of representation levels. All MRI modalities are used to train these networks. Scores are computed on the local validation set.

	Enhancing tumor	Whole tumor	Tumor core	# parameters
ResUNet	0.739 ± 0.207	0.892 ± 0.064	0.785 ± 0.200	2 M
ContextNet	**0.752 ± 0.207**	**0.897 ± 0.059**	**0.797 ± 0.195**	1.7 M
ContextNet - 3 RL	0.746 ± 0.217	0.894 ± 0.062	0.763 ± 0.225	1.6 M
ContextNet - 2 RL	0.751 ± 0.208	0.892 ± 0.064	0.750 ± 0.229	1.3 M

Figure 2 depicts the feature maps extracted from the residual layers and GPC modules on the ContextNet models with reduced representation levels, as explained in Sect. 2.3. We show only the first 4 feature maps with the highest mean absolute value at each representation level and stage, and we organize the information in a grid in which each row corresponds to a representation level and each column corresponds to a module type in the network (pre-GPC residual layer, GPC, and post-GPC residual layer).

Overall it is clear that the deeper the representation level, the coarser the features the network is able to extract. It can also be seen that, as we move from the pre-GPC residual layer to the GPC module and then the post-GPC residual layer, the features that the network extracts are increasingly abstract, even at the first representation levels. For instance, on the first level of both models the pre-GPC residual layer is enhancing fine-details of the images, such as the enhancing tumor ring; then the GPC module aggregates contextual information and captures global features such as the healthy part of the brain (the activation around whole tumor region is close to 0); finally the post-GPC residual layer combines the global features extracted by the GPC modules and refines their boundaries.

Predictions obtained from ResUNet, ContextNet constrained to three representation levels and full ContextNet are shown in Fig. 3. On one hand, the

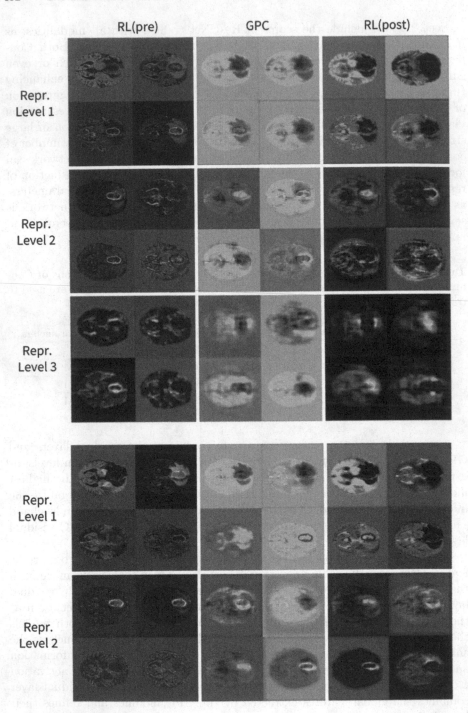

Fig. 2. Feature map visualization for GPC module interpretability. Top figure shows the activations of the ContextNet variant with 3 representation levels, while the bottom figure shows the activations of the ContextNet variant with 2 representation levels.

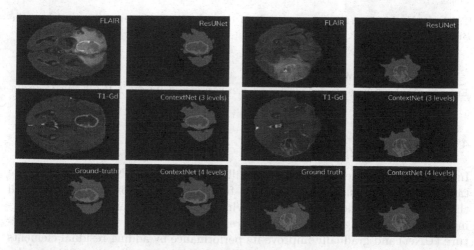

Fig. 3. From top to bottom, left to right: FLAIR and T1-Gd MR modalities, ground-truth labels and segmentations produced by ResUNet, ContextNet trained with 3 representation levels and ContextNet with all (4) representation levels of 2 subjects from the local validation split of the BraTS 2018 dataset.

ContextNet variant with three representation levels is shown to perform similarly to other networks on the first subject. On the other hand, it can be noticed that the depth reduction in this ContextNet variant affects the segmentation of the necrotic tissue on the second subject, in which both the full ContextNet and ResUnet are able to produce a more refined representation than the limited representation-level ContextNet.

Finally we evaluate the best performing ContextNet and ResUNet models and an ensemble of these two models on the BraTS 2018 validation data. As discussed in [9], model ensembling yields more robust segmentation maps by reducing the influence of the hyper-parameters and configurations of individual models. Specifically, we compute the average confidence score per class for each voxel across the models in the ensemble, and we obtain the final segmentation by assigning to each voxel the class with the highest average confidence score. As a consequence of model ensembling, we observe improved DICE scores and Hausdorff 95% quantile distances in practically all structures (shown in Table 3). Therefore, we submit this model ensemble to the BraTS 2018 challenge [4] and we report the resulting scores on the test set in Table 4.

Table 3. Evaluation scores (avg ± std) obtained on the BraTS 2018 validation set. The evaluated models are ContextNet trained on all MR image modalities, ResUNet trained on T1-Gd, T1 and FLAIR, and an ensemble of both models.

	Dice			Hausdorff 95		
	ET	WT	TC	ET	WT	TC
ResUNet	0.729 ± 0.279	0.882 ± 0.071	0.741 ± 0.256	5.578 ± 11.249	9.896 ± 16.803	9.532 ± 12.407
ContextNet	0.735 ± 0.281	0.883 ± 0.112	0.753 ± 0.269	7.004 ± 13.944	**7.594 ± 12.453**	9.505 ± 11.557
Ensemble	**0.758 ± 0.264**	**0.895 ± 0.07**	**0.774 ± 0.253**	**4.502 ± 8.227**	10.656 ± 19.286	**7.103 ± 7.084**

Table 4. Evaluation scores (avg ± std) of the submitted segmentation model on the BraTS 2018 test set.

	Dice			Hausdorff 95		
	ET	WT	TC	ET	WT	TC
Ensemble	0.694 ± 0.289	0.856 ± 0.147	0.754 ± 0.283	6.872 ± 13.21	9.676 ± 15.947	8.123 ± 12.713

4   Conclusion

In this work, we present several 3D fully-convolutional CNNs to address the task of automatic tumor segmentation from magnetic resonance images, with the objective of accelerating and improving radiotherapy planning and monitoring of patients with gliomas of varied grades. We start with a baseline architecture (UNet) and gradually improve its performance by adding residual elements (ResUNet) and enlarging the receptive field of its components via GPC modules (ContextNet).

We further investigate the behavior of the GPC modules by training networks with a limited number of representation levels and visualizing their intermediate representations, and show that equivalent performance can be achieved using GPC modules even when the number of representation levels (and consequently the depth and number of trainable parameters) of the network is considerably reduced.

Future work includes improving the performance of individual models by means of hyper-parameter optimization, uncertainty estimation via Monte-Carlo Dropout or related techniques, in-depth investigation of intermediate representations and use of other deep learning interpretability methods to better understand the behavior of the proposed GPC modules.

References

1. Bakas, S., et al.: Segmentation labels and radiomic features for the pre-operative scans of the TCGA-GBM collection. Cancer Imaging Archive **286** (2017)
2. Bakas, S., et al.: Segmentation labels and radiomic features for the pre-operative scans of the TCGA-LGG collection. Cancer Imaging Archive **286** (2017)
3. Bakas, S., et al.: Advancing the cancer genome atlas glioma MRI collections with expert segmentation labels and radiomic features. Sci. Data **4**, 170117 (2017)
4. Bakas, S., Reyes, M., et al.: Identifying the Best Machine Learning Algorithms for Brain Tumor Segmentation, Progression Assessment, and Overall Survival Prediction in the BRATS Challenge. arXiv preprint arXiv:1811.02629 (2018)
5. Çiçek, Ö., Abdulkadir, A., Lienkamp, S.S., Brox, T., Ronneberger, O.: 3D U-Net: Learning Dense Volumetric Segmentation from Sparse Annotation. CoRR abs/1606.06650 (2016). http://arxiv.org/abs/1606.06650
6. Clevert, D., Unterthiner, T., Hochreiter, S.: Fast and accurate deep network learning by exponential linear units (elus). CoRR abs/1511.07289 (2015). http://arxiv.org/abs/1511.07289

7. He, K., Zhang, X., Ren, S., Sun, J.: Deep residual learning for image recognition. Arxiv. Org **7**(3), 171–180 (2015). https://doi.org/10.3389/fpsyg.2013.00124, http://arxiv.org/pdf/1512.03385v1.pdf

8. Ioffe, S., Szegedy, C.: Batch normalization: accelerating deep network training by reducing internal covariate shift. CoRR abs/1502.03167 (2015). http://arxiv.org/abs/1502.03167

9. Kamnitsas, K., et al.: Ensembles of multiple models and architectures for robust brain tumour segmentation. CoRR abs/1711.01468 (2017). http://arxiv.org/abs/1711.01468

10. Kingma, D.P., Ba, J.: Adam: a method for stochastic optimization. CoRR abs/1412.6980 (2014). http://arxiv.org/abs/1412.6980

11. Long, J., Shelhamer, E., Darrell, T.: Fully convolutional networks for semantic segmentation. In: Proceedings of the IEEE Conference on Computer Vision and Pattern Recognition, pp. 3431–3440 (2015). https://doi.org/10.1109/CVPR.2015.7298965

12. Menze, B.H., et al.: The multimodal brain tumor image segmentation benchmark (BRATS). IEEE Trans. Med. Imaging **34**(10), 1993–2024 (2015). https://doi.org/10.1109/TMI.2014.2377694. http://ieeexplore.ieee.org/document/6975210/

13. Menze, B.H., et al.: The multimodal brain tumor image segmentation benchmark (brats). IEEE Trans. Med. Imaging **34**(10), 1993 (2015)

14. National Brain Tumor Society: Quick brain tumor facts (2018). http://braintumor.org/brain-tumor-information/brain-tumor-facts/. Accessed 12 June 2018

15. Ostrom, Q.T., et al.: The epidemiology of glioma in adults: a "state of the science" review. Neuro-oncology **16**(7), 896–913 (2014)

16. Peng, C., Zhang, X., Yu, G., Luo, G., Sun, J.: Large kernel matters - improve semantic segmentation by global convolutional network. CoRR abs/1703.02719 (2017). http://arxiv.org/abs/1703.02719

17. Ronneberger, O., Fischer, P., Brox, T.: U-Net: Convolutional Networks for Biomedical Image Segmentation, May 2015. http://arxiv.org/abs/1505.04597

18. Szegedy, C., et al.: Going deeper with convolutions. In: 2015 IEEE Conference on Computer Vision and Pattern Recognition (CVPR), pp. 1–9, June 2015. https://doi.org/10.1109/CVPR.2015.7298594

19. Weiss, E., Hess, C.F.: The impact of gross tumor volume (GTV) and clinical target volume (CTV) definition on the total accuracy in radiotherapy theoretical aspects and practical experiences. Strahlenther Onkol **179**(1), 21–30 (2003)

Automatic Brain Tumor Segmentation and Overall Survival Prediction Using Machine Learning Algorithms

Eric Carver[1,3], Chang Liu[1], Weiwei Zong[1], Zhenzhen Dai[1],
James M. Snyder[2], Joon Lee[1], and Ning Wen[1(✉)]

[1] Department of Radiation Oncology, Henry Ford Health System,
2799 W. Grand Blvd, Detroit, MI 48202, USA
nwen1@hfhs.org
[2] Department of Neurosurgery, Henry Ford Health System,
Detroit, MI 48202, USA
[3] Department of Oncology, Wayne State University, Detroit, MI 48201, USA

Abstract. Purpose: This study was designed to evaluate the ability of a U-net neural net-work to properly identify three regions of a brain tumor and an ELM for the prediction of patient overall survival after gross tumor resection using preoperative MR images.

Methods: 210 GBM patients were used for training, while 66 LGG and GBM patients were used for validation. Multiple preprocessing steps were performed on each patient's data before loading them into the model. The segmentation model consists of three different U-nets, one for each region of interest. These created segmentations were then analyzed by use of common quantitative metrics with respect to physician created contours. Regarding the patient overall survival prediction, 59 high grade glioma patients with gross total resection (GTR) were provided for training. 28 patients with GTR were used to validate the algorithm.

Results: The average [s.d] DSC for the whole tumor, enhanced tumor, and tumor core contours were 0.882 [0.080], 0.712 [0.294], and 0.769 [0.263], respectively. The average [s.d.] Hausdorff distance were 7.09 [11.57], 4.46 [8.32], and 9.57 [14.08], respectively. The average [s.d.] sensitivity for the whole tumor, enhanced tumor, and tumor core contours were 0.887 [0.126], 0.770 [0.245], and 0.750 [0.293], respectively. The average [s.d.] specificity was 0. 993 [0.005], 0.998 [0.003], 0.998 [0.002], respectively. The predictive power of patient overall survival is 0.607 using an extreme learning machine algorithm.

Conclusion: The U-Net model was very effective in determining accurate location of the whole tumor and segmenting the whole tumor, enhancing tumor and tumor core. The most predictive features of patient overall survival are both age and location of the tumor when all 163 validation cases were utilized.

Keywords: Magnetic resonance imaging · Neural network · U-net

---

Eric Carver and Chang Liu—The authors have equal contributions.

© Springer Nature Switzerland AG 2019
A. Crimi et al. (Eds.): BrainLes 2018, LNCS 11384, pp. 406–418, 2019.
https://doi.org/10.1007/978-3-030-11726-9_36

1 Introduction

Clinical care for patients with brain tumors such as glioblastoma (GBM) is ripe for advancement through the use of deep neural networks (DNNs) that augment imaging analysis and real-world clinical utility through automated segmentation and progression-free and overall survival prediction. GBM is the most common primary malignant brain tumor designated as a World Health Organization grade IV tumor with a median overall survival of less than two years with best available treatments [1]. The one- and three-year survival rates are 39.7% and 10.1%, respectively. An estimated 25,000 adults in the US are projected to be diagnosed with a primary malignant brain tumor in 2018 and the majority of those diagnosed with a GBM will not survive through 2020 [2]. Over 90% of GBMs develop in elderly patients without evidence of low grade precursor lesions (primary GBM). Other patients that harbor lower grade gliomas such as astrocytomas, oligodendrogliomas, and their anaplastic variants, will face disease recurrence despite treatment, often with malignant transformation and behavior that can manifest as GBM in astrocytic tumors (secondary GBM) [3]. In addition to aggressive intrinsic biology, primary brain tumors often result in profound functional deficits for patients due to neuroanatomic disruption that has implications for both quality of life and survival. Patients with GBM and other incurable brain tumors represent a group of patients with an unmet need that may benefit from refined prognostic tools and disease sub-classifications that ultimately result in more precise therapeutic interventions. The use of predictive analytics prior to surgical intervention for patients harboring malignant gliomas such as GBM presents such an opportunity.

Patients harboring malignant gliomas are monitored for disease progression with magnetic resonance imaging (MRI) studies of the brain. Initial symptoms of neurologic compromise are often investigated with a CT scan followed by an MRI. This process may include multiple MRI studies prior to surgery and ultimate tissue diagnosis which are used for complex intraoperative neuro-navigation and presurgical functional neuroanatomic investigation. Conventional MR imaging techniques contribute morphological information of the scanned area and they can generate images influenced by different types of tissue parameters including T1-weighted images (T1WI), T2-weighted images (T2WI) and fluid attenuation inversion recovery (FLAIR). The multimodal MRI sequences allows for complementary images to come together to describe the underlying heterogeneous anatomic and pathologic processes, including necrosis, enhancing and non-ET core and peritumoral edema. Diffusion and perfusion weighted images can be used to assess pathophysiological processes noninvasively. Radiomic features extracted from these MRI images that fuel precision medicine and contribute to multi-parametric knowledge networks such as those anchored in the genomic and molecular underpinnings of disease are refining our understanding of health and disease resulting to a seismic shift in healthcare delivery [4].

We participated in the 2018 International MICCAI BraTS challenge to segment heterogeneously-enhancing tumors, necrotic lesions and surrounding edema in preoperative MRI scans; and to predict overall survival from these scans. [12] It is imperative that neuroanatomic data is accurately reported to aid in therapeutic interventions such as surgical navigation. This process may be enhanced through DNNs linked to patient outcome that aid clinicians in therapeutic decision making. Over the last few years, DNNs have shown potential to aid physicians in tumor delineation and outcome prediction. The aim of this study was to employ a DNN to use automated image segmentation coupled with real world physician and scientific knowledge to predict survival outcomes in patients anticipated to undergo gross total resection (GTR) of a GBM.

2   Methods

2.1   Patient Population

All data was provided by the BraTS multimodal Brain Tumor Segmentation Challenge 2018. A total of 210 GBM patients were used during training, while 66 patients presenting with GBM and LGG were used for validation. Four conventional MRI sequences were provided for each patient case consisting of: T1WI, Gadolinium enhanced (T1CE), T2WI, and FLAIR. The MRI images were generated at 19 institutions utilizing multiple device manufacturers and clinical protocols to ensure variety of data. One to four raters, following the same annotation protocol, segmented enhancing tumor, peritumoral edema, and necrosis (non-enhancing tumor core) on each MRI sequence for every patient [5–8]. The annotations were approved by expert neuro-radiologists.

2.2   Image Preprocessing

All MRI modalities (T1, T2, FLAIR) were rigidly registered with the T1CE and resampled to a spatial resolution of $1 \times 1 \times 1$ mm^3. In addition to the registration and resampling, skull stripping and normalization were performed. Sixty-four slices image patches covering the enhanced tumor and edema were extracted from original 3D volumes ($240 \times 240 \times 155$) for each patient in order to speed up training/validation/testing and eliminate non-enhancing tumors and regions without surrounding edema. To create an image size conducive to the 2×2 max-pooling utilized by the U-net, zero margins were added to pad each 2D image to have image sizes of 256×256. Considering the brain exhibits marked symmetry across the sagittal plane, each slice was flipped left/right to decrease the dependence on location and increase the variability of the available dataset using data augmentation. In the validation phase, all 155 slices of each 3D volume went through the pipeline for segmentation.

2.3   Neural Network Architecture

Three U-nets were used in the study. One U-net was trained for each of the segmented areas in the MRI images: whole tumor (WT), enhancing tumor (ET), and tumor core (TC). Four scaling layers were used following Pelt *et al.'s* [9] study as the accuracy was improved by using a U-net with 3 or 4 scaling layers. The number of features doubled for each consecutive convolutional layer and halved for each consecutive up-sampling layer as suggested by Ronneberger *et al.* [10]. The merged layer was comprised of features from the up-sampling layers and the convolutional layers. The convolutional layers utilized batch normalization, which was placed right after every merged layer, as well as two more convolutional layers at the end of the model. [11] Soft Dice was utilized as the loss function and is explained by formula 1 with constant value chose to avoid division-by-zero singularities (Fig. 1).

$$\text{Dice Loss} = \frac{2 * \langle y_{true}, y_{pred} \rangle + c}{\langle y_{true}, y_{true} \rangle + \langle y_{pred}, y_{pred} \rangle + c},$$ (1)

$$\bar{y}_{true} = Clinical\,Contour, \bar{y}_{pred} = Model\,Contour\; c = 0.01\; threshold = 0.5$$

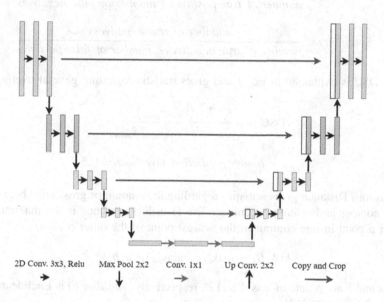

2D Conv. 3x3, Relu   Max Pool 2x2   Conv. 1x1   Up Conv. 2x2   Copy and Crop

Fig. 1. U-Net Model Architecture. (Blue: Res. 256 × 256, 64 Features; Red: Res 128 × 128, 128 Features; Orange: Res. 64 × 64, 256 Features; Yellow: Res 32 × 32, 512 Features) (Color figure online)

All four modalities were used for segmenting WT, TC, and ET. The U-net was run three different times, each of which had one type of the reference contours as the input. A corresponding mask was generated as the output for each input structure. In this way, a WT model, an ET model, and a tumor center model were trained respectively. The combination of these created masks during post-processing is crucial to the success of the study. It was surmised that these masks should be combined in a way that best mirrors the physicians contouring method, specifically for the enhancing tumor. This required that the intensity difference between T1CE and T1WI was calculated on a pixel level in each slice. The WT contour was generated by summing all three masks of each model. The TC contour was created by the combining the masks of the ET and TC models. The ET contour was the direct output of the ET model. The best model for each type of contour was chosen according to the validation loss within 100 epochs run on GPU (Titan XP, nVidia, Santa Clara, CA).

2.4   Evaluation Matrices

The metrics used to evaluate the agreement between the created U-net model and the given reference contours are the Dice similarity coefficient (DSC), Hausdorff distance (HD), sensitivity, and specificity. [12] Sensitivity is also known as a true positive. It is explained by Eq. 2. Specificity is also known as a true negative and explained by Eq. 3.

$$Sensitivity = \frac{number\ of\ true\ positives}{number\ of\ true\ positives + number\ of\ false\ negatives} \tag{2}$$

$$Specificity = \frac{number\ of\ true\ negativess}{number\ of\ true\ negatives + number\ of\ false\ postivies} \tag{3}$$

The DSC is explained in Eq. 4 and gives statistic regarding general overlap:

$$DSC = \frac{2 * \langle y_{true}, \bar{y}_{pred} \rangle}{\langle y_{true}, y_{true} \rangle + \langle \bar{y}_{pred}, \bar{y}_{pred} \rangle}, \tag{4}$$

$$\bar{y}_{pred} = binary\ prediction, threshold = 0.5$$

Hausdorff Distance gives statistics regarding the amount of gross error between the created contour and reference contour. The Hausdorff distance is the maximum distance of a point in one contour to the nearest point of the other contour:

$$h(A, B) = \max_{a \in A} \{\min_{b \in B} \{d(a, b)\}\} \tag{5}$$

where a and b are points of sets A and B, respectively, and $d(a, b)$ is Euclidean metric between these points [13].

2.5   Overall Survival Prediction

One hundred sixty-three high grade glioma (HGG) patients, 59 out of whom underwent GTR, were provided for training. Twenty-eight standalone HGG patients with GTR were used to validate the algorithm. Given the scale of training data, we proposed to represent patient condition with hand crafted features that utilized and integrated information about pre-operative MRI images, tumor and its sub-region segmentation masks, and patients' age. Those features were then fed into a "shallow" regression algorithm inspired by ELM [14].

ELM was proposed as a simplified substitute for back-propagation (BP) to train a single hidden layer feedforward neural network. In contrast to other popular regression algorithms, such as BP based neural network, support vector machine (SVM), solution of ELM was analytically computed instead of iteratively tuned, which makes its implementation outstandingly efficient. Kernelized version of ELM was used in our work for its stability. To be specific, an RBF kernel $(K(\mathbf{u}, \mathbf{v}) = \exp\left(-\gamma\|\mathbf{u} - \mathbf{v}\|^2\right)$ with kernel parameter γ to be tuned) was used to transform the data and then parameters of the model were computed as solution to an optimization problem (Eq. 6) that aims to minimize the training errors as well as the L-2 norm of the parameters to prevent overfitting.

$$Minimize : L_{ELM} = \frac{1}{2}\|\beta\|^2 + C\frac{1}{2}\sum_{i=1}^{N} \epsilon_i^2 \tag{6}$$

$$Subject\ to : h(x_i)\beta = t_i - \epsilon_i, \quad i = 1, \ldots, N,$$

where N is number of samples, β is the weight connecting hidden layer and output layer to be analytically computed, $h(x_i)$ is the hidden layer mapping of input x_i, ϵ_i is the error from predicted output $h(x_i)\beta$ and targeted output t_i. Here kernel is defined as multiplication of two mapping functions: $K(x_i, x_j) = h(x_i) \cdot h(x_j)$.

3 Results

3.1 The Validation of the Models

The model had different levels of success depending on the tumor area it was locating as well as tumor grade. Results show that the WT was most successfully segmented, specifically on GBM patients. This makes sense as GBM presents with increased tumor size and higher contrast with surrounding tissues. In general, the U-net performed worst when segmenting the ET in low grade gliomas. It is believed that the relatively diminutive size of the ET created problems in segmentation. Table 1 summarizes the Dice similarity coefficient, sensitivity, specificity and Hausdorff distances of the validation cases.

Table 1. The evaluation of the segmentation accuracy for 66 validation cases. WT had best results for DSC and Sens. ET had best HD and TC had best Spec.

	DSC	Sensitivity	Specificity	HD (mm)
ET	0.71 ± 0.29	0.77 ± 0.25	0.998 ± 0.003	4.46 ± 8.32
WT	0.88 ± 0.08	0.89 ± 0.13	0.993 ± 0.005	7.09 ± 11.57
TC	0.77 ± 0.26	0.75 ± 0.29	0.998 ± 0.002	9.57 ± 14.08

3.2 Overall Survival Prediction

The volume of the WT, TC, and enhanced tumor was calculated from ground truth masks. In addition to these volumes, the center-point of 3-D WT was found. Because in current stage segmentation of non-enhanced and enhanced TC isn't desirably accurate, information involving them was avoided in our work. Some experimental trials validated our hypothesis by showing that adding such information only deteriorated the performance.

Finally, our feature variables include, age, volume ratio of WT and brain, and 3-D coordinate of the center point of WT. Note that, among 163 HGG patients for training, only 59 underwent GTR with 104 other patients' resection status either partial or unknown. Although resection status played an important role in patients' outcome, we performed the experiment with merely GTR cases and compared that with all training data and found out that extra data can improve the performance to some extent (Table 2).

Table 2. Accuracy of overall survival prediction on standalone 28 validation patients. As shown in this table, single variable alone has very limited predicting power. And the best result was obtained when we used all 163 cases and considered both age and location of the tumor.

Accuracy	Age	Ratio of WT/Brain	WT Loc.	Age + WT/Brain	Age + Loc.	ALL
59 GTR	0.321	0.321	0.321	0.500	0.500	0.536
163 Case	0.321	0.25	0.393	0.571	0.607	0.357

4 Discussion

There have been significant research breakthroughs in deep machine learning in recent years, which has the potential to transform healthcare. Perhaps no area of healthcare is in greater need of rapid progress than malignant brain tumors which also harbor great potential for advancement through computational science owing to the wealth of available radiomics data provided by lifetime serial multimodality imaging and the existing genomic compendium provided by the cancer genome atlas and other investigations. The architecture of deep neural networks offers a powerful framework for multimodality image analysis and processing.

We have presented a deep U-net architecture to correctly detect and segment the following areas of brain lesions: ET, TC, and WT. Study specific modifications were employed in an attempt to delineate regions more accurately. The best and worst results are explored in the discussion, as well as ideas for potential future work.

4.1 Modifications

One such improvement on the basic U-net is to change the ET, WT, and TC contour definitions during post processing. Currently, the WT contour created by this model is defined as the sum of three masks created by the three neural networks; the ET contour created is the sum of the TC mask and the ET mask, and the TC contour is created by the TC mask. In an effort to improve the results, the definition of each created contour was modified. The purpose of this modification was to exclude any pixels in the contour that was outside of that specific mask. For example, the WT contour was originally created from the sum of the three masks (WT, ET, TC) if any of the other masks were outside of the WT mask, that mask's input in that area would be disregarded.

In this U-net, the input of each model consisted of all four MRI modalities in addition to just one contour, so that WT, ET, and TC each had their own specific U-net.

Further investigation is warranted to derive the optimal combination and weighting factor of each sequence. For example, FLAIR is the best modality to segment non-ET and edema, but FLAIR is a non-specific sequence and the data set did not provide enough data for the model to benefit from the FLAIR sequences. Ideally, the weighted values of each imaging sequence as determined by the context of the expected information from that sequence (i.e. enhancement to define necrotic tumor margin, FLAIR hyperintensity with T1 hypo-intensity represent non-ET) force the model to use the proper MRI modality for finding each GBM structural deviation structure in the context of the remaining MRI sequences and overlapping anatomy. This could potentially aid the model to overcome a smaller training dataset.

4.2 Encouraging Results

Both patients in Fig. 2 presented with GBM which is easily differentiated into WT, ET, and TC segmented areas. These patients benefitted from proper contrast agent uptake during the T1CE MRI. This level of success was achieved because the tumor was farther away from the midline of the brain which decreased the structural similarity, high contrast with surrounding healthy tissue, and similarity to many of the training images.

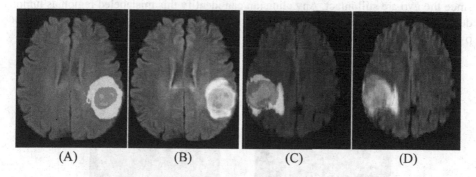

(A)   (B)   (C)   (D)

Fig. 2. Example of two good results from the validation set. (Green = WT, Blue = TC, Red = ET) B, D are shown to show correct identification of lesion. (FLAIR) (Color figure online)

4.3 Poor Results

The patient in Fig. 3 had ET, WT, and TC DSC values of 0, 0.884, 0.001, respectively. This patient presents with a large T2 lesion and limited enhancement (represents necrotic core) or absence of a cystic interior which does not indicate GBM. The poor results for this patient most likely came from the low ratio of ET and TC to WT as well as some artifacts in the image series. Currently methods are being investigated to utilize low grade gliomas during training to solve this issue

The patient in Fig. 4 had ET, WT, and TC DSC values of 0, 0.45863, and 0.45863, respectively. Encouragingly, this was the only patient to present with a WT DSC value less than 0.66. While (B) in Fig. 4 shows the left frontal lobe T2 hyperintensity shows agreement with physician review of segmentation, the image segmentation on the

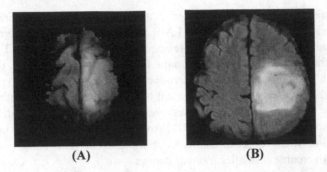

(A)   **(B)**

Fig. 3. Example patient presenting with poor results from the validation set. The contour is the blue outline for ease of viewing true tumor location. A show image slices with poor results, while B show image slices with encouraging results.

image "A" shows segmented tumor mask in the right superior-posterior frontal lobe raised concern with physician review. This case is interesting as the tumor shown in slice "A" maintains structural integrity as it is not filling. Naturally the U-net looks at structural integrity when defining a region, and this region was mislabeled as healthy since the gyri are still intact. Any clinician can identify this mislabeled region as tumor due to the pixel values. Future work will investigate placing priority of pixel values over structural similarities.

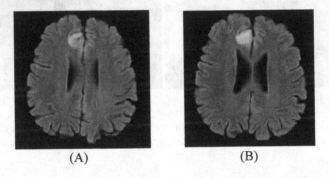

(A)   (B)

Fig. 4. Example patient presenting with poor results from the validation set. The contour is the blue outline for ease of viewing true tumor location. "A" shows an image slice with poor results, while "B" shows image slice with encouraging results.

In an attempt to fix this patient's problem, normalizing the pixel values per slice versus volumetric normalization was also investigated. It was found that volumetric normalization increases the DSC by about 0.8% while also increasing specificity. However, sensitivity increases when slice normalization is employed. Figure 5 "A & B" shows a specific case where volumetric normalization missed the target although the size of the false positive is also significantly smaller. It is also worth noting that when volumetric normalization was used performance decreased as it did not segment any part of the tumor in the slice discussed earlier as shown in Fig. 5 "C & D".

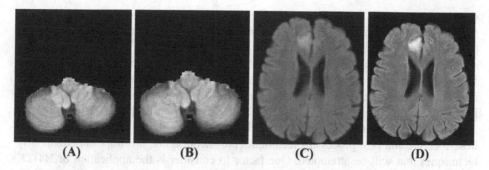

Fig. 5. Slice normalization(A, C) v Volumetric normalization (B, D)

4.4   Segmentation Study Comparison

It is important to compare these results with other methods, specifically the methods that also employed the U-net architecture. This architecture was created in 2015 by researchers at the University of Freiburg, Germany. Several competitors in BraTS have attempted to improve the basic architecture by making modifications. Comparison of our results from our modified U-net to some of the more successful papers in the BraTS preceding conference papers from 2017 and 2018 is necessary to discuss our contribution to this field. These results are compared in Table 3.

Table 3. Preliminary DSC Results from Similar Studies in BraTS

Study	ET	WT	TC
Our study	0.71	0.88	0.77
Varghese	0.69	0.83	0.69
Amorim	0.83	0.91	0.91
Beers	0.68	0.78	0.67
Feng	0.75	0.90	0.80
Rodriguez	0.47	0.82	0.57
Baid	0.75	0.88	0.83

Varghese et al. utilized a fully connected 2D U-net. This decreased computational time compared to patch-based techniques as it employs a single forward pass to classify pixels slice by slice. Z-score normalization was done for preprocessing. [15] Utilizing a 3D U-net allows for there to be features utilized along the z axis; however, it increases computational time. Amorim et al. studied the effects of utilizing 3D with $64 \times 64 \times 64$ patches. Similarities to our study's method is that one U-net was trained for each mask. Pre-processing included histogram equalization for T1 and T2 images, while standard scaling and normalization were performed on T1CE and FLAIR. [16] Beers et al. proposed the idea of employing sequential 3D U-nets. The idea centered on fully utilizing the GPU. This method was complex as the stacking of U-nets is complicated [17] Feng et al. utilized a patch-wise three-dimensional U-net. Class labels were generated by a sliding window approach. Pre-processing steps included normalization and patch (64 × 64 × 64) extraction. This patch extraction differs from conventional methods as there are many patches per image and each patch is input into the U-net. [18] Rodriquez et al. applied two 3D U-nets in a unique method. The first U-net found the WT. The found WT was then the input for the second U-net, which segmented the tumor into WT, ET, and TC. respectively. [19] Baid et al. utilized a 3D U-net with 64 × 64 × 64 patches. Preprocessing included normalization and N4ITK correction [20].

Although the results that our study achieved from utilizing three 2D U-nets with described pre and post processing techniques are encouraging; it is important to identify techniques that will be attempted. One factor to consider is the application of N4ITK correction during pre-processing. Another possible change to make is to utilize a 3D U-net. We preferred to utilize a 2D U-net instead of a 64 × 64 × 64 patch 3D U-net as the 3D U-net requires eithers down-sampling of the data and/or processing of the whole volume at multiple locations. The former potentially loses spatial information while the latter introduces complexity. The added complexity necessitates a theoretical computational time around 39 times longer than the 2D U-net as there is a need to process the image at multiple locations to cover the 240 × 240 × 155 region with a 64 × 64 × 64 patch without overlap. There must be further study performed to compare our 2D U-net with a 64 × 64 × 64 3D U-net with regards to finding specific training time differences, complexity, and size of data. We argue that by using a 2D patch of 256 × 256 we can quickly apply the segmentation slice by slice in a competitive manner.

4.5   Overall Survival Study Comparison

For years, leading consortiums in neuro-oncology and radiomics have attempted to provide guidelines for multimodality imaging to achieve standardized interpretation and data uniformity for clinical trials [21]. The use of DNNs to aid in the radiomic analysis of malignant brain tumors harbors tremendous opportunity to improve care. Radiomic tools that accurately predict patient survival or aid the clinical team in patient care and disease stratification are of great value, especially when used with other modalities that refine our understanding of the intrinsic biology of disease such as malignant brain tumors. Furthermore, use of a predictive tool in the presurgical navigation of brain tumor care has the potential to optimize resources, initiate precision medicine claims, and provide a framework for clinical evidence generation using real time data, which some argue is the near future of technology-enhanced healthcare delivery. In this study, we investigated the prediction power of overall survival using a DNN. However, the accuracy was very low due to the limited patient size and MR image sequences. Our study showed improved results using only a handful features such as age, ratio of WT to the brain and tumor location. We reviewed the results from other groups and noticed the similar findings.

Varghese et al. used features such as: age of patient, ratio of number of voxels of edema to number of voxels of lesion, ration of number of voxels of necrosis to number of voxels of lesion, and etc. These features were used as input to support vector machine utilizing a linear kernel. Preliminary reported accuracy is 60%. [15] It is very similar to our method, except that we avoided using segmentation of necrosis and enhanced TC due to its inaccuracy. We also added extra information about tumor location, which was shown to be important. Amorim et al. used a total of 120 features found from morphological and statistical features of the tumor regions and a XGBboost classifier. Preliminary reported accuracy is 48%. [16] Feng et al. used a linear regression model with features such as sum of the voxels, surface area, age, and resection status. Preliminary reported accuracy is 32%. [18] Baid et al. found 468 features for edema, TC, and ET by use of first order statistics, shape features, gray level co-occurrence matrix, and gray level run length matrix. This study achieved higher accuracy with multilayer perceptron than random forest. Preliminary reported accuracy is 57% [20].

Patients are evaluated with serial MRIs throughout their disease course often at short intervals of 2–3 months for the first several years. These MRI studies utilize a compendium of sequences described above in a multiparametric format to provide qualitative and quantitative information about the tumor and resultant neuroanatomic disruptions. The changes of radiomics features extracted from these serial MRIs may have better predictive power. Our next step is to increase the patient sample size with longitudinal MR scans and investigate the predictive power using delta radiomics.

5 Conclusion

U-net, a deep neural network, was used to automatically segment the brain into WT, TC, and ET. The difference one will expect to see when using volumetric versus slice normalization during pre-processing was investigated and discussed. These created segmentations were then analyzed by use of quantitative metrics (DSC, HD, sensitivity, and specificity) with respect to physician created contours. The results of these statistics were encouraging, especially for finding the WT. It also shows that if a 2D U-net can be competitive with many 3D U-nets if one implements a methodology during post-processing that mirrors the method that physicians utilize. Future study is still needed to investigate the impact of different scanners, protocols, sequences, and dataset volumes on the performance of segmentation. The opportunity to impact clinical care through refined radiomic tools such as this described application of a deep neural network for overall patient survival prediction after gross tumor resection is of particular interest in the management of GBM and other brain tumors which rely heavily on medical imaging.

Acknowledgement. This study was supported by a Research Scholar Grant, RSG-15-137-01-CCE from the American Cancer Society.

References

1. Stupp, R., Taillibert, S., Kanner, A., et al.: Effect of tumor-treating fields plus maintenance temozolomide vs maintenance temozolomide alone on survival in patients with glioblastoma: a randomized clinical trial. JAMA **318**, 2306–2316 (2017)
2. Ostrom, Q.T., Gittleman, H., Liao, P., et al.: Cbtrus statistical report: primary brain and other central nervous system tumors diagnosed in the united states in 2010–2014. Neuro Oncol. **19**, v1–v88 (2017)

3. Ohgaki, H., Kleihues, P.: The definition of primary and secondary glioblastoma. Clin. Cancer Res. **19**, 764–772 (2013)
4. Toward precision medicine: Building a knowledge network for biomedical research and a new taxonomy of disease. Washington (DC) (2011)
5. Menze, B.H., Jakab, A., Bauer, S., et al.: The multimodal brain tumor image segmentation benchmark (brats). IEEE Trans. Med. Imaging **34**, 1993–2024 (2015)
6. Bakas, S., Akbari, H., Sotiras, A., et al.: Advancing the cancer genome atlas glioma mri collections with expert segmentation labels and radiomic features. Sci. Data **4**, 170117 (2017)
7. Bakas, S.,et al.: Segmentation labels and radiomic features for the pre-operative scans of the TCGA-GBM collection. In: The Cancer Imaging Archive (2017)
8. Bakas, S., et al.: Segmentation labels and radiomic features for the pre-operative scans of the TCGA-LGG collection. In: The Cancer Imaging Archive (2017)
9. Pelt, D.M., Sethian, J.A.: A mixed-scale dense convolutional neural network for image analysis. Proc. Nat. Acad. Sci. **115**, 254 (2018)
10. Ronneberger, Olaf., Fischer, P., Brox, T.: U-net: convolutional networks for biomedical image segmentation. In: Navab, N., Hornegger, J., Wells, W.M., Frangi, A.F. (eds.) MICCAI 2015. LNCS, vol. 9351, pp. 234–241. Springer, Cham (2015). https://doi.org/10.1007/978-3-319-24574-4_28
11. Keras, C.F.: Keras. Github Repository (2015)
12. Bakas, S., Reyes, M., et al.: Identifying the best machine learning algorithms for brain tumor segmentation, progression assessment, and overall survival prediction in the brats challenge. arXiv, https://arxiv.org/abs/1811.02629 2018
13. Kumarasiri, A., Siddiqui, F., Liu, C., et al.: Deformable image registration based automatic CT-to-CT contour propagation for head and neck adaptive radiotherapy in the routine clinical setting. Med. Phys. **41**, 121712 (2014)
14. Huang, G.B., Zhou, H., Ding, X., Zhang, R.: Extreme learning machine for regression and multiclass classification. IEEE Trans. Syst. Man Cybern. Part B (Cybern.) **42**, 513–529 (2012)
15. Varghese Alex, M.S., Krishnamurthi, G.: Brain tumor segmentation from multi modal MR images using fully convolutional neural network. Spain (2017)
16. Amorim, P.H.A.C., Escudero, G.G., Oliveira, D.D.C., Pereira, S.M., Santos, H.M., Scussel, A.A.: 3D u-nets for brain tumor segmentation in MICCAI. Spain (2017)
17. Beers, A., et al.: Sequential 3D U-nets for brain tumor segmentation. Spain (2017)
18. Feng, X., Meyer, C.: Patch-based 3D U-net for brain tumor segmentation. Spain (2017)
19. Rodríguez Colmeiro, R.G., Verrastro, C.A., Grosges, T.: Multimodal brain tumor segmentation using 3D convolutional networks. In: Crimi, A., Bakas, S., Kuijf, H., Menze, B., Reyes, M. (eds.) BrainLes 2017. LNCS, vol. 10670, pp. 226–240. Springer, Cham (2018). https://doi.org/10.1007/978-3-319-75238-9_20
20. Baid, U., et al.: Gbm segmentation with 3D U-net and survivalprediction with radiomics. Spain (2018)
21. Wen, P.Y., Macdonald, D.R., Reardon, D.A., et al.: Updated response assessment criteria for high-grade gliomas: response assessment in neuro-oncology working group. J. Clin. Oncol. **28**, 1963–1972 (2010)

Deep Hourglass for Brain Tumor Segmentation

Eze Benson[1(✉)], Michael P. Pound[1], Andrew P. French[1,2],
Aaron S. Jackson[1], and Tony P. Pridmore[1]

[1] School of Computer Science, University of Nottingham, Nottingham, UK
{ezenwoko.benson,michael.pound,andrew.p.french,
aaron.jackson,tony.pridmore}@nottingham.ac.uk
[2] School of Biosciences, University of Nottingham, Nottingham, UK

Abstract. The segmentation of a brain tumour in an MRI scan is a challenging task, in this paper we present our results for this problem via the BraTS 2018 challenge, consisting of 210 high grade glioma (HGG) and 75 low grade glioma (LGG) volumes for training. We train and evaluate a convolutional neural network (CNN) encoder-decoder network based on a singular hourglass structure. The hourglass network is able to classify the whole tumour (WT), enhancing (ET) tumour and core tumour (TC) in one pass. We apply a small amount of preprocessing to the data before feeding it to the network but no post processing. We apply our method to two different unseen sets of volumes containing 66 and 191 volumes. We achieve an overall Dice coefficient of 92% on the training set. On the first unseen set our network achieves Dice coefficients of 0.66, 0.82 and 0.72 for ET, WT and TC. On the second unseen set our network achieves Dice coefficients of 0.62, 0.79 and 0.65 on ET, WT and TC.

Keywords: Convolutional neural network · Deep learning · Hourglass · Glioma

1 Introduction

Identifying regions of the brain which are tumourous is a task often carried out by medical professionals. Manually classifying segments of the tumour is a subset of a group of problems commonly referred to as semantic segmentation. Semantic segmentation is the task of assigning a class to each pixel within an image, modern automated solutions to this problem often use convolutional neural networks (CNN). The introduction of fully convolutional networks (FCN) [1] established a convolutional neural network architecture that is widely used for the task of semantic segmentation. Architectures such as U-NET [2] achieved success in biomedical imaging by adopting a similar architecture.

We propose the use of an adapted hourglass [3] network to solve the problem of tumour segmentation. The hourglass network improves on U-NET by using bottleneck blocks and adding convolutions to the skip connections. Training a CNN for this problem is a natural choice as they have demonstrated state-of-the-art performance on

© Springer Nature Switzerland AG 2019
A. Crimi et al. (Eds.): BrainLes 2018, LNCS 11384, pp. 419–428, 2019.
https://doi.org/10.1007/978-3-030-11726-9_37

semantic segmentation problems such as the widely used Pascal VOC2012 [9] and cityscapes [10] datasets.

2 Methods

2.1 Data

The dataset of BraTS 2018 [4–8] provides defined training and validation sets. The training set is composed of 210 MRI scans of high grade gliomas (HGG) and 75 MRI scans of low grade gliomas (LGG). Whilst the validation set is a group of 66 mixed HGG and LGG tumours. The MRIs are volumes in the format given by Eq. (1), in that format they have the dimensions 240 × 240 × 155. Each volume has four corresponding modalities FLAIR T1, T2 and T1CE.

$$X \times Y \times Z \tag{1}$$

Where x is the delineation between dimensions and X, Y and Z are the dimensions on a 3D coordinate system.

2.2 PreProcessing

A high variance in intensity in both validation and training set was observed this lead us normalise the training set to be centred around zero with a standard deviation of one. By normalizing the data, we found that the required training time was reduced and the accuracy of the network was increased. The formula for normalization is given in Eq. (2). Each modality was normalized separately due to the variance in intensity profile between modalities.

$$Z = \frac{x - \mu}{\sigma} \tag{2}$$

Where x is the current intensity, μ is the mean of the modality and σ is the standard deviation of the modality.

2.3 Hourglass Architecture

Our approach is to handle 2D slices of each volume separately, a 2D semantic segmentation problem. We performed additional experimentation using a volumetric encoder-decoder but found that the benefit of an end-to-end volumetric approach was outweighed by the significant necessary drop in features at each layer due to memory restrictions.

We design our network using an encoder-decoder structure, adapted from an hourglass network, popularized in the domain of human-pose estimation [3] The structure of the hourglass is similar to other encoder-decoder networks, but contains a denser use of residual blocks throughout.

The encoder starts with an input of 4 channels and contains 7 residual bottleneck blocks [14], after each a max-pooling layer performs spatial downsampling. A further three residual blocks at the lowest spatial resolution derive higher-level features before a series of bilinear upsampling operations return the network to the original spatial resolution. As in the encoder, all upsampling operations of the decoder are interleaved with residual blocks. Skip layers are added between each matching resolution of the encoder and decoder, with each containing an additional residual block to learn an appropriate mapping.

The first residual block has 64 filters, the filter amount doubles after each pooling operation up to a maximum of 512 filters in a convolution.

In order to improve the network's results for the final test set we made architectural changes to improve accuracy whilst keeping memory consumption to a minimum. We found that the choice of upsampling layer (e.g. bilinear, max-unpooling [11]) made little difference to the performance of the network. Unlike the original work [3] we chose not to stack hourglass networks sequentially and perform intermediate supervision, we found this too had a negligible effect on performance. The number of spatial-downsampling layers, 7 in total, were originally chosen based on the input resolution. However, through experimentation we found that using 5 downsampling layers was optimal and save memory. Only one residual block is used at each depth because adding two at all depths immediately doubles memory consumption which surpasses current memory constraints. We also found that replacing elementwise summation with concatenation followed by a 1×1 convolution improved results noticeably. Despite the additional memory consumption of the concatenation and convolutional layer, the increase in performance boost makes the change worthwhile.

2.4 Training

The training was split into two phases pre and post true validation set release. In the first phase the dataset was split into a test set, validation set and training set where each set was 10%, 10% and 80% of the original training set respectively. The data provided is treated as though it is the entire dataset so that our training can be validated and tested in preparation for the true validation set. This allows the network to avoid overfitting and approximate the results expected on the release of the second dataset. Later the network is retrained using a 10% test set and 90% training set split in order to obtain test results on the original data whilst maximizing the training set size. The network is trained for the same number of epochs for all training. The second phase is conducted post true validation set release. In this phase the BraTS dataset is split into 10% validation and 90% training.

The network is trained using an identical training scheme for both the natural and augmented dataset.

The hourglass network implemented in this paper only uses spatial convolutions, to accommodate this we convert MR volumes into a set of 155 images of spatial resolution 240^2. To do this we separate the volume along the depth dimension. For convenience we pad the images to the new resolution 256^2, this allows us to perform pooling operations where the output resolution of a feature map is always 2^x. In turn this allows us to perform concatenations or elementwise summations in the decoder network without a resolution difference between two feature maps. The 285 volumes therefore become a dataset of 44175 images. All four modalities are used for training and are given to the network as 4 input channels of a single image.

The hourglass network was chosen because it has been successful in other tasks such as human pose estimation [3] and allows the stacking of the network. Stacking the network multiple times sequentially can give performance boosts as shown before [3]. Spatial convolutions were chosen instead of volumetric convolutions because they consume much less memory, volumetric convolutions would exceed available memory if a stacked network was used. In addition, volumetric convolutions are so memory intensive that they do not allow a network to be trained on the entire MR volume at the same time as having a rich set of filters in a deep network. An alternative to this volumetric network is a volumetric network with a subset of a volume included E.g. A $32 \times 32 \times 32$ chunk. However, the problems remain largely unsolved, the performance boost given by depth context is potentially outweighed by the larger number of filters available in a spatial network. This multitude of reasons led to the choice of a spatial network which would be deeper and wider than the equivalent volumetric network given the same memory constraints.

The hourglass is trained on a NVIDIA TITAN X GPU using a cross entropy loss function with a learning rate of 10^{-5} which is decreased by a factor of 10 every 30 epochs. A batch size of 8 is used and the network is trained for a total of 50 epochs therefore the learning rate is only adapted once. The adaptive gradient descent algorithm, RMSProp is used to train the network faster than the typical stochastic gradient descent.

2.5   Data Augmentation

Two methods of data augmentation are used in this paper vertical flipping and random intensity variation. Vertical flipping is used because it matches the natural symmetrical shape of the brain.

Random intensity variation is used because the intensity between MRI scans varies significantly. This is shown by the fact that the standard deviation of the FLAIR modality in the dataset is greater than the mean by almost a factor of 10. E.g. The standard deviation and mean for the FLAIR modality are 529.2 and 61.8 respectively. The T1, T1CE and T2 modalities have similar standard deviations. Intensity variation is performed on the normalised dataset by first rescaling the standard deviation of the dataset and then shifting the mean. This allows the dataset to include image intensities which are not present in the original dataset but could appear on an MRI volume. The range for randomly changing the standard deviation is between zero and two. The mean is shifted between values of 0.4 and –0.4. Values above a standard deviation of two were experimented with but lead to a significant decrease in accuracy. Shifting the mean by over 0.5 and under –0.5 were trialed but also caused an accuracy decrease. The network is trained with and without data augmentation to experimentally ascertain whether augmentation gives any performance increase when using this network on the dataset.

3 Results and Discussion

The results are split into three sections, the results on the training data set, the results on the later released validation set and the results on the final test set. Results are shown for networks trained on the standard data and on augmented data in the validation set.

3.1 Training Dataset

We trained the network on 90% of the data leaving 10% for testing purposes. The network achieved a Dice coefficient of 92% with an IOU of 86%. We find that IOU approximates the network's worst performance on the test set in contrast to Dice which gives an approximate representation of the average case.

3.2 Validation Dataset

The results presented in this section are those achieved when segmenting the validation set using the network trained in Sect. 3.1. Table 1 shows the results of the segmentation without augmentation and Table 2 shows the results with flipping and intensity variation. The metrics provided in both tables are the standard metrics output by the BraTS automatic online evaluation server. Some metrics have been omitted to save space, only the most important evaluation metrics have been included.

Table 1. The results of the hourglass network segmenting the unseen validation set without augmentation in the training data.

	Dice ET	Dice WT	Dice TC	Hausdorff ET	Hausdorff WT	Hausedorff TC
Mean	0.59	0.82	0.64	18.12	94.28	130.70
Std	0.28	0.12	0.24	26.62	50.15	42.40
Median	0.71	0.86	0.71	5.732	97.13	132.59
25 quantile	0.48	0.78	0.51	3.162	52.72	103.36
75 quantile	0.80	0.90	0.83	20.03	135.81	163.39

Table 2. The results of the hourglass network segmenting unseen validation set where the network has been trained with augmented data

	Dice ET	Dice WT	Dice TC	Hausdorff ET	Hausdorff WT	Hausedorff TC
Mean	0.56	0.82	0.61	14.29	13.57	17.95
Std	0.29	0.13	0.22	23.26	15.32	18.14
Median	0.67	0.87	0.67	5.92	6.59	11.18
25 quantile	0.40	0.78	0.50	2.83	4.18	8.30
75 quantile	0.80	0.90	0.79	12.56	14.97	18.79

After comparing the metrics between a dataset with augmentation and one without we find that in this challenge augmentation appears to give a small increase in accuracy for Dice coefficient and improves the Hausdorff accuracies significantly. It is likely the case that the frequency at which the network misclassifies pixels remains similar but the network's ability to localize the pixels is increased.

Overall the network segments the whole tumour more accurately than it does the core tumour or enhancing tumour, from the results in previous challenges this result is expected. Naturally the enhancing and core tumour are much more difficult to segment due to the similarity between all classes.

Tables 1 and 2 both show a large disparity between the median and mean accuracy especially with results for the enhancing tumour where the difference is around 10%. The difference is caused by the difficulty of detecting the enhancing tumour and core tumour in some volumes. In most volumes the Dice coefficients are well above the mean however some outliers achieve a score of 0 therefore reducing the mean significantly. When removing these cases the mean Dice coefficient increases by 4% showing that the disparity can be explained by a few very difficult volumes. Some examples of the metrics achieved on these volumes are shown in Table 3.

Table 3. Segmentation results for very difficult volumes using a network trained with augmented data

	Dice ET	Dice WT	Dice TC	Hausdorff ET	Hausdorff WT	Hausedorff TC
TCIA09_248_1	0	0.79	0.63	0	14.18	10.82
TCIA10_195_1	0	0.80	0.63	0	15.23	25.98
TCIA11_612_1	0	0.74	0.60	0	52.78	48.52
TCIA12_613_1	0	0.69	0.26	0	49.97	9.00
TCIA13_646_1	0	0.90	0.40	0	35.83	6.48

3.3   Test Dataset

Before the release of the final evaluation dataset we train our network using 95% of the training data. The remaining 5% of the training data is used for on the fly validation of the network to monitor training and prevent overfitting. The network architecture has been adapted to improve the results on the validation set, these architectural changes are discussed in Sect. 2.3. We present the new validation set results along with the test set results. Section 3.2 showed that the network has an increase in Hausdorff95 accuracy when data augmentation was used. The network used for the results in this section was trained using data augmentation.

Table 4. The results of the hourglass network segmenting unseen the validation set

	Dice ET	Dice WT	Dice TC	Hausdorff ET	Hausdorff WT	Hausdorff TC
Mean	0.66	0.82	0.72	15.94	26.41	18.87
Std	0.27	0.10	0.23	25.56	23.61	20.56
Median	0.79	0.84	0.80	4.69	17.32	12.47
25 quantile	0.56	0.78	0.62	2.45	7.19	6.61
75 quantile	0.84	0.89	0.89	17.60	38.13	19.60

Table 5. The results of the hourglass network segmenting unseen test set

Label	Dice ET	Dice WT	Dice TC	Hausdorff ET	Hausdorff WT	Hausdorff TC
Mean	0.62	0.79	0.65	47.48	13.54	31.58
Std	0.32	0.25	0.34	113.76	23.51	83.24
Median	0.77	0.88	0.82	3.00	5.00	6.40
25 quantile	0.47	0.80	0.48	1.73	3.00	3.32
75 quantile	0.85	0.92	0.90	9.84	9.72	14.80

Table 4 shows the results of the hourglass network on the validation set. The dice scores for the validation set increase by 10% for both ET and TC whilst remaining approximately the same for the whole tumour segmentation. Conversely the Hausdorff scores increase (where a higher score is a decrease in performance) by 1, 13 and 2 for ET, WT and TC respectively. The increase in dice score indicates that the total number of pixels that are being classified correctly has increased but the decrease in Hausdorff score shows that the largest error in the shape of the classified pixels is much higher. The qualitative analysis presented in Sect. 3.4 shows that this may be because misclassification of background pixels far away from the site of the tumour.

The median Hausdorff distance and dice score are significantly better than the mean indicating that the mean results are being distorted by a small subset of difficult to segment brain tumour volumes. This is discussed in Sect. 3.2. The std of both metrics is also very high showing that the networks performance varies largely between volumes.

The network shows a significant improvement in the most problematic volumes highlighted in Table 3. Table 6 shows the modified network's performance on the selected examples. The average Hausdorff distance for the selected examples indicates an overall performance decrease however performance on individual volumes varies significantly when dice scores are compared. The network architecture was modified in order to increase performance on the enhanced tumour, Table 6 shows that on 3 out of 5 selected cases there is an increase of between 4.6% and 38% for the enhancing tumour dice score. The variability in dice score amongst the other two metrics indicates that the training scheme has altered the networks ability to classify the tumour in these volumes.

Table 6. The modified network's segmentation results on a subset of problematic volumes

	Dice ET	Dice WT	Dice TC	Hausdorff ET	Hausdorff WT	Hausedorff TC
TCIA09_248_1	0.00	0.80	0.48	0.00	61.26	12.41
TCIA10_195_1	0.00	0.86	0.71	0.00	22.67	30.23
TCIA11_612_1	0.38	0.63	0.40	98.47	59.87	98.25
TCIA12_613_1	0.06	0.94	0.94	58.26	4.12	2.83
TCIA13_646_1	0.05	0.70	0.61	111.19	87.68	15.13

The test set results show a decrease in performance on both dice score and hausdorff distance when compared to validation set results. The median scores for both metrics are noticeably better. This indicates that the validation set contains easier to segment volumes but the ratio between difficult and easy volumes is higher. The test set appears to have much more difficult volumes, this is corroborated by the very high standard deviation values. The results suggest that the percentage of easily segmented volumes in the test set is higher than the validation set.

Despite the differences between the network's performance on the validation and test sets both Tables 4 and 5 indicate the same overall strengths and weaknesses of the network as well as the difficulties within the dataset.

3.4   Qualitative Analysis

In this section we present singular slices taken from the network output. The output has 4 classes which are represented by 4 different colours in the segmentation map. Black, yellow, blue and red represent background, whole tumour, core tumour and enhancing tumour.

The network makes many mistakes when segmenting unseen volumes, most often these errors are misclassifying healthy brain tissues as tumourous. Often the mistakes are of a small area which does not affect the dice score significantly but has a noticeable impact on the hausdorff distance. These errors are important and can be improved upon however for brevity this section will focus on the largest errors associated with the problematic volumes highlighted in Sect. 3.3. Figure 1 shows large errors in classification. The largest errors the network makes occur when the input image has large errors of darkness within the tumour caused by necrosis or an irregular tumour shape. It is unclear why this occurs but could be because the training set contains mostly tumour which have small amounts of necrosis which are masses enveloped by the whole tumour. Therefore when given to the network it is unable to deal with the variance.

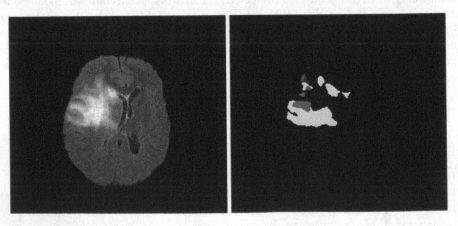

Fig. 1. Left, a FLAIR volume slice containing both brain and tumour tissues. Right, a slice from the network output showing erroneous segmentation results. Two similar looking dark regions on the left side of tumour have been classified different despite having largely the same appearance. These are the most error prone areas for the network.

4 Conclusion

We propose a solution which achieves a 92% Dice coefficient on the training set and 0.66, 0.82 and 0.72 on the validation set. On the test set the network achieves 0.62,0.79 and 0.65 Dice scores. Although the network underperforms on Dice score it can achieve a competitive Hausdorff distance.

Much of the network's underperformance is related to outliers in the set which could be mitigated in future with better preprocessing techniques. Future networks should train more on these difficult volumes using wider public datasets or through synthetic images generated by a CNN. Memory consumption is often a problem when using CNNs, to combat this we plan to add residual blocks in depths which increase the overall accuracy of the network the most. We also plan to add skip connections with an inception block structure [12] as shown in [13] to increase accuracy further.

We show that 2D architectures can segment 3D volumes with success but require fine tuning and a deeper architecture to achieve better results. An approach to bridge the gap may between 2D and 3D may be required. 3D networks outperform 2D networks when depth context is key, how much context is required in most tasks remains unclear. In future works we plan to use a 2.5D approach where each slice has an accompanying adjacent slice either side to provide some depth context.

References

1. Long, J., Shelhamer, E., Darrell, T.: Fully convolutional networks for semantic segmentation. In: Proceedings of the IEEE Conference on Computer Vision and Pattern Recognition, pp. 3431–3440 (2015)
2. Ronneberger, O., Fischer, P., Brox, T.: U-Net: convolutional networks for biomedical image segmentation. In: Navab, N., Hornegger, J., Wells, W.M., Frangi, A.F. (eds.) MICCAI 2015. LNCS, vol. 9351, pp. 234–241. Springer, Cham (2015). https://doi.org/10.1007/978-3-319-24574-4_28
3. Newell, A., Yang, K., Deng, J.: Stacked hourglass networks for human pose estimation. In: Leibe, B., Matas, J., Sebe, N., Welling, M. (eds.) ECCV 2016. LNCS, vol. 9912, pp. 483–499. Springer, Cham (2016). https://doi.org/10.1007/978-3-319-46484-8_29
4. Menze, B.H., et al.: The multimodal brain tumor image segmentation benchmark (BRATS). IEEE Trans. Med. Imaging 34(10), 1993 (2015)
5. Bakas, S., et al.: Advancing the cancer genome atlas glioma MRI collections with expert segmentation labels and radiomic features. Sci. data. 5(4), 170117 (2017)
6. Bakas, et al.: Segmentation labels and radiomic features for the pre-operative scans of the TCGA-GBM collection. The Cancer Imaging Archive, p. 286 (2017)
7. Bakas, S., et al.: Segmentation labels and radiomic features for the pre-operative scans of the TCGA-LGG collection. The Cancer Imaging Archive, p. 286 (2017)
8. Bakas, S., Reyes, M., Menze, B.: Identifying the Best Machine Learning Algorithms for Brain Tumor Segmentation, Progression Assessment, and Overall Survival Prediction in the BRATS Challenge. arXiv preprint arXiv:1811.02629 (2018)
9. Everingham, M., Van Gool, L., Williams, C.K., Winn, J., Zisserman, A.: The pascal visual object classes (voc) challenge. Int. J. Comput. Vis. 88(2), 303–338 (2010)

10. Cordts, M., et al.: The cityscapes dataset for semantic urban scene understanding. In: Proceedings of the IEEE Conference on Computer Vision and Pattern Recognition, pp. 3213–3223 (2016)
11. Badrinarayanan, V., Kendall, A., Cipolla, R.: Segnet: a deep convolutional encoder-decoder architecture for image segmentation. arXiv preprint arXiv:1511.00561, 2 November 2015
12. Szegedy, C., et al.: Going deeper with convolutions. In: Proceedings of the IEEE Conference on Computer Vision and Pattern Recognition, pp. 1–9 (2015)
13. Bulat, A., Tzimiropoulos, G.: Binarized convolutional landmark localizers for human pose estimation and face alignment with limited resources. In: The IEEE International Conference on Computer Vision (ICCV), vol. 1, no. 2, p. 4, 1 October 2017
14. He, K., Zhang, X., Ren, S., Sun, J.: Deep residual learning for image recognition. In: Proceedings of the IEEE Conference on Computer Vision and Pattern Recognition, pp. 770–778 (2016)

Deep Learning Versus Classical Regression for Brain Tumor Patient Survival Prediction

Yannick Suter[1]([⊠]), Alain Jungo[1], Michael Rebsamen[1], Urspeter Knecht[1], Evelyn Herrmann[2], Roland Wiest[3], and Mauricio Reyes[1]

[1] Institute for Surgical Technology and Biomechanics,
University of Bern, Bern, Switzerland
yannick.suter@istb.unibe.ch
[2] University Clinic for Radio-Oncology, Inselspital, Bern University Hospital,
University of Bern, Bern, Switzerland
[3] Support Center for Advanced Neuroimaging,
University Institute of Diagnostic and Interventional Neuroradiology, Inselspital,
University of Bern, Bern, Switzerland

Abstract. Deep learning for regression tasks on medical imaging data has shown promising results. However, compared to other approaches, their power is strongly linked to the dataset size. In this study, we evaluate 3D-convolutional neural networks (CNNs) and classical regression methods with hand-crafted features for survival time regression of patients with high-grade brain tumors. The tested CNNs for regression showed promising but unstable results. The best performing deep learning approach reached an accuracy of 51.5% on held-out samples of the training set. All tested deep learning experiments were outperformed by a Support Vector Classifier (SVC) using 30 radiomic features. The investigated features included intensity, shape, location and deep features.

The submitted method to the BraTS 2018 survival prediction challenge is an ensemble of SVCs, which reached a cross-validated accuracy of 72.2% on the BraTS 2018 training set, 57.1% on the validation set, and 42.9% on the testing set.

The results suggest that more training data is necessary for a stable performance of a CNN model for direct regression from magnetic resonance images, and that non-imaging clinical patient information is crucial along with imaging information.

Keywords: Brain tumor · Survival prediction · Regression ·
3D-Convolutional Neural Networks

1 Introduction

High-grade gliomas are the most frequent primary brain tumors in humans. Due to their rapid growth and infiltrative nature, the prognosis for patients with

© Springer Nature Switzerland AG 2019
A. Crimi et al. (Eds.): BrainLes 2018, LNCS 11384, pp. 429–440, 2019.
https://doi.org/10.1007/978-3-030-11726-9_38

gliomas ranking at grade III or IV on the Word Health Organization (WHO) grading scheme [17] is poor, with a median survival time of only 14 months. Finding biomarkers based on magnetic resonance (MR) imaging data could lead to an improved disease progression monitoring and support clinicians in treatment decision-making [10].

Predicting the survival time from pre-treatment MR data is inherently difficult, due to the high impact of the extent of resection (e.g., [18,23]) and response of the patient to chemo- and radiation therapy. The progress in the fields of automated brain tumor segmentation and radiomics have led to many different approaches to predict the survival time of high-grade glioma patients. Further, the introduction of the survival prediction task in the BraTS challenge 2017 [4,19] makes a direct performance comparison of methods possible. The current state-of-the-art approaches can roughly be classified into

1. Classical radiomics: Extracting intensity features and/or shape properties from segmentations and use regression techniques such as random forest (RF) regression [6], logistic regression, or sparsity enforcing methods such as LASSO [25].
2. Deep features: Neural networks are used to extract features, which are subsequently fed into a classical regression method such as logistic regression [7], support vector regression (SVR), or support vector classification (SVC) [14].
3. A combination of classical radiomics and deep features (e.g., [15]).
4. Survival regression from MR data using deep convolutional neural networks (CNNs) with or without additional non-imaging input (e.g., [16]).

Our experiments with 3D-CNNs for survival time regression confirmed observations made by other groups in last year's competition (e.g., [16]), that these models tend to converge and overfit extremely fast on the training set, but show poor generalization when tested on the held-out samples. The top-ranked methods of last year's competition were mainly based on RF. A reason for this may be the relatively few samples to learn from. Classical regression techniques typically have fewer learnable parameters compared to a CNN and perform better with sparse training data.

We present experiments ranging from simple linear models to end-to-end 3D-CNNs and combinations of classical radiomics with deep learning to benchmark new, more sophisticated approaches against established techniques. We believe that a thorough comparison and discussion will provide a good baseline for future investigations of survival prediction tasks.

2   Methods

2.1   Data

The provided BraTS 2018 training and validation datasets for the survival prediction task consist of 163 and 53 subjects, respectively. The challenge ranking is based on the performance on a test dataset with 77 subjects with gross total resection (GTR).

A subject contains imaging and clinical data. The imaging data includes images from the four standard brain tumor MR sequences (T1-weighted (T1), T1-weighted post-contrast (T1c), T2-weighted, and T2-weighted fluid-attenuated inversion-recovery (FLAIR)). All images in the datasets are resampled to isotropic voxel size ($1 \times 1 \times 1 \, \text{mm}^3$), size-adapted to $240 \times 240 \times 155 \, \text{mm}^3$, skull-stripped, and co-registered. The clinical data comprises the subject's age and resection status. The three possible resection statuses are: (a) gross total resection (GTR), (b) subtotal resection (STR), and (c) not available (NA).

Segmentation: For our experiments, we rely on segmentations of the three brain tumor sub-compartments (i.e., enhancing tumor, edema, and necrosis combined with non-enhancing tumor). In the validation and testing dataset, the segmentation is not provided due to the overlap with the data of the BraTS 2018 segmentation task. To obtain the required segmentations, we thus employ the cascaded anisotropic CNN by Wang et al. [26]. Their method is publicly available[1] and contains pre-trained models on the BraTS 2017 training dataset, which is identical to the BraTS 2018 [2,3,5] training dataset. This enables us to compute the segmentations with the available models without retraining a new segmentation network.

2.2 Deep Survival Prediction and Deep Features

Two different CNNs are built for the survival regression task (see Fig. 1). CNN1 consists of five blocks with an increasing number of filters, each block has two convolutional layers and a max pooling operation. The last block is connected to two subsequent fully connected layers. CNN2 consists of three convolutional layers with decreasing kernel sizes with intermediary max-pooling, followed by fully-connected layers connected to the single value regression target. To include clinical information into the CNN2, the age and resection status were appended to the first fully-connected layers of CNN2, which we refer to as CNN2+Age+RS.

Both CNN variants take the four MR sequences and additionally the corresponding segmentation (see Sect. 2.1) as input, and output the predicted survival in days. We observed no performance gain by the additional segmentation input but it improved the training behavior of the network. Instead of regressing the survival days, we also tested direct classification in long-, mid-, and short-term survival, but without improvements.

We trained the CNNs with the Adam optimizer [13] and a learning rate of 10^{-5}, and performed model selection based on Spearman's rank coefficient on a held-out set. Batch normalization and more dropout layers did not lead to improvements, neither on the training behaviour nor the results.

Deep Feature Extraction: For the extraction of deep features, the size of the two last fully connected layers are decreased to 100 and 20 elements. The activations of these two layers serve as deep feature sets.

---

[1] https://github.com/taigw/brats17.

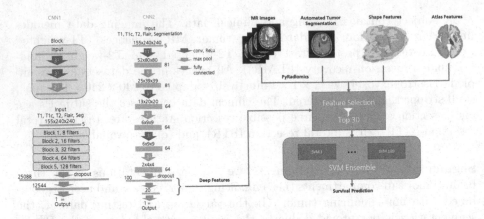

Fig. 1. Summary of the tested methods for GBM patient survival predictions. Left: The architectures of our CNNs for direct survival regression. CNN2 was additionally used to extract deep features from the fully connected layers. For direct regression, the two last fully connected layers of CNN2 had 2048 and 384 elements. Right: Combination of classical radiomics, shape, and atlas features. The top 30 features were used to predict survival classes with a SVC.

2.3   Classical Survival Prediction

Feature Extraction: We extract an initial set of 1353 survival features from the computed segmentation together with the four MR images (i.e., T1, T1c, T2, and FLAIR).

Gray-Level and Basic Shape: 1128 intensity and 45 shape features are computed with the open-source Python package *pyradiomics*[2] version 2.2.0 [11]. It includes shape, first-order, gray level co-occurrence matrix, gray level size zone matrix, gray level run length matrix, neighbouring gray tone difference matrix, and gray level dependence matrix features. Z-score normalization and a Laplacian of Gaussian filter with $\sigma = 1$ is applied to the MR images before extraction. A bin width of 25 is selected and the minimum mask size set to 8 voxels. The features are calculated from all MR images and for all tumor sub-compartments provided by the segmentation (i.e., enhancing tumor, edema, necrosis combined with non-enhancing tumor).

Shape: 15 additional enhancing tumor shape features previously used as predictors for survival [12,21] complement the basic shape features from *pyradiomics*. These features are the rim width of the enhancing tumor, geometric heterogeneity, combinations of rim width quartiles and volume ratios of all combinations of the three tumor compartments.

---

[2] https://github.com/Radiomics/pyradiomics.

Atlas Location: Tumor location has previously been used for survival prediction (e.g., [1]), therefore atlas location features are included. Affine registration is used to align all subjects to FreeSurfer's [8] *fsaverage* subject and its subcortical segmentation (*aseg*) is used as the atlas. The volume fraction of each anatomical region occupied by the contrast enhancing tumor is used as a feature, resulting in 43 features in total.

Clinical Information: The two provided clinical features resection status and age are further added to the feature set.

Feature Selection: Since the number of extracted features ($n = 1353$) is much higher than the available samples ($n = 163$), a subset of features needs to be used. Apart from being necessary for many machine learning methods, a reduction of the feature space improves the interpretability of possible markers regarding survival [20].

We analyzed the following feature selection techniques to find the most informative features: (a) step wise forward/backward selection with a linear model, (b) univariate feature selection, and (c) model-based feature selection by the learned feature weights or importances. We observed a rather low overlap among the selected features by the different techniques, or even the parameterization of the techniques. Consequently, we chose the feature subsets according to their performance on the training dataset for different classical machine learning methods (e.g., linear regression, SVC, and RF). The best results were obtained by the feature subset produced by the model-based feature selection from a sparse SVC model, which consists of the features listed in Table 1.

Our model-based feature selection identified age by far as most important feature. Additionally, a majority of the 30 selected features are intensity-based, but the subset also contains shape and atlas features. We note that none of the 120 deep features was retained.

Feature-Based Models: Although the BraTS survival prediction task is set up as a regression task, the final evaluation is performed on the classification accuracy of the three classes: short-term (less than 10 months), mid-term (between ten and 15 months), and long-term survivors (longer than 15 months). As a consequence, we include classification models in addition to the regression models in our experiments. Since the prediction is required in days of survival, the output of the classifiers needs to be transformed from a class (i.e., short-term, mid-term, long-term) to a day scalar. We do this by replacing each class by its mean time of survival (i.e. 147, 376, 626 days).

For our experiments, we consider the following feature-based regression and classification models [9]:

- Linear and logistic regression
- RF regression and classification
- SVR and SVC
- SVC ensemble

Table 1. Selected feature set with feature category (Cat.), tumor sub-compartment (Comp.), and the MR image (Img.) in decreasing order of importance. ED: Edema, ET: Enhancing tumor, NCR/NET: Necrosis and non-enhancing tumor. Feature importance is decreasing from top to bottom.

Feature	Cat.	Comp.	Img.
Age	Clinical		
Sphericity	Shape	ED	
Optic chiasm	Atlas		
Small area low gray level emphasis	Intensity	ED	T2
Correlation	Intensity	CE	Flair
Cluster shade	Intensity	NCR/NET	T1c
Small dependence high gray level emphasis	Intensity	CE	T1c
Correlation	Intensity	NCR/NET	T1
Maximum	Intensity	ED	T1
Maximum	Intensity	ED	T1c
Left amygdala	Atlas		
Information measure of correlation 1	Intensity	NCR/NET	T1
Large dependence low gray level emphasis	Intensity	NCR/NET	T1c
Cluster shade	Intensity	ED	T2
Inverse variance	Intensity	ED	T1
Small dependence high gray level emphasis	Intensity	CE	T2
Median	Intensity	CE	T2
Busyness	Intensity	ED	T1
Correlation	Intensity	NCR/NET	T1c
Right-vessel	Atlas		
Large area low gray level emphasis	Intensity	ET	Flair
Right caudate	Atlas		
Difference variance	Intensity	ED	Flair
Right cerebellum cortex	Atlas		
Cluster prominence	Intensity	ED	T2
Maximum 2D diameter slice	Intensity	ED	
Inverse difference normalized	Intensity	CE	Flair
Skewness	Intensity	ED	Flair
Median	Intensity	ET	T1
Right ventral diencephalon	Atlas		

We use 50 trees and an automatic tree depth for the RF models and linear kernels for the support vector approaches, SVR, and SVC. To handle the multi-class survival problem we employ the *one-versus-rest* binary approach for SVC and logistic regression. The ensemble method consists of 100 SVC models that are separately built on random splits of 80% of the training data. The final class prediction is performed by majority vote. We choose an ensemble to increase robustness against outliers or unrepresentative subjects in the training set. All classical feature-based models are implemented with *scikit-learn*[3] version 0.19.1.

2.4   Evaluation

We evaluated the classical feature-based approaches by 50 repetitions of a stratified five-fold cross-validation on the BraTS 2018 training dataset. These repetitions allowed us to examine the models' robustness besides their average performance. The CNN approaches were evaluated on a randomly defined held-out split of the training set, consisting of 33 subjects. This held-out set was also used to evaluate a subset of the feature-based methods in order to compare classical approaches to the CNN approaches. Moreover, the classical and CNN models were evaluated on the BraTS 2018 validation set. This dataset contains 53 subjects but only the 28 subjects with resection status GTR are evaluated. Finally, we selected the best-performing model to predict survival on the BraTs 2018 challenge test dataset, which consists of 77 evaluated subjects with GTR resection status (out of 130 subjects).

3   Results

In this section, we compare the performance of the CNN to the classical feature-based machine learning models on the BraTS 2018 training and validation datasets, and present the BraTS 2018 test set results. We introduced a reference baseline for the comparison of the different models. This baseline consists of a logistic regression model solely trained on the age feature. This minimal model provides us with a reference for the training and validation set.

Table 2 lists the results of the different models on the training dataset. To ensure a valid comparison, the table is subdivided by the two evaluation types, repeated cross-validation (CV) and hold-out (HO) (see Sect. 2.4). The results from the CV analysis highlights that by far the best results are achieved by the logistic regression, SVC, and ensemble SVC models, which performed very similarly. Except for the RF model, the classification models clearly outperformed their regression counterparts. The results from the HO analysis (Table 2, bottom) additionally reveals that well-performing classical methods (logistic regression and SVC) outperform all three CNN approaches (CNN1, CNN2, CNN2+Age+RS) by a large margin.

Table 3 presents the results obtained on the validation dataset. We can observe similar patterns as for the training set results: the classification models outperform the regression models with respect to the accuracy (except the RF), the SVC models (i.e., SVC ensemble and SVC) achieve the best performances,

---

[3] http://scikit-learn.org/stable/index.html.

Table 2. Results achieved on the BraTS 2018 training dataset by 100 stratified five-fold cross-validation (CV) runs (reported as mean ± standard deviation) and on one split with 33 held-out (HO) samples. The baseline consists of a logistic regression model with age as single feature. Best results per metric and evaluation type (Eval.) are presented in bold. Acc.: Accuracy, MSE: Mean squared error, r_S: Spearman's rank coefficient, RS: Resection status.

Eval.	Method	Acc.	MSE/days^2	r_S
CV	Baseline	0.489 ± 0.06	136323 ± 44378	0.300 ± 0.14
	Linear regression	0.552 ± 0.08	260706 ± 647176	0.573 ± 0.12
	SVR	0.554 ± 0.08	257542 ± 637062	0.574 ± 0.12
	RF regression	0.444 ± 0.08	117320 ± 42503	0.332 ± 0.17
	Logistic regression	0.721 ± 0.07	$\mathbf{93158 \pm 35665}$	$\mathbf{0.617 \pm 0.12}$
	SVC	$\mathbf{0.722 \pm 0.07}$	93571 ± 35861	0.612 ± 0.12
	RF	0.512 ± 0.07	136334 ± 47392	0.324 ± 0.15
	SVC ensemble	0.720 ± 0.07	93485 ± 35652	$\mathbf{0.617 \pm 0.12}$
HO	Logistic regression	0.697	30756	0.579
	SVC	$\mathbf{0.727}$	$\mathbf{28226}$	$\mathbf{0.616}$
	CNN1	0.515	50598	0.298
	CNN2	0.424	56496	0.235
	CNN2+Age+RS	0.394	61798	-0.194

Table 3. Results achieved on the BraTS 2018 validation dataset (28 samples). The baseline consists of a logistic regression model with age as single feature. Best results per metric are presented in bold. Acc.: Accuracy, MSE: Mean squared error, r_S: Spearman's rank coefficient, RS: Resection status.

Method	Acc.	MSE/days^2	r_S
Baseline	0.464	128841	0.288
Linear regression	0.464	89059	0.426
SVR	0.464	89035	0.426
RF regression	0.393	80980	0.342
Logistic regression	0.5	90791	0.393
RF	0.357	169782	-0.058
SVC	0.536	85471	0.501
SVC ensemble	$\mathbf{0.571}$	$\mathbf{79381}$	$\mathbf{0.556}$
CNN1	0.370	172821	0.104
CNN2	0.394	157617	-0.112
CNN2+Age+RS	0.444	137912	-0.005

and the CNNs remain behind the feature-based methods and the baseline. Additionally, we observe an overall decrease in performance compared to the training set results.

The results of CNN1 on our validation split (accuracy of 0.515) could not be replicated on the BraTS validation set, where it performed poorly with an accuracy of 0.37. CNN2 showed worse results on our validation split than the deeper CNN1, but performed better on the BraTS validation set.

Overall, the SVC ensemble performed best on the training and validation set and we consequently selected it for the challenge, where our method achieved an accuracy of 0.429, a mean squared error of 327725 days^2 and a Spearman's rank coefficient of 0.172.

4 Discussion

In this section, we discuss the presented results and highlight findings from the deep learning, and classical regression and classification experiments.

CNNs: The two CNNs overfit very fast on the training data, and showed highly variable performance between epochs. Model selection during training was therefore challenging, since both the accuracy and Spearman's rank coefficient were very unstable.

We postulate that more data would be needed to fully benefit from direct survival estimation with 3D-CNNs. When inspecting the filters of CNN1 and CNN2, most of the learning took place at the fully connected layers and almost none at the first convolutions layer. This effect and the fast overfitting of the CNN models indicate the lack of samples and are reasons for the poor performance on unseen data.

Classical Regression and Classification: Using classical regression techniques with hand-crafted features has the advantage of better interpretability. Models with fewer learnable parameters, such as the classical regression methods we tested, typically achieve more robust results on unseen data when only few training samples are available.

The atlas used for feature extraction most likely has too many regions for the number of training samples. Small anatomical structures, such as the optic chiasm, may not be accurately identified by simple registration to an atlas. Figure 2 shows the distribution of the contrast enhancing tumor segmentation per survival class. The short survivors with large contrast enhancing tumor loads contribute highly to the overall cumulative occurrence in the training data. The class-wise occurrence maps suggest that more training samples are needed to detect predictive location patterns (e.g. as reported in [22,24]). Additionally, a coarser atlas subdivision driven by clinical knowledge is in order. In the light of this caveat, the location features used here should be seen as approximate localization information with limited clinical interpretability.

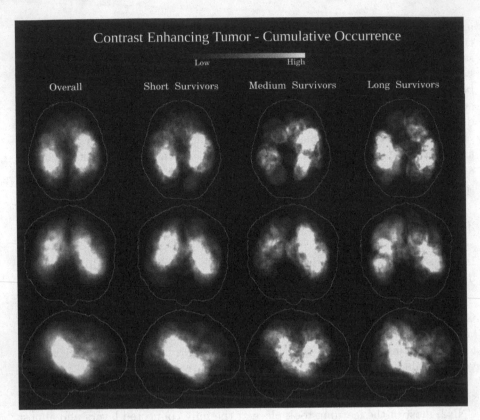

Fig. 2. Cumulative occurrence of the contrast enhancing tumor. Columns, from left to right: Overall across all three survival classes, short survivors (<10 months), medium survivors (≥10 months and ≤15 months), and long survivors (>15 months). Rows: Projection along axial, coronal and sagittal axes.

Performance on the Testing Data: The accuracy of 72.2% and 57.1% on the training and validation set could not be maintained on the testing data. The large performance drop might be caused by still too many features compared to the training set size. Other possible reasons may include a lack of feature robustness or different class distribution compared to the training data. Moreover, the survival time distributions within classes do not drop at the class boundaries, such that a small shift in the prediction can cause a large accuracy difference because ending up in a different class.

In conclusion, classical machine learning techniques using hand-crafted features still outperform deep learning approaches with the given data set size. The robustness of features regarding image quality and across MR imaging centers needs close attention, to ensure that the performance can be maintained on unseen data. We hypothesize that adding post-treatment imaging data and more clinical information to the challenge dataset would boost the performance of the survival regression.

Acknowledgements. We gladly acknowledge the support of the Swiss Cancer League (grant KFS-3979-08-2016) and the Swiss National Science Foundation (grant 169607). We are grateful for the support of the NVIDIA corporation for the donation of a Titan Xp GPU. Calculations were partly performed on UBELIX, the HPC cluster at the University of Bern.

References

1. Awad, A.W., et al.: Impact of removed tumor volume and location on patient outcome in glioblastoma. J. Neuro Oncol. **135**(1), 161–171 (2017). https://doi.org/10.1007/s11060-017-2562-1
2. Bakas, S., et al.: Segmentation labels and radiomic features for the pre-operative scans of the TCGA-GBM collection. Cancer Imaging Arch. (2017). https://doi.org/10.1038/sdata.2017.117
3. Bakas, S., et al.: Segmentation labels and radiomic features for the pre-operative scans of the TCGA-LGG collection. Cancer Imaging Arch. (2017). https://doi.org/10.1038/sdata.2017.117
4. Bakas, S., Reyes, M., et al.: Identifying the best machine learning algorithms for brain tumor segmentation, progression assessment, and overall survival prediction in the BRATS challenge. ArXiv e-prints, November 2018
5. Bakas, S., et al.: Advancing The Cancer Genome Atlas glioma MRI collections with expert segmentation labels and radiomic features. Sci. Data **4**, 170117 (2017). https://doi.org/10.1038/sdata.2017.117
6. Breiman, L., Friedman, J.H., Olshen, R.A., Stone, C.J.: Classification and regression trees (1984)
7. Cox, D.R.: The regression analysis of binary sequences. J. R. Stat. Society Ser. B (Methodol.), **20**(2), 215–242 (1958)
8. Fischl, B.: Freesurfer. Neuroimage **62**(2), 774–781 (2012). https://doi.org/10.1016/j.neuroimage.2012.01.021
9. Hastie, T., Friedman, J., Tibshirani, R.: The Elements of Statistical Learning. SSS, vol. 1. Springer, New York (2001). https://doi.org/10.1007/978-0-387-21606-5
10. Gillies, R.J., Kinahan, P.E., Hricak, H.: Radiomics: images are more than pictures, they are data. Radiology **278**(2), 563–577 (2015). https://doi.org/10.1148/radiol.2015151169
11. van Griethuysen, J.J., et al.: Computational radiomics system to decode the radiographic phenotype. Cancer Res. **77**(21), e104–e107 (2017). https://doi.org/10.1158/0008-5472.CAN-17-0339
12. Jungo, A., et al.: Towards uncertainty-assisted brain tumor segmentation and survival prediction. In: Crimi, A., Bakas, S., Kuijf, H., Menze, B., Reyes, M. (eds.) BrainLes 2017. LNCS, vol. 10670, pp. 474–485. Springer, Cham (2018). https://doi.org/10.1007/978-3-319-75238-9_40
13. Kinga, D., Adam, J.B.: A method for stochastic optimization. In: International Conference on Learning Representations (ICLR), vol. 5 (2015)
14. Lampert, C.H., et al.: Kernel methods in computer vision. Found. Trends® Comput. Graph. Vis. **4**(3), 193–285 (2009). https://doi.org/10.1561/0600000027
15. Lao, J., et al.: A deep learning-based radiomics model for prediction of survival in glioblastoma multiforme. Sci. Rep. **7**(1), 10353 (2017). https://doi.org/10.1038/s41598-017-10649-8

16. Li, Y., Shen, L.: Deep learning based multimodal brain tumor diagnosis. In: Crimi, A., Bakas, S., Kuijf, H., Menze, B., Reyes, M. (eds.) BrainLes 2017. LNCS, vol. 10670, pp. 149–158. Springer, Cham (2018). https://doi.org/10.1007/978-3-319-75238-9_13

17. Louis, D.N., et al.: The 2016 world health organization classification of tumors of the central nervous system: a summary. Acta Neuropathol. **131**(6), 803–820 (2016). https://doi.org/10.1007/s00401-016-1545-1

18. Meier, R., et al.: Automatic estimation of extent of resection and residual tumor volume of patients with glioblastoma. J. Neurosurg. **127**(4), 798–806 (2017). https://doi.org/10.3171/2016.9.JNS16146

19. Menze, B.H., Jakab, A., Bauer, S., et al.: The multimodal brain tumor image segmentation benchmark (BRATS). IEEE Trans. Med. Imaging **34**(10), 1993–2024 (2015). https://doi.org/10.1109/TMI.2014.2377694

20. Pereira, S., et al.: Enhancing interpretability of automatically extracted machine learning features: application to a RBM-random forest system on brain lesion segmentation. Med. Image Anal. **44**, 228–244 (2018). https://doi.org/10.1016/j.media.2017.12.009

21. Pérez-Beteta, J., et al.: Glioblastoma: does the pre-treatment geometry matter? A postcontrast T1 MRI-based study. Eur. Radiol. (2017). https://doi.org/10.1007/s00330-016-4453-9

22. Rathore, S., et al.: Radiomic MRI signature reveals three distinct subtypes of glioblastoma with different clinical and molecular characteristics, offering prognostic value beyond idh1. Sci. Rep. **8**(1), 5087 (2018). https://doi.org/10.1038/s41598-018-22739-2

23. Sanai, N., Polley, M.Y., McDermott, M.W., Parsa, A.T., Berger, M.S.: An extent of resection threshold for newly diagnosed glioblastomas. J. Neurosurg. **115**(1), 3–8 (2011). https://doi.org/10.3171/2011.2.JNS10998

24. Steed, T.C., et al.: Differential localization of glioblastoma subtype: implications on glioblastoma pathogenesis. Oncotarget **7**(18), 24899 (2016). https://doi.org/10.18632/oncotarget.8551

25. Tibshirani, R.: Regression shrinkage and selection via the lasso. J. R. Stat. Society Ser. B (Methodol.), **58**(1), 267–288 (1996)

26. Wang, G., Li, W., Ourselin, S., Vercauteren, T.: Automatic brain tumor segmentation using cascaded anisotropic convolutional neural networks. In: Crimi, A., Bakas, S., Kuijf, H., Menze, B., Reyes, M. (eds.) BrainLes 2017. LNCS, vol. 10670, pp. 178–190. Springer, Cham (2018). https://doi.org/10.1007/978-3-319-75238-9_16

Semi-automatic Brain Tumor Segmentation by Drawing Long Axes on Multi-plane Reformat

David Gering(✉), Kay Sun, Aaron Avery, Roger Chylla,
Ajeet Vivekanandan, Lisa Kohli, Haley Knapp, Brad Paschke,
Brett Young-Moxon, Nik King, and Thomas Mackie

HealthMyne, Madison, WI 53717, USA
david.gering@healthmyne.com

Abstract. A semi-automatic image segmentation method, called SAMBAS, based on workflow familiar to clinical radiologists is described. The user initializes 3D segmentation by drawing a long axis on a multi-plane reformat (MPR). As the user draws, a 2D segmentation updates in real-time for interactive feedback. When necessary, additional long axes, short axes, or other editing operations may be drawn on one or more MPR planes. The method learns probability distributions from the drawing to perform the MPR segmentation, and in turn, it learns from the MPR segmentation to perform the 3D segmentation. As a preliminary experiment, a batch simulation was performed where long and short axes were automatically drawn on each of 285 multi-spectral MR brain scans of glioma patients in the 2018 BraTS Challenge training data. Average Dice coefficient for tumor core was 0.86, and the Hausdorff-95% distance was 4.4 mm. As another experiment, a convolution neural network was trained on the same data, and applied to the BraTS validation and test data. Its outputs, computed offline, were integrated into the interactive method. Ten volunteers used the interface on the BraTS validation and test data. On the 66 scans of the validation data, average Dice coefficient for core tumor improved from 0.76 with deep learning alone, to 0.82 as an interactive system.

Keywords: Brain tumor · Image segmentation ·
Semi-automatic · Machine learning

1 Introduction

Evidence from cancer researchers suggests that extraction of quantitative variables from medical images can contribute more information for decision support in management of cancer patients. Specifically, quantitative metrics can improve both (1) diagnostic and prognostic accuracy; as well as (2) longitudinal monitoring of patient response [1]. Criteria for monitoring radiographic brain tumor progression include the Macdonald criteria [2], Response Evaluation Criteria in Solid Tumors (RECIST) [3, 4], WHO criteria [5], and RANO criteria [6].

Currently, radiological studies are generally limited to detection and staging along with qualitative descriptions. Quantitative descriptors are not yet in the standard of care

© Springer Nature Switzerland AG 2019
A. Crimi et al. (Eds.): BrainLes 2018, LNCS 11384, pp. 441–455, 2019.
https://doi.org/10.1007/978-3-030-11726-9_39

primarily due to a lack of infrastructure and tools to derive, test, and deploy these quantitative metrics at the point-of-care for all patients. Currently available tools to do this are limited to research or clinical trials, and have not been widely deployed as they lack the speed, precision and consistency required for wider clinical use [7]. The amount of time required to delineate lesion boundaries correctly could be intrusive to the radiologist's workflow. Delineation can be performed by manually drawing the tumor boundary on each image slice, by semi-automatically guiding an algorithm, or by fully automated methods. In either the semi-automated or fully automated methods, editing is necessary. Although manual delineation offers complete control to the user, humans exhibit great variability and the process is very time-consuming. Even if an automatic or semi-automatic method were to suffer a shortcoming in accuracy, as long as there is consistency in defining the boundary, then the volume change or change in a quantitative feature may be tracked more precisely.

For MR brain tumors, recent research with fully automated segmentation, especially based on deep neural networks, has been promising [8]. SAMBAS (Semi-Automated Map-BAsed Segmentation) differs from CAD (Computer Aided Detection) because it relies on a radiologist to make an indication. The motivation is adoption by clinical radiologists who desire full-control over the segmentation, real-time feedback, an algorithm that is ready to run immediately without the need to first be trained on a large database from their site, and an algorithm whose rationale behind decisions is explainable. We expect that real-time guidance of a semi-automated approach may often have faster workflow than editing of a fully automated method.

The vital part of any measurement tool is an interface that is both familiar and effortless. Drawing the longest axis across a lesion is a natural choice for initiating contours because radiologists are already accustomed to drawing the long axis. Oncologists participating in clinical trials follow published international criteria for objectively gauging the extent and progression of disease. The Macdonald, RECIST, and WHO criteria each incorporate long axis measurements. However, inherent challenges with axis-based criteria have been reported for aggressive brain tumors [9], thus motivating the discovery of volumetric-based criteria with similar familiarity as axis-based criteria.

Besides familiarity, there are several more goals of volumetric contouring. One goal is to achieve inter-observer consistency, while also catering to individual preferences for accuracy and style. Consistency results from initialization strategies that are reproducible, such as generating 3D volumetric contours from a straight stroke rather than free-form drawing. Tailoring to individual preferences is accomplished by editing tools prepared for whenever the initial contours may be unsatisfactory. Another goal is to provide a contingency plan in case the radiologist is both unsatisfied with the contours, and unwilling to invest the requisite time to edit them. Radiologists should be given the choice of confirming either the contours (thereby enabling volumetric measures), or just the long axis, which has already been drawn, and is held in reserve as an instant alternative. Yet another goal is automatic, large-scale, quantitative validation. Given hundreds of datasets that have been manually contoured, batch processing can be implemented by calculating the long axis from each expert's contours, and employing the long axis as the simulated user input. Yet another goal is to alleviate the need to select tools from a confusing suite of options. Ideally, there is exactly one tool in a reading room, generally applicable to all organs, yet simultaneously specialized with organ-specific features. The organ is automatically identified upon tool initialization.

SAMBAS aims to satisfy all the aforementioned goals, namely familiarity, consistency, individualism, contingency, automatic validation, and general applicability yet specialization. While advancements in processing speed have propelled deep learning (DL) in various fields, medical image analysis is missing the mass quantities of new labeled data needed for training artificial intelligence networks [10]. The multimodal Brain Tumor Segmentation challenge (BraTS) represents a pioneering step in this direction [11, 12]. One of the goals of the software was to generate such contours on new scans, at the point of read, which in turn, can serve as the labeled image data for DL in subsequent clinical application.

2   Methods

The proposed system consists of an interactive algorithm and two compute-intensive components, which are whole-brain tissue segmentation and deep learning. Each has a run-time of roughly one minute on a typical PC, so they are run offline prior to a user's interaction with the system. While the interactive algorithm is employed by a user to segment only the core tumor, the offline components support partitioning of the tumor into its constituent parts: edema, necrosis, and actively enhancing regions.

The integration of all elements into one system is presented first, followed by the offline elements and the interactive algorithm, along with associated experiments.

2.1   System that Integrates Offline Components with User Interaction

Figure 1 presents a system flowchart. User interaction occurs in real-time because only a portion of the image is being segmented since whole-brain analysis occurred earlier.

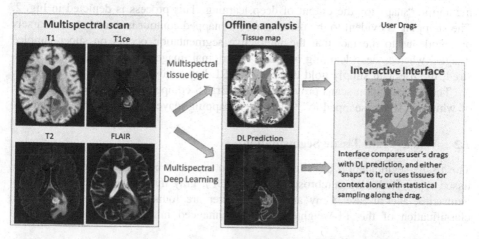

Fig. 1. Two components run offline prior to the user interacting with the system.

As the user draws a long axis on an image, only the core tumor is segmented in 2D with real-time feedback. Figure 2 displays a screenshot of the red 2D contour responding interactively to the drawn blue axis. When the user clicks a button to indicate all drawing is complete, then a 3D segmentation process runs for roughly 1–3 s, relying on the output of deep learning and tissue segmentation to find the edema associated with that particular core tumor. When a scan contains multiple distinct tumors, the user must draw a separate long axis for each tumor.

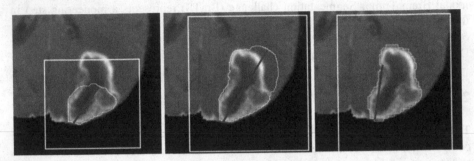

Fig. 2. These are three screenshots taken during a user's real-time interaction while drawing the long axis (blue). LEFT: The user has started drawing a long axis, but has only partially traversed the tumor at the time of the screen capture. MIDDLE: The user has over-drawn the lesion to show how the red contour always presents a reasonable result given strong image contrast in some areas, and little to none in others. RIGHT: The user has placed both endpoints of the long axis on the boundary of the output of deep learning. Consequently, the red contour "snaps to" deep learning's output contour even though the true longest axis in the plane was not indicated. (Color figure online)

As the user draws a long or short axis, whenever all endpoints of the axes are proximal to the boundary of core tumor, as found by deep learning, then the segmentation "snaps to" the output of deep learning. This process is depicted in Fig. 2. The snapping is evident to the user because a snapped contour is drawn more coarsely pixilated due to the fact that the interactive segmentation occurs on super-sampled images, whereas deep learning occurred on original images. When snapping is undesired, the user can simply hold down the CTRL key to disable it while drawing.

The 66 validation scans provided by the BraTS competition contained 89 tumors, of which 35 were "snapped to". Therefore, snapping played a role on 39% of tumors.

2.2 Whole-Brain Tissue Segmentation

The tissue segmentation classifies every brain voxel as belonging to one of several tissue types, including cerebrospinal fluid (CSF), gray matter, white matter, vessels, ventricles, and disease. Gray and white matter are found by performing Bayesian classification of the T1-weighted, contrast-enhanced image using the Expectation

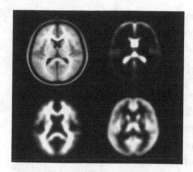

Fig. 3. The SPM atlas features an average of 305 scans (upper right) and probability maps for CSF (upper right), white matter (lower left), and gray matter (lower right).

Maximization (EM) algorithm [21]. One element of Bayesian classification is the probability that a voxel belongs to a certain tissue class prior to observing its brightness. When this prior probability varies across the image, it is referred to as a spatially-varying prior (SVP). The SVP is estimated through affine registration of the SPM atlas, as shown in Fig. 3.

Rules of logic are applied to the set of all four MR spectra to derive the other tissues. For example, enhancing tumor is described by areas that show hyper-intensity under contrast-enhancement when compared to the non-enhanced image, but also when compared to healthy white matter.

The resultant tissue segmentation will be used by the integrated system for anatomic context. For example, it will know to exclude vessels and ventricles from tumors.

2.3 Deep Learning Segmentation

Following the recent increasing successes of deep learning approaches in automated organ and tumor segmentations [13–19], a convolution neural network (CNN) was used. The CNN is based on the 3D U-Net architecture by Isensee et al. [15] (Fig. 4), which had one of the top scores in the 2017 BraTS Challenge [20]. Briefly, the input

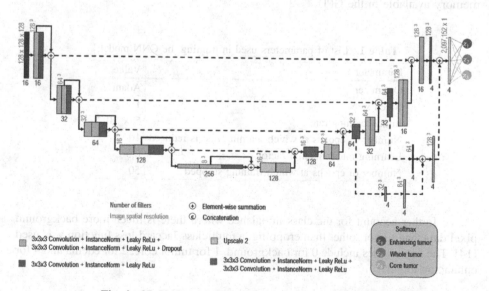

Number of filters   ⊕ Element-wise summation
Image spatial resolution ⓒ Concatenation

3x3x3 Convolution + InstanceNorm + Leaky ReLu +
3x3x3 Convolution + InstanceNorm + Leaky ReLu + Dropout

3x3x3 Convolution + InstanceNorm + Leaky ReLu

Upscale 2

3x3x3 Convolution + InstanceNorm + Leaky ReLu +
3x3x3 Convolution + InstanceNorm + Leaky ReLu

Softmax
① Enhancing tumor
② Whole tumor
③ Core tumor

Fig. 4. 3D U-Net architecture based on Isensee et al. [3].

image data is set to 128 × 128 × 128 voxels, constrained by the limited memory in the GPU. Processing from left to right, the 3D image volume is sequentially reduced in spatial resolution with multiple 3 × 3 × 3 convolution layers while increasing the number of filters or feature maps as the levels move deeper. Once the lowest level is reached, the extracted feature maps are then upsampled to sequentially restore the spatial resolution at each level, concatenating with feature maps preserved during the downsampling to help restore lost information. The Softmax function classifies the 3 tumor classes. Dropouts with probability 0.3 are included to minimize overfitting.

A set of MR training data consisting of brain scans of 210 subjects with high grade glioma (HGG) and another 75 with low grade glioma (LGG) was provided by the BraTS competition. Each subject has a T1 weighted, a post-contrast T1-weighted, a T2-weighted, and a FLAIR MR image. In addition, a segmented tumor mask that contains demarcations for whole tumor, core tumor and enhancing tumor, manually demarcated by expert physicians, was also provided as the ground truth for evaluation. The images were preprocessed prior to supervised training by cropping to remove the extraneous background and preserve only the brain, resizing to 128 × 128 × 128, and normalizing the MR intensities of each modality by subtracting the mean and dividing by the standard deviation.

During the supervised training, the 285 subject data were randomly split into training and validation dataset (80%–20%), each consisted of image volumes from the four different modalities and segmented truths. The training dataset was fed into the 3D U-Net model for optimization and the segmented truths were used for evaluations during the backpropagation. The model was tested at each step with the validation dataset, on which the model had not been trained. Table 1 lists the parameters used in the training of the CNN model. Batch size of one was used, despite the lower performance, in order to load the entire 285 subject image volumes into the limited memory available in the GPU.

Table 1. List of parameters used in training the CNN model.

Parameter	Value
Optimizer	Adam
Batch size	1
Initial learning rate	5×10^{-4}
Number of epochs at which learning rate is reduced	10
Learning rate reduction factor	0.5
Number of epochs at which training stopped	50

To further account for the class imbalance where there is much more background pixel data than tumor, other than cropping, a multiclass Jaccard loss function was used [14]. The four classes include 0 for background, 1 for tumor core, 2 for edema and 4 for enhancing tumor.

$$loss = -\frac{1}{K} \sum_{k \in K} \frac{\sum_i u_i^k v_i^k}{\sum_i u_i^k + \sum_i v_i^k - \sum_i u_i^k v_i^k} \tag{1}$$

The loss function is expressed in Eq. 1, where u is the prediction of the CNN and v is from the ground truth segmentation value, i is the pixel number, and k is each class in all $K = 4$ classes. The Jaccard coefficient is a measure of similarity between the segmented prediction and truth image volumes, where higher value indicates greater overlap. The multiclass version is the intersection over union of the two volumes averaged over the four classes. A negative term was added to the loss function to ensure the minimum loss function was optimized. CNN development used the open-source machine learning library, TensorFlow, and neural networking API, Keras.

2.4 Interactive MPR Segmentation

The interactive algorithm is implemented as a probabilistic framework with efficient user control. Like a digital simulation of a traditional light box on which radiologists used to view film, the 3D volume is visualized by displaying 2D planes. A Multi-Plane Reformat (MPR) refers to reformatting more than one plane, and we display a trio of planes side-by-side, such that there are axial, coronal, and sagittal orientations.

The user initializes the segmentation process by drawing a long axis on one plane of the MPR. As the user draws the long axis, a 2D segmentation updates in real-time for interactive feedback. The feedback has proven to be very helpful for the user to know precisely where to place the endpoint of the axis. Upon release of the mouse, 2D segmentation occurs immediately on the other MPR planes.

When the 2D contour is unsatisfactory, an optional short axis may be drawn perpendicular to the long axis. Other editing operations are available, such as a "ball tool" for drawing with a digital brush. A correct 2D segmentation is important since probability distributions are learned from the 2D segmentation to be employed in segmenting the other MPR planes.

When the contours on other MPR planes are unsatisfactory, then the user can draw there with the same editing tools, along with the option for drawing a long axis and short axis. This is especially useful for lesions which are irregularly shaped or oriented obliquely. Once satisfied, the user clicks a button to initiate 3D segmentation.

2.5 3D Segmentation

Multivariate Bayesian classification [21] labels image voxels as belonging to one of two classes, Background or Foreground. Classification combines the *likelihood* of class membership based on voxel brightness, with the probability of membership *prior* to observing brightness. The *likelihoods* are conditional probability distributions that do not vary across the image, while the *prior* probabilities are spatially varying, and a function of distance from region boundaries.

The user directly drives the segmentation process by manipulating four types of regions, where some regions govern the likelihoods, while some regions govern the prior probabilities. Various regions are described in Table 2, and illustrated in Fig. 5.

Table 2. Regions which drive probabilities.

Region	Color in Fig. 1	Description
Inclusion	Green	All voxels within belong to the Foreground class, and statistically sample it
Containment	Yellow	All voxels outside belong to the Background class
Background	Blue & pink	Statistically typify Background class
Avoidance	Not shown	Spatially prohibit Foreground without affecting statistics

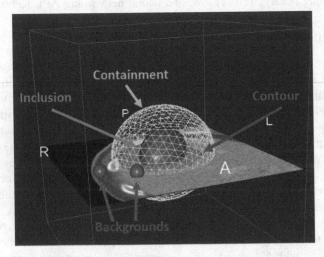

Fig. 5. Some regions are initially configured as ellipsoids, and then become warped. The image shown is a CT since the interactive algorithm was designed to be general purpose.

The sizes and poses of regions are automatically derived from the long axis. While the long axis describes lesion extent along one dimension, an initialization stage estimates lesion extent along other dimensions by analyzing orthogonal scout planes given statistical sampling along the long and short axes. Probability distributions are modeled parametrically as Gaussian Mixture Models (GMM) [22] while placing the Inclusion and Containment regions, and as non-parametric distributions thereafter.

Background regions are automatically placed by searching the vicinity outside the Containment region, and within the body outline, while maximizing the Mahalanobis distance [21] from the Inclusion region. Once Background and Inclusion regions are initialized, the voxels within are used to perform Parzen windowing [21] to estimate the likelihoods for Bayesian classification.

Noise and artifacts in CT vary by dose and choice of reconstruction type and kernel, and in MR by field strength, RF coil configuration, and protocol parameters, so Bayesian classification is augmented with a Markov Random Field [23] with 3 iterations of mean-field approximation.

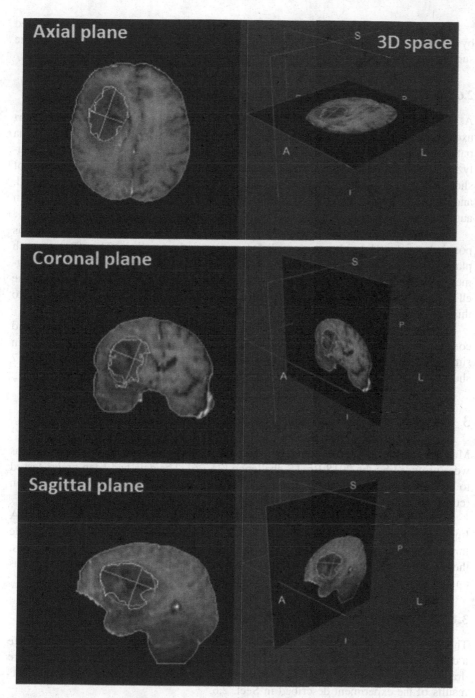

Fig. 6. Axial, coronal, and sagittal planes of MPR are shown from top to bottom. The blue ellipse was fit to the yellow contour of ground-truth in order to generate the green long and short axes. This process simulated a human user manually drawing on MPR. (Color figure online)

The output is a 3D mesh fit to voxel classification by adapting vertices connected by virtual springs to their neighbors to provide a regularizing force that smooths the surface. The true long axis is measured, which may not lie in any orthogonal plane.

2.6 Experiment with Simulated Drawing of Long Axes

As a preliminary experiment, a batch simulation was performed where long and short axes were automatically drawn on each of 285 multispectral MR brain scans of glioma patients in the BraTS 2018 training data. To achieve this, the ground-truth was analyzed to find an appropriate slice on which to draw the long axis. Given the range of slices that contained any ground-truth, the central third of this range was considered, and from that subset, the slice with the largest area of ground-truth was chosen. An automatic process then drew the long axis across the ground-truth on that slice.

In order to simulate the type of long axis that a human user might draw, the axis position was favored to be more medial than the true longest axis. Therefore, on each plane, an ellipse was fit by Principle Component Analysis (PCA) [21] to the segmentation on that slice. The long axis with the same orientation as the major axis of the ellipse was found. The short axis was then found as the longest axis perpendicular to this, as shown in Fig. 6.

The center of the long axis was used for the center of the reformatted sagittal and coronal planes to comprise a 3-plane MPR. Then long and short axes were drawn in similar manner on all planes. The drawn axes precipitate MPR segmentation. Figure 7 shows a few examples.

3 Results

Multi-institutional, routine clinically-acquired pre-operative multispectral MR scans were provided by the 2018 BraTS challenge [24–26]. The data had been preprocessed to be co-registered to the same anatomical template, interpolated to the same resolution (cubic mm), and skull-stripped.

Segmentation accuracy was computed by uploading labeled images to the CBICA Image Processing Portal, which measured statistics for active (enhancing) tumor, whole tumor, and tumor core. While the ground truth was available for the 285 training cases, there were an additional 66 validation cases, and 191 test cases where ground truth was unavailable to participants.

3.1 Experiment with Simulated Drawing of Long Axes

The T1-weighted post-contrast scan was combined with the T2-weighted scan to create a dual-spectra image that was input to the interactive algorithm. Since long axes where drawn on core tumor, this experiment segmented only that structure. Table 3 lists results of the experiment described in Sect. 2.6.

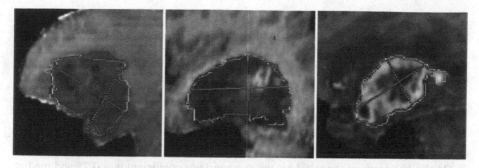

Fig. 7. MPR segmentation (red) depicted relative to ground-truth contours (yellow) and long/short axes (green) on a reformatted sagittal slice. MPR segmentations (not final 3D) were measured to have 0.90 average Dice, compared to ground-truth, for 855 planes of 285 cases. (Color figure online)

Table 3. BraTS 2018 validation results with *simulated* user interaction (core tumor only).

	Dice	Sensitivity	Hausdorff-95%
Mean	0.862	0.893	4.38
Std. Dev	0.060	0.081	3.18
Median	0.873	0.915	3.61

3.2 Experiment with Deep Learning Alone

All four MR modalities of the 66 validation cases were presented to the trained CNN model, and Table 4 presents the results.

Table 4. BraTS 2018 validation results with *no* user interaction.

	Dice			Sensitivity			Hausdorff-95%		
	Active	Whole	Core	Active	Whole	Core	Active	Whole	Core
Mean	0.696	0.878	0.763	0.823	0.893	0.883	6.92	10.1	8.35
Std. Dev	0.200	0.0378	0.256	0.134	0.102	0.127	11.0	16.7	11.5
Median	0.848	0.894	0.918	0.836	0.933	0.938	1.41	3.00	2.83

3.3 Experiment with User Interaction

Ten volunteers used the interface on the BraTS validation and test data. On the 66 scans of the validation data, average Dice coefficient on tumor core improved from 0.76 with deep learning alone, to 0.82 as an integrated, interactive system. The Hausdorff-95% distance improved from 8.4 to 7.5 mm, as detailed in Table 5. Progress since the challenge has improved scores further to 0.87 Dice and 4.9 mm, which is presently the lowest Hausdorff distance on the BraTS leaderboard.

Table 5. BraTS 2018 validation results with *real* user interaction.

	Dice			Sensitivity			Hausdorff-95%		
	Active	Whole	Core	Active	Whole	Core	Active	Whole	Core
Mean	0.730	0.890	0.823	0.752	0.885	0.791	4.37	5.53	7.53
Std. Dev	0.264	0.057	0.173	0.244	0.0865	0.203	6.86	7.06	10.9
Median	0.828	0.904	0.885	0.807	0.904	0.870	2.50	3.87	3.74

On the 191 scans of the test data, the average Dice coefficient was 0.75 and median 0.86. Compared with the simulation experiment where mean and median were quite similar, the disparity between mean and median here suggest that human volunteers and ground-truth disagreed, curtailing certain scores. Table 6 presents the details.

Table 6. BraTS 2018 *test* results with real user interaction.

	Dice			Hausdorff-95%		
	Active	Whole	Core	Active	Whole	Core
Mean	0.643	0.852	0.750	4.88	8.15	8.44
Std. Dev	0.300	0.135	0.267	5.77	12.9	12.7
Median	0.765	0.896	0.864	3.00	4.12	4.24

The Hausdorff-95% distance was nearly a factor of two lower on the simulation experiment, when compared with the other three experiments, which shows the value of knowing precisely where to draw axes. An interesting observation is that the sensitivity for core tumor was higher without user interaction. The median was even slightly higher even though the mean was much lower. This suggests DL fared better than humans on obvious tumors, but users provided essential aid when DL missed badly.

4   Discussion

In comparison with other semi-automatic tools, products from Invivo [27] and Mirada [28] feature initialization by a single click, whereas the additional information contained in SAMBAS' long axis bolsters reliability relative to a click. Perhaps the most similar algorithm to SAMBAS is the GrowCut algorithm [29, 30] implemented in the 3D Slicer [31]. Both have general applicability, and a concept of Background and Foreground regions. However, GrowCut is not initiated as quickly as a drag across the long axis, and one study measured lung lesion contouring to require an average of 10 min [32], whereas a clinical goal is sub-minute. Perhaps the most similar initialization method is [33] for the Random Walker algorithm [34], because a clicked point or stroke commences 2D segmentation from which Background and Foreground seeds are generated for 3D segmentation. However, the SAMBAS approach intentionally seeks statistical separation rather than a simple circumscribed shape for Background. GrowCut and the Random Walker both lack the two additional regions that SAMBAS adds, Containment and Avoidance, which make editing expeditious. Furthermore, SAMBAS differs by its Bayesian framework, which in conjunction with the added regions, make it possible to seamlessly incorporate organ-specific processing, and to employ DL-based CAD to derive additional SVP probability maps.

Quantitative results were promising, while leaving ample opportunity for improvement. During the interactive experiment, the long axis was drawn manually by human users, with the guidance of real-time MPR segmentation as constructive feedback. The advantage of feedback did not produce better quantitative scores than the batch-generated long and short axes of the simulation experiment. The drop-off in scores between the simulation experiment on training data, and the interactive experiment on validation data, indicates the value of knowing where to draw.

The novel "snap to" feature introduced here may offer a solution to the problem of false positives with CAD systems. Only those CAD findings which are drawn on by the user will be output. The other CAD findings could be withheld from clinicians to avoid biasing their judgment.

The interactive algorithm was developed to be general-purpose, and is well-suited for CT lung and liver lesions. The MR-specific and brain-specific enhancements presented herein are a new addition, which is a work in progress, and we look forward to upgrading the tissue segmentation and deep learning components to improve the overall system. The fact that the integrated system outperformed deep learning alone on the validation data bodes well for interfaces which unite neural networks with expert users.

References

1. Nordstrom, R.: The quantitative imaging network in precision medicine. Tomography 2(4), 239–241 (2016)
2. Macdonald, D.R., Cascino, T.L., Schold Jr., S.C., Cairncross, J.G.: Response criteria for phase II studies of supratentorial malignant glioma. J. Clin. Oncol. 8, 1277–1280 (1990)
3. Therasse, P., et al.: New guidelines to evaluate the response to treatment in solid tumors. J. Natl. Cancer Inst. 92(3), 205–216 (2000)
4. Sorensen, A.G., Batchelor, T.T., Wen, P.Y., Zhang, W.T., Jain, R.K.: Response criteria for glioma. Nat. Clin. Pract. Oncol. 5, 634–644 (2008)
5. Suzuki, C., et al.: Radiologic measurements of tumor response to treatment: practical approaches and limitations. Radiographics 28, 329–344 (2008)
6. Wen, P.Y., et al.: Updated response assessment criteria for high-grade gliomas: response assessment in neuro-oncology working group. J. Clin. Oncol. Off. J. Am. Soc. Clin. Oncol. 28, 1963–1972 (2010)
7. Yankeelov, E., Mankoff, D., Schwartz, L., Rubin, D.: Quantitative imaging in cancer clinical trials. Clin. Cancer Res. 22, 284–290 (2016)
8. https://www.cbica.upenn.edu/sbia/Spyridon.Bakas/MICCAI_BraTS/MICCAI_BraTS_2017_proceedings_shortPapers.pdf

9. Mehta, A.I., Kanaly, C.W., Friedman, A.H., Bigner, D.D., Sampson, J.H.: Monitoring radiographic brain tumor progression. Toxins **3**(3), 191–200 (2011). https://doi.org/10.3390/toxins3030191

10. Freedman, D.H.: A reality check for IBM's AI ambitions. MIT Technol. Rev. (2017). https://www.technologyreview.com/s/607965/a-reality-check-for-ibms-ai-ambitions/

11. Menze, B.H., et al.: The multimodal brain tumor image segmentation benchmark (BRATS). IEEE Trans. Med. Imaging **34**(10), 1993–2024 (2015). https://doi.org/10.1109/TMI.2014.2377694

12. Bakas, S., et al.: Advancing The Cancer Genome Atlas glioma MRI collections with expert segmentation labels and radiomic features. Nat. Sci. Data **4**, 170117 (2017). https://doi.org/10.1038/sdata.2017.117

13. Çiçek, Ö., Abdulkadir, A., Lienkamp, S.S., Brox, T., Ronneberger, O.: 3D U-Net: learning dense volumetric segmentation from sparse annotation. In: Ourselin, S., Joskowicz, L., Sabuncu, M.R., Unal, G., Wells, W. (eds.) MICCAI 2016. LNCS, vol. 9901, pp. 424–432. Springer, Cham (2016). https://doi.org/10.1007/978-3-319-46723-8_49

14. Kayahbay, B., Jensen, G., Van Der Smagt, P.: CNN-based segmentation of medical imaging data. arXiv Prepr. arXiv:1701.03056 (2017)

15. Isensee, F., Kickingereder, P., Wick, W., Bendszus, M., Maier-Hein, K.H.: Brain tumor segmentation and radiomics survival prediction: contribution to the BRATS 2017 challenge. In: Crimi, A., Bakas, S., Kuijf, H., Menze, B., Reyes, M. (eds.) BrainLes 2017. LNCS, vol. 10670, pp. 287–297. Springer, Cham (2018). https://doi.org/10.1007/978-3-319-75238-9_25

16. Ronneberger, O., Fischer, P., Brox, T.: U-Net: convolutional networks for biomedical image segmentation. In: Navab, N., Hornegger, J., Wells, W.M., Frangi, A.F. (eds.) MICCAI 2015. LNCS, vol. 9351, pp. 234–241. Springer, Cham (2015). https://doi.org/10.1007/978-3-319-24574-4_28

17. Roth, H.R., et al.: An application of cascaded 3D fully convolutional networks for medical image segmentation. Comput. Med. Imaging Graph. **66**, 90–99 (2018)

18. Jackson, P., Hardcastle, N., Dawe, N., Kron, T., Hofman, M.S., Hicks, R.J.: Deep learning renal segmentation for fully automated radiation dose estimation in unsealed source therapy. Front. Oncol. (2018)

19. Wang, G., Li, W., Ourselin, S., Vercauteren, T.: Automatic brain tumor segmentation using cascaded anisotropic convolutional neural networks. In: Crimi, A., Bakas, S., Kuijf, H., Menze, B., Reyes, M. (eds.) BrainLes 2017. LNCS, vol. 10670, pp. 178–190. Springer, Cham (2018). https://doi.org/10.1007/978-3-319-75238-9_16

20. Multimodal brain tumor segmentation challenge 2017 rankings. https://www.med.upenn.edu/sbia/brats2017/rankings.html

21. Duda, R.O., Hart, P.E., Stork, D.G.: Pattern Classification, 2nd edn. Wiley, New York (2001)

22. Press, W.H., Teukolsky, S.A., Vetterling, W.T., Flannery, B.P.: Numerical Recipes: The Art of Scientific Computing, 3rd edn. Cambridge University Press, Cambridge (2007)

23. Li, S.Z.: Markov Random Field Modeling in Image Analysis. Advances in Computer Vision and Pattern Recognition. Springer, London (2009). https://doi.org/10.1007/978-1-84800-279-1

24. Bakas, S., et al.: Segmentation labels and radiomic features for the pre-operative scans of the TCGA-GBM collection. Cancer Imaging Arch. (2017). https://doi.org/10.7937/K9/TCIA.2017.KLXWJJ1Q

25. Bakas, S., et al.: Segmentation labels and radiomic features for the pre-operative scans of the TCGA-LGG collection. Cancer Imaging Arch. (2017). https://doi.org/10.7937/K9/TCIA.2017.GJQ7R0EF

26. Bakas, S., et al.: Identifying the best machine learning algorithms for brain tumor segmentation, progression assessment, and overall survival prediction in the BRATS challenge. arXiv preprint arXiv:1811.02629 (2018)
27. Invivo DynaCAD. http://www.invivocorp.com/solutions/lung-cancer-screening/
28. Mirada XD3. http://www.mirada-medical.com
29. Vezhnevets, V., Konouchine, V.: GrowCut: interactive multi-label ND image segmentation by cellular automata. In: Proceedings of Graphicon, vol. 1, pp. 150–156 (2005)
30. Zhu, L., Kolesov, I., Gao, Y., Kikinis, R., Tannenbaum, A.: An effective interactive medical image segmentation method using fast growcut. In: MICCAI Workshop on Interactive Medical Image Computing, Boston (2014)
31. D Slicer. http://www.slicer.org
32. Velazquez, E.R., et al.: Volumetric CT-based segmentation of NSCLC using 3D-Slicer. Sci. Rep. **3**, 3529 (2013)
33. Jolly, M.P., Grady, L.: 3D general lesion segmentation in CT. In: 5th IEEE International Symposium on Biomedical Imaging (ISBI): From Nano to Macro, pp. 796–799. IEEE (2008)
34. Grady, L.: Random walks for image segmentation. IEEE Trans. Pattern Anal. Mach. Intell. **28**(11), 1768–1783 (2006)

Ensembles of Densely-Connected CNNs with Label-Uncertainty for Brain Tumor Segmentation

Richard McKinley[⊠], Raphael Meier, and Roland Wiest

Support Centre for Advanced Neuroimaging,
University Institute of Diagnostic and Interventional Neuroradiology,
Inselspital, Bern University Hospital, Bern, Switzerland
richard.mckinley@gmail.com

Abstract. We introduce a new family of classifiers based on our previous DeepSCAN architecture, in which densely connected blocks of dilated convolutions are embedded in a shallow U-net-style structure of down/upsampling and skip connections. These networks are trained using a newly designed loss function which models label noise and uncertainty. We present results on the testing dataset of the Multimodal Brain Tumor Segmentation Challenge 2018.

1 Introduction

We present a network architecture for semantic segmentation, heavily inspired by the recent Densenet architecture for image classification [7], in which pooling layers are replaced by heavy use of dilated convolutions [16]. Densenet employs dense blocks, in which the output of each layer is concatenated with its input before passing to the next layer. A typical Densenet architecture consists of a number of dense blocks separated by transition layers: the transition layers contain a pooling operation, which allows some degree of translation invariance and downsamples the feature maps. A Densenet architecture adapted for semantic segmentation was presented in [8], which adopted the now standard approach of U-net [15]: a downsampling path, followed by an upsampling path, with skip connections passing feature maps of the sample spatial dimension from the downsampling path to the upsampling path.

In a previous paper [12], we described an alternative architecture adapting Densenet for semantic segmentation: in this architecture, which we called Deep-SCAN, there are no transition layers and no pooling operations. Instead, dilated convolutions are used to increase the receptive field of the classifier. The absence of transition layers means that the whole network can be seen as a single dense block, enabling gradients to pass easily to the deepest layers. While we believe that this approach offers many advantages over U-net, by avoiding pooling and upscaling, this comes at the price of very high memory consumption, since all feature maps are present at the resolution of the final segmentation image. This restricts the possible depth, batch size, and input patch size of the network.

© Springer Nature Switzerland AG 2019
A. Crimi et al. (Eds.): BrainLes 2018, LNCS 11384, pp. 456–465, 2019.
https://doi.org/10.1007/978-3-030-11726-9_40

In this paper we describe a family of CNN models for segmentation which represent a continuum from our previously described DeepSCAN models to U-net-like models, in which a pooling-free dense net is embedded inside a U-net style network. This allows the dense part of the network to operate at a lower resolution, improving memory efficiency while maintaining many good properties of the original DeepSCAN architecture.

We describe the general architecture of the family of DeepSCAN models, plus the particular features of the network as applied to brain tumor segmentation, including pre-processing, data augmentation, and a new uncertainty-motivated loss function. We report preliminary results on the validation portion of the BRATS 2018 dataset.

2 The DeepSCAN Family of Models

We describe here the constituent parts of the DeepSCAN family of models.

2.1 Densely Connected Layers and Densenet

Densenet [7] is a recently introduced architecture for image classification. The fundamental unit of a Densenet architecture is the densely connected block, or dense block. Such a block consists of a number of consecutive dense units, as pictured in Fig. 1. In such a unit, the output of each convolutional layer (where a layer here means some combination of convolutional filters, non-linearities and batch normalization) is concatenated to its input before passing to the next layer. The goal behind Densenet is to build an architecture which supports the training of very deep networks: the skip connections implicit in the concatenation of filter maps between layers allows the flow of gradients directly to those layers, providing an implicit deep supervision of those layers.

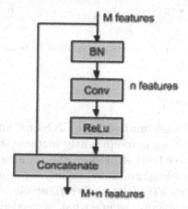

Fig. 1. A Dense unit, as used in the DeepSCAN architecture

In the original Densenet architecture, which has state-of-the-art performance on the CIFAR image recognition task, dense blocks are combined with transition blocks: non-densely connected convolutional layers, followed by a maxpooling layer. This helps to control parameter explosion (by limiting the size of the input to each dense block), but also means that the deep supervision is not direct, at the lowest layers of the network. This Dense-plus-transition architecture was also adopted by Jegou et al. [8], whose Tiramisu network is a U-net-style variation of the Densenet architecture designed for semantic segmentation.

In our previous paper [12], we dispensed with the transition layers: this means, in effect that the whole network (except for the final one by one convolutions) is a single dense block. This led to networks which were highly parameter efficient, but which had a very large memory footprint. In the current paper we hybridize this approach with the down/up-sampling approach of U-net [15].

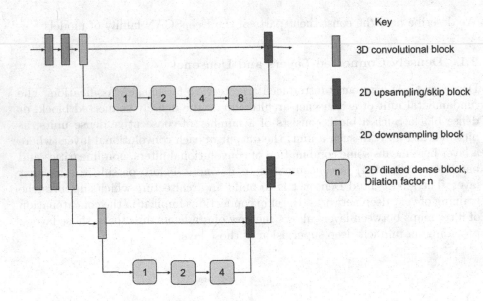

Fig. 2. Two DeepSCAN architectures, as applied to brain tumor segmentation

2.2 Dilated Convolutions

Some kind of pooling is found in almost all CNNs for image classification. The principal reason to use pooling is to efficiently increase the receptive field of the network at deeper levels without exploding the parameter space, but another common justification of pooling, and maxpooling in particular, is that it enables some translation invariance. Translation invariance is of course undesirable in semantic segmentation problems, where what is needed is instead translation *equivariance*: a translated input corresponding to a translated output. To that end, we use layers with dilated convolutions to aggregate features at multiple

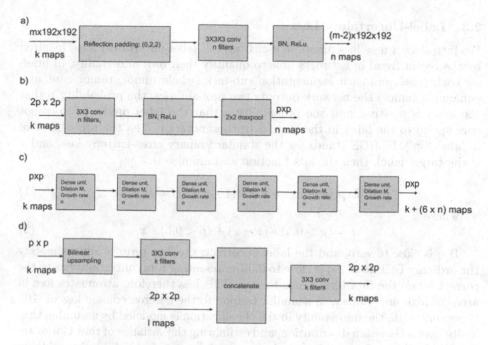

Fig. 3. Units of the DeepSCAN architecture: (a) 3D convolutional blocks, (b) Down-sampling block, (c) Dense block, with dilation M, (d) upsampling block. Except in the 3D block, all convolutions are preceded by 2 by 2 reflection padding.

scales. Dilated convolutions, sometimes called atrous convolutions, can be best visualized as convolutional layers "with holes": a 3 by 3 convolutional layer with dilation 2 is a 5 by 5 convolution, in which only the centre and corner values of the filter are nonzero, as illustrated in Fig. 4. Dilated convolutions are a simple way to increase the receptive field of a classifier without losing spatial information.

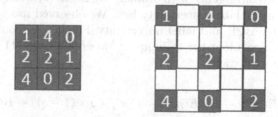

Fig. 4. Left, a 3 by 3 kernel. Right, a 3 by 3 kernel with dilation 2, visualized as a 5 by 5 kernel

2.3   Label-Uncertainty Loss

We introduce a new loss function, which we call label-uncertainty loss, inspired by the recent trend in networks able to quantify their own uncertainty. In brief, for each voxel, and each segmentation sub-task (whole tumor, tumor core, and enhancing tumor) the network outputs two probabilities: the probability p that the label is positive, and the probability q that the label predicted does not correspond to the label in the ground-truth annotation (i.e., the probability of a 'label flip'). IF BCE stands for the standard binary cross-entropy loss, and x is the target label, then the loss function we minimize is:

$$BCE(p, (1-x)*q + x*(1-q)) + BCE(q, z) \qquad (1)$$

where

$$z = (p > 0.5)*(1-x) + (p < 0.5)*x \qquad (2)$$

If q is close to zero, and the label is correct, the first term is approximately the ordinary BCE loss: if q is close to 0.5 (representing total uncertainty as to the correct label) the first term tends to zero. This loss therefore attenuates loss in areas of high uncertainty, in a similar fashion to the heteroscedastic loss of [10]. However, in [10] the uncertainty in the classification is modeled by assuming that logits have a Gaussian distribution, and estimating the variance of that Gaussian: this cannot be performed directly by gradient descent, instead requiring Monte Carlo sampling of the Gaussian distribution to perturb the output of the network. By contrast, label-uncertainty can be incorporated directly into the loss-function of the network. In fact, the label-uncertainty q can also be viewed as a variance: if we assume that the logit of p follows not a Gaussian but a logistic distribution (as is the standard assumption in classical statistical learning) with mean logit(p), then if the probability that a sample from that distribution is below zero is q, the variance of the logistic distribution is abs(logit(p)/logit(q)).

Since the label-uncertainty loss incorporates the current prediction in evaluating the probability of a label flip, it is important to apply the loss to a network which has already been pre-trained with ordinary BCE loss: for each of our networks we trained to convergence with ordinary BCE loss (typically 10–20 epochs) then switched to using label uncertainty loss. We observed more stability when using both ordinary BCE and label uncertainty. Further, to counter the effects of label imbalance, we adopt the technique of *focal loss* from [11]: therefore, the final loss function used was

$$(1 - p_x)^\gamma (BCE(p, x) + BCE(p, (1-x)*q + x*(1-q)) + BCE(q, z)) \qquad (3)$$

where p_x is p if x is 1 and (1-p) otherwise. For our experiments the value of γ used was 2.

2.4   The DeepSCAN Architecture

The design principles of the DeepSCAN models are (i) non-isotropic input volumes, with one dimension being rather small (in this case, 5 by 192 by 192)

(ii) initial application of enough 3D convolutions to reduce the short dimension to length 1, and (iii) a subsequent hybrid of 2D U-net and 2D Densenet, in which one or steps of convolution and maxpooling are followed by a number of densely connected blocks of dilated convolutions, with the dilation factor increasing with increasing depth, and then finally U-net-style upsampling blocks with skip connections from the previous downward path. The building blocks of these networks are shown in Fig. 3, and two architectures built from these blocks are shown in Fig. 2.

3   Initial Application to Brain Tumor Segmentation

Brain Tumor segmentation has become a benchmark problem in medical image segmentation, due to the existence since 2012 of a long-running competition, BRATS, together with a large curated dataset [1–3,13] of annotated images. Both fully-automated and semi-automatic approaches to brain-tumor segmentation are accepted to the challenge, with supervised learning approaches dominating the fully-automated part of the challenge. A good survey of approaches which dominated BRATS up to 2013 can be found here [5]. More recently, CNN-based approaches have dominated the fully-automated approaches to the problem [6,9,14].

We trained two networks, as pictured in Fig. 2. The networks were built using Pytorch, and trained using the Adam optimizer. Rather than using a softmax layer to classify the three labels (edema, enhancing, other tumor) we employ a multi-task approach to hierarchically segment the tumor into the three overlapping targets: whole tumor, tumor core and enhancing: thus the output of the network is three logits, one for each target. In addition, as per the label uncertainty loss, for each target the network outputs one label-flip logit.

3.1   Data Preparation and Homogenization

The raw values of MRI sequences cannot be compared across scanners and sequences, and therefore a homogenization is necessary across the training examples. In addition, learning in CNNs proceeds best when the inputs are standardized (i.e. mean zero, and unit variance). To this end, the nonzero intensities in the training, validation and testing sets were standardized, this being done across individual volumes rather than across the training set. This achieves both standardization and homogenization.

4   Cascaded Non-brain-tissue Removal

The BRATS dataset was assembled from a large number of data sources, and does not comprise raw imaging data: the volumes are re-sampled to 1 mm isovoxels, and in addition have been automatically skull-stripped. Unfortunately, the results of this skull-stripping vary: see Fig. 5 for an example with large amounts of residual skull tissue. Other examples have remnants of the dura or optic nerves. This remaining tissue can confound classification in two ways: it can be misidentified by the classification algorithm (though this is increasingly less likely as classifiers improve) and it can affect the distribution of the intensities in a volume, adversely impacting the global standardization of voxel values. To combat this effect, we used a cascade of networks to first segment the parenchymia from the poorly skull-stripped images, followed by a second network which identifies the tumor compartments as above. The ground truth for the brain mask was obtained by applying FSL-FAST to the T1 post Gadolinium imaging, as this tended to have the best definition in all three planes. The brain tissue label was assembled by taking the union of tumor, white matter and grey matter labels, and then taking the largest connected component.

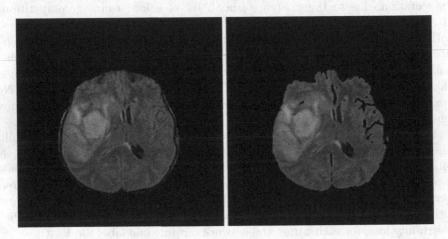

Fig. 5. A FLAIR image from the BRATS2018 testing dataset before (Left) and after (Right) additional brain extraction by our method

This brain-mask tissue label was used during training to ensure the training of networks robust to the presence or absence of non-brain tissue. In addition, we added a brain-mask label to the existing labels in the ground-truth for training, so that during testing a brain-mask for additional skull-stripping could be generated.

4.1 Data Augmentation

During training, we applied the following data augmentation: randomly flipping along the midline, random rotations in a randomly chosen principal axis, and random shifting and scaling of the standardised intensity values. In addition, the classifier was randomly shown either the original images, or images masked with the brain-mask generated as above.

4.2   Training

The network segments the volume slice-by slice: the input data is five consecutive slices from all four modalities, Ground truth for such a set of slices is the lesion mask of the central slice. Input images were initially cropped to remove as much empty space as possible. Batch size during training was 2. As a result, the input tensor to the model has dimensions 2 * 4 * 5 * 192 * 192. Models were trained using a cosine-annealing learning rate schedule, in which the learning rate was varied between 1e−5 and 1e−9 during each epoch.

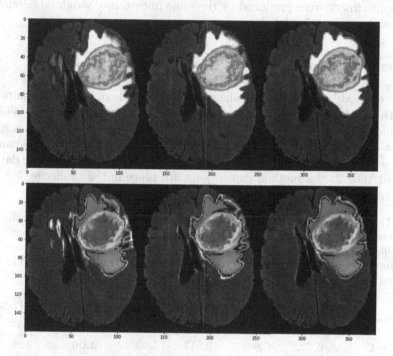

Fig. 6. Above: Whole tumor classified in the (sagittal, coronal, axial) plane. Below: Label-flip probability of the (sagittal, coronal, axial) segmentations

Slices from all three directions (sagittal, axial, coronal) were fed to the classifier for training. Examples of the different segmentations in those three directions (just for the whole tumor label) can be seen in Fig. 6.

4.3   Application of the Classifier

The initial application of the classifier is as follows: the volume is classified in the axial, sagittal and coronal planes separately, by both trained networks. This yields six logit maps, and six label-flip logit maps, for each target label. The logit maps were binarized with a threshold of 0 (corresponding to a standard threshold of 0.5 on the sigmoid of the logit).

Voxel-wise label confidence weights were then derived from the label-flip logits as the minimum of 0 and the negative of the label-flip logit, so that very confident classifications (corresponding to very large negative label-flip logits) contribute more than less-confident classifications. These weights were than used to ensemble the binarized maps.

The brain-mask label from this ensembled classification was then used to mask the input modalities, and the volume was again classified by both networks in all three directions. This yielded another six logit maps (with corresponding label-flip logit maps) for each tissue compartment. The final segmentations for each compartment were produced by the same uncertainty weighted ensembling as above, over all twelve label maps.

5   Results

Results of an ablation study are shown in Table 1, where we show results with and without label-uncertainty-based ensembling and brain extraction. While no single model showed dominance, the model with both novel features achieved the best results in one of Dice or Hausdorff distance for all three compartments, so was selected as the final model. Results on the BRATS 2018 testing data are shown in Table 2: this method gained joint 3rd place in the challenge [4].

Table 1. Results on the BRATS 2018 validation set using the online validation tool. Base denotes the ensemble of two DeepSCAN models over three directions, where ensembling is achieved by averaging logits. "+ U" denotes using averaging over label uncertainty instead of logits. "+ BE" denotes averaging over both original and brain-extracted inputs.

	Dice-ET	Dice-WT	Dice-TC	HD95-ET	HD95-WT	HD95-TC
Base	0.795	0.901	**0.854**	3.61	4.26	5.37
Base + U	0.792	0.901	0.847	3.60	**4.06**	4.99
Base + BE	**0.797**	0.901	0.851	3.60	4.41	5.58
Base + U + BE	0.796	**0.903**	0.847	**3.55**	4.17	**4.93**

Table 2. Results of the ensemble with brain extraction and uncertainty-driven ensembling on the BRATS 2018 testing set

Label	Dice-ET	Dice-WT	Dice-TC	HD95-ET	HD95-WT	HD95-TC
Mean	0.73189	0.88593	0.79926	3.48082	5.5185	5.5347
StdDev	0.27443	0.10182	0.26008	5.52176	9.34294	8.14881
Median	0.83199	0.91786	0.90847	1.73205	3.0	2.82843
25quantile	0.73922	0.87113	0.82327	1.41421	2.23607	1.73205
75quantile	0.88342	0.9396	0.93653	2.82843	5.09902	5.52101

References

1. Bakas, S., et al.: Advancing the cancer genome atlas glioma MRI collections with expert segmentation labels and radiomic features. Nature Sci. Data **4**, 170117 (2017)
2. Bakas, S., et al.: Segmentation labels and radiomic features for the pre-operative scans of the TCGA-GBM collection. In: The Cancer Imaging Archive (2017)
3. Bakas, S., et al.: Segmentation labels and radiomic features for the pre-operative scans of the TCGA-LGG collection. In: The Cancer Imaging Archive (2017)
4. Bakas, S., Reyes, M., Menze, B., et al.: Identifying the best machine learning algorithms for brain tumor segmentation, progression assessment, and overall survival prediction in the brats challenge. 1811.02629 (2015). https://arxiv.org/abs/1811.02629
5. Bauer, S., Wiest, R., Nolte, L.L., Reyes, M.: A survey of mri-based medical image analysis for brain tumor studies. Phys. Med. Biol. **58**(13), R97–129 (2013)
6. Havaei, M., et al.: Brain tumor segmentation with deep neural networks. Med. Image Anal. **35**, 18–31 (2017)
7. Huang, G., Liu, Z., van der Maaten, L., Weinberger, K.Q.: Densely connected convolutional networks. In: Proceedings of the IEEE Conference on Computer Vision and Pattern Recognition (2017)
8. Jégou, S., Drozdzal, M., Vázquez, D., Romero, A., Bengio, Y.: The one hundred layers tiramisu: Fully convolutional densenets for semantic segmentation. vol. abs/1611.09326 (2016). http://arxiv.org/abs/1611.09326
9. Kamnitsas, K., et al.: Efficient multi-scale 3D CNN with fully connected crf for accurate brain lesion segmentation. Med. Image Anal. **36**, 61–78 (2017)
10. Kendall, A., Gal, Y.: What uncertainties do we need in bayesian deep learning for computer vision? In: NIPS (2017)
11. Lin, T.Y., Goyal, P., Girshick, R.B., He, K., Dollár, P.: Focal loss for dense object detection. In: 2017 IEEE International Conference on Computer Vision (ICCV), pp. 2999–3007 (2017)
12. McKinley, R., Jungo, A., Wiest, R., Reyes, M.: Pooling-free fully convolutional networks with dense skip connections for semantic segmentation, with application to brain tumor segmentation. In: Crimi, A., Bakas, S., Kuijf, H., Menze, B., Reyes, M. (eds.) BrainLes 2017. LNCS, vol. 10670, pp. 169–177. Springer, Cham (2018). https://doi.org/10.1007/978-3-319-75238-9_15
13. Menze, B.H., et al.: The multimodal brain tumor image segmentation benchmark (BRATS). IEEE Trans. Med. Imaging **34**(10), 1993–2024 (2015)
14. Pereira, S., Pinto, A., Alves, V., Silva, C.A.: Brain tumor segmentation using convolutional neural networks in MRI images. IEEE Trans. Med. Imaging **35**(5), 1240–1251 (2016)
15. Ronneberger, O., Fischer, P., Brox, T.: U-Net: convolutional networks for biomedical image segmentation. In: Navab, N., Hornegger, J., Wells, W.M., Frangi, A.F. (eds.) MICCAI 2015. LNCS, vol. 9351, pp. 234–241. Springer, Cham (2015). https://doi.org/10.1007/978-3-319-24574-4_28
16. Yu, F., Koltun, V.: Multi-scale context aggregation by dilated convolutions. In: Proceedings of International Conference on Learning Representations (ICLR 2017) (2017)

Brain Tumor Segmentation Using Bit-plane and UNET

Tran Anh Tuan[1(✉)], Tran Anh Tuan[1], and Pham The Bao[2]

[1] Faculty of Math and Computer Science, University of Science,
Vietnam National University, Ho Chi Minh City, Vietnam
tatuan@hcmus.edu.vn, tuantran261083@gmail.com
[2] Department of Computer Science, Sai Gon University,
Ho Chi Minh City, Vietnam
ptbao2005@gmail.com

Abstract. The extraction of brain tumor tissues in 3D Brain Magnetic Resonance Imaging plays an important role in diagnosis gliomas. In this paper, we use clinical data to develop an approach to segment Enhancing Tumor, Tumor Core, and Whole Tumor which are the sub-regions of glioma. Our proposed method starts with Bit-plane to get the most significant and least significant bits which can cluster and generate more images. Then U-Net, a popular CNN model for object segmentation, is applied to segment all of the glioma regions. In the process, U-Net is implemented by multiple kernels to acquire more accurate results. We evaluated the proposed method with the database BRATS challenge in 2018. On validation data, the method achieves a performance of 82%, 68%, and 70% Dice scores and of 77%, 48%, and 51% on testing data for the Whole Tumor, Enhancing Tumor, and Tumor Core respectively.

Keywords: 3D brain MRI · Brain tumor · Bit-plane ·
2D U-Net · CNN · BRATS challenge in 2018

1 Introduction

Accurate extraction of brain tumor types plays an important role in diagnosis and treatment planning. Neuro-imaging methods in Magnetic Resonance Imaging (MRI) provide anatomical and pathophysiological information about brain tumors and aid in diagnosis, treatment planning and follow-up of patients. Manual segmentation of brain tumor tissue is a difficult and time-consuming job. Therefore, brain tumor segmentation from 3D Brain MRI automatically can solve these problems. Among many types of brain tumor, Gliomas are the most common primary brain malignancies, with different degrees of aggressiveness, variable prognosis and various heterogeneous histological sub-regions. In this paper, we focus on Enhancing Tumor, Tumor Core, and Whole Tumor segmentation which are the sub-regions of gliomas segmentation.

Segmentation of brain tumors in multimodal MRI scans is one of the most challenging tasks in medical image analysis. Currently, there are many methods related to brain tumor segmentation have been proposed [1, 2]. In this paper, we divide these methods into two categories: mathematical methods and machine learning methods

© Springer Nature Switzerland AG 2019
A. Crimi et al. (Eds.): BrainLes 2018, LNCS 11384, pp. 466–475, 2019.
https://doi.org/10.1007/978-3-030-11726-9_41

- In mathematical methods: the tumor can be segmented by using threshold, Edge-Based Method [3], Atlas [4]. In Dubey et al. [5], rough set based fuzzy clustering is proposed to segment the tumor.
- In machine learning methods: Traditionally, many features are extracted manually from image and given to the classifier. However, in recent years, Convolution Neural Networks (CNNs) which have been shown to excel learning a hierarchy task-adapted complex feature are seen prominent success in image classification, object detection and image semantic segmentation [6–8]. Many of the brain tumor segmentation methods based on CNNs or combining CNNs with the traditional method are also proposed [9–11].

In this study, we combine the Bit-plane method [12] and U-Net architecture [13] for tumor segmentation. First, we use Bit-plane to divide images into many images by determining significant bits. Second, the images with the significant bits can be used to segment the object boundary. Finally, original images and images with least significant bits can be used to determine tissues inside the boundary. Both stages used the U-Net with multiple kernels to segment the tissues more accurately.

The rest of the paper is organized as follows: in the next Sect. 2, we present our proposed method for brain tumor segmentation and the experimental results are shown in Sect. 3. We give the conclusion and discussion in Sect. 4.

2 Our Method

The proposed method is illustrated in Fig. 1. There are three main stages: preprocess, object boundary segmentation, and tissues segmentation. As shown in Fig. 1, after converting to 2D images and grouping images, the first U-Net predicts object boundary of the Whole Tumor and the other U-Net utilizes features to predicts the label of all pixel inside the boundary.

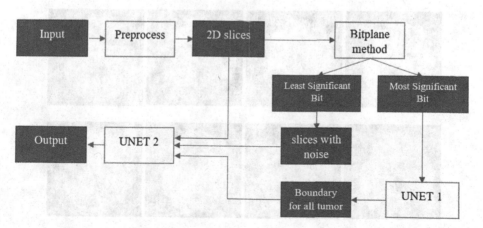

Fig. 1. The overall of our proposed method for brain tumor segmentation

2.1 Preprocessing

The preprocessing is the necessary stage before any tissue segmentation. We implement three main steps

- Normalization: each individual 3D image is scaled to the range [0–255].
- Brain Slice Category [14]: We group the slices which can contain the tumors together to get the accuracy better. The implementation can be done automatically by learning feature or set manually by omitting some first and end slices. Here, we detect the tumor from slices 40–140.
- Object Region: each 2D image can be cropped to implement deep learning effectively. Here, we cropped the image size from (256, 256) to (176, 176).

Majority of the volumes in the dataset were acquired along the axial plane and hence had the highest resolution this plane. Therefore, all 3D brain MRI is transformed to 2D brain slices on axial slice extracted from all four sequences. After the preprocessing stage, all the 2D slices is from (155, 256, 256) to (100, 176, 176) with value range (0–255).

2.2 Boundary for All Tumors

The bit plane method [12] is based on decomposing a multilevel image into a series of binary images. The intensities of an image are based on the Eq. (1):

$$a_{m-1}2^{m-1} + a_{m-2}2^{m-2} + \ldots + a_1 2^1 + a_0 2^0 \qquad (1)$$

We realize that the final plane contains the most significant bit. In order to segment the boundary of the object, we proposed using k significant bits to eliminate the noise which can affect the image. Instead of using a single plane, we can combine multiple planes together. We represented the slice by keeping from one-bit to eight-bit planes in Fig. 2.

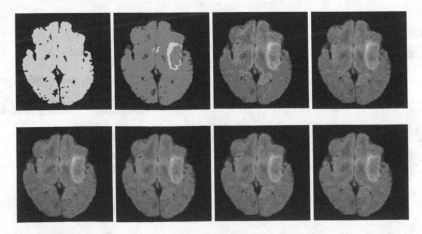

Fig. 2. An example of most significant bits from Flair image. In the first row, from the left to the right are images keeping one-bit to four-bit plane. In the second row, from the left to the right are images keeping five-bit to eight-bit plane.

In this study, we eliminate the last 6 bits to remove the noise and only used the first 2 bits to keep the significant data to generate images for training to detect the object boundary. After getting the images, U-Net is used to segment the background and the Whole Tumor by using the 2D slices input and the image which contains 2 significant bits.

2.3 Tissues Segmentation

After segmenting the tumor boundary, different types of tumors inside the boundary can be segmented by using other U-Net. The input data is the data which is preprocessed from the first stage. However, to get a better result, we suggest two contributions to enhance the segmentation:

- Another training data are the images with noises which are generated from the least significant bits. In this study, we implement the noise from three last bits of each image. The example of the implementation is shown in Fig. 3 with the input from Flair image.
- Implementing U-Net with multiple kernel size to get the better segmentation [15]. Let $K = \{(K_1, (a_1, b_1)), \ldots, (K_n, (a_n, b_n))\}$ is the set of n filters K with size (a, b). The output of layer i is the merge of feature maps that the layer i generate $\bigcup_{j=1}^{n} K_j$. In this study, the numbers within each Conv block comprises of 2 sets of convolutions by 3×3 kernels and 2 sets of convolutions by 5×5 kernel as shown in Fig. 3.

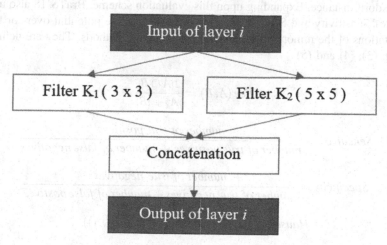

Fig. 3. Example using multiple kernels in each convolution for segmentation model

3   Results

We use BraTS'2018 training data [16–19], consisting of 210 pre-operative MRI scans of subjects with glioblastoma (HGG) and 75 scans of subjects with lower grade glioma (LGG). These multimodal scans describe (a) native (T1) and (b) post-contrast T1-weighted (T1Gd), (c) T2-weighted (T2), and (d) T2 Fluid Attenuated Inversion Recovery (FLAIR) volumes and were acquired with different clinical protocols and various scanners from multiple (n = 19) institutions. Ground truth annotations comprise the GD-Enhancing Tumor (ET—label 4), the peritumoral edema (ED—label 2), and the necrotic and non-enhancing tumor core (NCR/NET—label 1)

Our proposed method is implemented based on a Keras library [20] with backend Tensorflow [21]. 'Adam' optimizer [22] and 'binary_crossentropy' loss [23] are used in UNET. We run the method with 50 epochs on Ge-force GTX980 graphics card. Figure 4 shows the result from an example of experiments in the samples of image scans on the real data of the BraTS'18. The top row of Fig. 4 are the original images, from the left to the right: FLAIR, T1, T1ce and T2. The second row contains images from two most significant bits. The third row contains images with noise from three least significant bits. The fourth and the last row is the result of segmentation for each stage.

Tables 1 and 2 show the average performance for each label and score for all the validation patients and all the testing patients [24]. The BraTS'18 competition has four metrics to assess the accuracy of segmentation results and to measure the similarity between the segmentations A and B. For the segmentation task, and for consistency with the configuration of the previous BraTS challenges, we will use the Dice score and the Hausdorff distance. Expanding upon this evaluation scheme, BraTS'18 also use the metrics of Sensitivity and Specificity, allowing to determine potential over- or under-segmentations of the tumor sub-regions by participating methods. They are defined as Eqs. (2), (3), (4) and (5)

$$Dice\,(A,B) = \frac{2|A \cap B|}{|A| + |B|} \tag{2}$$

$$Sensitivity = \frac{number\ of\ true\ positives}{number\ of\ true\ positives + number\ of\ false\ negatives} \tag{3}$$

$$Specificity = \frac{number\ of\ true\ negatives}{number\ of\ true\ negatives + number\ of\ false\ postives} \tag{4}$$

$$Hausdorff\,(A,B) = \max\,(h(A,B), h(B,A)) \tag{5}$$

The Dice metric is the similarity between two volumes A and B, corresponding to the output segmentation of the model and clinical ground truth annotations, respectively. Sensitivity and Specificity are statistical measures employed to evaluate the behavior of the predictions and the proportions of True Positives, False Negatives, False Positives, and True Negatives voxels. Hausdorff(A, B) is the Hausdorff distance between the two surfaces of A and B where $h(A,B) = max_{a \in A} min_{b \in B} d(a,b)$. Here, $d(a,b)$ is the Euclidean distance between a and b. This metric indicates the

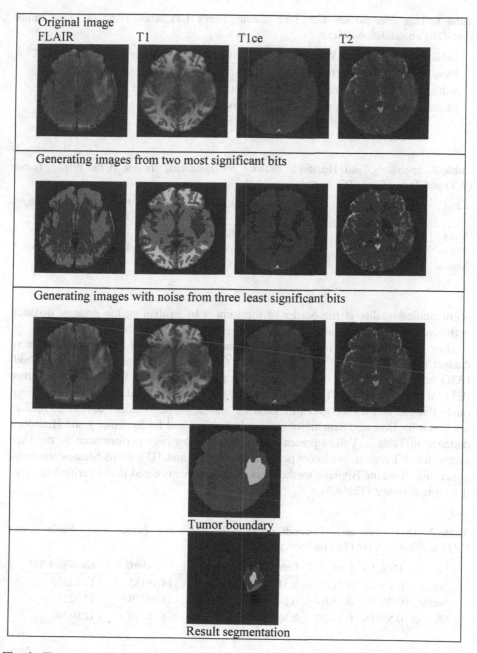

Fig. 4. The results from an example of brain tumor segmentation on the real data of the BraTS'18.

Table 1. Dice score and Sensitivity for Enhancing Tumor (ET), Whole Tumor (WT) and Tumor Core (TC) on validation data

Label	Dice_ET	Dice_WT	Dice_TC	Sensitivity ET	Sensitivity WT	Sensitivity TC
Mean	0.68252	0.81871	0.69986	0.70254	0.77338	0.64729
StdDev	0.28138	0.16968	0.2913	0.25413	0.20257	0.30542
Median	0.80902	0.88296	0.82567	0.7804	0.83364	0.75828

Table 2. Specificity, and Hausdorff distance for Enhancing Tumor (ET), Whole Tumor (WT) and Tumor Core (TC) on validation data

Label	Specificity ET	Specificity WT	Specificity TC	Hausdorff95 ET	Hausdorff95 WT	Hausdorff95 TC	Specificity ET
Mean	0.99783	0.99525	0.99862	7.01652	9.42113	12.46282	0.99783
StdDev	0.00403	0.00589	0.00197	9.53618	9.74773	14.68491	0.00403
Median	0.9989	0.9967	0.99905	2.82843	6.04138	6.16441	0.9989

segmentation quality at the border of the tumors by evaluating the greatest distance between the two segmentation surfaces and is independent of the tumor size.

For our participation in BraTS'2018 competition, we used 100% of the training dataset (285 subjects) for training purpose. Our model was trained to segment both HGG and LGG volumes. The result of the proposed method for Enhancing Tumor (ET), Whole Tumor (WT) and Tumor Core (TC) segmentation using the four previously defined metrics are given in Tables 1 and 2. Mean, standard deviation, median are given for Dice and Sensitivity metrics in Table 1 and for Specificity and Hausdorff distance in Table 2. Values presented in Table 1 show high performance on the Dice metric for WT region, but lower performance for ET and TC regions because the noise generating from the Bitplane method has a small difference and is not verified to make it as a real image (Table 3).

Table 3. Dice score, and Hausdorff distance for Enhancing Tumor (ET), Whole Tumor (WT) and Tumor Core (TC) on testing data

Label	Dice_ET	Dice_WT	Dice_TC	Hausdorff ET	Hausdorff WT	Hausdorff TC
Mean	0.47623	0.77338	0.51291	12.3933	14.19183	15.62507
StdDev	0.26239	0.15914	0.24294	12.33002	16.98779	14.32559
Median	0.55018	0.83227	0.56504	8.11168	8.06226	12.08305

4   Conclusions and Discussion

Nowadays, generating data is a good approach for segmentation. In this paper, we propose using Bit-plane to generate more image remaining significant features. Besides, we also implement the U-Net with multiple kernels to get better performance. The result is evaluated without additional data and is shown with a promising performance. In the future, we can concentrate on two main aspects:

- Using type of image
  As shown in Fig. 5, every type of image has specific characteristics. Therefore, instead of using all 4 types of images as an input for all stages, we can use a suitable type of image for each stage to get the better result.
- Using GAN
  Generative Adversarial Networks (GAN) [25] is one of the most promising recent developments in deep learning. GAN solve the problem of unsupervised learning by training two deep networks, called Generator and Discriminator, that compete and cooperate with each other. If we can combine GAN with Bitplane to generate more real images, the result segmentation will be better.

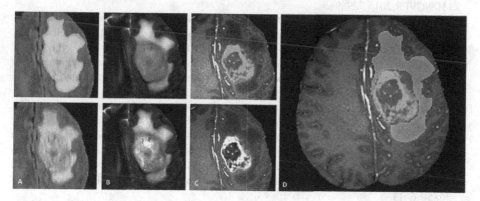

Fig. 5. Glioma sub-regions. Shown are image patches with the tumor sub-regions that are annotated in the different modalities (top left) and the final labels for the whole dataset (right). The image patches show from left to right: the Whole Tumor (yellow) visible in T2-FLAIR (Fig. A), the Tumor Core (red) visible in T2 (Fig. B), the Enhancing Tumor structures (light blue) visible in T1Gd, surrounding the cystic/necrotic components of the core (green) (Fig. C). The segmentations are combined to generate the final labels of the tumor sub-regions (Fig. D): edema (yellow), non-enhancing solid core (red), necrotic/cystic core (green), enhancing core (blue) [16–19]. (Color figure online)

Acknowledgement. We would like to thank Business Intelligence LAB at University of Economics and Law for supporting us throughout this paper. The study was supported by Science and Technology Incubator Youth Program, managed by the Center for Science and Technology Development, Ho Chi Minh Communist Youth Union, 2018.

References

1. Gordillo, N., Montseny, E., Sobrevilla, P.: State of the art survey on MRI brain tumor segmentation. IEEE (2013). https://doi.org/10.1016/j.mri.2013.05.002
2. Angulakshmi, M., Lakshmi Priya, G.G.: Automated brain tumor segmentation techniques—a review. Int. J. Imaging Syst. Technol. **27**, 66–77 (2017). https://doi.org/10.1002/ima.22211
3. Aslam, A., Khan, E., Beg, M.M.S.: Improved edge detection algorithm for brain tumor segmentation. Procedia Comput. Sci. **58**, 430–437 (2015). https://doi.org/10.1016/j.procs.2015.08.057
4. Bauer, S., Seiler, C., Bardyn, T., Buechler, P., Reyes, M.: Atlas-based segmentation of brain tumor images using a Markov Random Field-based tumor growth model and non-rigid registration. In: Proceedings of IEEE EMBC, pp. 4080–4083 (2010). https://doi.org/10.1109/IEMBS.2010.5627302
5. Dubey, Y.K., Mushrif, M.M., Mitra, K.: Segmentation of brain MR images using rough set based intuitionistic fuzzy clustering. Biocybern. Biomed. Eng., 413–426 (2016). https://doi.org/10.1016/j.bbe.2016.01.001
6. Lecun, Y., Bottou, L., Bengio, Y., Haffner, P.: Gradient-based learning applied to document recognition. Proc. IEEE, 2278–2324 (1998). https://doi.org/10.1109/5.726791
7. Shelhamer, E., Long, J., Darrell, T.: Fully convolutional networks for semantic segmentation. IEEE Trans. Pattern Anal. Mach. Intell. **39**(4), 640–651 (2017). https://doi.org/10.1109/CVPR.2015.7298965
8. Badrinarayanan, V., Kendall, A., Cipolla, R.: SegNet: a deep convolutional en-coder-decoder architecture for image segmentation. IEEE Trans. Pattern Anal. Mach. Intell. **39**(12), 2481–2495 (2017). https://doi.org/10.1109/TPAMI.2016.2644615
9. Guo, L., et al.: A fuzzy feature fusion method for auto-segmentation of gliomas with multimodality diffusion and perfusion magnetic resonance images in radiotherapy. Scientific Reports, vol. 8, Article number: 3231 (2018)
10. Shukla, G., et al.: Advanced magnetic resonance imaging in glioblastoma: a review. Chin. Clin. Oncol. **6**(4), 40 (2017). https://doi.org/10.21037/cco.2017.06.28
11. Bakas, S., et al.: GLISTRboost: combining multimodal MRI segmentation, registration, and biophysical tumor growth modeling with gradient boosting machines for glioma segmentation. In: Crimi, A., Menze, B., Maier, O., Reyes, M., Handels, H. (eds.) BrainLes 2015. LNCS, vol. 9556, pp. 144–155. Springer, Cham (2016). https://doi.org/10.1007/978-3-319-30858-6_13
12. Gonzalez, R.C., Woods, R.E.: Digital Image Processing. Prentice Hall Inc., Upper Saddle River (2002)
13. Ronneberger, O., Fischer, P., Brox, T.: U-Net: convolutional networks for biomedical image segmentation. In: Navab, N., Hornegger, J., Wells, W.M., Frangi, A.F. (eds.) MICCAI 2015. LNCS, vol. 9351, pp. 234–241. Springer, Cham (2015). https://doi.org/10.1007/978-3-319-24574-4_28
14. Tuan, T.A., Kim, J.Y., Bao, P.T.: 3D brain magnetic resonance imaging segmentation by using bitplane and adaptive fast marching. Int. J. Imaging Syst. Technol. **28**, 223–230 (2018). https://doi.org/10.1002/ima.22273
15. Szegedy, C., et al.: Going deeper with convolutions. In: IEEE Conference on Computer Vision and Pattern Recognition (CVPR) (2015). https://doi.org/10.1109/CVPR.2015.7298594
16. Menze, B.H., et al.: The multimodal brain tumor image segmentation benchmark (BRATS). IEEE Trans. Med. Imaging **34**(10), 1993–2024 (2015). https://doi.org/10.1109/TMI.2014.2377694

17. Bakas, S., et al.: Advancing The Cancer Genome Atlas glioma MRI collections with expert segmentation labels and radiomic features. Nat. Sci. Data **4**, 170117 (2017). https://doi.org/10.1038/sdata.2017.117
18. Bakas, S., et al.: Segmentation labels and radiomic features for the pre-operative scans of the TCGA-GBM collection. The Cancer Imaging Archive (2017). https://doi.org/10.7937/k9/TCIA.2017.KLXWJJ1Q
19. Bakas, S., et al.: Segmentation labels and radiomic features for the pre-operative scans of the TCGA-LGG collection. The Cancer Imaging Archive (2017). https://doi.org/10.7937/K9/TCIA.2017.GJQ7R0EF
20. Chollet, F., et al.: Keras (2015). https://keras.io
21. Abadi, M., et al.: TensorFlow: a system for large-scale machine learning. In: OSDI, vol. 16, pp. 265–283 (2016)
22. Kingma, D.P., Ba, L.J.: Adam: a method for stochastic optimization. In: International Conference on Learning Representations (2015)
23. Bishop, C.M.: Pattern Recognition and Machine Learning. Springer, New York (2006)
24. Bakas, S., Reyes, M., et al.: Identifying the best machine learning algorithms for brain tumor segmentation, progression assessment, and overall survival prediction in the BRATS challenge. arXiv preprint arXiv:1811.02629 (2018)
25. Goodfellow, I.J., et al.: Generative adversarial nets. In: Proceedings of the 27th International Conference on Neural Information Processing Systems, NIPS 2014 (2014)

Glioma Segmentation and a Simple Accurate Model for Overall Survival Prediction

Evan Gates[1], J. Gregory Pauloski[1], Dawid Schellingerhout[2], and David Fuentes[1(✉)]

[1] Department of Imaging Physics, University of Texas
MD Anderson Cancer Center, Houston, TX, USA
{EGates1,DTfuentes}@mdanderson.org
[2] Department of Cancer Systems Imaging and Diagnostic Radiology,
University of Texas MD Anderson Cancer Center, Houston, TX, USA

Abstract. Brain tumor segmentation is a challenging task necessary for quantitative tumor analysis and diagnosis. We apply a multi-scale convolutional neural network based on the DeepMedic to segment glioma subvolumes provided in the 2018 MICCAI Brain Tumor Segmentation Challenge. We go on to extract intensity and shape features from the images and cross-validate machine learning models to predict overall survival. Using only the mean FLAIR intensity, nonenhancing tumor volume, and patient age we are able to predict patient overall survival with reasonable accuracy.

Keywords: Glioblastoma · Segmentation · Neural network · Quantitative imaging

1 Introduction

Gliomas are highly malignant primary brain tumors that carry a dismal median overall survival of 15 months for high grade tumors [1]. One characteristic that contributes to this poor survival is the substantial heterogeneity. Spatial heterogeneity within a tumor implicitly increases the chances that a therapy resistant tumor subpopulation exists and thus frequently indicates poor clinical prognosis [2]. Successful and automated detection of distinct subvolumes (enhancing, nonenhancing, and necrotic regions, etc.) is a key step in quantitative analysis towards patient risk stratification and computer aided diagnosis. In recent years, convolutional neural networks (CNNs) are the undisputed champions of biomedical segmentation tasks [3]. Quantitative measurements of these subvolumes are likely to provide insight into patient's prognosis.

In this work we use a multi-scale convolutional neural network to segment glioma sub-volumes in multi-contrast MRI images. We go on to extract shape and intensity features from the sub-volumes to predict patient overall survival.

© Springer Nature Switzerland AG 2019
A. Crimi et al. (Eds.): BrainLes 2018, LNCS 11384, pp. 476–484, 2019.
https://doi.org/10.1007/978-3-030-11726-9_42

Results on the 2018 MICCAI Brain Tumor Segmentation (BraTS) Challenge [4–7] are provided. Final challenge rankings relative to other contest entries are available online [8].

2 Segmentation

2.1 Network Structure

Data Preprocessing. The BraTS 2018 Training set contains 285 multi-contrast MRI (T1, T1ce, T2, FLAIR) scans of high and low-grade gliomas. 75 of the 285 patients are labeled low-grade (LGG) and the remaining are high-grade (HGG). The imaging data is brain extracted, registered, and resampled to 1 mm isotropic voxel size. Each subject has a ground truth segmentation with four labels, non-tumor (label 0), necrotic and nonenhancing tumor core (label 1), peritumoral edema (label 2), and Gadolinium-enhancing tumor (label 4). The BraTS 2018 Validation set contains a mix of 66 HGG and LGG patients equivalently pre-processed and does not have ground truth segmentations.

All MRI scans were normalized by subtracting the mean intensity and dividing by the standard deviation. A binary brain mask for each patient was also created using the T1 scan, and this mask is used by the CNN to focus sampling on only the brain. The same preproccesing steps were also applied to the validation data set before segmenting.

Convolutional Neural Network. We used a 3-dimensional CNN built using the DeepMedic architecture created by Kamnitsas et al. [9]. DeepMedic has consistently produced high performing image segmentations in previous BraTS challenges. Sampling was used to produce image sub-volumes of size 37^3, and an equal number of sub-volumes centered on the foreground and background was taken to reduce class imbalance. Our CNN implementation contains three pathways consisting of eleven convolutional layers each. The pathways include one normal resolution and two downsampled where one was downsampled by a factor of 3^3 and the other by 5^3. The first seven layers use a 3^3 kernel with 30 to 50 features per layer. After the first seven layers, the downsampled pathways are upsampled to match the normal resolution pathway and all three pathways are concatenated. The concatenated features are then fed into two fully connected layers with 250 features and a kernel size of 3^3 and 1^3 respectively. The final layer is a fully connected layer with kernel of size 1^3 and four features. Dropout rates of 50% were used on the final two layers to prevent overfitting. The four features in the last fully connected layer are the output of the network and represent binary masks for each of the four segmentation labels.

Initially the CNN was trained on 80% of the BraTS 2018 Training Data and the remaining 20% was reserved for model validation. We tuned the batch size, learning rate, and optimizer until we found a set of parameters that gave the most accurate validation results and efficient use of our hardware. The final network was trained for 50 epochs with a batch size of 10, and the RMSprop optimizer was used with an initial learning rate of 0.001 and lowered throughout training. Before performing inference on the validation set we retrained the network with 95% of the BraTS 2018 Training Data and 5% reserved for validation. This training on a Nvidia Kepler Titan 6 GB took 96 h and, using the trained CNN, we performed a full inference on the BraTS 2018 Validation Data to produce binary mask for each of the four segmentation labels.

2.2   Training and Validation Set Results

The segmentation results from the full inference on the validation set were uploaded to the BraTS Challenge portal where the Dice score, sensitivity, specificity, and Hausdorff distance were calculated. Results for the 66 patients in the validation set are shown in Table 1. We also performed a full inference on the 5% of training set cases which were excluded from the model training process, so that we can compare the performance of the model against the ground truth segmentations. This comparison is shown in Table 2. From these samples, we can see that the CNN classifies the overall tumor well but has greater difficulty classifying regions with a dense mix of lower and higher grades.

Table 1. Mean values for metrics from segmentations on the training and validation data sets.

Data set	Label	Dice	Sensitivity	Specificity	Hausdorff
Training	Enhancing tumor	0.7332	0.84265	0.99781	6.20545
	Whole tumor	0.89633	0.88636	0.99531	5.14866
	Tumor core	0.75292	0.73297	0.99833	8.47618
Validation	Enhancing tumor	0.67831	0.72923	0.99611	14.52297
	Whole tumor	0.80558	0.81374	0.98703	14.415
	Tumor core	0.6852	0.68018	0.99619	20.01745

3   Survival Analysis

Of the 285 training cases, 163 cases had age (range 19–86 years) and overall survival data (range 5–1767 days) provided. We used this clinical data and features extracted from segmentations to predict the overall survival in days.

After determining the best model, we applied it to the 28 challenge validation set cases with patient age and gross total resection status.

The challenge assesses the predictions based on the accuracy: total number of cases correctly assigned a survival <10 months, between 10 and 15 months, and >15 months. The mean-square-error (MSE) is also used as a performance metric.

Table 2. Ground truth versus CNN segmentation samples from the validation data excluded from training. Red is nonenhancing tumor and necrosis (tumor core), green is edema, and yellow is enhancing tumor. The T1 weighted image for each patient is shown for reference.

Patient ID	Brats18_CBICA _AQQ_1	Brats18_CBICA _AXL_1	Brats18_CBICA _AYA_1

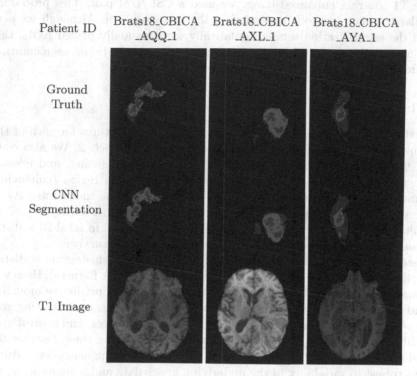

Ground Truth

CNN Segmentation

T1 Image

3.1   Image Processing

The format of the provided imaging data is described in Sect. 2.1. For the survival task, we further pre-processed the data by normalizing based on reference tissue intensities. Creating a consistent intensity scale between patients allows images features to discriminate short and long survival patients. Note, this is different than the normalization used in the segmentation task where each image had mean zero and standard deviation one. To apply this normalization, we placed small regions of interest for each patient in the gray matter (GM) of the lentiform nucleus, the cerebrospinal fluid (CSF) of the ventricles, and the normal appearing white matter (WM). Using the mean intensity for a pair of reference tissues, each voxel in the image was linearly scaled to map the mean intensities to 0 and 1 respectively. For example, in the FLAIR image each voxel value x was transformed according to

$$y_{CSF/WM} = \frac{x - \overline{CSF}}{\overline{WM} - \overline{CSF}}$$

For the FLAIR image CSF and WM were chosen because they were the darkest and brightest reference tissues respectively. For sequences T1 and FLAIR we normalized using the CSF/WM pair, for T2 we used the WM/CSF pair, and for the T1 contrast enhanced image we used a CSF/GM pair. This procedure is similar to other methods presented in the literature [10]. Although we performed the normalization semi-automatically with manually placed ROIs, this procedure can be performed fully automated using brain tissue segmentation software applied to the non-tumor regions.

3.2   Image Features

To predict patient overall survival we calculated image features for each of the available image sequences and segmentation labels from Sect. 2. We also computed the union of the three regions (nonenhancing, enhancing, and edema) to generate a whole-tumor ROI for each patient. For each region (enhancing, nonenhancing, edema, and whole tumor) we computed the mean intensity of that region for each image. (T1, T1 contrast enhances, T2, FLAIR) as well as the volume using the *Pyradiomics* software package [11]. So, in total 20 features (16 means and 4 volumes) were used for predicting overall survival.

We experimented with features quantifying higher order histogram statistics (quantiles, skewness, etc) and complex shape descriptors (i.e. flatness). However, we found these features did not improve the performance of predictive modeling beyond using just mean values. Similarly, we quantified image texture using gray level co-occurrence matrices and gray level run length matrices, and nearest gray tone difference matrices [12] but again found that including these features did not substantially increase model performance. Since these higher-order features are less robust to variability in the underlying image data and segmentation, we chose to consider only mean intensity and volume features in our final analysis.

3.3   Survival Task

An overview of our model development approach to predict survival is shown in Fig. 1. A family of models was considered with distinct permutations of variable selection methods and machine learning prediction algorithm. The best model was used to make predictions on the provided validation data. Modeling for the survival task was implemented in R version 3.4.0.

We partitioned the training data into 80% training and 20% testing data with an approximately equal proportion of short, medium, and long survivors in each set. Using the 80% partition, we performed variable selection and trained several classes of predictive models including linear models, neural networks, and random forests using leave-one-out cross validation. We selected the model with the highest Pearson correlation (R^2) between predicted and observed overall survival within the cross validation and made predictions on the testing set to see how well the model generalized.

For feature selection we consecutively applied univariate, multivariate, and step-wise feature elimination. After each selection step the resulting variables

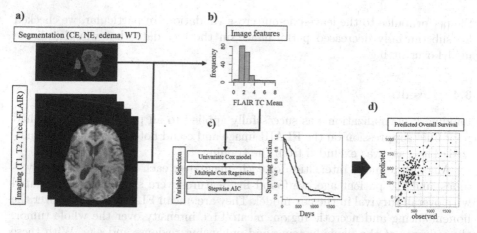

Fig. 1. Flowchart depicting the modeling process and variable selection methods for predicting overall survival. First the imaging data and computed segmentations (**a**) are used to extract mean intensity features and volumes. Panel (**b**) shows a histogram of the mean FLAIR intensity over the tumor core (TC) region. Several variable selection methods based on the cox model are used to generate input sets for predictive models (**c**). Features were first tested for significant association with overall survival using univariate cox models and discarding non-significant features. The set of univariate significant features was further reduced by constructing a multivariate cox model and again eliminating redundant (non-significant features), followed by stepwise AIC. The remaining image features and the patient's age, an important clinical factor, were used to predict overall survival (**d**). Before the variable selection 20% of the training data was held out for as an independent testing and the remaining 80% is used to select the best variable and model combination.

were stored as a possible set of inputs to predictive models. First, we used a Cox model to individually determine which image features were significantly associated with overall survival. Any feature with $p > 0.05$ for the Wald-test was discarded. Next, the remaining features were fed into a multivariate Cox model to reduce redundancy. Features with $p < 0.05$ in the multivariate Cox model were retained. Lastly, we further reduced the set of inputs using step-wise elimination based on Akaike Information Criteria (AIC) [13]. Starting with all variables, the stepwise AIC algorithm eliminates or replaces variables one at a time to maximize the AIC.

In addition, we applied the Boruta method [14] to select variables predictive of overall survival. The Boruta method is based on variable importance from the random forest algorithm, which has traditionally been a top performing machine leaning model.

Each variable subset was used to train several candidate models for predicting overall survival. We tested a linear model, random forest, and neural network and assessed the average cross-validation accuracy of each. After selecting the best model and variable combination, we trained a final model on all the training data, made predictions on the challenge-provided validation set, and compared the performance to the leave-one-out cross validation. In particular, we checked for substantially decreased performance on the test data that would indicate model over-fitting.

3.4   Results

Landmark normalization was successfully applied to all patients. One case had poor fluid suppression on the FLAIR image and could not be effectively normalized. This case was excluded from the training.

Among the mean intensities for each image over each region, the region volumes, and the patient age, we found five features were significantly associated with overall survival in the cox model. They are: mean FLAIR intensity over the nonenhancing and necrotic region, mean T1ce intensity over the whole tumor, the volumes of the nonenhancing and enhancing regions, and age. With these variables input into a multivariate Cox model only age, the FLAIR nonenhancing mean, and nonenhancing volume were independently significant. Applying stepwise AIC did not change the variable selections any further.

Among the candidate models we tested (random forest, neural network, linear model) the linear model performed best with $R^2 = 0.134$ and mean-square-error 114994 using the three inputs selected by the multivariate cox model. With the same model parameters fit to all 162 evaluable challenge cases the model to predict overall survival is given by.

$$\text{Survival} = 926.8 - 10.5 \cdot \text{Age} + 91.6 \cdot \text{FTCM} - 55.1 \cdot \text{TCV}$$

where Age is the patient's age in years, FTCM is the "FLAIR Tumor Core Mean" value on the landmark normalized scale, and TCV is the "Tumor Core Volume" in units of $\text{mm}^3/10000$ consisting of nonenhancing and necrotic areas. This volume scaling makes the range of values comparable to the other features. Surprisingly, this simple linear model performed substantially better on the testing data and on the challenge validation dataset. This strongly suggests the model is not over fitting the data. The metrics are provided in Table 3.

Table 3. Performance metrics for our linear model on the training data: (80% of 163 provided cases), testing data (20% of 163 provided cases), and validation data (26 cases without known survival). The Pearson R^2 for the validation data is not provided.

	R^2, predicted vs observed	Accuracy	MSE
Training data	0.134	44.5%	114994
Testing data	0.399	38.2%	55193
Validation data	-	53.6%	87998

4 Discussion

Brain tumor segmentation and prediction of overall survival are both challenging tasks. Despite good results, our segmentation model did not perform as well as the implementation of DeepMedic by Kamnitsas et al. that won the BraTS 2017 challenge [15]. Their model achieved better segmentation results by averaging results across an ensemble of six different models. The single model we used is not as robust as their ensemble method but provides satisfactory results without the high computational cost.

In the task of predicting patients as short, medium, or long survivors we achieved a validation accuracy of 54% with a MSE of 87998. In the training data the most frequent class is short survivors at 65 of 163 (39.9%) which means our models are performing better than chance. The root mean square error for continuous prediction is on the order of 300 days, which is comparable to the range seen among all patients. Overall survival is impacted by several factors, including age, treatment, and performance status (not provided) and the accuracy and MSE reflects the complexity of this task even when some variables are controlled for.

We were able to produce good results using two highly primitive image measurements (mean intensity and volume) and a linear regression model. Although vast numbers of higher-order texture features and nonlinear models are commonly employed to mine imaging data, we found they were not useful in predicting overall survival for this task. We suspect this is because these features are more sensitive to tumor segmentation (and segmentation error) as well as other variations in image quality and processing. Since predicting overall survival is already a highly uncertain task, it is easy for models to over-fit the higher order features. In other words, the simple and robust features more easy to generalize.

Our best performing model only included intensity information from one of the four magnetic resonance sequences available (FLAIR) and only one of the four segmentation labels used to extract features (enhancing tumor, tumor core consisting of nonenhancing tumor and necrosis, edema, and whole tumor). This may have happened for a few reasons: While the available image types (T1, T2, etc) contain different kinds of information about the tumors, there was a lot of variability between images of the same type from different patients. This intra-sequence variability reduces the impedes the models ability to predict overall survival based on the complementary nature of the different image contrasts.

5 Conclusion

We found we could segment glioma tumors with high accuracy using a multi-scale convolutional neural net. Using these segmentations and simple image features we were able to predict overall survival with reasonable accuracy.

References

1. Stupp, R., et al.: The European organisation for research and treatment of cancer brain tumor and radiotherapy groups, and "the national cancer institute of canada clinical trials group". radiotherapy plus concomitant and adjuvant temozolomide for glioblastoma. N. Engl. J. Med. **352**(10), 987–996 (2005)
2. Shipitsin, M., et al.: Molecular definition of breast tumor heterogeneity. Cancer cell **11**(3), 259–273 (2007)
3. Ronneberger, O., Fischer, P., Brox, T.: U-Net: convolutional networks for biomedical image segmentation. In: Navab, N., Hornegger, J., Wells, W.M., Frangi, A.F. (eds.) MICCAI 2015. LNCS, vol. 9351, pp. 234–241. Springer, Cham (2015). https://doi.org/10.1007/978-3-319-24574-4_28
4. Menze, B.H., et al.: The multimodal brain tumor image segmentation benchmark (BRATS). IEEE Trans. Med. Imaging **34**(10), 1993–2024 (2015)
5. Bakas, S., et al.: Advancing the cancer genome atlas glioma MRI collections with expert segmentation labels and radiomic features. Sci. Data **4**, 170117 (2017)
6. Bakas, S., et al.: Segmentation labels and radiomic features for the pre-operative scans of the TCGA-GBM collection, The Cancer Imaging Archive (2017). https://doi.org/10.7937/K9/TCIA.2017.KLXWJJ1Q
7. Bakas, S., et al.: Segmentation labels and radiomic features for the pre-operative scans of the TCGA-LGG collection, The Cancer Imaging Archive (2017). https://doi.org/10.7937/K9/TCIA.2017.GJQ7R0EF
8. Bakas, S., Reyes, M., et al.: Identifying the best machine learning algorithms for brain tumor segmentation, progression assessment, and overall survival prediction in the BRATS challenge. arXiv preprint arXiv:1811.02629 (2018)
9. Kamnitsas, K., et al.: Efficient multi-scale 3D CNN with fully connected CRF for accurate brain lesion segmentation. Med. Image Anal. **36**, 61–78 (2017)
10. Leung, K.K., et al.: Alzheimer's disease neuroimaging initiative. robust atrophy rate measurement in alzheimer's disease using multi-site serial MRI: tissue-specific intensity normalization and parameter selection. NeuroImage **50**(2), 516–523 (2010)
11. van Griethuysen, J.J.M., et al.: Computational radiomics system to decode the radiographic phenotype. Cancer Res. **77**(21), e104 LP–e107 (2017)
12. Haralick, R., Shanmugan, K., Dinstein, I.: Textural features for image classification (1973)
13. Akaike, H.: A new look at the statistical model identification. IEEE Trans. Autom. Control **19**(6), 716–723 (1974)
14. Kursa, M.B., Rudnicki, W.R., et al.: Feature selection with the Boruta package. J. Stat. Softw. **36**(11), 1–13 (2010)
15. Kamnitsas, K., et al.: Ensembles of multiple models and architectures for robust brain tumour segmentation. CoRR, abs/1711.01468 (2017)

Ensemble of Fully Convolutional Neural Network for Brain Tumor Segmentation from Magnetic Resonance Images

Avinash Kori[ID], Mehul Soni[ID], B. Pranjal[ID], Mahendra Khened,
Varghese Alex[ID], and Ganapathy Krishnamurthi[✉][ID]

Indian Institute of Technology Madras, Chennai 600036, India
gankrish@iitm.ac.in

Abstract. We utilize an ensemble of the fully convolutional neural networks (CNN) for segmentation of gliomas and its constituents from multimodal Magnetic Resonance Images (MRI). The ensemble comprises of 3 networks, two 3-D and one 2-D network. Of the 3 networks, 2 of them (one 2-D & one 3-D) utilize dense connectivity patterns while the other 3-D network makes use of the residual connection. Additionally, a 2-D fully convolutional semantic segmentation network was trained to distinguish between air, brain, and lesion in the slice and thereby localize the lesion the volume. Lesion localized by the above network was multiplied with the segmentation mask generated by the ensemble to reduce false positives. On the BraTS validation data ($n = 66$), the scheme utilized in this manuscript achieved a whole tumor, tumor core and active tumor dice of 0.89 0.76, 0.76 respectively, while on the BraTS test data ($n = 191$), our scheme achieved the whole tumor, tumor core and active tumor dice of 0.83 0.72, 0.69 respectively.

Keywords: Brain tumor · MRI · CNN · 3-D · Ensemble

1 Introduction

Manual tracing, detection of organs and tumor structure from medical images is considered as one of the preliminary step in diseases diagnosis and treatment planning. In a clinical setup this time-consuming process is carried out by radiologists, however, this approach becomes infeasible as the number of patients increases. This necessitates the scope of research in automated segmentation methods.

Diffused boundaries of the lesion and partial volume effects in the MR images makes automated segmentation of gliomas from MR volumes a challenging task. In the recent year's convolutional neural networks (CNN) have produced state of the art results for the task of segmentation of gliomas from MR images [6,9]. Typically, medical images are volumetric, organs being imaged are 3-D entities and henceforth we exploit the nature of 3-D CNN based architectures for segmentation task.

© Springer Nature Switzerland AG 2019
A. Crimi et al. (Eds.): BrainLes 2018, LNCS 11384, pp. 485–496, 2019.
https://doi.org/10.1007/978-3-030-11726-9_43

The segmentation generated by a trained network has an associated bias and variance. Ensembling the predictions generated by multiple models or networks aids in the reduction of the variance in the generated segmentation. In this manuscript, we make use of 3 networks (two 3-D networks and one 2-D network) for the task of segmentation of gliomas from MR volumes. Additionally, a 2-D fully semantic segmentation network was trained to delineate the air, brain, and lesion in a slice of the brain. The aforementioned network was used to reduce the false positive generated by the ensemble. The predictions were further processed by conditional random fields (CRF) & 3-D connected components analysis.

2   Materials and Methods

An ensemble of fully convolutional neural network were utilized to segment gliomas and its constituents from multi modal MR volume. The ensemble comprises of 3 networks (two 3-D networks and one 2-D network). Two networks (a 3-D and a 2-D network) utilizes dense connectivity patterns while the other 3-D network comprises of residual connection. The networks with dense connectivity pattern were semantic segmentation networks and predicts the class associated with all pixels or voxels that form the input to the network. The network with residual connectivity pattern was composed of inception modules so as to learn multi-resolution features. This multi-resolution network unlike the other networks in the ensemble classifies only a subset of voxels.

A 2-D fully convolutional semantic segmentation (Air-Brain-Lesion Network) was trained to delineate air, brain and lesion from axial slice of the MR volumes and thereby localize the lesion in the volume. The predictions generated by the ensemble were smoothened by using Conditional random fields. The smoothened prediction and the output generated by the Air-Brain-Lesion network were used in tandem to reduce the false positives in the prediction. The false positives in the predictions were further reduced by incorporating a class-wise 3-D connected component analysis in the pipeline. The pipeline utilised for segmentation of glioma is illustrated in Fig. 1.

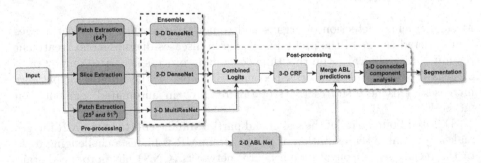

Fig. 1. Proposed pipeline for segmentation of Brain tumor and its constituents from Magnetic Resonance Images.

2.1 Data

Brats 2018 challenge data was used to train the networks [1–4, 8] was used in this manuscript for segmentation task. The training dataset comprises 210 high-grade glioma volumes and 75 low-grade gliomas along with expert annotated pixel level ground truth segmentation mask. Each subject comprises 4 MR sequences, namely FLAIR, T2, T1, T1 post contrast.

2.2 Data Pre-processing

As a part of pre-processing, the volumes were normalized to have zero mean and unit standard deviation.

2.3 Segmentation Network

The 3-D networks used in ensemble accepts 3-D patches as input while the 2-D network accepts an axial slice of the brain as the input. The architecture, training and testing regime associated with each network in the ensemble is explained in the following paragraphs.

3-D Densely Connected Semantic Segmentation Network

Architecture: The network is a fully convolutional semantic segmentation network. The network accepts input cubes of size 64^3 and predicts the class associated with all the voxels in the input cube fed to the network. The network is composed of an encoding and decoding section. The encoding section is composed of Dense blocks and Transition Down blocks. The Dense blocks are composed of a series of convolutions followed by non-linearity (ReLU) & each convolutional layer receives input from all the preceding convolutional layers in the block. This connectivity pattern leads to the explosion of a number of feature maps with the depth of the network which was circumvented by setting the number of output feature maps per convolutional layer to a small value ($k = 4$). The Transition down blocks are utilized in the network to reduce the spatial dimension of the feature maps.

The decoding or the up-sampling pathway in the network comprises of the Dense blocks and Transition Up blocks. The Transition Up blocks are composed of transposed convolution layers to upsample feature maps. The features from the encoding section of the network are concatenated with the up-sampled feature maps to form the input to the Dense block in the decoding section. The architecture of the network is given in Fig. 2.

Patch Extraction: Patches of size 64^3 were extracted from the brain. The class imbalance among the various classes in the data was addressed by extracting relatively more number of patches from lesser frequent classes such as necrosis. Figure 3 illustrates the number of patches extracted for each class.

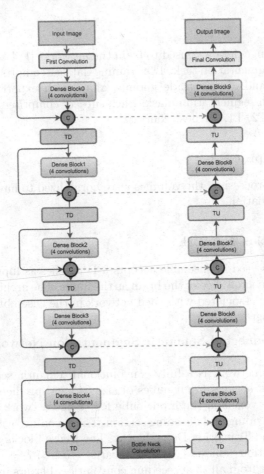

Fig. 2. Densely connected convolutional network used for segmentation task. TU: Transition Up block; TD: Transition Down block; C: Concatenation block

The 3-D dense fully connected network accepts an input of dimension 64^3 and predicts the class associated to all the voxels in the input. The network comprises 77 layers. The dense connection between the various convolutional layers in the network aids in the effective reuse of the features in the network. The presence of dense connections between layers increases the number of computations. This bottleneck was circumvented by keeping the number of convolutions to a small number say 4. Figure 2 shows the network architecture used in semantic segmentation task.

Training: Stratified sampling based on the grade of the gliomas was done to split the dataset into training, validation, and testing in the ratio 70: 20: 10. The network was trained and validated on 182 and 63 HGG & LGG volumes respectively. To further address the issue of class imbalance in the network, the parameters of the network were trained by minimizing weighted cross entropy.

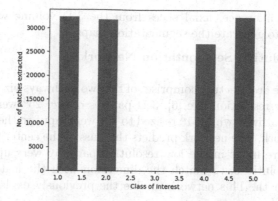

Fig. 3. Histogram of patches sampled surrounding certain class

The weight associated with each class was equivalent to the ratio of the median of the class frequency to the frequency of the class of interest [5]. The number of samples per batch was set at 4, while the learning rate was initialized to 0.0001 and decayed by a factor of 10% every-time the validation loss plateaued.

Testing: During inference, patches of the dimension of 64^3 were extracted from the volume and fed to the network with the stride of 32. CNN's being a deterministic technique is bound to generate predict the presence of the lesion in physiologically impossible place.

2-D Semantic Segmentation Network

Architecture: The architecture of this network is similar to that of the architecture of the 3-D network. The only difference between the networks is the usage of 2-D convolutions rather than 3-D convolutions. The network comprises 77 layers. The network accepts inputs of dimension 240×240 and predicts the class associated with all the pixels in the input.

Slice Extraction: In the given dataset, apart from the T1 post contrast, sequences such as FLAIR, T2 & T1 were 2-D sequences. Majority of the 2-D sequences in the given dataset were acquired axially and thus had good resolution along the axial plane. The 2-D network was trained on the axial slices of brain. The class imbalance in the dataset was addressed by extracting slices which comprise of at least one pixel of the lesion in it.

Training: The parameters of the network were initialized using Xavier initialization and the parameters of the network were learned by reducing the hybrid loss (cross entropy & dice loss). The imbalance among the various classes was further reduced by using weighted cross entropy rather than vanilla cross entropy. The weights assigned to each class were determined as explained earlier. Hyperparameters such as batch size, learning rate, and learning rate decay etc. were similar to the ones used to train the 3-D network.

Testing: During inference, axial slices from the 3-D volume were fed to the trained network to generate the segmentation maps.

3-D Multi-resolution Segmentation Network

Architecture: The architecture comprises of the two pathways viz high-resolution pathway and low resolution like [6]. 3-D patches of size 25^3 were input to the high-resolution pathway while 51^3 resized to 19^3 were input to the low-resolution path in the network. The network predicts the class of the center 9^3 voxels of the input. The feature maps in the low resolution pathway were upsampled using transposed convolutions, to match the dimension with the feature maps from high-resolution path. This network, unlike the previously explained two other networks, differs by:

1. Predicting the class associated to a subset of voxels in the input 3-D patch.
2. Making use of dual pathway to captures associated global and local features.
3. Making use of inception module [10] (3×3, 5×5 & 7×7) so as to learn multi-resolution features.

The architecture of the network is given in Fig. 4(a) and the building block of each unit in the network is illustrated in Fig. 4(b).

Patch Extraction: Patches of sizes 25^3 and 51^3 centered around voxels were extracted to form the training data to the network. The degree of class imbalance was reduced by extracting more patches from under-represented classes.

Training: Parameters in the network were initialized with Xavier initialization technique. The network was trained using the similar hyper-parameters that were used for the other two other networks proposed in the ensemble. The network was trained for 50 epochs and model that yielded lowest validation error was utilized for inference.

Testing: For testing, the stride was set to 9^3 and patches of 25^3 and 51^3 were extracted from the MR volume and input to the trained network to produce the segmentation mask.

2.4   Post-processing

Air-Brain-Lesion Network. The Air-Brain-Lesion (ABL Net) network was 2-D network densely connected the fully convolutional network. The network was trained to delineate lesion, air and the brain in a volume. The prediction made by this network was used to reduce the false positives generated by the segmentation network.

Architecture: The architecture of the network is similar to the 2-D network utilized in the segmentation ensemble model.

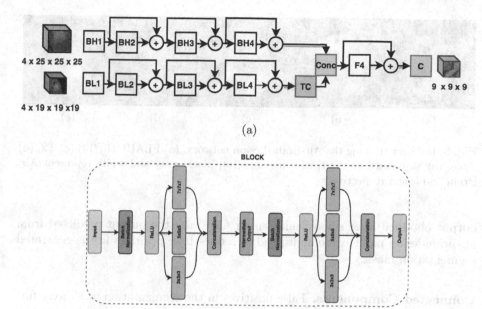

(a)

(b)

Fig. 4. 3-D Multi-Resolution Network for segmentation of gliomas from MR volumes. (a) The architecture of the network. The top portion of the network accepts high-resolution patches (25^3) while the bottom pathway accepts low-resolution input (51^3 patches resized to 19^3) as input. Both the high and low-resolution pathway is composed to inception modules so as learn multi-resolution features. TC in the network stands for transposed convolution and is used to match the spatial dimension of the features in low-resolution pathway with those learned in the high-resolution path. (b) The building block of the network. In the block, the dimension of the feature map in an inception module was maintained by setting the padding to 0, 1, 2 for 3×3, 5×5 & 7×7 respectively.

Slice Extraction: The Network was trained using axial slices as they correspond to the highest resolution. Various constituents of the lesion were clubbed to form the lesion while air and brain class labels were determined using a threshold on the volume Fig. 5 illustrates the slice of the brain with the aforementioned classes.

Training and Testing: The training & testing regime were similar to the ones used for the 2-D Densely connected segmentation network.

CRF. To the smoothen the segmentation predicted by the models a fully connected conditional random fields with Gaussian edge potentials as proposed by Krähenbühl et al. [7] was utilized. The posterior probabilities generated by each model in the ensemble were averaged to form the unary potentials for the CRF. The CRF was implemented by using open source code from the pydenscrf[1]. The

---

[1] pydensecrf: https://github.com/lucasb-eyer/pydensecrf.

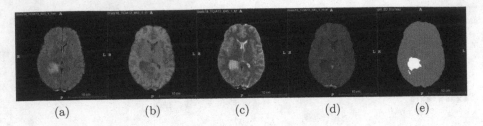

(a)   (b)   (c)   (d)   (e)

Fig. 5. Data for training the Air-Brain-Lesion network. (a) FLAIR, (b) T1, (c) T2, (d) T1ce, (e) Modified Ground truth. In image (e), black, gray and white represent Air, Brain and lesion respectively.

output obtained after smoothening using CRF and the output predicted from air-brain-lesion model were multiplied to reduce false positives in the generated segmentation mask.

Connected Components. False positives in the segmentation mask were further reduced by performing class-wise 3-D connected component analysis. All components within each class which composed more than 12,000 voxels were retained while the rest were discarded.

3   Results

The performance of the network was tested on 3 different namely: held out test data ($n = 40$), BraTS validation data ($n = 66$) & BraTS testing data ($n = 191$) (Table 1).

3.1   Performance of the Segmentation Networks on the Held Out Test Data

On the held out test data ($n = 40$), the performance of each of the network in the segmentation ensemble is given in Table 2(a, b, c). Table 2(d) showcases the performance on the held out test data post ensembling the networks. Comparing the whole tumor, tumor core and active tumor core dice score it was observed that ensembling of networks aided in reducing the variance and increasing the overall performance of the network. Figure 6 illustrates the segmentation generated by a trained network.

The post-processing which included CRFs & 3-D class-wise connected components aid in reducing the false positives generated by the networks. Figure 7 illustrates the effect post-processing on segmentation. The contribution of the various the components in the post processing pipeline (CRF, ABL Net, & Connected Components) are illustrated in Table 2.

Table 1. Performance of individual networks and ensemble on held out test data (n = 40). In the table WT, TC, AT stand for the whole tumor, tumor core & active tumor respectively.

3-D Densely connected network

	WT	TC	AT
Mean	0.88	0.78	0.72
Std	0.10	0.26	0.31
Median	0.92	0.92	0.86

2-D Densely connected network

	WT	TC	AT
Mean	0.88	0.72	0.69
Std	0.11	0.30	0.28
Median	0.91	0.88	0.82

3-D Multi-Resolution network

	WT	TC	AT
Mean	0.88	0.74	0.73
Std	0.21	0.26	0.32
Median	0.89	0.81	0.83

Ensemble of network

	WT	TC	AT
Mean	0.89	0.78	0.78
Std	0.08	0.21	0.20
Median	0.92	0.89	0.86

Table 2. The contribution of all the components used in post processing pipeline. (CC: 3-D Connected Components)

No post-processing

	WT	TC	AT
Mean	0.85	0.76	0.71
Std	0.10	0.18	0.29
Median	0.88	0.86	0.84

CRF post-processing

	WT	TC	AT
Mean	0.86	0.77	0.73
Std	0.09	0.19	0.28
Median	0.89	0.85	0.84

CRF + ABL Network

	WT	TC	AT
Mean	0.86	0.79	0.74
Std	0.12	0.19	0.28
Median	0.91	0.85	0.84

CRF + ABL + CC

	WT	TC	AT
Mean	0.89	0.78	0.78
Std	0.08	0.19	0.28
Median	0.92	0.89	0.86

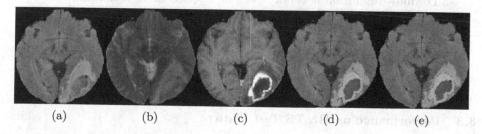

(a)   (b)   (c)   (d)   (e)

Fig. 6. (a) FLAIR, (b) T2, (c) T1c, (d) Prediction, (e) Segmentation. In images d and e, Green, Yellow & Red represent Edema, Enhancing Tumor and Necrosis present in the lesion. (Color figure online)

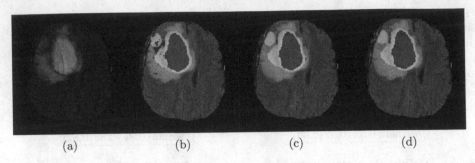

(a)   (b)   (c)   (d)

Fig. 7. (a) FLAIR, (b) Without Post-processing, (c) With Post-processing, (d) Ground truth. In images b, c and d, Green, Yellow & Red represent Edema, Enhancing Tumor and Necrosis present in the lesion. (Color figure online)

3.2 Performance on the BraTS Validation Data

On the BraTS validation data (n = 66), the performance of each of the networks that form the ensemble is listed in Table 3 respectively. Similar to the observation seen in the held out test data, it was observed that ensembling prediction from multiple networks helped in achieving better segmentation results by lowering variance in the predictions.

Table 3. Performance on validation data (n = 66)

3-D Densely connected network

	WT	TC	AT
Mean	0.85	0.74	0.71
Std	0.11	0.22	0.28
Median	0.88	0.82	0.83

2-D Densely connected network

	WT	TC	AT
Mean	0.87	0.73	0.71
Std	0.10	0.27	0.27
Median	0.90	0.86	0.82

3-D Multi-Resolution network

	WT	TC	AT
Mean	0.85	0.73	0.71
Std	0.17	0.26	0.30
Median	0.90	0.83	0.85

Ensemble of network

	WT	TC	AT
Mean	0.89	0.76	0.76
Std	0.07	0.23	0.25
Median	0.91	0.86	0.86

3.3 Performance on BraTS Test Data

The performance of the proposed scheme on the BraTS test data (n = 191) is illustrated in Table 4. It was observed that the network achieved good segmentation on unseen data.

Table 4. Performance of the Ensemble of Segmentation on the test data (n = 191)

	Whole tumor	Tumor core	Active tumor
Mean	0.83	0.72	0.69
Std	0.19	0.29	0.29
Median	0.90	0.87	0.80

4   Conclusion

We made use of an ensemble of convolutional neural networks for segmentation of gliomas. From the experiments carried out it was observed that the ensemble aids in reducing the variance associated in the prediction and also helped in increasing quality of the segmentation generated. The false positives generated by the network were minimized by using multiplying the predictions with network trained to delineate lesion from MR volumes. The segmentation was further post-processed by utilizing CRF & 3-D connected component analysis. On the BraTS 2018 validation data (n = 66), the network achieved a competitive dice score of 0.89, 0.76 and 0.76 for the whole tumor, tumor core and active tumor respectively. On the BraTS test data, the network used in the manuscript achieved a mean whole tumor, tumor core and active tumor dice of 0.83, 0.72 and 0.69 respectively.

References

1. Bakas, S., et al.: Segmentation labels and radiomic features for the pre-operative scans of the TCGA-GBM collection. Cancer Imaging Arch., 286 (2017)
2. Bakas, S.: Segmentation labels and radiomic features for the pre-operative scans of the TCGA-LGG collection. Cancer Imaging Arch. (2017)
3. Bakas, S., et al.: Advancing the cancer genome atlas glioma MRI collections with expert segmentation labels and radiomic features. Sci. Data **4**, 170117 (2017)
4. Bakas, S., Reyes, M., et al.: Identifying the best machine learning algorithms for brain tumor segmentation, progression assessment, and overall survival prediction in the brats challenge. arXiv preprint arXiv:1811.02629 (2018)
5. Eigen, D., Fergus, R.: Predicting depth, surface normals and semantic labels with a common multi-scale convolutional architecture. In: Proceedings of the IEEE International Conference on Computer Vision, pp. 2650–2658 (2015)
6. Kamnitsas, K., et al.: Efficient multi-scale 3D CNN with fully connected CRF for accurate brain lesion segmentation. Med. Image Anal. **36**, 61–78 (2017)
7. Krähenbühl, P., Koltun, V.: Efficient inference in fully connected CRFs with Gaussian edge potentials. In: Advances in Neural Information Processing Systems, pp. 109–117 (2011)
8. Menze, B.H., et al.: The multimodal brain tumor image segmentation benchmark (BRATS). IEEE Trans. Med. Imaging **34**(10), 1993 (2015)

9. Pereira, S., Pinto, A., Alves, V., Silva, C.A.: Brain tumor segmentation using convolutional neural networks in MRI images. IEEE Trans. Med. Imaging **35**(5), 1240–1251 (2016)

10. Szegedy, C., Ioffe, S., Vanhoucke, V., Alemi, A.A.: Inception-v4, inception-resnet and the impact of residual connections on learning. In: AAAI, vol. 4, p. 12 (2017)

Learning Contextual and Attentive Information for Brain Tumor Segmentation

Chenhong Zhou[1], Shengcong Chen[1], Changxing Ding[1($\boxtimes$)], and Dacheng Tao[2]

[1] School of Electronic and Information Engineering,
South China University of Technology, Guangzhou, China
chxding@scut.edu.cn

[2] UBTECH Sydney AI Centre, SIT, FEIT, University of Sydney, Sydney, Australia

Abstract. Thanks to the powerful representation learning ability, convolutional neural network has been an effective tool for the brain tumor segmentation task. In this work, we design multiple deep architectures of varied structures to learning contextual and attentive information, then ensemble the predictions of these models to obtain more robust segmentation results. In this way, the risk of overfitting in segmentation is reduced. Experimental results on validation dataset of BraTS 2018 challenge demonstrate that the proposed method can achieve good performance with average Dice scores of 0.8136, 0.9095 and 0.8651 for enhancing tumor, whole tumor and tumor core, respectively. The corresponding scores for BraTS 2018 testing set are 0.7775, 0.8842 and 0.7960, respectively, winning the third position in the BraTS 2018 competition among 64 participating teams.

1 Introduction

Brain tumor is one of the most fatal cancers, which consists of uncontrolled, unnatural growth and division of the cells in the brain tissue [1]. The most frequent types of brain tumors in adults are gliomas that arise from glial cells and infiltrating the surrounding tissues [2]. According to the malignant degree of gliomas and their origin, these neoplasms can be categorized into Low Grade Gliomas (LGG) and High Grade Gliomas (HGG) [2,3]. The former is slower-growing and comes with a life expectancy of several years, while the latter is more aggressive and infiltrative, having a shorter survival period and requiring immediate treatment [2]. Therefore, segmenting brain tumor timely and automatically would be of critical importance for assisting the doctors to improve diagnosis, perform surgery and make treatment planning.

In recent years, convolutional neural networks (CNNs) have been widely applied to automatic brain tumor segmentation tasks. Pereira et al. [15] and Havaei et al. [13] respectively trained a CNN to predict the label of the central

---

C. Zhou and S. Chen—Equal contribution.

© Springer Nature Switzerland AG 2019
A. Crimi et al. (Eds.): BrainLes 2018, LNCS 11384, pp. 497–507, 2019.
https://doi.org/10.1007/978-3-030-11726-9_44

voxel only within a patch, which causes that they suffer from high computational cost and time consumption during inference. To reduce the computational burden, Kamnitsas et al. [5] propose an efficient model named DeepMedic that can predict the labels of voxels within a patch simultaneously, in order to achieve dense predictions. Recently, fully convolutional networks (FCNs) have achieved promising results. Shen et al. [6] and Zhao et al. [11] allow end-to-end dense training and testing for brain tumor segmentation at the *slice* level to improve computational efficiency. With a large variety of CNN architectures proposed, the performance of automatic brain tumor segmentation from Magnetic Resonance Imaging (MRI) images has been improved greatly.

In this work, we construct multiple different CNN architectures and approaches to ensemble their prediction results, in order to produce stable and robust segmentation performance. We evaluate our approaches on the validation set of 2018 Brain Tumor Segmentation (BraTS) challenge, where we obtain the good performance with average Dice scores of 0.8136, 0.9095 and 0.8651 for enhancing tumor, whole tumor and tumor core, respectively. Correspondingly, we achieve promising scores for BraTS 2018 testing set are 0.7775, 0.8842 and 0.7960, respectively.

2   Data

We use the dataset of 2018 Brain Tumor Segmentation challenge [2,4,7,8,21] for experiments, which consists of the training set, validation set and testing set. The training set contains 210 HGG and 75 LGG cases whose corresponding manual segmentations are provided. As shown in Fig. 1, the provided manual segmentations include four labels: 1 for necrotic (NCR) and the non-enhancing (NET) tumor, 2 for edema (ED), 4 for enhancing tumor (ET), and 0 for everything else, i.e. normal tissue and background (black padding). The validation set and testing set contain 66 cases and 191 cases with unknow grade and hidden segmentations, respectively. Each case has four MRI sequences that are named T1, T1 contrast enhanced (T1ce), T2 and FLAIR, respectively. These datasets are provided after their pre-processing, i.e. co-registered to the same anatomical template, interpolated to the same resolution ($1\,mm^3$) and skull-stripped, where dimensions of each MRI sequence are $240 \times 240 \times 155$. Besides, the official evaluation is calculated by merging the predicted labels into three regions: whole tumor (1,2,4), tumor core (1,4) and enhancing tumor (4). The valuation for validation set is conducted via an online system[1].

3   Methods

3.1   Basic Networks

As is well known, brain tumor segmentation from MRI images is a very tough and challenging task due to the severe class imbalance problem. Following [14],

---

[1] https://ipp.cbica.upenn.edu/.

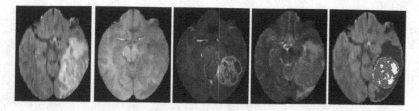

Fig. 1. Example of images from the BRATS 2018 dataset. From left to right: Flair, T1, T1ce, T2 and manual annotation overlaid on the Flair image: edema (green), necrosis and non-enhancing (yellow), and enhancing (red). (Color figure online)

we decompose the multi-class brain tumor segmentation into three different but related sub-tasks to deal with the class imbalance problem. (1) *Coarse segmentation to detect whole tumor*. In this sub-task, the region of whole tumor is located. To reduce overfitting, we define the first task being the five-class segmentation problem. (2) *Refined segmentation for whole tumor and its intra-tumoral classes*. The above obtained coarse tumor mask is dilated by 5 voxels as the ROI for the second task. In this sub-task, the precise classes for all voxels within the dilated region are predicted. (3) *Precise segmentation for enhancing tumor*. We specially design the third sub-task to segment the enhancing tumor, due to its high difficulty of segmentation.

Model Cascade. In view of the above three sub-tasks, it is probably easy to train a CNN individually for each sub-task, which is the currently popular Model Cascade (MC) strategy. We use a 3D variant of the FusionNet [10], as illustrated in Fig. 2. The network architecture consists of an encoding path (upper half of the network) to extract complex semantic features and a symmetric decoding path (lower half of the network) to recover the same resolution as the input to achieve voxel-to-voxel predictions. The network is constructed by four types of basic building blocks, as shown in Fig. 2. In addition, the network has not only the short shortcuts in residual blocks, but also three long skip connections to merge the feature maps from the same level in the encoding path during decoding by using a voxel-wise addition. We employ the identical network architecture for each sub-task, except for the final convolutional classification layer. The number of channels of last classification layer is equal to 5, 5 and 2 for the first, second and third sub-tasks, respectively. Besides, size of input patches for the network is $32 \times 32 \times 16 \times 4$, where the number 4 indicates the four MRI modalities. During inference, we adopt overlap-tile strategy in [9]. Thus, we abandon the prediction results of border region and only retain the predictions in the center region ($20 \times 20 \times 5$). This trick is also used in the following models. Different from [20] that is a typical example of model cascade strategy, we dilate the coarse tumor mask to prevent tumor omitting in the second sub-task and adopt the same 3D basic network architecture for each sub-task instead of sophisticated operations that design different networks for different sub-tasks.

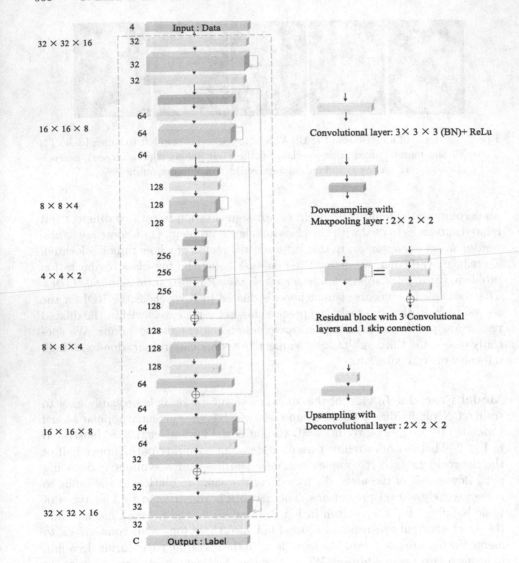

Fig. 2. Network architecture used in each sub-task. The building blocks are represented by colored cubes with numbers nearby being the number of feature maps. C equals to 5, 5, and 2 for the first, second, and third task, respectively. (Best viewed in color) This figure is reproduced from [14].

One-Pass Multi-task Network. The above proposed model cascade approach has obtained promising segmentation performance. To a certain extent, it alleviates the problem of class imbalance. However, model cascade approach needs to train a series of deep models individually for the three different sub-tasks, which leads to large memory cost and system complexity during training and testing. In addition, we have observed that the networks used for three sub-tasks are almost the same except for the training data and the classification layer. It is obvious that the three sub-tasks are relative to each other.

Therefore, we employ the one-pass multi-task network (OM-Net) proposed in [14], which is a multi-task learning framework that incorporates the three sub-tasks into a end-to-end holistic network, to save a lot of parameters and exploit the underlying relevance among the three sub-tasks. The OM-Net proposed in [14] is described in Fig. 3, which is composed of the sharable parameters and task-specific parameters. Specially, the shared backbone model refers to the network layers outlined by the yellow dashed line in Fig. 2, while three respective branches for different sub-tasks are designed after the shared parts.

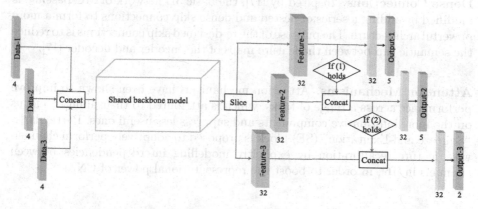

Fig. 3. Architecture of OM-Net. Data-i, Feature-i, and Output-i denote training data, feature, and classification layer for the i-th task, respectively. The shared backbone model refers to the network layers outlined by the yellow dashed line in Fig. 2. This figure is reproduced from [14].

In addition, inspired by the curriculum learning theory proposed by Bengio et al. [12] that humans can learn a set of concepts much better when the concepts to be learned are presented by gradually increasing the difficulty level, we adopt the curriculum learning-based training strategy in [14] to train OM-Net more effectively. The training strategy of our framework is to start training the network on the first easiest sub-task, then gradually add the more difficult sub-tasks and their corresponding training data to the model. This is a process from easy to difficult, highly consistent with the thought of manual segmentation of the tumor. Besides, the training data conforming to the sampling strategy of the other sub-tasks can be transferred to achieve data sharing. Eventually, the OM-Net is a single deep model to slove three sub-tasks simultaneously in one-pass. It is also significantly smaller in the number of trainable parameters than model cascade strategy and can be trained end-to-end using stochastic gradient descent to achieve data sharing and parameters sharing in a holistic network.

3.2   Extended Networks

In this section, we extend and improve the MC-baseline and OM-Net from four aspects to further promote the performance. The four aspects are elaborated in the following.

Deeper OM-Net. We deepen the OM-Net by appending a residual block (the violet block in Fig. 2 right) after each existing residual block of OM-Net, which is the easiest and most direct way to boost the performance.

Dense Connections. Inspired by [17], the basic 3D network of MC-baseline is modified by adding a series of nested and dense skip connections to form a more powerful architecture. The purpose of the re-designed skip connections is to reduce the semantic gap between the feature maps of the encoder and decoder [17].

Attention Mechanisms. Attention mechanisms have been shown to improve performance across a range of tasks, which is attributed to their ability to focus on the more informative components and suppress less useful ones. Particularly, "Squeeze-and-Excitation" (SE) block is proposed to adaptively perform channel-wise feature recalibration by explicitly modelling interdependencies between channels in [16], in order to boost the representational power of CNNs.

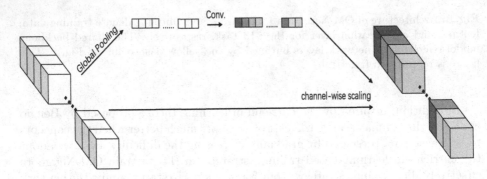

Fig. 4. The adopted "Squeeze-and-Excitation" (SE) block.

Inspired by it, we introduce SE blocks to OM-Net, in order to recalibrate the feature maps and further improve the learning and representational properties of OM-Net. The SE block is described in Fig. 4. Similar to [16], the SE block focuses on channels to adaptively recalibrate channel-wise feature responses in two steps, squeeze and excitation. It helps the network to increase the sensitivity to informative features and suppress less useful ones.

Multi-scale Contextual Information. To deal with the 3D medical scans, we employ the above 3D CNNs that process small 3D patches. However, small patches cause the network to lean the limited contextual information. It seems necessary to introduce larger patches, in order to provide larger receptive fields and more contextual information to the network. Therefore, inspried by [5], we design a two parallel pathway architecture that processes two scale input patches simultaneously. As shown in Fig. 5, we incorporate both local and larger contextual information to the model, which not only extracts semantic features at a higher resolution, but also considers larger contextual information from the lower resolution level. It can provide rich information to discriminate voxels that appear very similar when considering only local appearance, avoiding making wrong predictions.

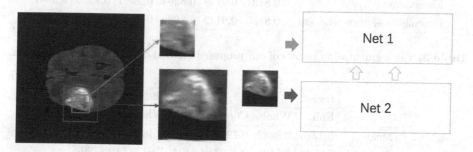

Fig. 5. The proposed network architecture to introduce multi-scale contextual information.

Table 1. Mean values of Dice and Hausdorff95 measurements on BraTS 2018 validation set (submission id DL-86-61).

Method	Dice			Hausdorff95		
	Enh.	Whole	Core	Enh.	Whole	Core
MC-Net	0.7732	0.9006	0.8232	4.1647	4.4849	7.6216
OM-Net	0.7882	0.9034	0.8273	3.1003	6.5218	7.1974
MC-Net (Dense connections)	0.7768	0.9049	0.8358	3.3994	4.2390	6.8503
MC-Net (Multi-scale)	0.7751	0.9059	0.8181	2.8192	4.0085	6.3437
OM-Net (Attention)	0.7792	0.8986	0.8329	3.8949	6.1926	8.3459
Deeper OM-Net	0.7882	0.8991	0.8405	2.7649	8.0177	7.3671
Deeper OM-Net (Attention)	0.7925	0.8948	0.8333	2.8099	4.8093	6.7755
Ensembles	**0.8137**	0.9092	0.8530	2.7092	4.4519	7.1535
Ensembles + post-processing	0.8136	**0.9095**	**0.8651**	2.7162	4.1724	6.5445

Table 2. Mean values of Sensitivity and Specificity measurements on BraTS 2018 validation set.

Method	Sensitivity			Specificity		
	Enh.	Whole	Core	Enh.	Whole	Core
MC-Net	0.8082	0.9118	0.7971	0.9980	0.9943	0.9981
OM-Net	0.7953	0.9084	0.7883	0.9984	0.9948	0.9985
MC-Net (Dense connections)	0.8144	0.9127	0.8323	0.9979	0.9947	0.9973
MC-Net (Multi-scale)	0.8095	0.9065	0.7970	0.9980	0.9951	0.9981
OM-Net (Attention)	0.8245	0.9078	0.8103	0.9977	0.9943	0.9979
Deeper OM-Net	0.7962	0.9059	0.8176	0.9984	0.9945	0.9979
Deeper OM-Net (Attention)	0.7995	0.8972	0.8159	0.9983	0.9946	0.9975
Ensembles	0.8137	0.9148	0.8294	0.9983	0.9950	0.9981
Ensembles + post-processing	0.8135	0.9142	0.8683	0.9983	0.9951	0.9968

Table 3. The segmentation results of our proposed method on BraTS 2018 testing set.

	Dice			Hausdorff95		
	Enh.	Whole	Core	Enh.	Whole	Core
Mean	0.7775	0.8842	0.7960	2.9366	5.4681	6.8773
StdDev	0.2533	0.1127	0.2593	4.6894	7.6479	10.1779
Median	0.8498	0.9183	0.9030	1.7321	3.1623	3.0000
25quantil	0.7596	0.8725	0.8062	1.4142	2.0000	1.7321
75quantil	0.8997	0.9437	0.9376	2.7337	5.3852	7.2798

3.3 Ensembles of the Above Multiple Models

Model ensembling is an effective method to improve performance, e.g. Kamnitsas et al. [19] ensembled DeepMedic [5], 3D FCN [18], and 3D U-Net [9] into EMMA. In this paper, we also adopt model ensembling to obtain more robust segmentation results. Above multiple models, including MC-Net, OM-Net and their variants are trained separately, and the predicted probabilities are averaged at testing time. Additionally, a simple yet effective post-processing method [14] is adopted to improve segmentation performance.

4   Experiments and Results

Pre-processing. We adopt the minimal pre-processing operation to the BraTS 2018 data. That is, each sequence is individually normalized by subtracting its mean value and dividing by its standard deviation of the intensities within the brain area in that sequence.

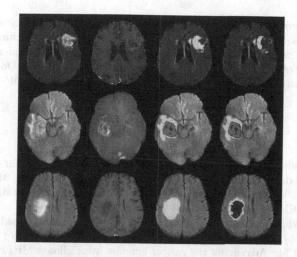

Fig. 6. Example segmentation results on the validation set of BraTS 2018. From left to right: Flair, T1ce, segmentation results using MC-Net only overlaid on Flair image, and segmentation results using the proposed method overlaid on Flair image; edema (green), necrosis and non-enhancing (blue), and enhancing (red). (Color figure online)

Segmentation Results. Table 1 presents the mean values of Dice and Hausdorff95 measurements of the different models on BraTS 2018 validation set, meanwhile Table 2 presents the corresponding mean values of Sensitivity and Specificity measurements. We can see that the OM-Net is superior to MC-Net, despite the fewer training parameters of OM-Net. Besides, the extended networks including MC-Net (Dense connections), MC-Net (Multi-scale), OM-Net (Attention), Deeper OM-Net and Deeper OM-Net (Attention) improve the segmentation performance to some extent. Finally, it shows that the proposed method achieves promising performance with average Dice scores of 0.8136, 0.9095 and 0.8651 for enhancing tumor, whole tumor and tumor core, respectively. In addition, we also provide qualitative comparisons in Fig. 6. From Fig. 6, we can see that model ensembling is much better and the effectiveness of the proposed method is justified.

Table 3 presents the segmentation results of our proposed method on BraTS 2018 testing set. It shows that the proposed method yields excellent performance, winning the third position in the BraTS 2018 competition.

5   Conclusion

In this work, we employ the OM-Net to obtain strong basic results, and then extend and improve MC-baseline and OM-Net from multiple aspects to further promote the performance. Eventually, the predictions of these models are ensembled to produce robust performance for brain tumor segmentation. The proposed method yields promising results, winning third place in the final testing stage of the BraTS 2018 challenge.

Acknowledgments. Changxing Ding was supported in part by the National Natural Science Foundation of China (Grant No.: 61702193), Science and Technology Program of Guangzhou (Grant No.: 201804010272), and the Program for Guangdong Introducing Innovative and Entrepreneurial Teams (Grant No.: 2017ZT07X183). Dacheng Tao was supported by Australian Research Council Projects (FL-170100117, DP-180103424 and LP-150100671).

References

1. Işın, A., Direkoğlu, C., Şah, M.: Review of MRI-based brain tumor image segmentation using deep learning methods. Procedia Comput. Sci. **102**, 317–324 (2016)
2. Menze, B.H., et al.: The multimodal brain tumor image segmentation benchmark (BRATS). IEEE TMI **34**(10), 1993–2024 (2015)
3. Bauer, S., Wiest, R., Nolte, L.P., Reyes, M.: A survey of MRI-based medical image analysis for brain tumor studies. Phys. Med. Biol. **58**, R97–R129 (2013)
4. Bakas, S., et al.: Advancing the cancer genome atlas glioma MRI collections with expert segmentation labels and radiomic features. Nat. Sci. Data **4**, 170117 (2017)
5. Kamnitsas, K., et al.: Efficient multi-scale 3D CNN with fully connected CRF for accurate brain lesion segmentation. Med. Image Anal. **36**, 61–78 (2017)
6. Shen, H., Wang, R., Zhang, J., McKenna, S.J.: Boundary-aware fully convolutional network for brain tumor segmentation. In: Descoteaux, M., Maier-Hein, L., Franz, A., Jannin, P., Collins, D.L., Duchesne, S. (eds.) MICCAI 2017. LNCS, vol. 10434, pp. 433–441. Springer, Cham (2017). https://doi.org/10.1007/978-3-319-66185-8_49
7. Bakas, S., et al.: Segmentation labels and radiomic features for the pre-operative scans of the TCGA-GBM collection. In: The Cancer Imaging Archive (2017)
8. Bakas, S., et al.: Segmentation labels and radiomic features for the pre-operative scans of the TCGA-LGG collection. In: The Cancer Imaging Archive (2017)
9. Ronneberger, O., Fischer, P., Brox, T.: U-Net: convolutional networks for biomedical image segmentation. In: Navab, N., Hornegger, J., Wells, W.M., Frangi, A.F. (eds.) MICCAI 2015. LNCS, vol. 9351, pp. 234–241. Springer, Cham (2015). https://doi.org/10.1007/978-3-319-24574-4_28
10. Quan, T.M., et al.: Fusionnet: a deep fully residual convolutional neural network for image segmentation in connectomics. arXiv preprint arXiv:1612.05360 (2016)
11. Zhao, X., Wu, Y., Song, G., et al.: A deep learning model integrating FCNNs and CRFs for brain tumor segmentation. Med. Image Anal. **43**, 98–111 (2018)
12. Bengio, Y., Louradour, J., Collobert, R., and Weston, J.: Curriculum learning. In: ICML, pp. 41–48. ACM (2009)
13. Havaei, M., et al.: Brain tumor segmentation with deep neural networks. Med. Image Anal. **35**, 18–31 (2017)
14. Zhou, C., Ding, C., Lu, Z., Wang, X., Tao, D.: One-pass multi-task convolutional neural networks for efficient brain tumor segmentation. In: Frangi, A.F., Schnabel, J.A., Davatzikos, C., Alberola-López, C., Fichtinger, G. (eds.) MICCAI 2018. LNCS, vol. 11072, pp. 637–645. Springer, Cham (2018). https://doi.org/10.1007/978-3-030-00931-1_73
15. Pereira, S., Pinto, A., Alves, V., Silva, C.A.: Brain tumor segmentation using convolutional neural networks in MRI images. IEEE Trans. Med. Imag. **35**(5), 1240–1251 (2016)
16. Hu, J., Shen, L., and Sun, G.: Squeeze-and-excitation networks. arXiv preprint arXiv:1709.01507 (2016)

17. Zhou, Z., Siddiquee, M.M.R., Tajbakhsh, N., and Liang, J.: UNet++: A nested U-Net architecture for medical image segmentation. arXiv preprint arXiv:1807.10165 (2018)

18. Long, J., et al.: Fully convolutional networks for semantic segmentation. In: CVPR, pp. 343–3440 (2015)

19. Kamnitsas, K., et al.: Ensembles of multiple models and architectures for robust brain tumour segmentation. In: Crimi, A., Bakas, S., Kuijf, H., Menze, B., Reyes, M. (eds.) BrainLes 2017. LNCS, vol. 10670, pp. 450–462. Springer, Cham (2018). https://doi.org/10.1007/978-3-319-75238-9_38

20. Wang, G., Li, W., Ourselin, S., Vercauteren, T.: Automatic brain tumor segmentation using cascaded anisotropic convolutional neural networks. In: Crimi, A., Bakas, S., Kuijf, H., Menze, B., Reyes, M. (eds.) BrainLes 2017. LNCS, vol. 10670, pp. 178–190. Springer, Cham (2018). https://doi.org/10.1007/978-3-319-75238-9_16

21. Bakas, S., Reyes, M., et al.: Identifying the best machine learning algorithms for brain tumor segmentation, progression assessment, and overall survival prediction in the BRATS challenge. arXiv preprint arXiv:1811.02629 (2018)

Glioblastoma Survival Prediction

Zeina A. Shboul[(✉)], Mahbubul Alam, Lasitha Vidyaratne,
Linmin Pei, and Khan M. Iftekharuddin

Vision Lab, Electrical and Computer Engineering,
Old Dominion University, Norfolk, USA
{zshbo001,malam001,lvidy001,lxpei001,
kiftekha}@odu.edu

Abstract. Glioblastoma is a high-grade invasive astrocytoma tumor. The highly invasive nature makes timely detection and characterization of the tumor critical for the survivability prediction of patients. This work proposes MRI- and clinical information-based automated pipeline that implements various state-of-the-art image processing, machine learning, and deep learning techniques to obtain robust tumor segmentation and patient survival estimation. We use 163 cases from the training dataset, and 28 cases from the validation dataset provided by the BraTS 2018 challenge for the evaluation of our model. We achieve an accuracy of 0.679 using the validation dataset and that of 0.519 for the test dataset.

1 Introduction

High-grade glioblastoma (HGG) or glioblastoma represents tumors arising from the gluey or supportive tissue of the brain. HGG is considered the most aggressive type of brain tumor. According to the American Brain Tumor Association (ABTA) [1] HGGs represent 74.6% of all malignant tumors and 24.7% of all primary brain tumors. World Health Organization (WHO) categorize HGGs as stage IV brain cancer [2]. Typically, the survival duration of patients with HGG tumor is less than two years [3, 4]. Therefore, accurate and timely detection of HGG tumor is essential for devising an appropriate treatment plan that may improve patient survival duration.

Recent works [5–10] have focused on developing automated survival prediction techniques for patients with HGG tumor. Different studies analyze tumor heterogeneity using different types of imaging [11] such as Magnetic Resonance Imaging (MRI) [12, 13]. This suggests MR as a potential non-invasive imaging biomarker for Glioblastoma diagnostic, prognostic and survival prediction. Jain et al. [14] extract morphological imaging features represented by Visually Accessible Rembrandt Images (VASRAI) [15] from the non-enhancing region of GBM and then correlate these features to the relative cerebral blood volume of a non-enhancing region and non-enhancing region crossing the midline. Gutman et al. [16] utilize four VASARI imaging features that describe the size of the contrast-enhanced, necrosis, non-enhance, edema, and the size

Z. A. Shboul, M. Alam, L. Vidyaratne and L. Pei—The authors have equal contribution.

© Springer Nature Switzerland AG 2019
A. Crimi et al. (Eds.): BrainLes 2018, LNCS 11384, pp. 508–515, 2019.
https://doi.org/10.1007/978-3-030-11726-9_45

of the whole tumor. Then these imaging features are associated with genetic mutation, survival prediction and Verhaak subtypes [17]. Nicolasjilwan et al. [18] combine clinical factors, VASARI imaging features, and genomics in a stepwise multivariate Cox model to predict overall survival time. Prasanna et al. [19] extract radiomic texture features that characterize three tumor regions; enhancing tumor, peritumoral brain zone, and necrosis from MR images. These features are assessed to overall survival prediction. Itakura et al. [20] utilize quantitative MRI features that describe tumor histogram statistics, texture, edge sharpness, compactness, and roughness. The authors cluster the features into three MRI phenotypic imaging subtypes; pre-multifocal, spherical, and rim-enhancing tumor. The three distinct subtypes are then correlated with overall survival and associated with molecular pathways.

Our proposed deep learning and machine learning based survival prediction technique have shown the best performance in the BraTS 2017 survival prediction challenge on the validation and test dataset, respectively [9]. This work proposes a sophisticated computational modeling-based survival prediction method. Specifically, we approach the survival prediction task as a two-step process, brain tumor segmentation followed by survival prediction, considering the segmentation output as an input to the second step. The brain tumor segmentation task is performed by utilizing two state-of-the-art convolutional neural networks (CNN) models, U-Net and fully convolutional neural network (FCN). The outcome of these two models is fused together to achieve the final segmentation output. This segmentation output along with the original MRI volumes are considered as input to the survival prediction step. Several radiomics features such as texture, topological, histogram etc. are extracted from the raw MRI sequences and the segmented tumor volume. Furthermore, a state-of-the-art 3D CNN architecture is utilized to extract additional features useful for the survival prediction task. These features are then processed using a gradient boosting-based regression technique known as extended gradient boosting (XGBoost) to obtain the survival estimation of patients.

2 Dataset

In this study, we use MR images of 163 high-grade GBM patients from BtaTS18 training dataset [21–25]. The dataset provides the ground truth segmentation of tumor tissues which comprises of enhancing tumor (ET), edema (ED), and the necrosis and non-enhancing tumor (NCR/NET). The training dataset provides age and overall survival (in days) data. The available scans of the MRI are native (T1), post-contrast T1-weighted (T1Gd), T2-weighted (T2), and T2 Fluid Attenuated Inversion Recovery (FLAIR) volumes. The dataset is co-registered, re-sampled to 1 mm^3 and skull-stripped. In addition, for overall survival validation proposes we use the 28 cases of BraTS18 validation dataset.

3 Methodology

3.1 Brain Tumor Segmentation

This work utilizes two state-of-the-art CNN architectures, U-Net and fully convolutional neural network (FCN) to perform the brain tumor segmentation task. The following two sub-sections provide a very brief outline of these two models.

Brain Tumor Segmentation Using Deep CNNs. This work utilizes a CNN based U-Net model [26, 27] to perform the brain tumor segmentation task. Unlike patch based CNN segmentation pipeline where the model only sees a localized region of the brain, the U-Net based segmentation model captures the global information from a different region of the brain tissues which is essential to achieve robust segmentation performance. Moreover, U-Net based model allows achieving an end-to-end segmentation framework rather than a pixel-wise classification technique. The U-Net architecture utilized in this work is implemented following [26]. Rather than using regular cross-entropy based loss function, we utilize a soft dice similarity coefficient (DSC) metric based loss function to train the U-Net model [28]. The U-Net model is trained using mini-batch gradient descent (GD) technique which minimizes the soft dice cost function. This work also uses a fully convolutional network (FCNs) [29, 30] for tumor segmentation. We adapt VGG-11 [31] as a pre-trained model. The overall FCN architecture contains an encode and a decode stage. The encoding stage contains convolution and max-pooling steps whereas the decoding stage contains a deconvolution step to obtain the same output size as the input. The final segmentation output is achieved by fusing the outcomes obtained from the above mentioned CNN based tumor segmentation pipelines as shown in Fig. 1.

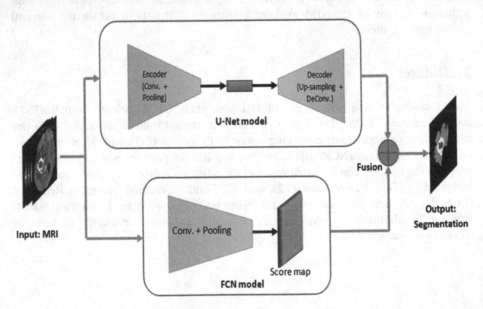

Fig. 1. Brain tumor segmentation pipeline using U-Net and FCN

3.2   Survival Prediction

The proposed survival prediction pipeline essentially involves three stages, (1) feature extraction stage, (2) feature selection stage, and (3) feature classification/regression stage.

Feature Extraction. Overall, approximately 31 thousand features representing texture, volume, area, and Euler characteristics are extracted from the tumor and the sub-regions (edema, enhancing tumor, and tumor core). Each feature type is selected to emphasize different characteristics of the tumor that may be relevant for survival prediction. For instance, texture features define the heterogeneity of the different tumor tissues, and Volumetric and Euler characteristic features define tumor shape.

We extract forty-one representative features [32] from three raw MRI sequences, and from eight texture representations of the tumor volume that includes several Texton filters [33], and fractal characterizations using algorithms such as PTPSA [34], mBm [35], and Holder exponent [36]. The features obtained from these representations include histograms, co-occurrence matrix, grey-tone difference matrix, and several other statistical measures. Furthermore, histogram-based features are extracted from the different modalities of different histogram graphs of the tumor tissue regions (edema, enhancing tumor, and necrosis). We also extract many representative volumetric features from the different tumor tissues with respect to the brain and whole tumor regions. In addition, we compute the Euler characteristics of the whole tumor, edema, enhancing and necrosis, for each slice as feature vectors. Euler characteristic [37] identifies the shape of a tumor by computing tumor vertices, edges, and faces. Combining feature vectors from all the above-mentioned methods constitute a total 31,000 feature points for the survival prediction task.

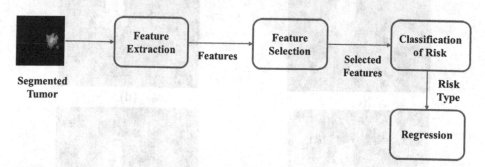

Fig. 2. Survival prediction pipeline

Feature Selection. We perform recursive feature selection on the Euler features alone, another recursive feature selection on the other features (texture, volumetric, histogram-graph based).

Feature Classification and Regression. Extreme Gradient Boosting (XGBoost) [38] is a tree boosting supervised machine learning technique that is highly effective. XGBoost is widely used in classification and regression tasks. In our study, XGBoost is utilized for classification and regression overall survival prediction on the selected features. The trained models are tuned to their optimized hyper-parameters when a tuned grid (search grid) is created by the different combination of the hyper-parameters. The complete pipeline for classification and regression overall survival is illustrated in Fig. 2.

4   Experimental Results

Following the proposed pipeline in Fig. 1, we first perform the FCN and U-Net fused tumor segmentation task. Figure 3 shows an example from the BraTS 2018 training dataset of the fused segmentation outcome of FCN and U-Net. We perform

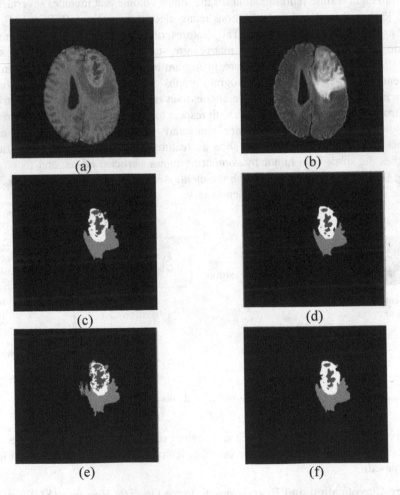

Fig. 3. Example input from training dataset and segmentation outcomes: (a) T1Gd sequence, (b) FLAIR sequence, (c) the segmentation outcome of FCN, (d) the segmentation outcome of U-Net, (e) ground truth (f) fused segmentation.

leave-one-out cross-validation analysis on the BraTS 2018 training dataset using the proposed survival prediction pipeline. We evaluate the performance of our proposed method using the root mean square error (RMSE), and the classification accuracy for a three-class setting defined as follows: (1) long – more than 15 months, (2) Medium – between 10 to 15 months, (3) short – less than 10 months. The best model is picked from the analysis of training data to be used to process the validation dataset of the BraTS 2018 competition. The ground truth is considered for the segmented tumor in the survival analysis of the training dataset, while the segmented tumor obtained in stage one of the proposed pipeline is used as input in the evaluation of the validation dataset. The leave-one-out cross-validation RMSE of the training dataset is 391.25, and the three-class leave-one-out cross-validated survival classification accuracy is 0.73. Table 1 shows the online evaluation results we achieve with the BraTS18 validation dataset and the test dataset.

Table 1. BraTS18 validation dataset and test dataset evaluation

Dataset	Accuracy	MSE	medianSE	stdSE	SpearmanR
Validation dataset	0.679	153600.515	36512.175	221707.697	0.204
Test dataset	0.519	367239.974	38416	945593.877	0.168

5   Conclusions

This work proposes a robust automated glioblastoma survival prediction using state-of-the-art computational modeling techniques. The survival prediction task is performed in two steps: tumor segmentation and survival prediction. A combination of hand-crafted and learned features are used in a regression technique to obtain the final survival prediction output. The performance of the proposed pipeline is evaluated using BraTS 2018 challenge training and validation datasets. Our results show a leave-one-out cross-validated classification accuracy of 0.679 for the validation dataset and that of 0.519 for the test dataset.

Acknowledgements. This work was funded by NIBIB/NIH grant# R01 EB020683.

References

1. A.B.T. Association: Brain tumor statistics. vol. 2 (2016). Accessed May
2. Louis, D.N., et al.: The 2007 WHO classification of tumours of the central nervous system. Acta Neuropathol. **114**(2), 97–109 (2007)
3. Holland, E.C.: Progenitor cells and glioma formation. Curr. Opin. Neurol. **14**(6), 683–688 (2001)
4. Ohgaki, H., Kleihues, P.: Population-based studies on incidence, survival rates, and genetic alterations in astrocytic and oligodendroglial gliomas. J. Neuropathol. Exp. Neurol. **64**(6), 479–489 (2005)

5. Lao, J., et al.: A deep learning-based radiomics model for prediction of survival in glioblastoma multiforme. Sci. Rep. **7**(1), 10353 (2017)
6. Nie, D., Zhang, H., Adeli, E., Liu, L., Shen, D.: 3D deep learning for multi-modal imaging-guided survival time prediction of brain tumor patients. In: Ourselin, S., Joskowicz, L., Sabuncu, M.R., Unal, G., Wells, W. (eds.) MICCAI 2016. LNCS, vol. 9901, pp. 212–220. Springer, Cham (2016). https://doi.org/10.1007/978-3-319-46723-8_25
7. Huang, C., Zhang, A., Xiao, G.: Deep Integrative Analysis for Survival Prediction (2017)
8. Chato, L., Latifi, S.: Machine learning and deep learning techniques to predict overall survival of brain tumor patients using MRI images. In: 2017 IEEE 17th International Conference on Bioinformatics and Bioengineering (BIBE), pp. 9–14. IEEE (2017)
9. Shboul, Z.A., Vidyaratne, L., Alam, M., Iftekharuddin, K.M.: Glioblastoma and survival prediction. In: Crimi, A., Bakas, S., Kuijf, H., Menze, B., Reyes, M. (eds.) BrainLes 2017. LNCS, vol. 10670, pp. 358–368. Springer, Cham (2018). https://doi.org/10.1007/978-3-319-75238-9_31
10. Vidyaratne, L., Alam, M., Shboul, Z., Iftekharuddin, K.: Deep learning and texture-based semantic label fusion for brain tumor segmentation. In: Medical Imaging 2018: Computer-Aided Diagnosis, vol. 10575, p. 105750D. International Society for Optics and Photonics (2018)
11. Cyran, C.C., et al.: Visualization, imaging and new preclinical diagnostics in radiation oncology. Radiat. Oncol. **9**(1), 3 (2014)
12. Ellingson, B.M.: Radiogenomics and imaging phenotypes in glioblastoma: novel observations and correlation with molecular characteristics. Curr. Neurol. Neurosci. Rep. **15**(1), 506 (2015)
13. Kickingereder, P., et al.: Radiogenomics of glioblastoma: machine learning–based classification of molecular characteristics by using multiparametric and multiregional MR imaging features. Radiology **281**(3), 907–918 (2016)
14. Jain, R., et al.: Outcome prediction in patients with glioblastoma by using imaging, clinical, and genomic biomarkers: focus on the nonenhancing component of the tumor. Radiology **272**(2), 484–493 (2014)
15. VASARI Research Project - Cancer Imaging Archive Wiki. https://wiki.cancerimagingarchive.net/display/Public/VASARI+Research+Project
16. Gutman, D.A., et al.: MR imaging predictors of molecular profile and survival: multi-institutional study of the TCGA glioblastoma data set. Radiology **267**(2), 560–569 (2013)
17. Verhaak, R.G., et al.: Integrated genomic analysis identifies clinically relevant subtypes of glioblastoma characterized by abnormalities in PDGFRA, IDH1, EGFR, and NF1. Cancer Cell **17**(1), 98–110 (2010)
18. Nicolasjilwan, M., et al.: Addition of MR imaging features and genetic biomarkers strengthens glioblastoma survival prediction in TCGA patients. J. Neuroradiol. **42**(4), 212–221 (2015)
19. Prasanna, P., Patel, J., Partovi, S., Madabhushi, A., Tiwari, P.: Radiomic features from the peritumoral brain parenchyma on treatment-naive multi-parametric MR imaging predict long versus short-term survival in glioblastoma multiforme: preliminary findings. Eur. Radiol. **27**(10), 4188–4197 (2017)
20. Itakura, H., et al.: Magnetic resonance image features identify glioblastoma phenotypic subtypes with distinct molecular pathway activities. Sci. Transl. Med. **7**(303), 303ra138 (2015)
21. Bakas, S., et al.: Advancing The Cancer Genome Atlas glioma MRI collections with expert segmentation labels and radiomic features. Sci. Data **4**, 170117 (2017)
22. Bakas, S., et al.: Segmentation labels and radiomic features for the pre-operative scans of the TCGA-GBM collection. The Cancer Imaging Archive (2017)

23. Bakas, S., et al.: Segmentation labels and radiomic features for the pre-operative scans of the TCGA-LGG collection. The Cancer Imaging Archive (2017)
24. Menze, B.H., et al.: The multimodal brain tumor image segmentation benchmark (BRATS). IEEE Trans. Med. Imaging **34**(10), 1993–2024 (2015)
25. Bakas, S., et al.: Identifying the Best Machine Learning Algorithms for Brain Tumor Segmentation, Progression Assessment, and Overall Survival Prediction in the BRATS Challenge. arXiv preprint arXiv:1811.02629 (2018)
26. Dong, H., Yang, G., Liu, F., Mo, Y., Guo, Y.: Automatic brain tumor detection and segmentation using U-Net based fully convolutional networks. In: Valdés Hernández, M., González-Castro, V. (eds.) MIUA 2017. CCIS, vol. 723, pp. 506–517. Springer, Cham (2017). https://doi.org/10.1007/978-3-319-60964-5_44
27. Ronneberger, O., Fischer, P., Brox, T.: U-Net: convolutional networks for biomedical image segmentation. In: Navab, N., Hornegger, J., Wells, W.M., Frangi, A.F. (eds.) MICCAI 2015. LNCS, vol. 9351, pp. 234–241. Springer, Cham (2015). https://doi.org/10.1007/978-3-319-24574-4_28
28. Milletari, F., Navab, N., Ahmadi, S.-A.: V-Net: Fully convolutional neural networks for volumetric medical image segmentation. In: 2016 Fourth International Conference on 3D Vision (3DV), pp. 565–571. IEEE (2016)
29. Long, J., Shelhamer, E., Darrell, T.: Fully convolutional networks for semantic segmentation. In: Proceedings of the IEEE Conference on Computer Vision and Pattern Recognition, pp. 3431–3440 (2015)
30. Zhao, X., Wu, Y., Song, G., Li, Z., Fan, Y., Zhang, Y.: Brain tumor segmentation using a fully convolutional neural network with conditional random fields. In: Crimi, A., Menze, B., Maier, O., Reyes, M., Winzeck, S., Handels, H. (eds.) BrainLes 2016. LNCS, vol. 10154, pp. 75–87. Springer, Cham (2016). https://doi.org/10.1007/978-3-319-55524-9_8
31. Simonyan, K., Zisserman, A.: Very deep convolutional networks for large-scale image recognition. arXiv preprint arXiv:1409.1556 (2014)
32. Vallières, M., Freeman, C.R., Skamene, S.R., El Naqa, I.: A radiomics model from joint FDG-PET and MRI texture features for the prediction of lung metastases in soft-tissue sarcomas of the extremities. Phys. Med. Biol. **60**(14), 5471 (2015)
33. Leung, T., Malik, J.: Representing and recognizing the visual appearance of materials using three-dimensional textons. Int. J. Comput. Vis. **43**(1), 29–44 (2001)
34. Iftekharuddin, K.M., Jia, W., Marsh, R.: Fractal analysis of tumor in brain MR images. Mach. Vis. Appl. **13**(5–6), 352–362 (2003)
35. Islam, A., Iftekharuddin, K.M., Ogg, R.J., Laningham, F.H., Sivakumar, B.: Multifractal modeling, segmentation, prediction, and statistical validation of posterior fossa tumors. In: Medical Imaging 2008: Computer-Aided Diagnosis, vol. 6915, p. 69153C. International Society for Optics and Photonics (2008)
36. Ayache, A., Véhel, J.L.: Generalized multifractional Brownian motion: definition and preliminary results. In: Dekking, M., Véhel, J.L., Lutton, E., Tricot, C. (eds.) Fractals, pp. 17–32. Springer, London (1999). https://doi.org/10.1007/978-1-4471-0873-3_2
37. Turner, K., Mukherjee, S., Boyer, D.M.: Persistent homology transform for modeling shapes and surfaces. Inf. Infer. J. IMA **3**(4), 310–344 (2014)
38. Chen, T., Guestrin, C.: XGBoost: a scalable tree boosting system. In: Proceedings of the 22nd ACM SIGKDD International Conference on Knowledge Discovery and Data Mining, pp. 785–794. ACM (2016)

Author Index

Printed in the United States
By Bookmasters

Printed in the United States
By Bookmasters